TREES

OF THE
CALIFORNIA LANDSCAPE

TREES
OF THE
CALIFORNIA LANDSCAPE

A Photographic Manual of Native and Ornamental Trees

CHARLES R. HATCH

A PHYLLIS M. FABER BOOK

UNIVERSITY OF CALIFORNIA PRESS

For my parents, Betty and Charles Hatch

University of California Press, one of the most distinguished university presses in the United States, enriches lives around the world by advancing scholarship in the humanities, social sciences, and natural sciences. Its activities are supported by the UC Press Foundation and by philanthropic contributions from individuals and institutions. For more information, visit www.ucpress.edu.

University of California Press
Oakland, California

Produced by Phyllis M. Faber Books, Mill Valley, California
Layout by Charles R. Hatch
Design and typesetting by Beth Hansen-Winter
Editing by Nora Harlow

Library of Congress Cataloging-in-Publication Data

Hatch, Charles R.
    Trees of the California landscape / Charles R. Hatch.
        p. cm.
    Includes bibliographical references and index.
    isbn-13: 978-0-520-25124-3 (cloth : alk. paper)
    1. Trees—California—Identification.   I. Title.

QK149.H38 2007
582.16'09794—dc22   2006051410

Manufactured in China

26 25 24 23 22 21
10 9 8 7 6 5 4 3

The paper used in this publication meets the minimum requirements of ansi/ niso z39.48-1992 (r 1997) (*Permanence of Paper*).

Cover: Native western sycamores (*Platanus racemosa*) in Laguna Canyon riparian habitat of southern California.
Half-title page: *Lagerstroemia* sp. in a garden setting.
Title page: *Acer macrophyllum* and *Arbutus menziesii* in riparian habitat in El Dorado County, California.
This page: Deciduous trees, possibly *Cercis canadensis*, at Sunset Books, Menlo Park, California.

TABLE OF CONTENTS

FOREWORD

The eminent American conservationist Aldo Leopold, in *A Sand County Almanac*, reminds us that "To keep every cog and wheel is the first precaution of intelligent tinkering."[1] One might add a second precaution—that it is necessary to know the identity and characteristics of each cog and wheel. *Trees of the California Landscape* is an excellent source for learning those cogs and wheels that we commonly know as trees.

California has long needed a comprehensive treatment of trees found in the natural landscape combined with trees used in designed landscapes. The last book to approach this on a statewide basis was Howard E. McMinn and Evelyn Maino's *An Illustrated Manual of Pacific Coast Trees*.[2] Their book was published in 1935 and described 445 taxa of which 22 were varieties or cultivars. *Trees of the California Landscape* presents 419 taxa of which 89 are varieties or cultivars. While these total numbers are similar, the listings in this book reflect the growth that has taken place in the numbers of certain tree species, especially in terms of varieties and cultivars that have been made available to landscape architects and home gardeners in the last 70 years. The ability to compare and contrast the 419 taxa described and illustrated makes *Trees of the California Landscape* an exceptional reference book. Recent books on California trees have focused, for the most part, on either native or introduced species. I often find myself fumbling between three or four tree books to find an adequate description of a particular landscape species. *Trees of the California Landscape* will eliminate the need to search through multiple books.

The author's extensive photography is one of the outstanding features of this book. The photographs illustrate the form of each tree (both in summer and winter for deciduous species) and the characteristics of leaves, fruit, flowers, and bark. They clearly illustrate the key characteristics of each species. The photographs of a leaf of each species, presented in the section on taxonomy, provide an excellent reference for identifying specimens and add greatly to the value of the book.

The introduction to the book brings together much information on the natural and the urban environments of California. The sections on topography, geography, and climate provide a concise framework for understanding the distribution of native species and geographic factors that influence the growth of introduced trees. Incorporation of the "Climatic Zones of California" map from Sunset Western Garden Book[3] and reference to the Sunset zones in the compendium of trees enables readers to cross-reference the information provided here with that of Sunset's widely used volume. Likewise, the use of codes from WUCOLS (Water Use Classification of Landscape Species) developed by Costello and Jones[4] and updated by the California Department of Water Resources is helpful in selecting trees with regard to irrigation needs in the Mediterranean climate of California.

The introduction also provides background information on the evolution of classification systems for natural vegetation in California, including *A California Flora* by Philip A. Munz (1968) and *Terrestrial Vegetation of California* by Michael Barbour and Jack Major (1997). It is a comprehensive treatment of information not previously synthesized in any publication to my knowledge. John O. Sawyer assisted in the adaptation of the series classification of tree species developed by Sawyer and Keeler-Wolf and also in applying a similar approach to the ordered listing of 19 native vegetation types and habitats with associated species. Including the recent classification scheme of Sawyer and Keeler-Wolf[5] in the context of older vegetation classification systems makes this book uniquely valuable to ecologists and to students of California vegetation.

The section on the use of trees in the interface between natural and urban environments. describes the physical aspects of these environments and discusses design concepts for using trees in designed landscapes. History, design concepts, codes and ordinances, and horticultural aspects are considered at a level that will be appreciated by landscape architects and designers, environmental specialists, planners, and home gardeners.

The focus on the California environment provides an understanding not commonly gained from more general treatments of these topics. In particular, the information on codes and ordinances, tree growth rates, and planting and irrigation methods have tailored the book to California. However, the species presented and much of the information on the use of trees will be equally applicable to portions of Washington, Oregon, Nevada, Arizona, and Baja California.

---

[1] Leopold, A. 1966, *A sand county almanac.* Oxford University Press, New York NY, p. 177.

[2] McMinn, H.E. and E. Maino, 1935, *An illustrated manual of Pacific Coast trees.* University of California Press, Berkeley CA.

[3] Sunset Books, 2007, *Sunset Western Garden Book,* Sunset Books, Menlo Park CA.

[4] Costello, L. and K. S. Jones, 1992, *WUCOLS Project* (Water Use Classification of Landscape Species), University of California Cooperative Extension, San Francisco and San Mateo County Office.

[5] Sawyer, J. O. and T. Keeler-Wolf, 1995, *A manual of California vegetation.* California Native Plant Society, Sacramento CA.

The lists of trees for special applications assist in the selection of trees based on different characteristics and functions. The lists are organized in categories such as leaf persistence, tree longevity, planting sites (e.g., streets, parking areas, court-yards, etc.) fruit and seed types, flower color, fall color, and trees that respond to special types of pruning. As the author points out, these lists are not meant to be definitive in choosing a tree for a particular function or setting. One should carefully read the compendium descriptions, examine the photographs, and then find and observe a specimen growing locally before making a selection.

Each entry in the compendium of trees is a gold mine of information. The author has formatted each page to allow for easy comparison of different species on many characteristics and has indicated various ways in which trees can be used in the landsacpe. The consistent format for each entry greatly facilitates identification, as well as comparison of differences even between quite similar trees.

The keys to tree genera and species, both native and introduced, are something not seen in their breadth of coverage in a California tree book since the now out-of-print *An Illustrated Manual of Pacific Coast Trees.* The keys are economical and efficient and can be used by readers with little knowledge of botanical vocabulary, although there is a fine glossary of terms included in the book. The illustrations and photographs throughout the book make the keys especially easy to use.

The author should be congratulated for producing a unique volume, rich in information on California trees, their presence in native plant communities, and their uses in designed landscapes. Publication of this book greatly enriches our understanding of these trees and broadens the possibilities for the wise use of trees adapted to and suitable for the California landscape.

Joe R. McBride, Professor
Department of Landscape Architecture
University of California, Berkeley, CA

INTRODUCTION

What is the California landscape? To paraphrase Garrett Eckbo in *The Landscape We See* (1969), the California landscape surrounds us everyday—at work, at play, while traveling from place to place or enjoying a vacation in the scenic outdoors. Humankind and nature coexist in this landscape not as separate entities but together, forming the whole. We see this most vividly when we fly over the state of California, looking down upon the cities and towns that grow out of the natural terrain. From this vantage point, the landscape is a mosaic of geographic features and native vegetation patterns interwoven with the organic and inorganic forms and stylized landscapes of cities, towns, and rural areas. The California landscape provides many scenic images of our state, recognizable even to those who have not visited California. The iconic lone cypress on the Monterey coastline, the palm trees of Beverly Hills, California missions from San Diego to the San Francisco Bay Area, and awe-inspiring coast redwoods are all trademarks of the California landscape.

Trees are one of the most dominant features of the California landscape. This book presents both the native and urban aspects of trees in the California landscape, looking at them as distinctly different but on an equal basis, together forming the overall landscape—native species associated with various habitats, ornamental species planted in urban landscapes, and orchard trees in agricultural areas, which often become integrated into expanding urban and suburban landscapes. Infor-

© Carr Clifton

California poppy, the state flower, is a familiar sight carpeting rolling valley grasslands and oak woodlands in springtime in many areas of the state.

mation about the wide range of trees we have to work with provides insight into some of the considerations involved in the design of effective and successful landscapes. Studying the types and extents of native trees and evaluating the current use and suitability of various ornamental trees facilitates the use of both in the California landscape. The diverse distribution of native and nonnative tree species, covering a wide range of climate zones, represents both a challenge and offers extraordinary opportunities for horticultural use as well. Knowledge of both native and nonnative trees also encourages the use of appropriate species in designed landscapes that reflect the context of natural species associations and the preservation of existing vegetation and habitats in ecologically sound landscape designs. Photographs and illustrations provide a means of identifying trees included in the compendium and differentiating native from ornamental species.

The ability to visualize design elements that we wish to create in a landscape brings relevance to the adage "a picture is worth a thousand words." The photographic details and graphic features of this book enable the reader to create a useful mental picture of the trees included in the compendium. This volume is intended primarily as a handy desk reference, rather than a scientific manual, though much botanical information is provided. The information is presented in a format intended to be useful to both horticulture and design professionals, possibly bridging the gap between the two areas of expertise. Urban planners, students in related fields, homeowners, gardeners, and landscape maintenance personnel may find the book useful as well. Every effort has been made to present information in a user-friendly manner and to make the identification, selection, arrangement, planting, and care of trees less intimidating for all. Information about the wide range of trees in California provides insight into some of the considerations involved in the design and planning of the landscape we see about us every day, but may otherwise take for granted.

THE CALIFORNIA LANDSCAPE

Trees and other vegetation associated with California sea-

© Mike Perry

This trademark "lone cypress" in Monterey is an instantly recognizable image of one of the many facets of the California landscape. Monterey cypress is one of California's unique native species.

© Larry Ulrich

The influence of early Spanish settlers and missionaries has remained strong, with this unique style relating to the Mediterranean climate of California. Missions from San Diego to Santa Barbara and other areas throughout the state, with their landscaped courtyards and surroundings, reflect the seasonal dryness occurring in a large portion of the state.

coasts, mountains, valleys, and deserts provide the settings for our towns and cities. The domesticated landscape of ornamental species becomes more apparent in urban areas, often with dramatic effect. The wide-ranging regional qualities of California's landscape make it quite unique, with recognizable identities reflecting geographic areas of the state from north to south and east to west.

The geographic effects of topography, climate, and resulting vegetation patterns have a direct bearing on the California landscape and have been well documented in other publications. These publications have become established standards applied by horticulturists, landscape architects, planners, and students in research and practice. The authors and publishers of several publications have graciously allowed the use of adaptations and extracts of their work, providing a means of maintaining consistency with the standards they have developed and encouraging their use in practical applications. Here they can be compared in a consistent side-by-side format, offering a visual means of correlating the separate maps of topographic, geographic, and climatic zones of California.

The large color maps in this section include a rendered relief map generated by the United States Geological Survey (USGS) Flagstaff Center. Also included are adaptations of "Geographic Subdivisions of California" developed for *The Jepson Manual: Higher Plants of California*, Hickman (1993) and "The West's Climate Zones" from the *Sunset Western Garden Book*, Sunset Publishing Corporation (2007). Viewing these maps together is useful in understanding the overall context of environmental factors affecting the California landscape.

TREES IN THE NATIVE LANDSCAPE

The native landscape reflects the topographic, geographic, and climatic environment of specific areas. The resulting species distribution and patterning throughout the state has slowly changed over the ages into the vegetation we see today. These natural processes are still at work, but accelerated human intervention has brought more complicated environmental issues to bear in the preservation of native landscapes around us while allowing cities and towns to thrive.

Regional landscapes are important to cities and suburban areas, with images depicting or suggesting what makes them desirable places to live. In most regional landscapes, trees are the dominant feature. Trees do not occur randomly or as a homogeneous feature throughout the state. Instead, they form recognizable vegetation types specific to certain areas, and without them regional identity may be lost.

To better understand the diversity of California's native vegetation, various means of categorizing plant associations and distributions have been developed and gradually perfected since the early 1900s. More detailed scientific, botanical, and environmental studies utilize a system of classifying native vegetation by series. The system currently in use was developed by the California Native Plant Society (CNPS) in *A Manual of California Vegetation*, Sawyer & Keeler-Wolf (1995). An adapted summary of its development and application is offered here to clarify the purposes of this newer approach and extend familiarity with its use. The abbreviated codes indicated with individual species in the compendium of trees have been combined with a condensed listing of CNPS series dominated by trees, including subordinate tree associations. The compendium more often illustrates native trees in parks

© Mike Perry

Agricultural land use associated with urban California landscapes provides scenic settings that often become extensions of valley grasslands with a carpet of wildflowers. Yellow-flowering mustard is a common sight in orchards throughout California's Central Valley.

© Ron and Patty Thomas

Cool, moist redwood forests are an easily recognizable feature of California's northwest coastal regions, representing the wettest climate in the state. Coast redwood is one of California's unique native species and a highly desirable addition to urban landscape plantings.

or landscape settings, since this book emphasizes pictorial aspects of landscape use for both native and nonnative trees. However, understanding common vegetation associations in nature is important to the preservation of native habitats as well as the use of native landscape materials in the context of designed landscapes.

Prior to the establishment of the series method of classification, a more pragmatic approach incorporating vegetation types or "plant communities" was used. The older system is still relevant today in the study of habitat and resource management and provides descriptive terms that remain familiar to many people. The color map of native vegetation types and habitats includes codes used in the compendium for trees native to California. These have been adapted from a broad listing of plant communities in *A California Flora*, Philip Munz (1959) and *Terrestrial Vegetation of California*, Barbour & Major (1977). These are condensed into 19 major vegetation zones representing general native plant associations and indicating their natural range throughout the state. While not precise, this map is an approximated graphic representation of how these zones might fit together and serves as a visual means of relating this map to the topographic, geographic, and climatic zone maps pictured elsewhere in this book. There is no established reference for this purpose.

Knowledge of plant habitats and communities can be useful to highway travelers, homeowners in rural or mountain areas, and recent migrants to California. Landscape architects and others may find this reference useful in incorporating native plant materials in regionally appropriate landscapes. Environmental consultants often apply these principles in the analysis and development of habitat and resource inventory studies and in vegetation restoration.

A summary of issues involved in the preservation of the native landscape and maintaining a transition to the urban landscape is provided in the section on trees in the urban landscape. This section explores the many ways that native habitats and vegetation can be integrated into suburban and rural landscapes while still remaining viable ecosystems that enhance the living environment of existing and new developments. The native landscape surrounding urban environments may also reveal natural factors affecting the success of ornamental landscapes that replace them, including geology (poor, unstable, or rocky soils), hydrology (soggy soils or areas prone to flooding), or species indicating microclimates. Understanding the native landscape can be an important step toward integrating and maintaining cultivated landscapes, especially in new developments.

TREES IN THE URBAN LANDSCAPE

Trees have an important role in sustaining the urban and suburban environment, which most often utilizes ornamental species. Trees and other types of vegetation adaptable to domestication are used within a landscape of artificial design, usually replacing the natural landscape to suit human needs. Trees create settings for the buildings, plazas, roads, parks, homes, and open spaces of our cities and towns, and agricultural orchards are also often included.

Various types of urban landscapes serve different purposes. In highly developed urbanized settings, space for buildings and

© Ron and Patty Thomas

Representing the driest regions of the United States, the desert landscape is often highlighted with a brilliant floral carpet in late spring, when winter rains offer just enough moisture to bring out these brief flowering displays.

streets is at a premium. This restricts the use of generous land-scaped areas, and trees become the major element of a land-scape often limited to tree wells, planters, or narrow planting strips between curbs and sidewalks or near buildings. More extensive open space elements such as freeway plantings, park-way corridors, parks, and plazas provide additional softening effects and offer places for people to gather and enjoy the out-doors in an active or passive setting. Residential landscapes may offer an opportunity to experiment with trees and other land-scape elements in a more personal or even whimsical manner.

The arrangement of trees and their use in various con-texts is an important tool of landscape design. Often a land-scape design is enhanced by organizing the spatial qualities and accentuating the aesthetic qualities of trees. Local codes and ordinances often determine or restrict the selection and placement of trees. However, the design elements of shape, space requirements, height, growth rate and longevity, den-sity, texture, and evergreen or deciduous qualities remain pri-mary considerations in the selection and placement of trees.

General planting and irrigation methods used in Califor-nia often differ from those in other parts of the country. Com-mercial nursery stock is most often grown in containers to maintain uniform moisture.

"Bare root" or "balled-in-burlap" stock tends to dry out too quickly in California. Irrigation is essential in most parts of California, and trees planted in narrow tree wells or plant-ers often require specialized treatment. Height and space re-quirements, including root space, must allow for the mature size of the tree. There are many options available for tree plant-ing and irrigation, depending on the situation, and special ef-

© Chuck Place

Palm trees are a trademark of the southern California landscape. They grace many skylines, notably Beverly Hills, Hollywood, Palm Desert, and other cit-ies and resorts throughout the southland area.

fects can also be achieved with the creative use of trees in un-usual landscape situations.

Culture, maintenance, and the control of disease are ma-jor factors affecting trees once they are planted. Selecting trees that may do best in cases where, for various reasons, they will not receive special care and treatment becomes very impor-tant. Trees that are more disease resistant, as well as those with fewer undesirable characteristics, are preferable in such situa-tions, though there is no tree without some sort of drawback. Selecting the "right" tree often becomes a question of how much maintenance a particular tree can or will be provided.

The Special Use Lists included in this section provide a means of selecting trees that might best fulfill the requirements of a particular situation. It is important to recognize that these are not all-inclusive lists, nor are the trees listed necessarily approved for the suggested use. The lists are intended to spark ideas and narrow the search for a particular type of tree.

TREES LISTED IN THIS BOOK

The U.S. Forest Service defines a tree as "woody plants having one erect perennial stem or trunk at least 3 inches in diameter at 4-1/2 feet, a more or less definitely formed crown of foliage, and a height of at least 13 feet." For the purposes of this book, this definition has been extended to include shrub-like species and dwarf cultivars. Some large ornamental shrubs are often trained and pruned into multi-trunked "tree" forms for their accent specimen value in landscape applications. Others are often trained into single-trunk "standards." These are important as small trees in urban situations where planter or canopy space is restricted.

© The Sacramento Bee/Manny Crisostomo

New housing developments with newly planted landscapes are a common sight throughout California. It usually takes about 15 years for new trees to soften the rather stark initial appearance and roughly 40-60 years for trees to reach sizable maturity.

© Marc Muench

Giant sequoia is a species unique to California, growing in moderately dry Sierra Nevada mountain regions. Shown here in the Calaveras Big Trees grove, giant sequoias represent some of the largest and oldest trees on earth, some as old as 2,500 years.

Compact or dwarf cultivars of tree species commonly used as small patio or garden specimens have been included as well. Orchard trees used for agricultural purposes also are included where these are commonly used in ornamental landscapes, such as apple, pear, plum, and cherry, or are often preserved from pre-existing orchards, such as pecan and walnut. Large native shrubs such as elderberry and willow are included. While some willows are definitely treelike, others are dominant shrubs in their native habitats.

Other publications are available with more complete listings than those covered in this book for the numerous cultivars, varieties, and subspecies of trees. The extensive listings in *Sunset Western Garden Book* include newly introduced species and varieties successfully tested for landscape use that are cultivated in sufficient quantity to provide general nursery availability. *North American Landscape Trees*, by Arthur Lee Jacobsen, also lists a wealth of useful varieties for special landscape situations, though some are available only through specialty nursery sources. The book also clarifies common misnomers regarding current botanical listings, synonyms, plant patents, and where a cultivar was developed. The most comprehensive listing of native plant species of California is *The Jepson Manual*, which represents a definitive work in regard to the ongoing identification and classification of California's native vegetation.

New cultivars and varieties of native and ornamental species are continually being introduced. Some quickly become popular, while others are found to be less desirable than expected. Newer introductions often require many years to evaluate their shortcomings, which are not noticeable until the tree matures. Availability of new introductions at first may be limited, and to list them here might be misleading. Readers are encouraged to investigate availability and determine which ones can be expected to grow successfully in a particular locale. The compendium in this book includes the most commonly found trees. The 78 cultivars and varieties included represent only those available for photographic study.

THE COMPENDIUM OF TREES

The compendium lists a total of 419 taxa, with 309 introduced nonnative ornamental species and 110 California natives. However, native trees outnumber nonnative trees in quantity throughout the state.

Trees in the compendium are featured in a full-page format for each tree, with consistently detailed text and a full photographic representation of its various aspects. Notable cultivars and varieties are represented in an abbreviated half-page format. Trees of average maturity are generally pictured, rather than those of exceptional size or ones not fully grown, giving a better representation of commonly seen shape and size, which may vary greatly.

Consistency of visual representations, the order of the items described in the text, and the alphabetical listing of botanical names in the compendium facilitate cross-referencing, which is necessary for proper evaluation when one is comparing similarities and differences between trees.

Coded abbreviations are at the top of each heading in the compendium to facilitate quick referencing. These headings begin with: **Continents & Regions** of the world in which the species is native, specifically: **c** (central), **e** (eastern), **s** (southern), **n** (northern), and **w** (western) **Africa, Asia, Asia Minor, Canary Islands, Australia, Europe, N.A.** (North America), **N.Z.** (New Zealand), **Medit.** (Mediterranean), **Pacific Is.** (Pacific Islands), **South Africa** and **South America. Cultivars** are listed without reference to regional origin, as they are horticulturally developed and propagated varieties.

Native trees are designated in the same manner, with an additional notation **California Native**, and (**CA only**) for native endemic species that occur only within California. Coded abbreviations indicating **Native Vegetation Zones** featured on pages 35–45 are as follows:

CO (Coastal Strand), **CM** (Coastal Marsh), **CP** (Coastal Prairie), **VG** (Valley Grassland), **AL** (Alpine), **CS** (Coastal Scrub), **MD** (Mojave Desert Scrub), **SD** (Sonoran Desert Scrub), **SS** (Sagebrush Scrub), **CH** (Chaparral), **OW** (Oak Woodland), **FW** (Foothill Woodland), **JP** (Juniper-Pine Woodland), **CC** (Closed Cone Forest), **ME** (Mixed Evergreen Forest), **RW** (Redwood Forest), **CF** (Klamath Mixed Conifer Forest) and **FF** (Montane & Subalpine Fir Forest). These are followed by codes and additional notations (*listed in italics*) for the *Jepson Manual* **Geographic Regions** featured on page 15. As an additional visual reference, a small **map** is also provided on each native species page to suggest general extents of the species within California.

Codes for **Landscape Use** follow, for both ornamentals and cultivated natives. These include: **ACC** Accent, **CNF** Conifer,

CLR or COL Color, DES Desert, EVG Evergreen, FAL Fall, FLW Flowering, FOL Foliage, FRAG Fragrance, FRU Fruit, NAT Native, ORN Ornamental, SCR Screen, SHD Shade, SPC Specimen, STR Street tree, VERT Vertical, and WPG Weeping.

A coded six-zone slashed entry such as (M/M/M/H/H/H) indicates WUCOLS water use requirements for each species. Details of WUCOLS (*Water Use Classification of Landscape Species*) are explained on page 59. WUCOLS requirements for native species are indicated in parentheses, since they may require supplemental watering until they become well established.

Scientific names in the botanical headings for all trees are bold and italicized, denoting genus and species, consistent with *Sunset Western Garden Book*, reflecting the most widespread usage by which the nursery trade generally makes them available. Varieties are denoted as var., subspecies as ssp. Cultivars are are in single quotes and capitalized.

Synonyms and Plant Patents are listed beneath where they apply, in smaller bold text. Synonyms preceded by an equal sign (=) are considered equal and valid, with slightly different *Jepson Manual* subspecies or other general use variety listings included beneath. Synonyms shown in parentheses are provided to clarify botanical name changes or other latinized names by which a plant is or has been known. For further reference on synonyms, see *Trees of North America*, by Arthur Lee Jacobsen.

Common names in the compendium include only those in most general use. For additional common names, see Index to Common Names.

Descriptive text for all trees begins with **Family**, followed by **Sunset** Climate Zones, where available, featured on page 17, excluding zones for Hawaii and Alaska. For native species, these are followed by ***Jepson Manual* Horticultural Entry Codes**, where available, with an asterisk (*) representing the Jepson "flower symbol," showing optimal *Sunset* climate zones in bold, and/or parenthesized abbreviated codes, with especially appropriate *Sunset* climate zones indicated in bold. **Horticultural Entry Codes** used in this book are: (**CVS**) cultivar(s) available, (**DFCLT**) difficult, requires special care, (**DRN**) requires excellent drainage, (**DRY**) summer water intolerant, (**INV**) invasive, (**IRR**) requires moderate summer watering, (**SHD**) best in full or part shade, (**STBL**) good soil stabilizer, (**SUN**) best in sun or nearly full sun, (**WET**) wet or continually moist soil required.

Trees are defined in the compendium text as evergreen or deciduous, followed by more specific information regarding the tree's origin and range. Details of growth rate, height, spread, character, and habit are followed by information regarding leaves, flowers, fruit, twigs, and bark, as applicable. A final summation lists general observations, beneficial aspects, significant drawbacks, susceptibility to disease or other items useful for consideration in determining suitability or adaptability in the landscape. Estimated longevity is stated in general terms. Native species also have a listing of their series associations, derived from *A Manual of California Vegetation*.

THE TAXONOMY SECTION

Taxonomy involves the classification of plants and animals. Classification involves an arrangement in a hierarchy of classes, according to a defined system, or grouping them by shared characteristics. The generalized section in this book clarifies many of the specific characteristics of trees listed in the compendium, both verbally and visually, as an overview. It also helps in understanding terms used in descriptions encountered in the compendium, or in other more detailed classification manuals. The drawings and text of this section begin with general characteristics of tree structure, bark types, habit, leaf veining, and leaf types. The pictorial collection of leaves is arranged alphabetically in a classification of simple, compound, needlelike or scalelike, to facilitate its use. Various types of flowers and fruit are also pictured for visual identification and comparison.

The glossary of terms included in this section serves as a dictionary of terminology encountered in more detailed manuals that may be used in conjunction with this book. The distribution of trees is listed here according to standard botanical classification, followed by a listing of genera by family, providing an overall view and means of relating the various trees included.

The identification keys are not as cumbersome as they may appear. They can also be a useful time-saving tool when they are thought of simply as an outline, with major headings clearly marking the main types. The key for tree genera precedes the one for tree species. One can quickly skip to the heading that applies, and then to the next subheading, where similar types are closely grouped, thus narrowing the search. While the keys follow the general structure of horticultural classification, the main differentiation is made between simple and compound

A dramatic specimen of *Dacrydium cupressimum*, or rimu, originally brought from New Zealand for the Panama-Pacific International Exposition in 1915. The tree was later transplanted to Golden Gate Park's Strybing Arboretum, where it has long been admired for its unusual beauty.

leaves. This easily discernible feature becomes a logical point to begin, and narrows the choices considerably, as in the case when one has only a single leaf to use for identification, without having seen the tree itself.

AUTHOR'S COMMENTS

In summation, there is no perfect tree, that is, one with no undesirable characteristics, nor is there a checklist for which tree is best for a certain situation. Instead, it becomes a task of selecting the tree species with the most desirable characteristics and the least number of drawbacks. Some of the questions we must first ask ourselves are whether the attractiveness of an especially desirable tree is worth the added effort and care, or whether one that will merely provide dependable shade with minimal care once it is established is a better choice. Personal preferences vary from person to person, and it is often difficult to arrive at a choice that pleases everyone, as landscape architects and designers know well. For some, the personal garden is a chance to enjoy the freedom to experiment with certain trees that are exotic or unusual, as specimens to be admired.

The pictorial aspects of this book give an in-depth look at the various aspects of individual trees. The text also looks at the background issues and techniques effective in understanding the current use of trees to create landscapes that are both attractive and enduring. While urban and native landscapes are often quite distinct and separate, there are other instances where they blend together. Trees from around the world enhance and beautify our surroundings, with stunning results. They often form an identity or sense of community as important features of the landscape, which make it distinct from other cities. Native trees tie in the surrounding geographical and regional qualities imparted by the setting.

In choosing species for tree plantings, it can be a much more rewarding process with sufficient background knowledge one can apply to the type of landscape intended. Then it becomes a matter of flipping pages, becoming familiar with the trees that stand out, and choosing. Again, there is no simple checklist or database, though these are becoming quite common. These aids are useful to those who are already familiar with most of the species, and can picture them in their minds as they run down a list of names, rather than searching through a book.

As one's knowledge of trees increases, the horizons broaden and possibilities for creative and innovative use in landscapes follows, not only in personal gardens, but in urban, rural, and suburban landscapes as well. One can also begin to appreciate the unique beauty of the native landscape that California has to offer, with its many distinctive and rare species, some of which are found only in this state.

The topics covered in this book were an important part of the curriculum offered to us as students of landscape architecture. Knowledge in these subjects has proven to be very useful through the years, and the desire here is to keep them current as relevant materials for students and for those in related professions. Their inclusion here may promote renewed study where these subjects have not been further developed or used, or it may serve as a refresher in cases where some of it may have been forgotten. While researching materials for this book, I have gained much valuable additional knowledge. I hope this information will be as useful to others as it has been for me. I encourage those interested in learning more to visit and support the arboretums listed in the back of the book, to which I am indebted.

ACKNOWLEDGMENTS

I would like to thank the many people who have helped make this book a reality. Without their encouragement and assistance from the beginning, it might never have been completed. In particular, the following people have been unfailingly supportive: Monty Knudsen, Nicholas DeLorenzo ASLA, Denis M. Kurutz, Michael Barbour, Joe R. McBride, John O. Sawyer, Elly Bade, Warren Roberts, James Harding, and Zona Noren. I am indebted to each of them for their generosity and help. I am also grateful to UC Press and co-publisher, Phyllis M. Faber, and for the extraordinary efforts of her team, Beth Hansen-Winter, designer, and Nora Harlow, copy editor. Many thanks also to Ed and Fran Egan, Ted and Wendy Rybicki, and my colleagues Gordon Bradley ASLA, William Andersen ASLA, S. Wayne Kelly ASLA, and Steve Fuhrman ASLA. Much of the inspiration for this book came from ideas and insight into landscape architecture imparted to me long ago by Robert Royston FASLA, Edgar C. Haag ASLA FAAR, James Degen, and John N. Reader AIA. Thank you all.

The strikingly lifelike artificial trees in the foreground, though of sculpted metal, are quite dramatic in this median landscape, and complement the real grass and redwoods behind.

THE CALIFORNIA LANDSCAPE

Topography, Geography, and Climate

One of the most noticeable features of the California landscape is its varied topography. The scenic beauty of California is heightened by dramatic elevation changes, often in close proximity. Rugged coastlines, flat expansive valleys, rolling foothills, snowcapped mountain peaks, and granitic escarpments carved out by glacial action or worn down by erosion are all part of the California landscape. The rocky Monterey coastline, Half Dome overlooking Yosemite Valley, the majestic San Gabriel Mountains, the Hollywood Hills surrounding the Los Angeles Basin, and the desert hills around Palm Springs are all landmarks of the California landscape easily recognized by millions of people who have never visited the state.

One of the steepest gradient changes in North America is found in California. In less than 80 miles distance between the top of Mount Whitney and the lowest point in Death Valley, the elevation difference is almost 14,800 feet. These two locations are the highest and lowest elevations in the contiguous United States, with the summit of Mount Whitney at 14,494' elevation and Badwater in Death Valley at 282' below sea level. It is not surprising that this considerable variety in relief has pronounced effects on regional climate and specifically determines geographical regions of this state.

Most states may have one mountain range, whereas California has five. The Klamath Ranges, including the Trinity Alps, occur in the northwestern corner of the state. The Cascade Ranges occur in the northeastern corner, extending into Oregon and Washington. Southward, the Sierra Nevada and associated ranges define the eastern edge of the state. The Coast Ranges border the entire length of the coastline. The southern portion of the state is defined by the Transverse Ranges, which are aligned latitudinally, or perpendicular to the coast. The other mountain ranges in the state run longitudinally and have a more significant impact on climate and rainfall. As continental air masses flow eastward from the coast, rainfall is produced by condensation at colder temperatures as air rises and crosses over these mountain ranges.

Elevation and the placement of landforms greatly affect climate and rainfall. The Coast Ranges soften the mild coastal influence of eastward-flowing air masses, with less rainfall occurring on the leeward side. The inland mountain ranges are much higher in elevation, with more significant effect. South of the Transverse Ranges, air masses flow relatively unimpeded to the east, and moisture is dispersed evenly until dissipated, usually by the time it reaches the desert regions. Rainfall in these warmer southern latitudes occurs less often than in northern regions of the state.

Much of California has a Mediterranean climate characterized by seasonal rainfall and temperature changes, which are most affected by latitude and elevation. This becomes apparent when we study variations in both native vegetation and the domesticated trees and shrubs that thrive in ornamental landscapes throughout the state. Upright conifers, their excurrent branching adapted to withstand heavy snows, predominate in the higher mountain ranges. Conifers accompany broadleaf evergreen trees found along windswept coastlines, as their tough-needled foliage can withstand harsh winds. Their rather low transpiration rate also allows some to withstand dry desert conditions. Broadleaf trees with decurrent branching gradually become more common below elevations of lasting winter snow cover, and eventually predominate in lower-elevation valley and foothill grasslands.

Temperature constraints have a significant effect on the cultivation of trees and other plants from the eastern U.S. and around the world. At higher elevations, trees such as Colorado blue spruce and sugar maple are quite at home, though they also tolerate valley conditions with less stunning results and reduced longevity. Trees such as these, as well as fruit trees such as apple and pear, require significant winter chill to produce their best growth. Trees grown at lower elevations with less winter freezing often display less brilliant fall color than they do in areas where seasonal change is greater. However, at lower elevations the list of trees for landscape use expands, especially along the temperate coast and in subtropical areas of southern California, where elevation and seasonal changes are less pronounced.

Elevation directly influences temperature, as seen in such features as the "snow belt," where snowfall often remains through the winter season. Red fir predominates, its excurrent branching structure bending rather than breaking with heavy snows. Throughout the Central Valley, the extents of the "fog belt" can be seen at roughly 1,100' elevation, where *Quercus lobata* is gradually replaced by *Q. kelloggii* above elevations with extended periods of winter fog.

TOPOGRAPHY

United States Geological Survey (USGS)

A map showing mountain ranges in rendered relief, with elevation differences indicated by color, offering a realistic representation of these elements over the entire state. The influence of the location of major mountain ranges and the valleys between, as well as the contrasting elevations can be seen, corresponding to distinct geographic regions and climate patterns which affect native vegetational patterns as well as ornamental plantings in the landscape. The smaller contour map is a two-dimensional means of conveying the three-dimensional aspects of topography, with steeper slopes occurring where contour lines are closest.

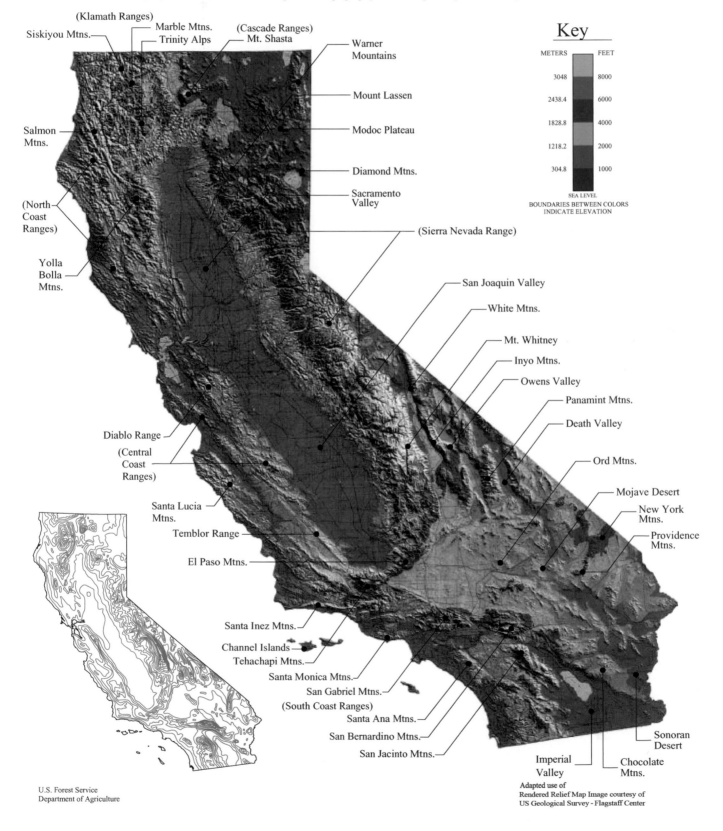

Key

METERS		FEET
3048		8000
2438.4		6000
1828.8		4000
1218.2		2000
304.8		1000

SEA LEVEL
BOUNDARIES BETWEEN COLORS
INDICATE ELEVATION

(Klamath Ranges)
Marble Mtns.
Siskiyou Mtns.
Trinity Alps
(Cascade Ranges)
Mt. Shasta
Warner Mountains
Mount Lassen
Modoc Plateau
Salmon Mtns.
Diamond Mtns.
Sacramento Valley
(North Coast Ranges)
(Sierra Nevada Range)
Yolla Bolla Mtns.
San Joaquin Valley
White Mtns.
Mt. Whitney
Inyo Mtns.
Owens Valley
Panamint Mtns.
Death Valley
Diablo Range
(Central Coast Ranges)
Ord Mtns.
Mojave Desert
Santa Lucia Mtns.
New York Mtns.
Temblor Range
Providence Mtns.
El Paso Mtns.
Santa Inez Mtns.
Channel Islands
Tehachapi Mtns.
Santa Monica Mtns.
San Gabriel Mtns.
(South Coast Ranges)
Santa Ana Mtns.
San Bernardino Mtns.
San Jacinto Mtns.
Sonoran Desert
Imperial Valley
Chocolate Mtns.

U.S. Forest Service
Department of Agriculture

Adapted use of
Rendered Relief Map Image courtesy of
US Geological Survey - Flagstaff Center

California's overall landscape is defined by latitudinal and longitudinal extents within which the noticeable topographic elements of mountain ranges, valleys, and coastlines create distinguishable geographic regions. These different regions have their own unmistakable identity, as evidenced in distinctly visible qualities and recognizable character. While county lines provide a means of arbitrarily dividing the state into locales, geographic regions are indicative of areas exhibiting common characteristics that reflect local climate, topography, and vegetation. Based on *The Jepson Manual — Higher Plants of California* (Hickman, 1993), the state can be subdivided geographically both east/west and north/south. A longitudinal line divides the main portion of the state, known as the California Floristic Province to the west, from the visibly drier regions of the Desert and Great Basin provinces, which extend eastward from the leeward side of the inland north-south mountain ranges. The latitudinal distinction between the north and south portions of the state occurs roughly at the southern tip of the Tehachapi Mountains toward Point Conception on the coast. Seasonal changes are more evident north of this line, and rainfall is more abundant.

The Jepson Manual introduced a map recognizing the geographic subdivisions of California. Much as the *Sunset* "Western Climate Zones" replaced the USDA system based solely on minimum temperatures, *The Jepson Manual's* subdivisions replaced county lines with recognizable geographic regions. These regions provide a consistent terminology for describing the locations of species, habitats, and plant communities and set a standard for defining the range and extents of California's native flora within geographic regions. The system is organized within a hierarchy of three broadly defined provinces, 10 regions, 20 subregions, and 17 districts. This offers a descriptive means of efficient definition, as in the differences evidenced between the southern and northern extents of the Sierra Nevada mountain ranges. The main subdivisions are based on recognizable features of the native California landscape, with lesser subregions reflecting local geologic, topographic, and climatic variation seen in specific vegetation types.

The Jepson Manual was published as a comprehensive up-

date to the earlier works of Willis Jepson, who devoted his life to defining the extensive native vegetation of California. The *Manual* was a major accomplishment, involving nearly 200 botanist-authors and illustrators who donated their time and expertise in assembling the information. The most complete inventory to date on the subject, this work set a standard for use accepted by the horticultural and botanical sciences.

With its comprehensiveness, conciseness, and accessibility, *The Jepson Manual* has made extensive botanical information available for a wide range of practical uses. The conventions used for the botanical listings, nomenclature, descriptions, and terms provide welcome consistency for practical use and identification purposes. The detailed information about California's topography, geography, and geology provides background and insight into the many significant changes that have taken place in the evolutionary process over eons of time, resulting in the California landscape we see today.

The codes provided in *The Jepson Manual* convey numerous aspects about species, including range and extents within and outside California, rarity, invasiveness, horticultural adaptability, and climatic and cultural requirements in cultivation. While the *Manual's* zones may not coincide exactly with those of *Sunset,* they indicate where particular species might be tried, where they can be expected to do best (shown in bold type), and where supplemental watering is needed to hasten establishment. Watering requirements are often quite different for native and for ornamental species. While focusing on native species, the *Manual* also lists certain other species that have been introduced to California and subsequently have become naturalized, persisting and surviving on their own. This information is helpful in identifying species that have ornamental value and others that may have become weedy and detrimental to the native landscape and native habitats.

The classification keys in *The Jepson Manual* are organized according to family, under which genus and species keys are grouped. While alphabetized listings are more convenient for indexing, listings by family following the format of classification keys are more practical for specialized botanical study. The botanical names in the compendium of this book follow *Sunset,* as they reflect commercial nursery listings in landscape use. *Jepson Manual* listings for native species are included where they differ.

About the Jepson Herbarium. Willis Jepson published many scientific papers and books on the California flora and natural history. In 1925, his *Manual of the Flowering Plants of California* was the first attempt to provide a comprehensive manual on the wild plants of California. This volume was instrumental in various movements to document and preserve California's environmental and botanical heritage. When Jepson died in 1946, his estate passed to the University of California at Berkeley, where his work is continued through the Jepson Herbarium. The services of this organization include many publications and research projects, educational services, and resources, including its herbarium, library, botanical workshops, and courses for scientific study. The recently published *The Jepson Desert Manual,* designed as a compact, easy-to-use manual, is a definitive work focusing on the vegetation of southeastern California deserts.

Arid desert regions are a distinctly recognizable geographic feature of the California landscape. With their fragile ecosystems and endemic species, these regions exhibit a harsh environment of singular beauty and efficiency.

GEOGRAPHY

A description of the various geographical areas, in terms of the topography and general location, and the major mountain ranges, foothills, valleys and desert regions throughout the state, allowing organized study of these regions according to their distinct charcteristics, and the extents to which they occur.

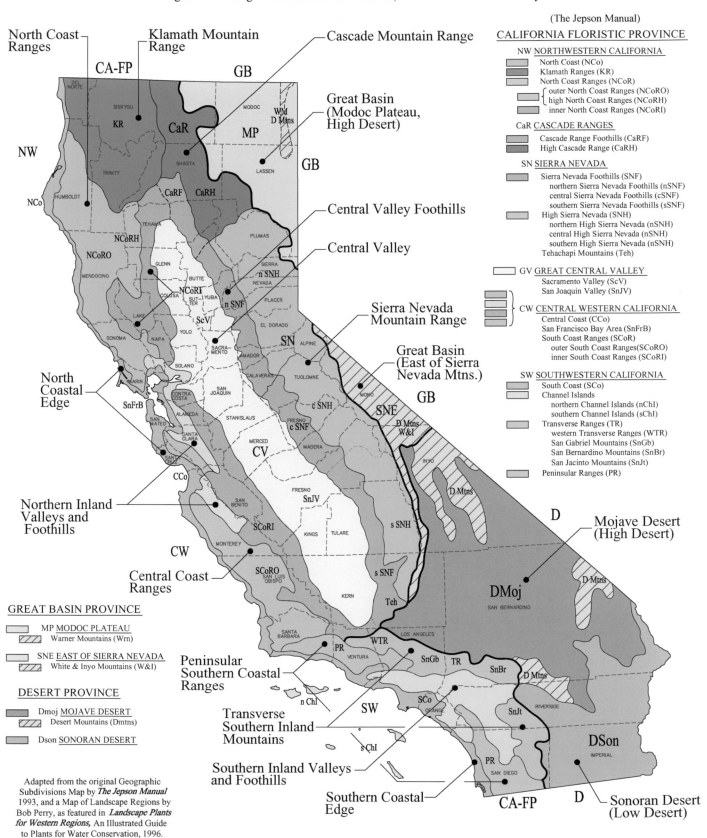

CALIFORNIA FLORISTIC PROVINCE (The Jepson Manual)

NW NORTHWESTERN CALIFORNIA
- North Coast (NCo)
- Klamath Ranges (KR)
- North Coast Ranges (NCoR)
  - outer North Coast Ranges (NCoRO)
  - high North Coast Ranges (NCoRH)
  - inner North Coast Ranges (NCoRI)

CaR CASCADE RANGES
- Cascade Range Foothills (CaRF)
- High Cascade Range (CaRH)

SN SIERRA NEVADA
- Sierra Nevada Foothills (SNF)
  - northern Sierra Nevada Foothills (nSNF)
  - central Sierra Nevada Foothills (cSNF)
  - southern Sierra Nevada Foothills (sSNF)
- High Sierra Nevada (SNH)
  - northern High Sierra Nevada (nSNH)
  - central High Sierra Nevada (nSNH)
  - southern High Sierra Nevada (nSNH)
- Tehachapi Mountains (Teh)

GV GREAT CENTRAL VALLEY
- Sacramento Valley (ScV)
- San Joaquin Valley (SnJV)

CW CENTRAL WESTERN CALIFORNIA
- Central Coast (CCo)
- San Francisco Bay Area (SnFrB)
- South Coast Ranges (SCoR)
  - outer South Coast Ranges(SCoRO)
  - inner South Coast Ranges (SCoRI)

SW SOUTHWESTERN CALIFORNIA
- South Coast (SCo)
- Channel Islands
  - northern Channel Islands (nChI)
  - southern Channel Islands (sChI)
- Transverse Ranges (TR)
  - western Transverse Ranges (WTR)
  - San Gabriel Mountains (SnGb)
  - San Bernardino Mountains (SnBr)
  - San Jacinto Mountains (SnJt)
- Peninsular Ranges (PR)

GREAT BASIN PROVINCE
- MP MODOC PLATEAU
- Warner Mountains (Wrn)
- SNE EAST OF SIERRA NEVADA
- White & Inyo Mountains (W&I)

DESERT PROVINCE
- Dmoj MOJAVE DESERT
- Desert Mountains (Dmtns)
- Dson SONORAN DESERT

North Coast Ranges

Klamath Mountain Range

Cascade Mountain Range

Great Basin (Modoc Plateau, High Desert)

Central Valley Foothills

Central Valley

Sierra Nevada Mountain Range

Great Basin (East of Sierra Nevada Mtns.)

Mojave Desert (High Desert)

North Coastal Edge

Northern Inland Valleys and Foothills

Central Coast Ranges

Peninsular Southern Coastal Ranges

Transverse Southern Inland Mountains

Southern Inland Valleys and Foothills

Southern Coastal Edge

Sonoran Desert (Low Desert)

Adapted from the original Geographic Subdivisions Map by *The Jepson Manual* 1993, and a Map of Landscape Regions by Bob Perry, as featured in *Landscape Plants for Western Regions,* An Illustrated Guide to Plants for Water Conservation, 1996.

Climate is the main limiting factor in the California landscape, especially with regard to the hardiness of plants. Moisture is the next the most decisive limiting factor, although it usually can be supplemented by artificial means in ornamental landscapes. In California, mean low temperatures range from 0 to -20 degrees F. in the coldest climates to 30 to 40 degrees in more temperate southern regions. Mean high temperatures may range from 100-115 degrees in desert regions to between 75-85 degrees along the coast.

"The West's 24 Climate Zones" in *Sunset Western Garden Book* (Lane Publishing Co., 1967) was developed in association with climatologists, bioclimatologists, meteorologists, and horticulturists and based on *California's Plantclimate Map* (University of California Agricultural Extension Service, 1967). The plant climate map was a departure from the USDA Hardiness Map, which was based on average minimum low temperatures and applied primarily in the agricultural context for which it was intended. The *Sunset* method initiated the term "plant climate," defined as an area in which a common set of temperature ranges, humidity patterns, and other geographic and seasonal characteristics combine to allow certain plants to succeed and cause others to fail. This provides a more accurate means of applying temperature constraints for specialized landscapes and bears a close resemblance to naturally occurring vegetation patterns, which exhibit the influence of climatic subtleties.

This system of plant climate zones incorporated the effects of latitude, elevation, ocean influence, continental air mass, mountains and valleys, and local terrain as determinants of localized climate. Generally winters are longer and colder and daylight hours increase in summer and decrease in winter the further the latitude north. Higher elevations experience colder winters and have comparatively colder night temperatures throughout the year. The greater the influence of the Pacific Ocean, the moister the comparative atmosphere is during all seasons, and the winters are milder and the summers cooler.

Coastal fog plays an important role in California's climate by bringing moisture inland to areas that do not receive summer rainfall. Nature's air conditioning draws in cooler moist air, in onshore breezes, with the resulting rivulets of fog a familiar sight along the coastline as warmer inland air rises in interior valleys.

The continental air mass produces colder winters and hotter, drier summers. Mountains and hills lessen or eliminate the marine influence of the continental air mass. The Coast Ranges lessen it, and the southern Cascade Ranges and Sierra Nevada eliminate it. Local terrain affects cold air and frosts during fall, winter, and spring, caused by thermal belts (cold air flowing downward along sloping terrain) and cold basins (lowlands into which cold air descends to the lowest points). Quite simply, warm air rises and cold air sinks. However, it is important to distinguish between weather (atmospheric conditions at a particular moment) and climate (the all-season accumulation and resulting overall effect of weather).

Sunset Western Garden Book has been a recognized authority in western gardening since the first edition appeared more than fifty years ago. *Sunset*'s close affiliation with horticulturists, botanical gardens, and the nursery trade keep it at the forefront of testing and evaluation of new plant introductions. The composite map presented here allows side-by-side comparison of climate with topography, geography, and vegetation maps, as well as with references in the compendium. For specific areas, readers should refer to more detailed climate zone maps contained in *Sunset Western Garden Book*. Updated in 2007, Sunset Western Climate Zone listings include climate zones for 11 western states, including Alaska, Hawaii, and portions of southwestern Canada, with California exhibiting 20 of these 32 climate zones.

Zones 1, 2, 3 – *Snowy regions of the west* – Northern and southern latitude areas of extreme winter cold, affected by high elevation and continental air masses.

Zones 7, 8, 9 – *The Central Valley and surrounding foothills* – Interior cold air basins and thermal belts, influenced by surrounding mountain ranges and continental air masses.

Zones 10, 11, 13 – *Desert regions* – Including California and extending eastward into high desert areas of Nevada and Utah in the north, with hot summers, cold winters, and cold night temperatures. Low desert areas extend into Colorado, Arizona, and New Mexico in the southwest, with hot summers, cool winters, and cool night temperatures.

Zone 14 – *Inland regions of coastal influence* – Northern and central California areas that experience the effects of cool ocean air or sinking cold air from surrounding mountains or foothills.

Zones 15, 16, 17 – *Areas of direct coastal influence* – Temperate but cool, windy, or foggy areas along the coast, including north coast cold-winter areas, central coast thermal belts, and the north coast marine belt.

Zones 18, 19 – *Southern latitude interior valleys* – Valleys and hilltops, influenced by thermal belts of surrounding mountain ranges and continental air masses.

Zones 20, 21 – *Southern latitude inland regions of coastal influence* – Cold air collecting basins, hilltops, and valleys inland from the coast, influenced by marine and continental air masses.

Zones 22, 23, 24 – *Southern latitude regions of direct coastal influence* – Cold-winter areas of the southern coastline, influenced by thermal belts and marine air masses.

Headings above are grouped by region on the map opposite. Zones 15/16, 18/19, 20/21, and 22/23 have been combined to enhance clarity.

CLIMATE

SUNSET WESTERN GARDEN BOOK
CLIMATE ZONES OF CALIFORNIA

An overall climate zone composite for California, from *Sunset Western Garden Book*, widely accepted as a planting guide for the west coast region, reflecting microclimates affecting plant growth, besides just high and low temperature ranges. It is shown here as a basic overall general reference guide and to note the similarities it shares with the maps of Topography, page 13, Geography, page 15 and Vegetation and Habitats, page 35.

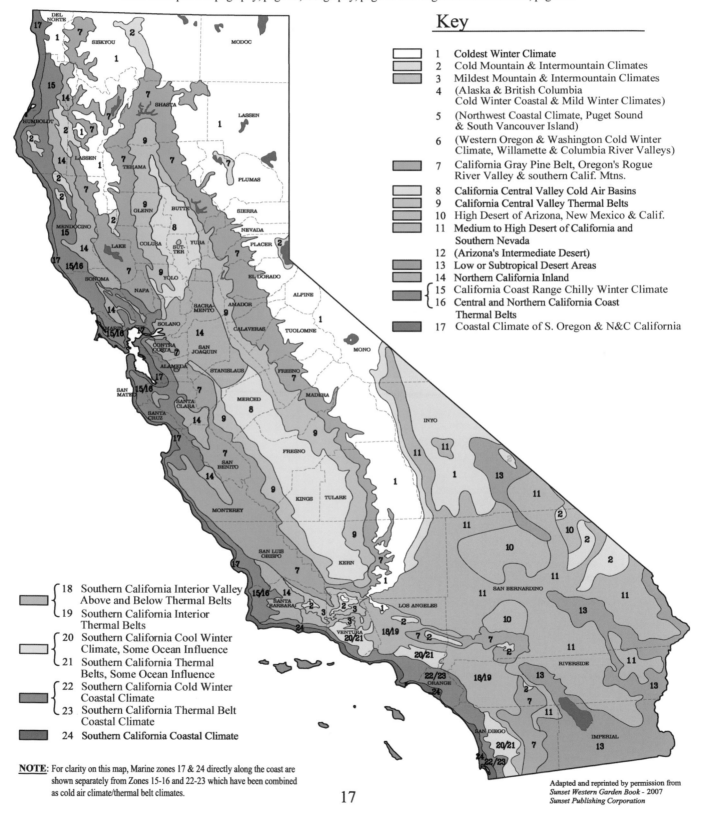

Key

1 Coldest Winter Climate
2 Cold Mountain & Intermountain Climates
3 Mildest Mountain & Intermountain Climates
4 (Alaska & British Columbia Cold Winter Coastal & Mild Winter Climates)
5 (Northwest Coastal Climate, Puget Sound & South Vancouver Island)
6 (Western Oregon & Washington Cold Winter Climate, Willamette & Columbia River Valleys)
7 California Gray Pine Belt, Oregon's Rogue River Valley & southern Calif. Mtns.
8 California Central Valley Cold Air Basins
9 California Central Valley Thermal Belts
10 High Desert of Arizona, New Mexico & Calif.
11 Medium to High Desert of California and Southern Nevada
12 (Arizona's Intermediate Desert)
13 Low or Subtropical Desert Areas
14 Northern California Inland
15 California Coast Range Chilly Winter Climate
16 Central and Northern California Coast Thermal Belts
17 Coastal Climate of S. Oregon & N&C California

18 Southern California Interior Valley Above and Below Thermal Belts
19 Southern California Interior Thermal Belts
20 Southern California Cool Winter Climate, Some Ocean Influence
21 Southern California Thermal Belts, Some Ocean Influence
22 Southern California Cold Winter Coastal Climate
23 Southern California Thermal Belt Coastal Climate
24 Southern California Coastal Climate

NOTE: For clarity on this map, Marine zones 17 & 24 directly along the coast are shown separately from Zones 15-16 and 22-23 which have been combined as cold air climate/thermal belt climates.

Adapted and reprinted by permission from *Sunset Western Garden Book* - 2007 Sunset Publishing Corporation

EFFECTS OF TOPOGRAPHY & GEOGRAPHY ON CALIFORNIA CLIMATE

Sectional diagrams through the northern, central and southern latitudes of the state, to visualize the wide range
of differences in elevation, topography and resulting climatic factors affecting the vegetation of California.

NORTHERN SECTION

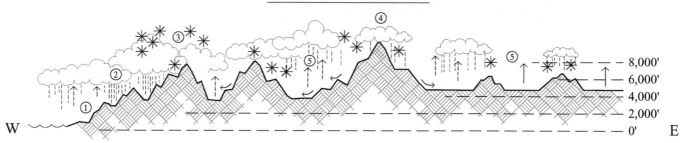

1. Marine air brings in frequent moisture and fog, maintaining moderate temperatures along the coastline, rarely freezing.
2. Outer coastal mountain ranges receive the most rainfall, losing condensed moisture as clouds push inland.
3. Higher coastal peaks receive periods of winter snow.
   Interior valleys between mountains with varied microclimates due to variable topography and elevation. Warm, dry summers with mild , but pronounced seasonal change, with relatively mild winters.

4. As coastal influence decreases, inland mountain ranges and foothills receive lasting winter snows, with warm, dry summers, cold winters.
5. High inland valleys and plains, receiving seasonal rainfall and snows, with warm summers and coldest winter temperatures west of the Great Divide. High summer transpiration, short growing season where temperatures are above freezing.

CENTRAL SECTION

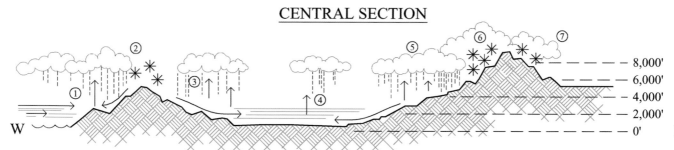

1. Marine air brings in moisture and maintains moderate temperatures along the coastline.
2. Outer coastal mountain ranges receive substantial rainfall as clouds push inland.
3. Inner coast ranges receive less rainfall, losing moisture to transpiration as clouds move westward. Thermal belts of sinking cool air from mountains.
4. Moderate transpiration in central valley. Thermal belts of sinking cold air slides down along side slopes, causing moisture in the air to condense, resulting in thick continual daytime "tule fog" in winter.

Rising hot air on summer days draws in cooler sea breezes through nature's "air conditioning" in the central valley, less so at the extreme ends, as sea breezes are blocked by mountains.
5. As elevation increases and temperature falls, rainfall increases in the forest belt of the state.
6. Winter snow belt, generally above 6,000', intermittently down to 3,500', representing colder temperature zones.
7. Eastward side of Sierra Nevada receives less rainfall, high transpiration. Cold winter temperatures.

SOUTHERN SECTION

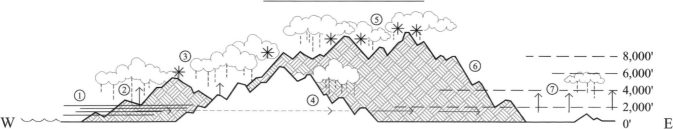

1. Marine air brings in frequent moisture and fog, maintaining moderate temperatures along the coastline, with little or no winter freezing.
2. Outer coastal mountain ranges receive the most rainfall, losing condensed moisture as clouds push inland.
3. Higher coastal peaks receive periods of winter snow.
   Interior valleys between mountains with varied microclimates due to variable topography and elevation. Warm, dry summers with little seasonal change, long growing season, mild winters with intermittent rainfall.
   Offshore sea breezes flow relatively unimpeded through the valley between the mountains running parallel to the south and north. Moisture is dissipated

relatively evenly through its course, diminishing as it passes beyond the inland mountains to the desert. (Onshore breezes flow from east to west, in reverse, drawing dry warm desert air toward the coast).
5. Highest peaks often receive lasting winter snows, contrasting with the flat valleys below.
6. Snows and rainfall diminish on leeward sides of mountain ranges.
7. Flat, rolling terrain of deserts may receive only brief intermittent rainfall, occasionally producing periods of heavy runoff during cloudbursts, followed by extremely dry climate with high transpiration rate.

TREES IN THE NATIVE LANDSCAPE

Native Vegetation Types and Habitats

Pinus ponderosa, with *Quercus kelloggii* understory, is typical of the Pacific montane forest of California. This is the timber belt of the Sierra Nevada, sustaining the timber industry, rural towns, and recreation areas. Western El Dorado County.

The California landscape is not homogeneous, where one part of the state looks like any other. The cool, moist regions of the north coast, the arid deserts of the southeastern portion of the state, the seasonally rainy Mediterranean central section, and the drier south coast regions present a wide variation. Coniferous forests predominate in higher mountains and along the coast, while broadleaf deciduous trees predominate in lower-elevation valley and foothill woodlands. Even within similar vegetation types, species in the north may differ significantly from those in the south, varying not only with topography and climate but with latitude and longitude. The map of vegetation patterns on page 35 shows how closely the native landscape reflects California's topography, geography, and climate.

Exposure, soils, and access to water are other factors that produce variations in these general vegetation types. Predominant tree species as well as subordinate and understory trees and vegetation types are affected by such factors as whether the site is a north- or south-facing hillside or a shaded ravine and dry rocky soil or alluvial floodplain where soil is fertile and the water table is relatively near the surface.

The native landscape has evolved over millions of years through species adaptation and succession, and it continues to evolve today. Receding species or those that have not continued to advance, such as *Abies bracteata, Torreya californica,* and *Sequoiadenron giganteum,* may become more scarce or even rare with time. Conversely, *Pseudotsuga menziesii* and *Pinus ponderosa* may become more prevalent due to their vigor and adaptability as well as the fact that they are used in reforestation plantings. Diseases are a major factor in species decline, currently affecting such species as *Chamaecyparis lawsoniana* and *Pinus radiata. Quercus agrifolia* has recently been affected by viral and fungal diseases that can decimate large numbers of trees in native forests as well as in ornamental landscapes. Fire has always been a part of plant succession, with certain species well suited to reproduction after fire. Cones of some pines remain closed indefinitely, bursting open under the intense heat of fire and releasing seeds for germination with the first rains. *Pseudotsuga macrocarpa* has the ability to sprout new growth from the trunk, replacing limbs burned in fires and eventually producing enough seeds to establish new stands of trees.

The native landscape sustains our livelihood. Forests help to retain water in watershed areas, and runoff provides water for cities, towns, and agricultural areas in valleys below. Timber production in this state is crucial to employment and the economic viability of towns associated with it. Forest land comprises 32% of California's 101 million acres. In timber production, California is second only to Oregon. More than 35 million seedlings are planted each year, replacing trees removed by timber harvesting with "new growth" forests. This effort is made to achieve sustainability, while attempting to allow portions of "old growth" forests and genetically pure stands to remain undisturbed.

Careful management is required to maintain the viability of California's forests while fulfilling the demands of one of the largest housing markets in the nation and providing for the livelihood of the many people employed by the forest products industry. Balancing the sometimes conflicting needs of healthy, productive forests, a growing population and

Oak woodland in low rolling foothill terrain rising from the northern Central Valley floor, commonly used for cattle grazing. Many small rural towns dot the landscape, and these areas are becoming quite appealing for residential expansion. Western Amador County.

economy, the diversity of natural species, and the recreational and restorative values of open space and scenic beauty has become a real concern in recent years that directly or indirectly affects us all.

REGIONAL NATIVE TREES

The Jepson Manual subdivides the state of California and adjacent geographic areas into three "provinces," including the California Floristic Province, the Great Basin, and Desert. Each province is divided into regions and subregions. These physiographic subdivisions are utilized here to highlight variations in climate, topography, and vegetation throughout the state, with an emphasis on the occurrence and range of native trees. Codes used in *The Jepson Manual* are provided to facilitate cross reference with that publication and with listings in the compendium of trees in this book.

According to *The Jepson Manual*, the California Floristic Province (Jepson CA-FP) contains all of the state of California west of the dry provinces of the Great Basin (Jepson GB) and Desert (Jepson D). The six regions of the CA-FP are Northwestern California, the Cascade Ranges, the Sierra Nevada, the Great Central Valley, Central Western California, and Southwestern California. The regions of the GB are the Modoc Plateau and East of the Sierra Nevada. The Desert regions include the Mojave Desert and the Sonoran Desert.

Northwestern California (Jepson NW) is the most heavily forested region of the state, with the most species diversification and the wettest and most predicable climate. Vegetation includes low-elevation and subalpine forests, mixed evergreen and hardwood forests, closed-cone pine forests, oak/pine woodland, oak woodland, chaparral, prairie, and sagebrush scrub. Species that range throughout this region include *Arbutus menziesii, Cercis occidentalis, Cercocarpus betuloides, Chrysolepis chrysophylla, Fraxinus latifolia, Lithocarpus densiflorus, Pinus attenuata, P. lambertiana, Quercus chrysolepis, Q. garryana, Salix scouleriana, Taxus brevifolia,* and *Umbellularia californica. Aesculus californica* occurs only in the central and southern portions of this region.

The **North Coast** (Jepson NCo) extends along the coast from Oregon to Bodega Bay. This subregion is heavily for-

Coastal salt and fresh water marshes in northern California's Humboldt Bay offer spectacular views of vast expanses of marsh grasses dotted with sparse stands of Monterey pine and beach pine. Various willow species extend into the salt zone. Central western Humboldt County.

ested with *Abies grandis, Chamaecyparis lawsoniana, Picea sitchensis, Thuja plicata, Torreya californica, Tsuga heterophylla,* with closed-cone pine forests of *Cupressus macrocarpa, Pinus muricata,* and *P. contorta* ssp. *contorta.* Groves of deciduous *Alnus oregona* and *Salix hookeriana* are mixed among evergreen forest trees along ravines, streams, and the coast.

The **North Coast Ranges** (Jepson NCoR) occur south of the Klamath Ranges, east of the coastal region, and west of the Central Valley. The region is moister to the west, becoming drier to the east.

A northern riparian zone, typical of the Klamath Mixed Coniferous Forest and Redwood Forests, with a dense background of evergreen conifers. Narrow bands of sparse *Populus, Salix, Alnus,* and *Acer macrophyllum* occur along sandy riverbanks where conifers cannot tolerate seasonal flooding. Central Humboldt County.

The **Outer North Coast Ranges** (Jepson NCoRO) receive the highest amount of rainfall. They are characterized by redwood, mixed evergreen, and mixed hardwood forests. Evergreens include *Abies concolor, A. grandis, Chamaecyparis lawsoniana, Pseudotsuga menziesii, Quercus garryana, Sequoia sempervirens, Thuja plicata, Torreya californica,* and *Tsuga heterophylla.* Deciduous species include *Alnus oregona* and *Salix hookeriana.* Evergreen hardwood species include *Arbutus menziesii* and *Q. agrifolia.*

The **High North Coast Ranges** (Jepson NCoRH) receive heavy snows. They typify high-elevation montane and subalpine fir forests, which include *Abies concolor, A. magnifica, Pinus jeffreyi,* and *P. ponderosa. P. lambertiana* and *Tsuga mertensiana* occur sparsely.

The **Inner North Coast Ranges** (Jepson NCoRI) are the driest portion of the North Coast Ranges, on the leeward side of the mountains, characterized by hot, dry summers. Chaparral vegetation includes *Cupressus macnabiana, C. sargentii,* and *Juniperus californica,* with pine/oak woodlands of *Pinus sabiniana, Quercus lobata,* and *Q. douglasii. Fraxinus dipetala* occurs in canyons, and *Salix gooddingii* is common along rivers and streams. Scattered groves of *Sequoia sempervirens* are found in Napa, Solano, and Sonoma counties.

The **Klamath Ranges** (Jepson KR) of the higher peaks to the east are bounded by the Klamath and South Fork Trinity rivers, Interstate 5 to the east, extending north to Oregon. This subregion is heavily forested, predominantly with conifers. *Abies concolor* and *Pseudotsuga menziesii* occur throughout. In the west, these are mixed with *Calocedrus decurrens, Pinus lambertiana,* and *P. ponderosa.* Less common are *Abies magnifica, Pinus contorta* ssp. *murrayana, Tsuga mertensiana, Pinus*

Plant competition is evidenced by spruces and firs mixed with red alder in the north coastal coniferous forest. Thickets form in small riparian ravines among the conifers. Willows and grasses extend toward an open marsh. Del Norte County near Crescent City.

albicaulis, P. attenuata, and *P. jeffreyi.* The less common *Abies amabilis, A. lasiocarpa, A. procera, Chamaecyparis lawsoniana, C. nootkatensis, Picea breweriana,* and *P. engelmannii* occur in a few isolated locations. Deciduous understory and riparian trees include *Acer glabrum, Betula occidentalis,* and *Populus tremuloides.*

Drier inland and eastern slopes of the northern Coast Ranges, with mixed chaparral and foothill woodland, as they rise above the flat grassland of the interior valleys. Though reasonably cool, the dry conditions limit dense forests, and usually sparse groves of trees are surrounded by dense shrub cover. Central Mendocino County.

The **Cascade Ranges** (Jepson CaR) are evidence of the great amount of volcanic activity that formed this region. Mount Shasta and Mount Lassen stand far above the surrounding flatlands. This region is bounded by Interstate 5 and the Great Central Valley on the west and the Modoc Plateau on the east, extending from Oregon to slightly north of the north fork of the Feather River, ending where the Sierra Nevada ranges begin. The chaparral and oak/pine woodland of the western slopes contrast with the grassy plains of the Central Valley, and the montane fir/pine forests stand out against the juniper savanna of the Great Basin to the east.

Conifers of the region include *Pinus attenuata, P. lambertiana, P. jeffreyi, Taxus brevifolia,* and *Calocedrus decurrens.* Deciduous and evergreen broadleaf trees include *Cercocarpus betuloides* and *Quercus garryana. Aesculus californica* occurs

in the southern portions and *Chrysolepis chrysophylla* occurs in northern portions. *Cercis occidentalis, Fraxinus latifolia,* and *Populus fremontii* are often found in ravines and washes.

In addition to trees that extend throughout the Cascade Ranges, the vegetation of the **Cascade Range Foothills** (Jepson CaRF) includes *Fraxinus dipetala, Pinus sabiniana, Quercus douglasii, Q. lobata, Q. wislizeni, Salix gooddingii, Torreya californica,* and *Umbellularia californica.*

The **High Cascade Range** (Jepson CaRH) becomes more heavily forested with coniferous trees at higher elevations receiving the heaviest winter snowfalls. *Abies concolor, A. magnifica, A. procera, Cupressus macnabiana, Juniperus occidentalis, Pinus albicaulis,* and *P. monticola* are predominant. *Acer glabrum, Arbutus menziesii,* and *Salix scouleriana* are subordinate understory and riparian species, in addition to other species that extend throughout the area.

The **Sierra Nevada** (Jepson SN) region extends from the Cascade Range southward to the Tejon Pass of southwestern California. The western slopes of the Sierra Nevada are bordered by the Central Valley. The eastern slopes are considered part of the Desert Province. Trees and vegetation show pronounced differences between northern and southern latitudes. *Calocedrus decurrens* and *Pinus ponderosa* occur throughout the region. Certain species are found in various locations throughout, namely *Pinus attenuata, P. jeffreyi, P. lambertiana,* and *Torreya californica. Taxus brevifolia* occurs only in the northern and central portions. Deciduous and evergreen broadleaf trees and shrublike trees include *Cercis occidentalis, Cercocarpus betuloides, Fraxinus dipetala, F. latifolia, Lithocarpus densiflorus,* and *Salix scouleriana. Fraxinus velutina* occurs only in the southern portions.

The **Sierra Nevada Foothills** (Jepson SNF) subregion is characterized by oak/foothill-pine woodlands, which extend to an elevation of roughly 2,500', where the forest becomes significantly more dense. Foothill woodlands commonly include other indicative species, such as *Aesculus californica, Arbutus menziesii, Pinus sabiniana, Quercus douglasii, Q. garryana, Q. lobata, Q. wislizeni, Salix gooddingii,* and *Umbellularia californica. Cupressus macnabiana* occurs sparsely in the northern Sierra Nevada foothills. *Platanus racemosa* and *Sequoiadendron giganteum* occur sparsely in the central and southern Sierra Nevada foothills. *Celtis reticulata* occurs sparsely in the northern and southern Sierra Nevada foothills, in association with other species that typify the area.

The blue hues of the central Sierra Nevada, where *Pinus ponderosa* and *Pseudotsuga menziesii* predominate in the timber belt, provide a foreground for the snowcapped peaks in the distance. Central Placer County.

Higher subalpine portions of the Sierra Nevada, bordering the alpine zone at treeline, exhibit exposed granitic slopes with sparse conifers and scattered shrub massings on steep slopes with rocky soils. Western El Dorado County.

The **High Sierra Nevada** (Jepson SNH) is a heavily forested subregion with *Abies magnifica, Juniperus occidentalis, Pinus albicaulis, P. contorta* ssp. *murrayana, P. flexilis, P. jeffreyi, P. monticola,* and *Tsuga mertensiana.* Higher elevations are dotted with groves of *Populus tremuloides* together with other species of the area. *Pseudotsuga menziesii* occurs in the northern and central High Sierra Nevada. *Pinus monophylla* occurs in the central and southern High Sierra Nevada.

The **Tehachapi Mountains** (Jepson Teh) are at the southern tip of the Sierra Nevada, where *Abies concolor, Cupressus arizonica,* and *Pinus monophylla* are common conifers in addition to other trees that extend throughout the region. Oaks such as *Quercus douglasii, Q. garryana, Q. lobata,* and *Q. wislizeni* are also commonly found here. *Aesculus californica, Celtis reticulata,* and *Platanus racemosa* occur in riparian habitats.

Inland valley wetlands, with *Salix gooddingii, Acer negundo, Fraxinus latifolia,* tules, and reeds, contrast with the dry fields beyond. These moist areas form essential wildlife habitats. Vegetation of this type also has been found to remove toxic agricultural residues from ground and surface water through photosynthesis, although this is an extremely slow process. Sacramento County.

The **Great Central Valley** (Jepson GV) includes the Sacramento Valley and the San Joaquin Valley. This region extends to where oak/pine woodlands or mixed-hardwood forests begin in the surrounding foothills. The region was once primarily grassland, with marshes in the delta and extensive riparian woodlands along the numerous rivers, creeks, and streams. Much of the native grassland has been converted to agricultural use for grazing, crops, and orchards, and urban development is now replacing much of the farmland.

The **Sacramento Valley** (Jepson ScV), in the northern portion of the Great Central Valley, receives moderate seasonal rainfall and is slightly cooler than the southern portion of the valley. Scattered groves of *Quercus lobata* are characteristic of grasslands of the area. *Salix gooddingii* is also a familiar sight, dotting the flat expanses of moist agricultural lands. *Platanus racemosa* is commonly found in riparian woodlands bordering streams and rivers. *Juglans hindsii,* northern black walnut, occurs in the southern end of this region and has mostly hybridized with *J. nigra,* an eastern U.S. species. Pure stands are limited to only a few isolated areas.

The **San Joaquin Valley** (Jepson SnJV), in the larger southern portion of the Great Central Valley, is hotter and drier than the valley to the north. Here, *Quercus douglasii* becomes more common. *Q. palmeri* occurs in the northwestern portion, though rarely. *Prosopis glandulosa* extends sparsely into this subregion, along with the introduced nonnative species *P. velutina,* which has become naturalized in some locations.

Central Western California (Jepson CW) is bounded by the Pacific Ocean on the west and the Central Valley on the east and extends from just north of the San Francisco Bay Area southward to Point Conception. The region is characterized by meandering valleys and low mountain ranges containing assorted grassland/woodland habitats. *Quercus agrifolia* and *Lithocarpus densiflorus* are predominant evergreens that typify the region. *Arbutus menziesii, Cercocarpus betuloides, Fraxinus dipetala, Platanus racemosa,* and *Prunus ilicifolia* are common understory and riparian trees. *Aesculus californica* occurs in the northern and central portions. Small groves of *Chrysolepis chrysophylla* occur as well, but not in the southern inland regions. *Pinus coulteri* occurs in scattered groves throughout the region.

The **Central Coast** (Jepson CCo) is defined in a narrow band directly along the coastline. This subregion includes the San Francisco Bay, from Bodega Bay in the north, southward to Point Conception. *Pseudotsuga menziesii, Sequoia sempervirens,* and *Cupressus macrocarpa* are most common in the northern section. Occasional groves of *Pinus muricata, Alnus oregona,* and *Salix scouleriana* occur in riparian habitats.

The **San Francisco Bay Area** (Jepson SnFrB) is quite mild in climate and also exhibits a wide diversity of vegetation types. These range from moist redwood forests and groves to dry oak/pine woodlands, chaparral, and coastal scrub. This variation is due to a diverse topography, which includes surrounding ranges and peaks. Assorted groves of *Pseudotsuga menziesii, Sequoia sempervirens,* and *Cupressus macrocarpa* are commonly intermixed with *Quercus agrifolia, Q. douglasii, Q. garryana, Q. lobata,* and *Umbellularia californica. Alnus oregona* and *Fraxinus latifolia* occur frequently in riparian and wetland habitats. Isolated groves of *Juglans hindsii,* hybridized with *J. nigra,* sparsely dot the landscape in grassland areas. *Cupressus sargentii* occurs in isolated groves on Mount Tamalpais. *Taxus brevifolia* is also found here, though sparsely. *Pinus attenuata* occurs in small groves in eastern inland portions.

The **South Coast Ranges** (Jepson SCoR) extending southward are typically drier. This subregion includes conifers such as *Cupressus sargentii* and *Pinus attenuata.* Broadleaf trees in-

Coastal oak woodland savannas create interesting patterns in the rolling terrain of central and southern inland coastal mountains and valleys, which are heavily influenced by the cool, moist air from the ocean. Contra Costa County.

clude *Quercus engelmannii, Q. lobata, Q. palmeri,* and *Q. wislizeni.*

The **Outer South Coast Ranges** (Jepson SCoRO) are the southernmost range of *Sequoia sempervirens.* This area also represents the northernmost range of *Pseudotsuga macrocarpa* and *Juglans californica. Abies bracteata* is an uncommon California species found only in the northern portion in the Santa Lucia Range. *Torreya californica* and *Umbellularia californica* occur intermittently throughout.

In the higher **Inner South Coast Ranges** (Jepson SCoRI), *Juniperus californica, Pinus jeffreyi,* and *Quercus douglasii* are common. *Pinus monophylla* occurs only in southeastern portions.

Southwestern California (Jepson SW), including the Channel Islands, comprises the area from Point Conception and the Tejon Pass southward to Mexico from the Pacific Ocean and extending eastward to the Desert Province. This region includes southern inland and coastal valleys, foothills, and mountains. The surrounding mountains provide a spectacular setting for the Los Angeles Basin, which is the most densely populated region of the state. *Pinus coulteri* and *P. lambertiana* occur only at higher elevations. *Quercus agrifolia* and *Q. wislizeni* are found at lower elevations along with *Platanus racemosa* and *Prunus ilicifolia.*

Coastal scrub, chamise, and chaparral of the rolling terrain of inland southern California among patches of oak woodland provide habitats for wildlife and scenic vistas from urban areas. Eastern Los Angeles County.

The **South Coast** (Jepson SCo) vegetation of coastal sage scrub and chaparral remains on hillsides that surround cities and towns at the lowest elevations. Assorted groves of *Quercus agrifolia, Q. dumosa,* and *Juglans californica* occur. *Q. lobata* is found in the northwest portion of this subregion. *Pinus torreyana* occurs in a few sparse groves in southern coastal areas. *Fraxinus velutina* and *Salix gooddingii* are common along washes and in other riparian habitats.

The **Channel Islands** (Jepson ChI) are home to many unique species, including *Prunus lyonii, Lyonothamnus floribundus,* and *Quercus tomentella.* Other common species include *Cercocarpus betuloides, Lithocarpus densiflorus,* and *Q. lobata. Arbutus menziesii* and *Pinus muricata* are found on the northern Channel Islands. *Pinus torreyana* is found only on Santa Rosa Island. *Q. engelmannii* occurs on Santa Catalina Island of the southern Channel Islands.

The **Transverse Ranges** (Jepson TR) include the mountains that run in an east/west orientation along the northern edge bordering the Los Angeles Basin. The climate becomes progressively hotter and drier to the east. Included in the chaparral vegetation at the lowest elevations are *Cercocarpus betuloides, Fraxinus dipetala, F. velutina,* and *Juglans californica* with sparse occurrences of *Umbellularia californica.* In the southern oak forest and dry montane forests at higher elevations, coniferous trees include *Abies concolor, Pinus flexilis, P. monophylla, P. jeffreyi,* and *Pseudotsuga macrocarpa.*

The **Western Transverse Ranges** (Jepson WTR) are nearest the coastline, north of the San Fernando Valley. *Arbutus menziesii, Lithocarpus densiflorus,* and *Quercus lobata* are found here. *Q. douglasii* also occurs on the northern slopes. *Q. garryana* occurs only in the northeastern portions.

The **San Gabriel Mountains** (Jepson SnGb) are slightly higher, with typically higher-elevation trees such as *Juniperus occidentalis, Pinus contorta* ssp. *murrayana,* and *P. jeffreyi. Quercus engelmannii, Q. palmeri,* and *Salix scouleriana* occur at lower elevations. *Q. lobata* is found only in the western portions.

The **San Bernardino Mountains** (Jepson SnBr), being the farthest east, are the driest. In these mountains *Juniperus occidentalis, Pinus contorta* ssp. *murrayana, Populus tremuloides,* and *Salix scouleriana* are found at the highest elevations. *Pinus attenuata* occurs at lower elevations. *Celtis reticulata* occurs in low-elevation ravines and washes.

The **Peninsular Ranges** (Jepson PR) include the various ranges and mountains occurring throughout the southland, south of the Transverse Ranges. *Abies concolor, Pinus monophylla, P. jeffreyi, P. attenuata,* and *Pseudotsuga macrocarpa* are found at the highest elevations. *Pinus quadrifolia* occurs only sparsely, in the southernmost mountains. *Arbutus menziesii, Celtis reticulata, Cercocarpus betuloides, Cupressus arizonica, Fraxinus dipetala, F. velutina, Prosopis glandulosa, Quercus engelmannii,* and *Salix gooddingii* are common at lower elevations. *Juglans californica* is found in the Santa Ana Mountains. *Q. palmeri* occurs in the eastern ranges. *Umbellularia californica* occurs sparsely in scattered locations.

The **San Jacinto Mountains** (Jepson SnJt) vary specifically in the presence of species such as *Pinus contorta* ssp. *murrayana, P. quadrifolia, P. flexilis,* and *Salix scouleriana.*

The **Great Basin Province** (Jepson GB) occurs along the eastern portion of the state, and extends southward to the Desert Province, east of the California Floristic Province. This area is characterized by low rainfall. Summers are hot to very hot, as reflected in the native vegetation. Trees are more sparse than in the heavily forested mountains to the west. Similar species occur here, however, such as *Pinus contorta* ssp. *murrayana* and *P. jeffreyi*. *P. lambertiana* and *P. sabiniana* occur on the western slopes. Riparian trees such as *Populus tremuloides*, *Populus fremontii*, and *Sambucus mexicana* are commonly found in ravines and washes. However, *P. fremontii* does not occur in the Modoc Plateau.

The **Modoc Plateau** (Jepson MP) covers most of the northeastern corner of the state. This high plateau extends north from Lake Tahoe. The characteristic vegetation is juniper savanna and sagebrush steppe. These areas also include mon-

The northern high desert, or Modoc Plateau, east of the Sierra Nevada and extending eastward to the Great Divide, is dotted with sagebrush and sparse groves of *Pinus ponderosa* and *Juniperus californica*. Eastern Placer County.

tane pine/fir forests in the few short, isolated mountain ranges. Coniferous trees include *Abies concolor*, *Juniperus occidentalis*, *Pinus attenuata*, *P. jeffreyi*, *P. monticola*, *P. ponderosa*, and *Tsuga mertensiana*. *Cercocarpus betuloides* occurs as a shrublike tree. *Acer glabrum*, *Alnus rhombifolia*, *Crataegus douglasii*, *Fraxinus latifolia*, and *Salix scouleriana* are common riparian species.

The **Warner Mountains** (Jepson Wrn), cold in winter, are home to *Pinus albicaulis*, picturesque pines famous for their windswept silhouette of gnarled and contorted branching. Some are the oldest living trees on earth.

The region **East of the Sierra Nevada** (Jepson SNE) extends southward from Lake Tahoe, at high elevations and slopes along the eastern Vegetation is a mosaic of sagebrush steppe, pinyon/juniper woodland, and riparian cottonwood-dominated habitats. *Abies magnifica*, *Pinus monophylla*, *P. flexilis*, and *P. monticola* are common. *Tsuga mertensiana* occurs only in the northernmost areas. *Celtis reticulata* and *Fraxinus velutina* are found with cottonwoods in riparian canyons and washes.

The **White and Inyo Mountains** (Jepson W&I), the southernmost mountains east of the Sierra Nevada, are unique in their subalpine bristlecone pine and limber pine communities, as well as their striking treeless alpine woodlands.

The **Desert Province** (Jepson D) covers the southeastern portion of the state, where creosote bush scrub is the predominant vegetation. This area is characterized by low rainfall and extremely hot summers. Few trees occur in this landscape and those that are present occur sparsely and are often shrublike, such as *Prosopis glandulosa*. *Chilopsis linearis* is commonly found in washes, where water may be present briefly after rainfall, and extends into bordering regions of the Transverse and

The subalpine forest along the eastern slopes of the Sierra Nevada is quite dry, with gravelly soils and exposed, open meadow terrain between sparse forests of the slopes. Western Inyo County.

Peninsular ranges. *Populus fremontii* occurs frequently in riparian locations of canyons and slopes.

The **Mojave Desert** (Jepson DMoj) occupies the northern two-thirds of the Desert Province. This region has more elevation changes and greater temperature ranges than the flatter Sonoran Desert to the south. The region is characterized by Mojave creosote-bush scrub, with areas of saltbrush scrub indicative of alkaline basins. *Yucca brevifolia* stands out in this desolate landscape. Other familiar species such as *Fraxinus velutina* also occur here, though not in any great quantity.

The low **Desert Mountains** (Jepson DMtns) are only high enough to subtly alter the uniformity of this expansive landscape. Coniferous trees such as *Abies concolor*, *Juniperus californica*, *J. occidentalis*, and *Pinus monophylla* occur sparsely in these mountains. *P. edulis* occurs only in the New York Mountains. *Quercus palmeri* and *Q. turbinella* are native oaks in this region, with the latter found only in the New York Mountains. *Celtis reticulata* occurs in riparian washes.

The **Sonoran Desert** (Jepson DSon) extends south of the Mojave Desert. With little transition, it becomes a lower, flatter, and warmer terrain, with a slightly different vegetation of Sonoran creosote-bush scrub. Only isolated oases where desert springs surface, highlighted by *Washingtonia filifera*, provide a hospitable environment for life. Resort communities such as Palm Springs, in contrast, have become quite lush urban oases in their own right.

Mojave creosote-bush scrub predominates in the foreground of the San Gorgonio Mountains. Central eastern Riverside County.

California contains the most diverse and complicated vegetation patterning of any area of comparable size in North America. In comparison to other states, California has the greatest number of plant taxa and, not surprisingly, the greatest number of endangered species. The complex and highly varied terrain and climatic conditions, and the resulting high species diversity, have presented a monumental task of developing an organized means for the study and evaluation of species and vegetation types in California.

The history of vegetation classification in California, as excerpted from *A Manual of California Vegetation* (Sawyer & Keeler-Wolf, 1995) traces the development of early pragmatic approaches to the more specific classification methods being developed today. This history is summarized as follows.

Early in the 1900s, distinct vegetation patterns were beginning to be categorized across the U.S. These distinctions were made according to the most obvious, easily recognizable characteristics. While somewhat simplistic, this approach set the groundwork for recognizing basic vegetation associations in terms of dominant species on a broad, generalized basis. While not identifying species associations at the smallest level of detail, it did establish a basic framework of vegetation patterns. This eventually led to the development of a Vegetational Type Map (VTM) Survey of California (Weislander 1935, Critchfield 1971).

Native vegetation adds subtle beauty to the backdrop of the foothill woodlands in the Coast Range mountains of Marin County. Here, the flowers of *Holodiscus discolor*, pink oceanspray, complement the blue hues of the coastal mountains in the background.

Mapping for these VTM vegetation types was based largely on U.S. Forest Service and individual field studies. After 1940, with the advent of aerial photography, the information gathering process became greatly enhanced. At first, maps of broadly defined plant associations were developed for use in practical applications of wildlife management and forest inventory. Later, more detailed versions were developed that included subtypes. These were more useful for botanical applications and introduced the concept of single species, mixed species, mosaic mixtures, and random isolated species. Weislander & Jensen (1946) synthesized their research into a statewide vegetation map based on the USGS mapping scale.

Detailed maps such as these continue to be important for vegetation studies and now utilize computer-assisted interpretations of satellite imagery and aerial photos. These include the California Gap Analysis Project (GAP) by the Fish and Wildlife Service and the California Vegetation Map by the California Department of Forestry and Fire Protection. While they incorporate species dominance as a determinant, scale hinders depiction of smaller vegetation units. Other mapping systems are currently under development, which create more accurate inventories of habitats and vegetation, with the ultimate goal of creating a national database extending beyond the boundaries of California.

A California Flora (Philip Munz, 1959) established a hierarchical listing for the statewide distribution of plants. This included five provinces, 11 vegetation types, and 29 plant communities. In addition to coining the still often used term of plant communities, this approach also introduced many terms still commonly used in reference to habitat characteristics, such as chaparral and oak woodland. However, several important and ecologically meaningful habitats were missing. The combined works of Ornduff (1974), Barbour & Major (1977), and others refined this early approach, extending study to a much more detailed level, and initiated the development of different methods of categorizing California vegetation according to more specific species associations.

The Wildlife Habitat Relationships (WHR) System (Mayer & Laudenslayer, 1988) prepared in conjunction with the USFS and other agencies and organizations, has been used primarily as a system of classification of California wildlife habitats. This system is useful in evaluating habitat value to wildlife. While it is less successful in differentiating between vegetation types, it does address the problematic confusion among various vegetation/habitat type classifications available for diverse disciplines. These include the study of wildlife, forestry, and range management, as well as ecological and botanical applications. The WHR includes a "Classification Crosswalk," which provides a means of translating one method to another. This results in some measure of continuity among various terminologies that utilize broad-based terms with similar definitions. Methods listed in the Crosswalk include Cheatham & Haller (1975), CNDDB (1986), Cowardin (1979), Kuchler (1977), Munz, & Keck (1973), Parker & Matyas (1981), Paysen et al. (1980), Proctor et al. (1980), and Thorne (1976). The Crosswalk chart features a qualitative comparison between the broad-based components of these methods in a hierarchy of tree, shrub, herbaceous plant, aquatic, and developed habitats.

The development of a more consistent classification method that could be directly applied to research and management use did not occur until the early 1980s. This development coincided with a growing awareness of the need for nationwide study, which required compatible standards and scale for consistency. These standards were prerequisites for application towards an integrated classification system that could be used in a comprehensive approach toward national ecosystem conservation and management. The use of broad-based terms presented a problem in the arising awareness, concern, and urgency for a method to clearly define quantitative descriptions of California's vegetation types in order to distinguish rare and endangered plant communities from more common ones. A common system was also required to ascertain whether certain vegetation was in fact endangered or slowly disappearing. In order to address issues regarding habitat conservation, preservation, restoration, and ecosystems that applied to National Environmental Protection Act (NEPA) and California Environmental Quality Act (CEQA) legislation enacted about that time, a consistent method was required that did not contain ambiguities subject to interpretation. To accomplish this, the California Native Plant Society (CNPS) began development of a classification system for this purpose. This also represented a departure from mapped representations.

The preceding paraphrased excerpt from *A Manual of California Vegetation* has been included here to clarify the transition from earlier methods and terminologies universally in use only 40 years ago to those of the present. Differing terminologies have produced confusion and frustration for those not directly involved in their development or professional use. Some of the earliest "plant community" concepts may now be considered outdated for the critical assessment of natural resource inventories, and newer systems continue to be developed that combine wildlife and vegetation habitats as well as forestry resources. Depending on the particular type of study, individualized systems and terminologies are developed to meet specific needs. Though the systems may vary, most tend to agree at basic levels. It should be understood, however, that vegetation maps and systems using broad-based terminologies are different from vegetation classification systems.

Current standards and methods are continually being evaluated and refined. Rather than causing frustration and confusion due to inconsistencies between various specific terminologies, it may help to understand that one should not be discouraged from using whichever is most accepted and applicable in understanding the complexities of the native landscape.

THE SERIES METHOD OF CLASSIFICATION

The California Native Plant Society (CNPS) published *A Manual of California Vegetation* (Sawyer & Keeler-Wolf, 1995), which introduced the series or alliance approach to classification. The term "series" is analogous to the more currently accepted term "alliance," which is now used in lieu of "series" in keeping with the new national vegetation classification system. This is a definitive method, using a keyed system of series reflecting the basic floristic units, thus avoiding the ambiguity of higher vegetation levels containing many component species, which may not be consistent throughout. The series approach represents a departure from previous methods of

Photo: S. Wayne Kelly

A colorful aspen grove in a subalpine region of the northern Sierra Nevada, with mixed conifers in the background, exemplifying an aspen/sagebrush-mule's ears association. Series associations reflect the occurrence of certain species where specific trees may dominate, listing afterward the ground plane shrubs, groundcovers, or grasses that are present.

applying physiognomic units within ground plane boundaries. Instead, it focuses on identifying specific diverse associations in terms of the taxa contained in each association, which are defined within a series. The key is divided into stages of series dominated by trees, shrubs, and herbs, unique stands, habitats, and vernal pools. Individual series are keyed in sequence within each category. The extents of these series are an inventory of what occurs on the ground plane. This approach is not subject to graphic representations on overall maps, which cannot accurately represent detail at a minute level.

The table that follows lists "Vegetation Series Dominated by Trees" from *A Manual of California Vegetation*. This table also contains a listing of subordinate tree species that may or may not occur. Subordinate tree species are arranged alphabetically according to botanical name, with the tree designator highlighted in parentheses to differentiate dominant from subordinate species. This table provides a clearer picture of major native tree species and their less common subordinate counterparts in various associations. The table also facilitates the use of the series method of abbreviated codes that may accompany individual species in the compendium.

Variation in species composition often occurs among stands comprising a particular series. Associations extend the primary series-level listings to include incidental species that may occur in specific locations. These may include other vegetation series as well. An example from the Douglas-fir series would be the Douglas-fir-incense cedar/California fescue association. Common names are utilized consistently in series and associations, rather than scientific names. Differentiations between and within series and associations involve a hierarchy, represented in layers. That is, they appear in order of dominance: trees/shrubs and herbs/unique stands/habitats/vernal pools. The "/" separates these series hierarchy layers, and hyphens (-) are used between multiple species within a layer. An example of this would be a white fir-Douglas-fir/wild rose-twinflower association. Trees, being the dominant feature of the landscape, are listed first if they occur, followed by shrubs and grasses, etc.

A Manual of California Vegetation (Sawyer & Keeler-Wolf, CNPS, 1995)

Tree series, as defined by major species in a condensed listing adapted from *A Manual of California Vegetation*. Herb & shrub-dominated series and habitats & vernal pools are not included. The term "series" is analogous to the currently accepted term "alliance," which is now used in lieu of "series" in keeping with the new national vegetation classification system.

FORESTS WHERE A SINGLE CONIFER SPECIES DOMINATES

CONIFER FORESTS
Port Orford-cedar series (PorOrfced)
Bigcone Douglas-fir series (BigDoufir)
Douglas-fir series (Doufir)
Engelmann spruce series (Engspruce)
Sitka spruce series (Sitspruce)
Mountain hemlock series (Mouhemlock)
Western hemlock series (Weshemlock)
Incense-cedar series (Inccedar)
Giant sequoia series (Giasequoia)
Redwood series (Redwood)

CYPRESS WOODLANDS
McNab cypress series (McNcypress)
Pygmy cypress series (Pygcypress)
Sargent cypress series (Sarcypress)

JUNIPER WOODLANDS
California juniper series (Caljuniper)
Mountain juniper series (Moujuniper)
Utah juniper series (Utajuniper)
Western juniper series (Wesjuniper)

FIR FORESTS
Grand fir series (Grafir)
Red fir series (Redfir)
Santa Lucia fir series (SanLucfir)
Subalpine fir series (Subfir)
White fir series (Whifir)

CLOSED-CONE FORESTS
Beach pine series (Beapine)
Bishop pine series (Bispine)
Coulter pine series (Coupine)
Knobcone pine series (Knopine)
Monterey pine series (Monpine)

PINYON WOODLANDS
Parry pinyon series (Parpinyon)
Singleleaf pinyon series (Sinpinyon)

TWO-LEAVED PINE FORESTS
Beach pine series (Beapine)
Bishop pine series (Bispine)
Lodgepole pine series (Lodpine)

THREE-LEAVED PINE FORESTS
Coulter pine series (Coupine)
Foothill pine series (Foopine)
Jeffrey pine series (Jefpine)
Knobcone pine series (Knopine)
Monterey pine series (Monpine)
Ponderosa pine series (Ponpine)
Washoe pine series (Waspine)

FIVE-LEAVED PINE FORESTS
Bristlecone pine series (Bripine)
Foxtail pine series (Foxpine)
Limber pine series (Limpine)
Western white pine series (Weswhipine)

FORESTS WHERE A SINGLE NON-CONIFER SPECIES DOMINATES

DESERT WOODLANDS
Fan palm series (Fanpalm)
Shrub series containing trees:
Joshua tree series (Jostree)
Mesquite series (Mesquite)
Mojave yucca series (Mojyucca)
Ocotillo series (Ocotillo)
Foothill palo verde-saguaro series (Foopalversag)

DECIDUOUS FORESTS & WOODLANDS
Black oak series (Blaoak)
Blue oak series (Bluoak)
Oregon white oak series (Orewhioak)
Valley oak series (Valoak)
Red alder series (Redalder)
White alder series (Whialder)
Water birch series (Watbirch)
Aspen series (Aspen)
Black cottonwood series (Blacottonwoo)
Fremont cottonwood series (Frecottonwoo)
California walnut series (Calwalnut)
California sycamore series (Calsycamore)

WILLOW THICKETS
Arroyo willow series (Arrwillow)
Black willow series (Blawillow)
Hooker willow series (Hoowillow)
Pacific willow series (Pacwillow)
Red willow series (Redwillow)
Sitka willow series (Sitwillow)

DECIDUOUS WOODLANDS
California buckeye series (Calbuckeye)
California sycamore series (Calsycamore)
Mesquite series (Mesquite)

EVERGREEN OAK FORESTS & WOODLANDS
Canyon live oak series (Canlivoak)
Coast live oak series (Coalivoak)
Engelmann oak series (Engoak)
Interior live oak series (Intlivoak)
Island oak series (Isloak)
Tanoak series (Tanoak)

MOUNTAIN-MAHOGANY WOODLANDS
Birchleaf mountain-mahogany series (Birmoumah)
Curlleaf mountain-mahogany series (Curmoumah)

EVERGREEN FORESTS
California bay series (Calbay)
Eucalyptus series (Eucalyptus)
Fan palm series (Fanpalm)
Tanoak series (Tanoak)

FORESTS WITH TWO SIMILARLY IMPORTANT SPECIES

COTTONWOOD FORESTS
Aspen series (Aspen)
Black cottonwood series (Blacottonwoo)
Fan palm series (Fanpalm)

OAK FORESTS
Blue oak series (Bluoak)

MIXED LIVE-OAK FORESTS
Bigcone Douglas-fir - canyon live oak series (BigDoufircan)
Canyon live oak series (Canlivoak)
Coast live oak series (Coalivoak)
Coulter pine - canyon live oak series (Coupincanliv)
Engelmann oak series (Engoak)
Santa Lucia fir series (SanLucfir)
Sargent cypress series (Sarcypress)
Black oak series (Blaoak)
Douglas-fir series (Doufir)
Douglas-fir - ponderosa pine series (Doufirponpin)
Douglas-fir - tanoak series (Doufirtan)
Port Orford-cedar series (PorOrfced)

Black oak series (Blaoak)
Jeffrey pine - ponderosa pine series (Jefpinponpin)
Ponderosa pine series (Ponpine)
Singleleaf pinyon - Utah juniper series (SinpinUtajun)
Giant sequoia series (Giasequoia)
Port Orford-cedar series (PorOrfced)
Sitka spruce series (Sitspruce)

MIXED FORESTS & WOODLANDS
Sargent cypress series (Sarcypress)
Black oak series (Blaoak)
Douglas-fir series (Doufir)
Douglas-fir - ponderosa pine series (Doufirponpin)
Douglas-fir - tanoak series (Doufirtan)
Port Orford-cedar series (PorOrfced)
Black oak series (Blaoak)
Jeffrey pine - ponderosa pine series (Jefpinponpin)
Ponderosa pine series (Ponpine)
Singleleaf pinyon - Utah juniper series (SinpinUtajun)
Giant sequoia series (Giasequoia)
Port Orford-cedar series (PorOrfced)
Sitka spruce series (Sitspruce)

FORESTS WITH MORE THAN TWO SIMILARLY IMPORTANT SPECIES

MIXED FORESTS & WOODLANDS
Blue palo verde - ironwood - smoke tree series (Blupalveriro)
Mixed oak series (Mixoak)
Mixed willow series (Mixwillow)
Giant sequoia series (Giasequoia)
Mixed conifer series (Mixconifer)
Mixed subalpine series (Mixsubfor)

UNIQUE STANDS

ONE-OF-A-KIND STANDS
Enriched stands in the Klamath Mountains
Stands on San Benito Mountain

CYPRESS FORESTS
Baker cypress stands (Bakcypsta)
Cuyamaca cypress stands (Cuycypsta)
Gowen cypress stands (Gowcypsta)
Monterey cypress stands (Moncypsta)
Piute cypress stands (Piucypsta)
Santa Cruz cypress stands (SanCrucypsta)
Tecate cypress stands (Teccypsta)

DESERT WOODLANDS
All-thorn stands (Allthosta)
Crucifixion-thorn stands (Cruthosta)
Elephant tree stands (Eletresta)

DUNES
Stands at Antioch dunes (Antdunsta)

FOREST STANDS
Alaska yellow-cedar stands (Alayelcedsta)
Catalina ironwood stands (Catirosta)
Hinds walnut stands (Hinwalsta)
Pacific silver fir stands (Pacsilfirsta)
Torrey pine stands (Torpinsta)
Twoleaf pinyon stands (Twopinsta)

DISTINCTIVE STANDS
Hollyleaf cherry stands (Holchesta)

An alphabetical listing of series dominated by trees, to facilitate references listed in the compendium. In the following table the names in bold are at the series level in *A Manual of California Vegetation* (Sawyer & Keeler-Wolf, CNPS, 1995). Here they are listed as vegetation types such as forests, woodlands, groves, oases, and thickets rather than series, as they represent associations that may not be consistent throughout. Associated trees do not occur consistently within these listings, only where their range may overlap the range of the designator(s), which are indicated in bold type. One can refer to the compendium for range of occurrence maps for individual species to see where associated trees might overlap and certain tree associations might occur. Note: Some associated species not covered in the compendium have been included here for consistency with *A Manual of California Vegetation*.

Alaska yellow-cedar stands (Alayelcedsta)
(*Chamaecyparis nootkatensis*)
Abies concolor
Abies procera
Chamaecyparis lawsoniana
Picea breweriana
Pinus contorta ssp. *murrayana*
Pinus monticola
Pseudotsuga menziesii
Taxus brevifolia
Tsuga mertensiana

Arroyo willow thickets (Arrwillow)
Acer macrophyllum
Platanus racemosa
Populus balsamifera ssp. *trichocarpa*
Populus fremontii
Salix lasiolepis
Salix species
Sambucus mexicana

Aspen groves (Aspen)
(*Populus tremuloides*)
Abies concolor
Abies magnifica ssp. *magnifica*
Pinus contorta ssp. *contorta*

Baker cypress stands (Bakcypsta)
(*Cupressus bakeri*)
Abies concolor
Abies magnifica var. *magnifica*
Calocedrus decurrens
Juniperus occidentalis ssp. *occidentalis*
Pinus attenuata
Pinus jeffreyi
Pinus lambertiana
Pinus ponderosa
Pseudotsuga menziesii

Beach pine forests (Beapine)
(*Pinus contorta* ssp. *contorta*)
Abies grandis
Arbutus menziesii
Picea sitchensis
Pinus muricata
Pseudotsuga menziesii
Tsuga heterophylla

Bishop pine forests (Bispine)
(*Pinus muricata*)
Arbutus menziesii
Cupressus goveniana ssp. *pygmaea*
Pinus contorta ssp. *contorta*
Pinus contorta ssp. *bolanderi*
Pinus radiata
Pseudotsuga menziesii
Sequoia sempervirens

Bigcone Douglas-fir forests (BigDoufir)
(*Pseudotsuga macrocarpa*)
Abies concolor
Calocedrus decurrens
Pinus coulteri
Pinus lambertiana
Pinus monophylla
Pinus ponderosa
Quercus chrysolepis
Quercus kelloggii

Bigcone Douglas-fir - canyon live oak forests (BigDoufircan)
(*Pseudotsuga macrocarpa*)
(*Quercus chrysolepis*)
Juglans californica
Pinus attenuata
Pinus coulteri
Pinus monophylla
Pinus sabiniana
Quercus kelloggii
Umbellularia californica

Birchleaf mountain-mahogany woodlands (Birmoumah)
(*Cercocarpus betuloides*)
Juniperus californica
Pinus monophylla
Pinus sabiniana
Quercus berberidifolia

Black cottonwood forests (Blacottonwoo)
(*Populus balsamifera* ssp. *trichocarpa*)
Abies concolor
Acer macrophyllum
Acer negundo
Alnus rubra
Fraxinus latifolia
Juniperus occidentalis
Pinus jeffreyi
Pinus contorta ssp. *murrayana*
Populus tremuloides
Salix exigua
Salix hookeriana
Salix lutea
Salix scouleriana

Black oak forests and woodlands (Blaoak)
(*Quercus kelloggii*)
Acer macrophyllum
Arbutus menziesii
Calocedrus decurrens
Pinus attenuata
Pinus jeffreyi
Pinus ponderosa
Pseudotsuga menziesii
Quercus agrifolia
Quercus chrysolepis
Quercus garryana
Quercus lobata
Umbellularia californica

Black willow thickets (Blawillow)
(*Salix lasiolepis*)
Alnus rhombifolia
Platanus racemosa
Populus balsamifera ssp. *trichocarpa*
Populus fremontii
Salix gooddingii
Salix hookeriana
Salix laevigata
Salix lucida ssp. *lasiandra*
Salix sitchensis
Salix species
Sambucus mexicana

Blue oak forests and woodlands (Bluoak)
(*Quercus douglasii*)
Cercocarpus betuloides

Juniperus occidentalis ssp. *occidentalis*
Pinus sabiniana
Quercus agrifolia
Quercus lobata
Quercus wislizeni
Quercus x *alvordiana*

Blue palo verde - ironwood - smoke tree woodlands (Blupalveriro)
(*Cercidium floridum*)
Chilopsis linearis
Prosopis glandulosa
Prosopis pubescens
Salix gooddingii

Bristlecone pine forests (Bripine)
(*Pinus longaeva*)
Pinus flexilis

California bay forests (Calbay)
(*Umbellularia californica*)
Arbutus menziesii
Lithocarpus densiflorus
Quercus agrifolia
Quercus chrysolepis
Quercus wislizeni
Sequoia sempervirens

California buckeye woodlands (Calbuckeye)
(*Aesculus californica*)
Fraxinus dipetala
Pinus sabiniana
Prunus ilicifolia
Quercus wislizeni
Umbellularia californica

California juniper woodlands (Caljuniper)
(*Juniperus californica*)
Pinus monophylla
Pinus quadrifolia
Quercus turbinella
Yucca brevifolia

California sycamore woodlands (Calsycamore)
(*Platanus racemosa*)
Alnus rhombifolia
Populus fremontii
Quercus agrifolia
Quercus lobata
Salix gooddingii
Salix laevigata
Salix lasiolepis
Salix lutea
Umbellularia californica

California walnut forests (Calwalnut)
(*Juglans californica* var. *californica*)
Fraxinus dipetala
Quercus agrifolia
Sambucus mexicana
Umbellularia californica

Canyon live oak forests and woodlands (Canlivoak)
(*Quercus chrysolepis*)
Abies concolor
Acer macrophyllum

Arbutus menziesii
Calocedrus decurrens
Pinus coulteri
Pinus lambertiana
Pinus ponderosa
Pseudotsuga macrophylla
Pseudotsuga menziesii
Quercus garryana
Quercus kelloggii
Umbellularia californica

Catalina ironwood stands (Catirosta)
(*Lyonothamnus floribundus*)

Coast live oak forests and woodlands
(Coalivoak)
(*Quercus agrifolia*)
Acer macrophyllum
Acer negundo
Arbutus menziesii
Quercus douglasii
Quercus engelmannii
Quercus kelloggii
Umbellularia californica

Coulter pine forest (Coupine)
(*Pinus coulteri*)
Pinus ponderosa
Pinus sabiniana
Pseudotsuga macrocarpa
Quercus agrifolia
Quercus chrysolepis
Quercus kelloggii
Quercus wislizeni

Coulter pine - canyon live oak forests and
woodlands (Coupincanliv)
(*Pinus coulteri*)
(*Quercus chrysolepis*)
Pinus ponderosa
Pinus sabiniana
Pseudotsuga macrocarpa
Quercus agrifolia
Quercus kelloggii

Curlleaf mountain-mahogany woodlands
(Curmoumah)
(*Cercocarpus ledifolius*)
Juniperus occidentalis ssp. *australis*
Juniperus occidentalis ssp. *occidentalis*
Pinus albicaulis
Pinus balfouriana
Pinus contorta ssp. *murrayana*
Pinus jeffreyi
Pinus monophylla

Cuyamaca cypress stands (Cuycypsta)
(*Cupressus stephensonii*)
Pinus coulteri

Douglas-fir forests (Doufir)
(*Pseudotsuga menziesii*)
Abies concolor
Abies magnifica
Acer circinatum
Acer macrophyllum
Arbutus menziesii
Calocedrus decurrens
Chamaecyparis lawsoniana
Cornus nuttallii
Lithocarpus densiflorus
Pinus jeffreyi
Pinus lambertiana
Pinus ponderosa
Quercus chrysolepis
Quercus garryana
Quercus kelloggii
Sequoia sempervirens
Tsuga heterophylla
Umbellularia californica

Douglas-fir - ponderosa pine forests
(Doufirponpin)
(*Pseudotsuga menziesii*)
(*Pinus ponderosa*)
Acer macrophyllum
Calocedrus decurrens
Pinus jeffreyi
Pinus lambertiana
Quercus chrysolepis
Quercus garryana
Quercus kelloggii

Douglas-fir - tanoak forests (Doufirtan)
(*Pseudotsuga menziesii*)
(*Lithocarpus densiflorus*)
Acer circinatum
Alnus rhombifolia
Arbutus menziesii
Chamaecyparis lawsoniana
Chrysolepis chrysophylla
Pinus lambertiana
Pinus ponderosa
Quercus chrysolepis
Quercus kelloggii
Taxus brevifolia
Umbellularia californica

Engelmann oak forests (Engoak)
(*Quercus engelmannii*)
Juglans californica var. *californica*
Quercus agrifolia
Quercus berberidifolia
Quercus kelloggii

Engelmann spruce forests (Engspruce)
(*Picea engelmannii*)
Abies concolor
Abies magnifica var. *shastensis*
Calocedrus decurrens
Picea breweriana
Pinus contorta ssp. *murrayana*
Pinus lambertiana
Pinus monticola
Pinus ponderosa
Pseudotsuga menziesii
Taxus brevifolia
Tsuga mertensiana

Eucalyptus groves (Eucalyptus)
Indigenous nonnative species
Eucalyptus camaldulensis
Eucalyptus citriodora
Eucalyptus cladocalyx
Eucalyptus globulus
Eucalyptus polyanthemos
Eucalyptus pulverulenta
Eucalyptus sideroxylon
Eucalyptus tereticornis
Eucalyptus viminalis

Fan palm oases (Fanpalm)
(*Washingtonia filifera*)
Fraxinus velutina
Platanus racemosa
Populus fremontii
Prosopis glandulosa
Quercus chrysolepis
Salix exigua
Salix gooddingii
Salix lasiolepis

Foothill pine woodland (Foopine)
Aesculus californica
Cercocarpus betuloides
Juniperus occidentalis
Pinus coulteri
Pinus sabiniana
Quercus agrifolia
Quercus douglasii
Quercus kelloggii

Quercus lobata
Quercus wislizeni

Foothill palo verde-saguaro woodlands
(Foopalversag)
Cercidium microphyllum

Foxtail pine forests (Foxpine)
(*Pinus balfouriana*)
Abies magnifica var. *shastensis*
Pinus albicaulis
Pinus contorta ssp. *murrayana*
Pinus monticola
Pinus flexilis
Tsuga mertensiana

Fremont cottonwood forests
(Frecottonwoo)
(*Populus fremontii*)
Acer negundo
Fraxinus latifolia
Juglans californica var. *californica*
Juglans californica var. *hindsii*
Juglans nigra
Platanus racemosa
Salix exigua
Salix gooddingii
Salix lucida ssp. *lasiandra*
Salix laevigata
Salix lasiolepis
Salix lutea

Giant sequoia groves (Giasequoia)
(*Sequoiadendron giganteum*)
Abies concolor
Abies magnifica var. *magnifica*
Calocedrus decurrens
Pinus contorta ssp. *murrayana*
Pinus jeffreyi
Pinus lambertiana
Pinus ponderosa
Pseudotsuga menziesii
Quercus kelloggii

Gowen cypress stands (Gowcypsta)
(*Cupressus goveniana* ssp. *goveniana*)
Pinus muricata
Pinus radiata

Grand fir forests (Grafir)
(*Abies grandis*)
Abies concolor
Alnus rubra
Lithocarpus densiflorus
Picea sitchensis
Pinus muricata
Pseudotsuga menziesii
Sequoia sempervirens
Tsuga heterophylla

Hinds walnut stands (Hinwalsta)
(*Juglans californica* var. *hindsii*)
Juglans nigra

Hollyleaf cherry stands (Holchesta)
(*Prunus ilicifolia* ssp. *ilicifolia*)
Prunus ilicifolia ssp. *lyonii*

Hooker willow thickets (Hoowillow)
(*Salix hookeriana*)
Alnus oregona
Alnus rhombifolia
Populus balsamifera ssp. *trichocarpa*
Populus fremontii
Salix gooddingii
Salix laevigata
Salix lucida ssp. *lasiandra*
Salix sitchensis
Salix species

Incense-cedar forests (Inccedar)
(*Calocedrus decurrens*)

Abies concolor
Pinus contorta ssp. *murrayana*
Pinus lambertiana
Pinus ponderosa
Pseudotsuga menziesii
Quercus chrysolepis

Interior live oak forests and woodlands (Intlivoak)
(*Quercus wislizeni*)
Arbutus menziesii
Lithocarpus densiflorus
Pinus sabiniana
Quercus douglasii
Quercus kelloggii

Island oak forests (Isloak)
(*Quercus tomentella*)
Prunus ilicifolia ssp. *lyonii*
Quercus macdonaldii

Jeffrey pine forests & woodlands (Jefpine)
(*Pinus jeffreyi*)
Abies concolor
Abies magnifica var. *magnifica*
Abies magnifica var. *shastensis*
Calocedrus decurrens
Cercocarpus ledifolius
Chamaecyparis lawsoniana
Chrysolepis sempervirens
Pinus attenuata
Pinus balfouriana
Pinus contorta ssp. *murrayana*
Pinus monticola
Pinus ponderosa
Pseudotsuga menziesii
Quercus chrysolepis
Quercus kelloggii
Quercus wislizeni

Jeffrey pine - ponderosa pine forests (Jefpinponpin)
(*Pinus jeffreyi*)
(*Pinus ponderosa*)
Abies concolor
Calocedrus decurrens
Fraxinus latifolia
Juniperus occidentalis ssp. *occidentalis*
Quercus kelloggii
Quercus vaccinifolia

Joshua tree woodlands (Jostree)
(*Yucca brevifolia*)
Pinus monophylla
Quercus turbinella

Knobcone pine forests (Knopine)
(*Pinus attenuata*)
Lithocarpus densiflorus
Pinus contorta ssp. *murrayana*
Pinus coulteri
Pinus monticola
Pinus radiata
Pinus sabiniana
Quercus chrysolepis
Quercus wislizeni

Limber pine forests (Limpine)
(*Pinus flexilis*)
Abies concolor
Cercocarpus ledifolius
Pinus albicaulis
Pinus balfouriana
Pinus contorta ssp. *murrayana*
Pinus jeffreyi
Pinus longaeva

Lodgepole pine forests (Lodpine)
(*Pinus contorta* ssp. *murrayana*)
Abies concolor
Abies magnifica var. *magnifica*

Abies magnifica var. *shastensis*
Pinus albicaulis
Pinus balfouriana
Pinus flexilis
Pinus monticola
Tsuga mertensiana

McNab cypress woodlands (McNcypress)
(*Cupressus macnabiana*)
Cupressus sargentii
Pinus attenuata
Pinus sabiniana
Quercus douglasii
Quercus wislizeni

Mesquite bosques (Mesquite)
(*Prosopis glandulosa*)
Prosopis pubescens

Mixed conifer forests (Mixconifer)
Abies concolor
Calocedrus decurrens
Lithocarpus densiflorus
Pinus jeffreyi
Pinus lambertiana
Pinus ponderosa
Pseudotsuga macrocarpa
Pseudotsuga menziesii
Quercus chrysolepis
Quercus kelloggii
Quercus vaccinifolia

Mixed oak forests and woodlands (Mixoak)
Aesculus californica
Arbutus menziesii
Pinus ponderosa
Pinus sabiniana
Pseudotsuga menziesii
Quercus agrifolia
Quercus douglasii
Quercus garryana
Quercus kelloggii
Quercus lobata
Quercus wislizeni
Umbellularia californica

Mixed subalpine forests (Mixsubfor)
Abies magnifica var. *magnifica*
Abies magnifica var. *shastensis*
Pinus albicaulis
Pinus balfouriana
Pinus contorta ssp. *murrayana*
Pinus monticola
Tsuga mertensiana

Mixed willow thickets (Mixwillow)
Acer macrophyllum
Alnus rhombifolia
Alnus rubra
Platanus racemosa
Populus balsamifera ssp. *trichocarpa*
Populus fremontii
Salix exigua
Salix gooddingii
Salix hookeriana
Salix laevigata
Salix lasiolepis
Salix lucida ssp. *lasiandra*
Salix sitchensis

Monterey pine forests (Monpine)
(*Pinus radiata*)
Arbutus menziesii
Pinus attenuata
Pinus muricata
Pinus ponderosa
Pseudotsuga menziesii
Quercus agrifolia
Sequoia sempervirens

Monterey cypress stands (Moncypsta)
(*Cupressus macrocarpa*)

Mountain hemlock forests (Mouhemlock)
(*Tsuga mertensiana*)
Abies concolor
Abies magnifica var. *magnifica*
Abies magnifica var. *shastensis*
Pinus albicaulis
Pinus contorta ssp. *contorta*
Pinus monticola

Mountain juniper woodlands (Moujuniper)
(*Juniperus occidentalis* ssp. *australis*)
Abies concolor
Pinus contorta ssp. *murrayana*
Pinus jeffreyi
Pinus monophylla
Quercus chrysolepis
Quercus kelloggii

Ocotillo woodlands (Ocotillo)
Cercidium floridum

Oregon white oak woodlands (Orewhioak)
(*Quercus garryana* var. *garryana*)
Arbutus menziesii
Calocedrus decurrens
Juniperus occidentalis ssp. *occidentalis*
Pinus jeffreyi
Pinus ponderosa
Pseudotsuga menziesii
Quercus chrysolepis
Quercus kelloggii

Pacific silver fir stands (Pacsilfirsta)
(*Abies amabilis*)
Abies magnifica var. *shastensis*
Abies procera
Picea breweriana
Pinus contorta ssp. *murrayana*
Pinus monticola
Tsuga mertensiana

Pacific willow thickets (Pacwillow)
Acer macrophyllum
Alnus rhombifolia
Alnus oregona
Platanus racemosa
Populus balsamifera ssp. *trichocarpa*
Populus fremontii
Salix gooddingii
Salix hookeriana
Salix laevigata
Salix lucida ssp. *lasiandra*
Salix sitchensis
Salix species
Sambucus mexicana

Parry pinyon woodlands (Parpinyon)
(*Pinus quadrifolia*)
Pinus jeffreyi
Pinus monophylla

Piute cypress stands (Piucypsta)
(*Cupressus nevadensis*)
Juniperus californica
Pinus monophylla
Pinus sabiniana
Quercus douglasii

Ponderosa pine forests and woodlands (Ponpine)
(*Pinus ponderosa*)
Abies concolor
Calocedrus decurrens
Cercocarpus ledifolius
Pinus coulteri
Pinus contorta ssp. *murrayana*
Pinus jeffreyi

Pinus lambertiana
Pseudotsuga menziesii
Quercus chrysolepis
Quercus kelloggii

Port Orford-cedar forests (PorOrfced)
(*Chamaecyparis lawsoniana*)
Abies concolor
Abies magnifica var. *shastensis*
Acer circinatum
Lithocarpus densiflorus
Picea sitchensis
Pinus jeffreyi
Pinus lambertiana
Pinus monticola
Pinus ponderosa
Pseudotsuga menziesii
Quercus vaccinifolia
Umbellularia californica

Pygmy cypress forests (Pygcypress)
(*Cupressus goveniana* ssp. *pygmaea*)
Cupressus goveniana ssp. *goveniana*
Pinus contorta ssp. *bolanderi*
Pinus contorta ssp. *contorta*
Pinus muricata
Sequoia sempervirens

Red alder forests (Redalder)
(*Alnus rubra*)
Abies grandis
Acer circinatum
Picea sitchensis
Populus balsamifera ssp. *trichocarpa*
Pseudotsuga menziesii
Salix hookeriana
Salix lasiolepis
Sequoia sempervirens
Tsuga heterophylla

Red fir forests (Redfir)
(*Abies magnifica* var. *magnifica*)
(*Abies magnifica* var. *shastensis*)
Abies concolor
Abies procera
Calocedrus decurrens
Chamaecyparis lawsoniana
Chrysolepis sempervirens
Picea breweriana
Pinus albicaulis
Pinus contorta ssp. *murrayana*
Pinus jeffreyi
Pinus lambertiana
Pinus monticola
Tsuga mertensiana

Red willow thickets (Redwillow)
(*Salix laevigata*)
Alnus rhombifolia
Platanus racemosa
Populus fremontii
Salix species
Sambucus mexicana

Redwood forests (Redwood)
(*Sequoia sempervirens*)
Abies grandis
Acer macrophyllum
Arbutus menziesii
Lithocarpus densiflorus
Pseudotsuga menziesii
Tsuga heterophylla
Umbellularia californica

Santa Cruz cypress stands (SanCrucypsta)
(*Cupressus abramsiana*)
Pinus attenuata
Pinus ponderosa
Quercus chrysolepis

Santa Lucia fir forests (SanLucfir)
(*Abies bracteata*)
Arbutus menziesii
Lithocarpus densiflorus
Pinus coulteri
Pinus lambertiana
Pinus ponderosa
Quercus chrysolepis

Sargent cypress woodlands (Sarcypress)
(*Cupressus sargentii*)
Cupressus macnabiana
Pinus attenuata
Pinus sabiniana
Quercus wislizeni
Umbellularia californica

Singleleaf pinyon woodlands (Sinpinyon)
(*Pinus monophylla*)
Juniperus californica
Juniperus occidentalis ssp. *australis*
Juniperus osteosperma
Pinus jeffreyi
Pinus ponderosa
Quercus chrysolepis

Sitka spruce forests (Sitspruce)
(*Picea sitchensis*)
Abies grandis
Alnus rubra
Sequoia sempervirens
Tsuga heterophylla

Sitka willow thickets (Sitwillow)
(*Salix sitchensis*)
Acer macrophyllum
Alnus rhombifolia
Alnus oregona
Platanus racemosa
Populus balsamifera ssp. *trichocarpa*
Populus fremontii
Salix gooddingii
Salix hookeriana
Salix laevigata
Salix lucida ssp. *lasiandra*
Salix species
Sambucus mexicana

Subalpine fir forests (Subfir)
(*Abies lasiocarpa*)
Abies amabilis
Abies concolor
Abies magnifica var. *shastensis*
Picea breweriana
Picea engelmannii
Pinus contorta ssp. *murrayana*
Pinus monticola
Taxus brevifolia
Tsuga mertensiana

Tanoak forests (Tanoak)
(*Lithocarpus densiflorus*)
Arbutus menziesii
Pinus coulteri
Pinus lambertiana
Pseudotsuga macrocarpa
Pseudotsuga menziesii
Quercus agrifolia
Quercus chrysolepis
Quercus kelloggii
Umbellularia californica

Tecate cypress stands (Teccypsta)
(*Cupressus forbesii*)

Torrey pine stands (Torpinsta)
(*Pinus torreyana*)
Quercus dumosa

Twoleaf pinyon stands (Twopinsta)
(*Pinus edulis*)

Quercus chrysolepis
Quercus turbinella

Valley oak forests and woodlands (Valoak)
(*Quercus lobata*)
Fraxinus latifolia
Platanus racemosa
Quercus agrifolia (coastal transition zones)
Quercus douglasii
Quercus kelloggii

Washoe pine forests and woodlands (Waspine)
(*Pinus washoensis*)
Abies concolor
Abies magnifica var. *magnifica*
Pinus contorta ssp. *murrayana*
Pinus jeffreyi
Pinus monticola
Pinus ponderosa

Water birch thickets (Watbirch)
(*Betula occidentalis*)
Populus balsamifera ssp. *trichocarpa*
Populus fremontii
Salix exigua
Salix lasiolepis
Salix laevigata
Salix lucida ssp. *caudata*

White alder forests (Whialder)
(*Alnus rhombifolia*)
Acer macrophyllum
Chamaecyparis lawsoniana
Fraxinus latifolia
Lithocarpus densiflorus
Platanus racemosa
Pseudotsuga menziesii

Western juniper woodlands (Wesjuniper)
(*Juniperus occidentalis* ssp. *australis*)
(*Juniperus occidentalis* ssp. *occidentalis*)
Abies concolor
Cercocarpus ledifolius
Pinus jeffreyi
Pinus ponderosa
Pinus washoensis
Quercus garryana
Quercus kelloggii

Western white pine forests (Weswhipine)
(*Pinus monticola*)
Abies concolor
Abies magnifica var. *magnifica*
Abies magnifica var. *shastensis*
Pinus attenuata
Pinus balfouriana
Pinus contorta ssp. *murrayana*
Pinus jeffreyi
Pseudotsuga menziesii

White fir forests (Whifir)
(*Abies concolor*)
Abies grandis
Abies magnifica var. *magnifica*
Abies magnifica var. *shastensis*
Acer circinatum
Acer macrophyllum
Arbutus menziesii
Calocedrus decurrens
Chamaecyparis lawsoniana
Chrysolepis chrysophylla
Chrysolepis sempervirens
Cornus nuttallii
Picea breweriana
Pinus contorta ssp. *murrayana*
Pinus jeffreyi
Pinus lambertiana
Pinus monophylla

Pinus ponderosa
Pseudotsuga menziesii
Quercus chrysolepis
Quercus kelloggii
Quercus sadleriana
Quercus vaccinifolia

Whitebark pine forests (Whipine)
(*Pinus albicaulis*)
Abies magnifica var. *magnifica*
Abies magnifica var. *shastensis*
Pinus balfouriana
Pinus contorta ssp. *murrayana*

TREE SPECIES NOT ILLUSTRATED IN THE COMPENDIUM

Cercidium microphyllum
Cercocarpus ledifolius
Cupressus abramsiana
Cupressus forbesii
Cupressus nevadensis
Cupressus stephensonii
Cupressus goveniana ssp. *pygmaea*
Cupressus goveniana ssp. *goveniana*
Juniperus osteosperma
Pinus balfouriana
Pinus contorta ssp. *bolanderi*
Pinus longaeva
Pinus washoensis
Prosopis pubescens
Quercus x *alvordiana*
Quercus macdonaldii
Salix exigua
Salix lucida ssp. *caudata*
Salix lutea

In addition, *Lithocarpus densiflorus* var. *echinoides*, *Quercus berberidifolia*, *Q. cornelius-mulleri*, *Q. durata*, *Q. garryana* var. *breweri*, *Q. sadleriana*, and *Q. vaccinifolia* are not included in the compendium. They are considered shrubs and do not become tree-sized.

NATIVE TREE SPECIES ILLUSTRATED IN THE COMPENDIUM

As a reference, one may refer to the compendium in this book for many of the trees indicated in the preceding series classification and associations listings. Most are listed in alphabetical order below, except for a few species that were not available for photographic representation.

Abies amabilis, Pacific silver fir
Abies bracteata, bristlecone fir
Abies concolor, white fir
Abies grandis, grand fir
Abies lasiocarpa, subalpine fir
Abies magnifica, California red fir
Abies procera, noble fir
Acer circinatum, vine maple
Acer glabrum, mountain maple
Acer grandidentatum, bigtooth maple
Acer macrophyllum, bigleaf maple
Acer negundo, box elder
Aesculus californica, California buckeye
Alnus oregona, red alder
Alnus rhombifolia, white alder
Arbutus menziesii, madrone
Betula occidentalis, western (water) birch
Calocedrus decurrens, incense cedar
Celtis reticulata, netleaf hackberry
Cercidium floridum, palo verde
Cercis occidentalis, western redbud
Cercocarpus betuloides, mountain ironwood

Chamaecyparis lawsoniana, Port Orfordcedar
Chamaecyparis nootkatensis, Nootka cypress
Chilopsis linearis, desert-willow
Chrysolepis chrysophylla, golden chinquapin
Cornus nuttallii, western dogwood
Crataegus douglasii, black hawthorn
Cupressus arizonica, Arizona cypress
Cupressus macnabiana, McNab cypress
Cupressus macrocarpa, Monterey cypress
Cupressus sargentii, Sargent's cypress
Fraxinus dipetala, foothill ash
Fraxinus latifolia, Oregon ash
Fraxinus velutina, Arizona ash
Juglans hindsii, California black walnut
Juniperus californica, California juniper
Juniperus occidentalis, western juniper
Lithocarpus densiflorus, tanbark oak
Lyonothamnus floribundus, Catalina ironwood
Picea breweriana, weeping spruce
Picea engelmannii, Engelmann spruce
Picea sitchensis, Sitka spruce
Pinus albicaulis, whitebark pine
Pinus attenuata, knobcone pine
Pinus contorta ssp. *contorta*, shore pine
Pinus contorta ssp. *murrayana*, lodgepole pine
Pinus coulteri, Coulter pine
Pinus edulis, Colorado pinyon
Pinus flexilis, limber pine
Pinus jeffreyi, Jeffrey pine
Pinus lambertiana, sugar pine
Pinus monophylla, singleleaf pinyon
Pinus monticola, western white pine
Pinus muricata, Bishop pine
Pinus ponderosa, ponderosa pine
Pinus quadrifolia, Parry pinyon pine
Pinus radiata, Monterey pine
Pinus sabiniana, gray pine
Pinus torreyana, Torrey pine
Platanus racemosa, western sycamore
Populus fremontii, Fremont cottonwood
Populus tremuloides, quaking aspen
Populus trichocarpa, black cottonwood
Prosopis glandulosa, honey mesquite
Prunus ilicifolia, hollyleaf cherry
Prunus lyonii, Catalina cherry
Pseudotsuga macrocarpa, bigcone Douglasfir
Pseudotsuga menziesii, Douglas-fir
Quercus agrifolia, coast live oak
Quercus chrysolepis, canyon live oak
Quercus douglasii, blue oak
Quercus dumosa, Nuttall's scrub oak
Quercus engelmannii, Engelmann oak
Quercus garryana, Oregon oak
Quercus kelloggii, California black oak
Quercus lobata, valley oak
Quercus morehus, oracle oak
Quercus palmeri, Palmer's oak
Quercus tomentella, island oak
Quercus turbinella, shrub live oak
Quercus wislizeni, interior live oak
Salix gooddingii, Goodding's black willow
Salix hindsiana, sandbar willow
Salix hookeriana, Hooker willow
Salix laevigata, red willow
Salix lasiandra, yellow willow
Salix lasiolepis, arroyo willow
Salix scouleriana, Scouler's willow
Sambucus mexicana, blue elderberry
Sequoia sempervirens, coast redwood
Sequoiadendron giganteum, giant sequoia
Taxus brevifolia, western yew
Thuja plicata, western red cedar
Torreya californica, California nutmeg

A stately sugar pine, with its long extended branches and enormous hanging cones, is a strikingly beautiful subordinate forest tree in the Douglas-fir - ponderosa pine series.

Tsuga heterophylla, western hemlock
Tsuga mertensiana, mountain hemlock
Umbellularia californica, California bay
Yucca brevifolia, Joshua tree
Washingtonia filifera, California fan palm

NATURALIZED SPECIES

Eucalyptus camaldulensis, red gum
Eucalyptus citriodora, lemon-scented gum
Eucalyptus cladocalyx, sugar gum
Eucalyptus globulus, blue gum
Eucalyptus polyanthemos, silver dollar gum
Eucalyptus pulverulenta, silver mountain gum
Eucalyptus sideroxylon, red ironbark
Eucalyptus viminalis, manna gum
Parkinsonia aculeata, Mexican palo verde
Prosopis velutina, Arizona mesquite

(Plant Communities)

Establishment of the series method as a classification system for vegetation represents a major milestone for effective and precise vegetation assessment and botanical use. Earlier approaches still have profound relevance, even though they are not as precise, especially because of their widely recognized terms associated with plant communities and habitats, which remain in common use today, such as forest, woodland, chaparral, and scrub. These terms reflect a hierarchy of vegetation types commonly understood and used by many people. Much like common names associated with scientific or latinized botanical names of plants, common terms for vegetation types provide instantaneous general recognition for most people.

The General Vegetation Types Map presented here is a simplified representation that consolidates three major categories, six subcategories, and 19 smaller vegetation units representing familiar plant community names. The intent is to provide an overall picture of major vegetation types and how they might fit together, like pieces of a puzzle. This map also can be used in conjunction with other maps showing topography, geography, and climate. The 19 vegetation types and habitats are defined in the pages that follow, and diagrams convey general range and occurrence in greater detail.

A stately weathered western juniper, perhaps nearly a thousand years old, sits atop a granite knoll overlooking Lake Tahoe.

The map shows vegetation types organized according to whether tree canopy, shrub canopy, or ground layer canopy is the most developed part of the habitat's vegetation. These are divided into six subheadings:

Marsh & Dune – Coastal borders with few sparse adjacent evergreen tree groves and few if any deciduous trees. These areas reflect the influence of constant wind and salt spray. They contain tidal basins with surrounding reedlike vegetation and grasses or drifting sand with little or no vegetation except for scattered low perennials.

Prairie/Grassland – Areas with roughly less than 10% cover of trees, which are mainly deciduous. These areas are characterized by flat, nearly treeless expanses of native grasses and perennials, with seasonal vernal pools in low areas where water ponds briefly in winter and spring. Dense patches of bright annual color often occur in spring, turning brown during dry summers. Savanna and prairie are other terms sometimes used to designate grassland.

Scrub & Chaparral – Scrubby vegetation with few if any trees. These areas are covered with low, dense, stunted shrubs, which are mainly evergreen. Shrubs occur on intermittent gravelly bare soils, sometimes along with sparse grasses. These are usually true drought tolerant plant communities, though they may also occur along the coast where fog may be present and above timberline in alpine locations. The harsh conditions are typically reflected in densely mounded vegetation, which is slow growing and generally woody.

Desert Scrub – Arid, flat valley areas with few if any trees and sparse, low, dense, stunted, mainly evergreen shrubs. These occur on gravelly or sandy bare soils, with only a few sparse grasses, if any. This is the most drought tolerant plant community, and the harsh conditions are reflected in the slow-growing, woody, sparsely foliaged vegetation.

Woodland – Areas in rolling valleys and lower foothill regions, with roughly 30% or more cover of dense groves or scattered and solitary coniferous or deciduous trees with a high, broad canopy. There are few shrubs and usually native grassland and perennial growth beneath. These areas occur in regions having dry summer months, and are semi-moist along the coast or hotter and drier inland. These areas often transition to surrounding grassland or sparse evergreens in adjacent foothills and mountains.

Forest – Areas in coastal, foothill, or mountainous regions with densely spaced trees. There is a tall overhead canopy of predominant evergreen and coniferous species and often an understory of smaller, sparse deciduous trees and evergreen or deciduous shrubs tolerant of deeply shaded conditions. Grasses are few, occurring only when grassland interrupts where dense cover is not present and receding when trees reseed and reform dense cover.

In the first 10 specific vegetation types, tree cover is sparse or not present. Here, trees are emergent. Tree cover is most developed in the remaining 9, with an understory of smaller trees, shrubs, and herbaceous vegetation beneath. In climax forests, dense tree cover shades out understory growth of shrubs and herbs.

NATIVE VEGETATION TYPES AND HABITATS

A composite diagram of major vegetation types, represented as plant communities, for an overall picture of how they all roughly fit together. The extents are approximate, and greatly simplified, for clarity and simplicity in conveying an overall picture. The various entities may have significant areas of overlap, which cannot be accurately conveyed in this diagram, and smaller inclusions of lesser areas also are not indicated. The diagrams that follow on pages 36-45 are derived from *Terrestrial Vegetation of California* (Barbour & Major, 1977) and more accurately depict the actual extents of each of the individual plant communities listed on this map.

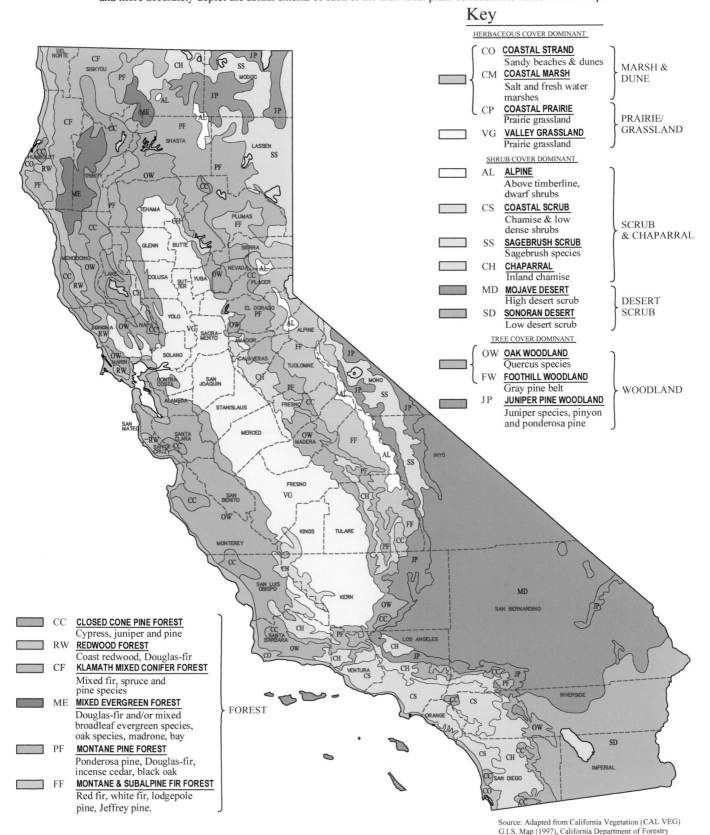

Key

HERBACEOUS COVER DOMINANT

CO **COASTAL STRAND** Sandy beaches & dunes
CM **COASTAL MARSH** Salt and fresh water marshes } MARSH & DUNE

CP **COASTAL PRAIRIE** Prairie grassland
VG **VALLEY GRASSLAND** Prairie grassland } PRAIRIE/ GRASSLAND

SHRUB COVER DOMINANT

AL **ALPINE** Above timberline, dwarf shrubs
CS **COASTAL SCRUB** Chamise & low dense shrubs
SS **SAGEBRUSH SCRUB** Sagebrush species
CH **CHAPARRAL** Inland chamise } SCRUB & CHAPARRAL

MD **MOJAVE DESERT** High desert scrub
SD **SONORAN DESERT** Low desert scrub } DESERT SCRUB

TREE COVER DOMINANT

OW **OAK WOODLAND** Quercus species
FW **FOOTHILL WOODLAND** Gray pine belt
JP **JUNIPER PINE WOODLAND** Juniper species, pinyon and ponderosa pine } WOODLAND

CC **CLOSED CONE PINE FOREST** Cypress, juniper and pine
RW **REDWOOD FOREST** Coast redwood, Douglas-fir
CF **KLAMATH MIXED CONIFER FOREST** Mixed fir, spruce and pine species
ME **MIXED EVERGREEN FOREST** Douglas-fir and/or mixed broadleaf evergreen species, oak species, madrone, bay
PF **MONTANE PINE FOREST** Ponderosa pine, Douglas-fir, incense cedar, black oak
FF **MONTANE & SUBALPINE FIR FOREST** Red fir, white fir, lodgepole pine, Jeffrey pine. } FOREST

Source: Adapted from California Vegetation (CAL VEG)
G.I.S. Map (1997), California Department of Forestry

CO COASTAL STRAND

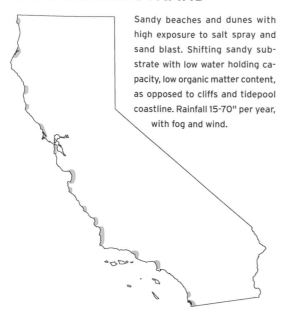

Sandy beaches and dunes with high exposure to salt spray and sand blast. Shifting sandy substrate with low water holding capacity, low organic matter content, as opposed to cliffs and tidepool coastline. Rainfall 15-70" per year, with fog and wind.

Bordering Trees

Cupressus macrocarpa (n,c)
Pinus contorta ssp. *contorta* (n)
Pinus radiata (c)
Pinus torreyana (s)
Salix hookeriana (n)

Naturalized Trees

Albizia distachya (c,s)
Myoporum laetum (c,s)

Canopy

Abronia maritima
Atriplex leucophylla
Elymus mollis
Myrica californica

n: northern
c: central
s: southern

Central western Humboldt County

Scattered locations along entire length of the state:

In n. Calif: San Francisco, Pt. Reyes, Dillon Beach, Bodega Bay, Point Arena, Fort Bragg, Humboldt Bay, Crescent City.

In s. Calif: Monterey, Morro Bay, Santa Maria River Bay, Los Angeles Bay, San Diego Bay.

CM COASTAL MARSH

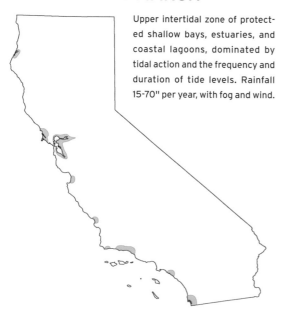

Upper intertidal zone of protected shallow bays, estuaries, and coastal lagoons, dominated by tidal action and the frequency and duration of tide levels. Rainfall 15-70" per year, with fog and wind.

Associated Trees

Alnus oregona (n)
Alnus rhombifolia (c)
Pinus radiata (c)
Salix hookeriana (n)
Salix scouleriana (n)

Canopy

Grindelia subterminalis
Salicomia virginica
Spartina foliosa

n: northern
c: central
s: southern

Central western Humboldt County

Scattered locations along entire length of the state:

In n. Calif: Humboldt Bay, San Francisco Bay.

In s. Calif: Santa Maria River Bay, Goleta Slough, Long Beach Bay, Newport Bay, San Diego Bay, Mission Bay.

CP COASTAL PRAIRIE

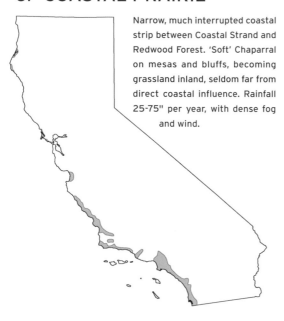

Narrow, much interrupted coastal strip between Coastal Strand and Redwood Forest. 'Soft' Chaparral on mesas and bluffs, becoming grassland inland, seldom far from direct coastal influence. Rainfall 25-75" per year, with dense fog and wind.

Scattered locations along entire length of the state:

In n. Calif: San Francisco, Pt. Reyes, Dillon Beach, Bodega Bay, Point Arena, Fort Bragg, Humboldt Bay, Crescent City.

In s. Calif: Monterey, Morro Bay, Santa Maria River Bay, Los Angeles Bay, San Diego Bay.

Associated Trees

Acer negundo (n,c)
* *Aesculus californica* (n,c)
* *Alnus oregona* (n)
* *Alnus rhombifolia* (c,s)
  Crataegus douglasiana (n)
* *Cupressus macrocarpa* (n,c)
  Myrica californica (c)
  Pinus radiata (n,c)
  Quercus agrifolia
  Quercus garryana (n)
* *Salix gooddingii* (s)
* *Salix hookeriana* (n)
* *Salix lasiolepis*
* *Salix scouleriana* (n,c)
* *Umbellularia californica* (n,c)

Canopy

Perennial grasses
  Danthonia, Festuca
  sedges and forbs
  Artemisia californica
  Baccharis pilularis
  Gaultheria shallon (n)

n: northern
c: central
s: southern

* riparian

Central western Humboldt County

VG VALLEY GRASSLAND

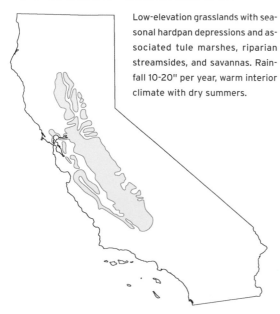

Low-elevation grasslands with seasonal hardpan depressions and associated tule marshes, riparian streamsides, and savannas. Rainfall 10-20" per year, warm interior climate with dry summers.

Low elevations of the Central Valley, extending into the Inner Coast Ranges, north and south.

Associated Trees

* *Acer negundo* (n,c)
* *Aesculus californica*
* *Alnus rhombifolia* (c,s)
* *Fraxinus latifolia* (n)
* *Juglans hindsii* (x *nigra*) (n)
* *Platanus racemosa* (c,s)
* *Populus fremontii* (n,c)
* *Populus trichocarpa*
  Quercus douglasii (n,c)
  Quercus lobata (n,c)
  Quercus wislizeni (n,c)
* *Salix gooddingii*
* *Salix hindsiana*
* *Salix laevigata*
* *Salix lasiandra*
* *Salix lasiolepis*
* *Sambucus mexicana*
  Umbellularia californica (n,c)

Canopy

Annual grasses and seasonal wildflowers

n: northern
c: central
s: southern

* riparian

Western central Colusa County

AL ALPINE

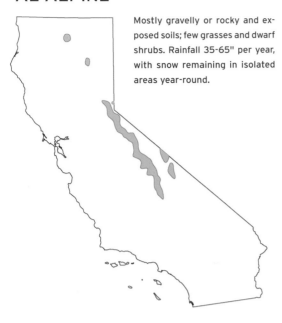

Mostly gravelly or rocky and exposed soils; few grasses and dwarf shrubs. Rainfall 35-65" per year, with snow remaining in isolated areas year-round.

Associated and Bordering Trees

Abies magnifica (n,c)
Juniperus occidentalis
Pinus albicaulis (n,c)
Pinus contorta ssp. m*urrayana*
Pinus flexilis (c,s)
Pinus monticola (n,c)
* *Populus tremuloides*
* *Salix* species
* *Sorbus* species
Tsuga mertensiana (n,c)

Canopy

Low shrubs, perennials, and bunchgrasses

n: northern
c: central
s: southern

* riparian

Eastern El Dorado County

Sierra Nevada mountaintops and ridges, above timberline at highest elevations, above 9,000'.

CS COASTAL SCRUB

Coastal cliffs, bluffs, and dry gravelly or rocky slopes, heavily influenced by wind exposure and low rainfall, below the chaparral region. Low shrubs, perennials, and grasses. Rainfall 10-20" per year with some summer fog.

Associated and Bordering Trees

Cupressus macrocarpa (n)
Pinus radiata (c)
Pinus torreyana (s)
* *Platanus racemosa* (s)
Quercus agrifolia (n,c)
Quercus chrysolepis
Quercus dumosa (c,s)

Canopy

Artemisia californica
Baccharis pilularis
Ceanothus leucodermis
Eriogonum fasciculatum
Salvia leucophylla

n: northern
c: central
s: southern

* riparian

Western San Mateo County

A much interrupted belt in the coastal regions of the Coast Ranges, south to Baja California.

In n. Calif: Point Reyes, Dillon Beach, Humboldt Bay, Crescent City and Monterey Bay.

In s. Calif: From Morro Bay to San Diego.

MD MOJAVE DESERT SCRUB

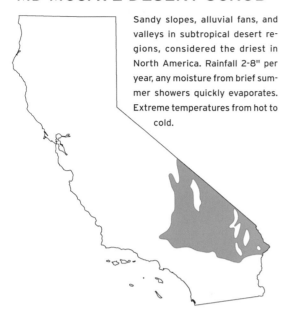

Sandy slopes, alluvial fans, and valleys in subtropical desert regions, considered the driest in North America. Rainfall 2-8" per year, any moisture from brief summer showers quickly evaporates. Extreme temperatures from hot to cold.

East of the Sierra Nevada Range from the south end of Owens Valley south to the southern California Transverse Ranges and eastward.

Associated Trees

Cercidium floridum
* *Chilopsis linearis*
Fraxinus velutina
Juniperus californica
Prosopis glandulosa
Prosopis velutina
Yucca brevifolia

Naturalized Trees

Parkinsonia aculeata

Understory

Ambrosia dumosa
Fouquieria splendens
Larrea tridentata
Olneya tesota

* riparian

Central Riverside County

SD SONORAN DESERT SCRUB

Dry desert washes with few scattered groves of low shrubby trees and scrub brush, mostly vast open sandy plains. Rainfall 2-4" per year or less.

From the south end of the Mojave Desert along the Colorado River to the south end of the Salton Sea.

Associated Trees

Cercidium floridum
* *Chilopsis linearis*
Parkinsonia aculeata
Prosopis glandulosa
Prosopis velutina
+ *Washingtonia filifera*

Naturalized Trees

Parkinsonia aculeata
* *Tamarix parvifolia*

Understory

Ambrosia dumosa
Fouquieria splendens
Larrea tridentata
Olneya tesota

* riparian washes
+ oases

Southern Riverside County

SS SAGEBRUSH SCRUB

High-elevation desert areas with few trees, dominated by low open shrub cover and sparse grasses. Rainfall 10-30" per year with some winter snow and few if any summer showers.

Northern and eastern exposure Sierra Nevada.

Associated Trees

* *Crataegus douglasiana* (n)
Juniperus occidentalis (n,c)
Pinus monophylla (c,s)
Pinus ponderosa
* *Populus trichocarpa*

Canopy

Artemisia arbuscula
Artemisia nova
Artemisia tridentata
Purshia tridentata

Various grass species

n: northern
c: central
s: southern

* riparian

Eastern Placer County

CH CHAPARRAL

Shrubby inland evergreen oak and shrub lands on dry hills and mountain slopes. Rainfall 14-25" per year, with hot dry summers and cool winters, light snows.

Dry slopes and ridges with rocky or gravelly to fairly heavy soils in the Coast Ranges from Shasta County south, below Yellow Pine Forest in the Sierra Nevada and southern ranges bordering Oak Woodland.

Associated Trees

Pinus coulteri (c,s)
Pinus sabiniana (c)
Quercus chrysolepis (nc)
Quercus dumosa (c,s)
Quercus lobata (c)
Quercus turbinella (s)
* *Quercus wislizeni* (c)
Aesculus californica (n,c)
* *Chrysolepis chrysophylla* (n,c)
* *Celtis reticulata* (s)
Cercis occidentalis
* *Cercocarpus betuloides*
Fraxinus dipetala (c,s)
Juniperus californica (c,s)
Pinus attenuata
Pinus muricata (n,c)
Pinus torreyana (s)
Prunus ilicifolia (s)
Quercus kelloggii, Q. morehus
Quercus palmeri (s)
* *Salix hindsiana*
* *Salix lasiandra* (n,c)
* *Salix lasiolepis*
Torreya californica

Canopy

Adenostoma fasciculatum
Ceanothus species
Arctostaphylos species
Quercus species

Western San Mateo County

n: nouthern
c: central
s: southern

* riparian

OW OAK WOODLAND

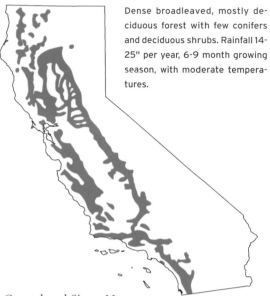

Dense broadleaved, mostly deciduous forest with few conifers and deciduous shrubs. Rainfall 14-25" per year, 6-9 month growing season, with moderate temperatures.

Coastal and Sierra Nevada foothills of northern and southern California, bordering and extending into Valley Grassland, below Foothill Woodland, at elevations from near sea level to 1,500'.

Common Trees

Quercus agrifolia
Quercus douglasii (n,c)
Quercus engelmannii (s)
Quercus lobata (n,c)
Quercus wislizeni

Other Trees

* Acer negundo
* Aesculus californica
* Alnus rhombifolia
Arbutus menziesii
* Cercis occidentalis
Cercocarpus betuloides
* Fraxinus dipetala
* Fraxinus velutina (s)
* Juglans hindsii (x nigra) (n)
* Juglans californica (s)
Lithocarpus densiflorus
Pinus muricata, P. sabiniana
Pinus radiata (c coast)
* Platanus racemosa (c,s)
* Populus fremontii
Prunus ilicifolia (s)
Quercus chrysolepis
Quercus garryana (n,c)
Quercus kelloggii
Quercus palmeri
* Salix species
* Sambucus mexicana
* Umbellularia californica

Understory

Arctostaphylos viscida
Ceanothus cuneatus
Heteromeles arbutifolia
Rhamnus californica

Central Sonoma County

n: northern
c: central
s: southern
* riparian

FW FOOTHILL WOODLAND

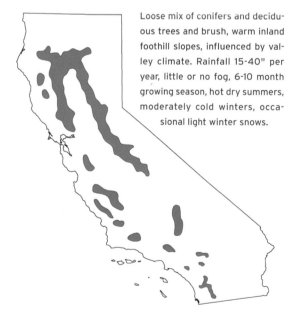

Loose mix of conifers and deciduous trees and brush, warm inland foothill slopes, influenced by valley climate. Rainfall 15-40" per year, little or no fog, 6-10 month growing season, hot dry summers, moderately cold winters, occasional light winter snows.

Foothills encircling the Central Valley and eastern exposure slopes of s. California Coast Ranges below 1,500' elevation.

Often referred to as the Gray Pine Belt.

Common Trees

Pinus coulteri (s)
Pinus sabiniana (n,c)
Quercus chrysolepis
Quercus douglasii (n,c)
Quercus lobata (n,c)
Quercus wislizeni (c)

Other Trees

* Acer negundo (n,c)
* Aesculus californica
* Alnus oregona (n)
* Alnus rhombifolia (c,s)
* Celtis reticulata (s)
* Cercis occidentalis
Fraxinus dipetala
* Fraxinus velutina (s)
Juniperus californica
* Populus fremontii (n,c)
* Populus trichocarpa
Prunus ilicifolia (s)
Quercus morehus
* Salix hindsiana, S. laevigata
* Salix lasiandra, S. lasiolepis
* Umbellularia californica (n,c)

Understory

Arctostaphylos viscida
Ceanothus cuneatus
Heteromeles arbutifolia
Rhamnus californica
Rhus diversiloba (s)

Western Placer County

n: northern
c: central
s: southern
* riparian

JP JUNIPER-PINE WOODLAND

Higher elevation plateau montane regions, with sparse forests of junipers and pines. Rainfall 0-30" per year, with winter snow, and few summer showers.

From the Great Basin Plateau, extending southward along the east base of the Sierra Nevada in the White & Inyo Mountains through the higher mountains of the Mojave Desert, below Yellow Pine Forest and above Joshua Tree Woodland of the Mojave Desert Scrub.

Common Trees

Juniperus occidentalis

Subdominant Trees

Abies concolor
Abies magnifica (n)
* *Acer glabrum*
Calocedrus decurrens (n,c)
Cercocarpus betuloides
Cupressus arizonica (s)
Pinus albicaulis (n,c)
Pinus flexilis (c,s)
Pinus ponderosa (n,c)
* *Populus tremuloides* (n,c)
Quercus chrysolepis
Quercus turbinella (s)
Quercus palmeri (s)

Understory

Amelanchier pallida
Artemisia arbuscula
Artemisia tridentata
Cercocarpus ledifolius
Purshia tridentata

n: northern
c: central
s: southern

* riparian

Eastern El Dorado County

CC CLOSED CONE FOREST

Coastal cliffs, bluffs, and dry gravelly or rocky slopes, heavily influenced by wind exposure and low rainfall, below the chaparral region. Low shrubs, perennials, and grasses. Rainfall 10-20" per year with some summer fog.

Interrupted forests scattered throughout the state, primarily in the Klamath Mountains, Coast Ranges in the north, and various isolated areas of the Sierra Nevada and Coast Ranges in the south, from near sea level to 1,000' elevation.

Common Trees

Cupressus macnabiana (n)
Cupressus macrocarpa (n,c)
Cupressus sargentii (n,c)
Juniperus californica (c,s)
Pinus attenuata
Pinus coulteri (s)
Pinus muricata
Pinus radiata

Other Trees

Abies concolor, A. magnifica
* *Acer macrophyllum*
Arbutus menziesii
Calocedrus decurrens (n,c)
Chrysolepis chrysophylla (n)
Cercocarpus betuloides
* *Fraxinus dipetala*
Lithocarpus densiflorus (n,c)
Pinus attenuata
Pinus jeffreyi, P. ponderosa
Pinus sabiniana (n,c)
Quercus chrysolepis (n,c)
Sequoiadendron giganteum (c)
* *Umbellularia californica*

Understory

Arctostaphylos species

Western San Mateo County

Ceanothus species
Rhamnus californica
Vaccinium ovatum (n)

n: northern
c: central
s: southern

* riparian

ME MIXED EVERGREEN FOREST

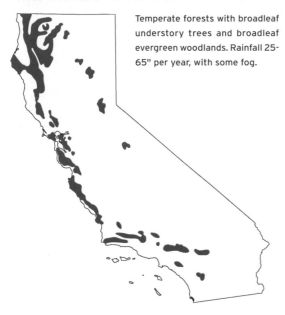

Temperate forests with broadleaf understory trees and broadleaf evergreen woodlands. Rainfall 25-65" per year, with some fog.

Along the inner edge and on the higher hills of Redwood Forest in Del Norte and Siskiyou counties, extending south along the western slopes of the central Coast Ranges and higher southern California mountains, at elevations 2,000-5,000'.

Common Trees

Arbutus menziesii
Lithocarpus densiflorus (n,c)
Pseudotsuga macrocarpa (s)
Pseudotsuga menziesii (n,c)

Other Trees

Abies concolor
Abies grandis (n)
* *Acer glabrum*
* *Acer macrophyllum* (n,c)
* *Alnus oregona* (n)
* *Alnus rhombifolia* (c,s)
Calocedrus decurrens (n,c)
Chrysolepis chrysophylla
Cercocarpus betuloides
* *Cornus nuttallii* (n)
* *Crataegus douglasii* (n)
Pinus contorta ssp. *murrayana* (n)
Pinus coulteri
Pinus jeffreyi
Pinus ponderosa (n,c)
Quercus agrifolia, Q. chrysolepis
Quercus garryana (n)
* *Salix lasiandra, S. lasiolepis*
* *Salix scouleriana*
* *Taxus brevifolia*
Tsuga heterophylla
* *Umbellularia californica*

Central Mendocino County

Understory

Arctostaphylos species
Mahonia aquifolium (n)
Vaccinium ovatum (n)

n: northern
c: central
s: southern

* riparian

RW REDWOOD FOREST

Dense Coast Range coniferous forest, including broadleaf trees and shrubs, on westerly seaward exposure, having a cool, moist climate. Rainfall 35-100" per year, with fog prevailing most of the year.

Seaward side of the Coast Ranges from Del Norte County south to San Luis Obispo County, ranging from near sea level to about 2,000' elevation.

Common Trees

Pseudotsuga menziesii
Sequoia sempervirens

Other Trees

* *Abies grandis*
* *Acer circinatum*
* *Acer macrophyllum*
* *Alnus oregona*
* *Arbutus menziesii*
Cornus nuttallii
Crataegus douglasii
Lithocarpus densiflorus
Pinus muricata
Quercus agrifolia
Quercus chrysolepis
Taxus brevifolia
Tsuga heterophylla
* *Umbellularia californica*

Understory

Gaultheria shallon
Polystichum munitum
Rhododendron macrophyllum
Vaccinium ovatum

* riparian

Central Mendocino County

CF KLAMATH MIXED CONIFER FOREST

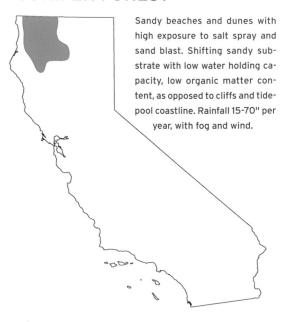

Sandy beaches and dunes with high exposure to salt spray and sand blast. Shifting sandy substrate with low water holding capacity, low organic matter content, as opposed to cliffs and tidepool coastline. Rainfall 15-70" per year, with fog and wind.

The Klamath Mountain region, Siskiyou Mountains, and the Trinity Alps, extending into southwestern Oregon, ranging in elevation from 4,500 to 6,000', separate from Redwood Forest.

Common Trees

Abies concolor (e,w)
Calocedrus decurrens (w)
Pinus lambertiana (c)
Pinus ponderosa (e)
Pseudotsuga menziesii (e,w)

Other Trees

Abies amabilis
Abies lasiocarpa
Chamaecyparis lawsoniana
Chamaecyparis nootkatensis
Chrysolepis chrysophylla
Picea breweriana
Pinus attenuata
Picea engelmannii
Pinus jeffreyi
Pinus monticola
Pinus contorta ssp. *murrayana*
Quercus chrysolepis
Quercus kelloggii
Taxus brevifolia
Tsuga mertensiana

Understory

Arctostaphylos nevadensis
Ceanothus prostratus
Chrysolepis sempervirens
Mahonia aquifolium

Central Del Norte County

Prunus emarginata   e: eastern
                             w: western
                             c: central

PF MONTANE PINE FOREST

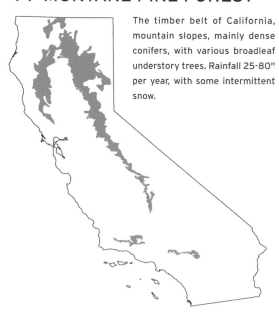

The timber belt of California, mountain slopes, mainly dense conifers, with various broadleaf understory trees. Rainfall 25-80" per year, with some intermittent snow.

Widespread throughout California at elevations 3,000-7,000' in northern California Coast, Cascade, and Sierra Nevada ranges, as far south as Kern County. Often referred to as Ponderosa Pine or Yellow Pine Forest.

Common Trees

Calocedrus decurrens (n,c)
Pinus ponderosa
Pseudotsuga menziesii (n,c)

Other Trees

Abies concolor (n,c)
* *Acer macrophyllum*
* *Alnus rhombifolia*
Arbutus menziesii
* *Cercis occidentalis* (n,c)
Cercocarpus betuloides
* *Cornus nuttallii* (n,c)
Lithocarpus densiflorus (n,c)
Pinus coulteri (s)
Pinus jeffreyi
Pinus lambertiana (n,c)
* *Populus trichocarpa*
Pseudotsuga macrocarpa (s)
Quercus chrysolepis
Quercus kelloggii (n,c)
* *Salix* species
Taxus brevifolia (n,c)

Understory

Arctostaphylos species
Ceanothus species
Chamaebatia foliolosa

Central western Humboldt County

Prunus virginiana   n: northern
Rhamnus califomica   s: southern

c: central   * riparian

FF MONTANE & SUBALPINE FIR FOREST

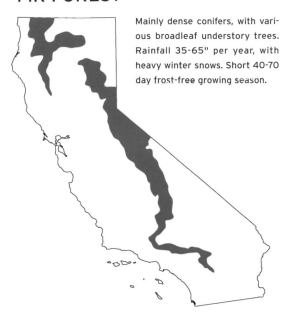

Mainly dense conifers, with various broadleaf understory trees. Rainfall 35-65" per year, with heavy winter snows. Short 40-70 day frost-free growing season.

Common Trees

Abies concolor
Abies magnifica
Pinus contorta ssp. *murrayana*
Pinus jeffreyi

Other Trees

* *Acer glabrum*
Juniperus occidentalis
Lithocarpus densiflorus
Pinus albicaulis
Pinus flexilis
Pinus monticola
* *Populus tremuloides*
Populus trichocarpa
Pseudotsuga menziesii (n,c)
Pseudotsuga macrocarpa (s)
* *Sambucus mexicana* (n,c)
Tsuga mertensiana (n,c)

Understory

Arctostaphylos nevadensis
Linnaea borealis
Symphoricarpos mollis

n: northern
c: central
s: southern

* riparian

Western central Nevada County

Above the Montane Coniferous Forest, at elevation 6,000' in the North Coast Ranges, elevations 5,500-7,500' in northern California, elevations 6,000-9,000' in the Sierra Nevada, and elevations 8,000-9,500' in southern California ranges. Often referred to as Red Fir or Jeffrey Pine Forest.

LAND USE

LEGEND

 Urban areas and agricultural use
National and State forests & parklands
Open space and rural communities

Source: USFS, and GIS Information Center

Three major land uses occur throughout the range of the 19 plant communities described. These use areas are fairly equally divided, comprising about one-third each: urban areas and agriculture, forest and parkland, and open space. Urban areas and agricultural and grazing lands generally occur in flatter regions that are arable and conducive to development and farming. Surrounding these uses, in more mountainous and desert areas of the state, are the National and State forest lands, including timberland, watersheds, recreational lands, and habitats that are preserved and managed. Remaining open space, including undeveloped land, riparian watersheds and rivers, and desert lands, do not fit into either category above and are either used for cattle grazing, where rural towns and scattered low density development has occurred without significant impact, or as yet are undeveloped. This fairly equal distribution represents a balance between what we need to live, what we need to sustain us, and what scenic qualities remain for us to enjoy. California's population is expected to grow from 35 million people to 67 million by the year 2050, making it increasingly difficult to maintain this balance.

THE INTERFACE

Transitions between Native and Urban Landscapes

In *The Landscape of Man (1975)* Geoffrey and Susan Jellicoe describe the unprecedented impact of man on the face of the earth. "There is a realization that the existing delicately balanced order of nature within the biosphere is being disturbed by the activities of man. Only his own exertions can restore a balance and ensure survival, in an efficient state of sustained existence and encouraged comprehensive ecological planning. However, international awareness has increased in the recent past, toward ecological soundness as a common goal. Design diversification has expanded to include habitat preservation and restoration as well as individualized stylized landscapes, with regard to how the urban landscape affects and interacts with the natural environment."

Regional and local planning involve assessment of zoning and density and evaluate other issues to achieve a balanced environment for inhabitants and the viability of the town, city, or region. Many people associate the California Environmen-

© Rex Babin/The Sacramento Bee

Suburban development can become a sprawling expanse of roads, paving, and buildings without sufficient open space and preservation of native vegetation. Where native vegetation is replaced by ornamental species, regional or local identity may be lost, with towns and suburban areas indistinguishable from each other except for architectural landmarks and distant vistas of remaining natural areas beyond.

tal Quality Act (CEQA) and the National Environmental Protection Act (NEPA) with protection of endangered species, but these State and Federal laws also focus on impacts that proposed development may have on existing communities. Existing infrastructure such as roads, water and sewer capabilities, noise levels, and traffic congestion are some of the issues assessed in environmental impact studies. Potential impacts of proposed developments are identified and addressed, with measures designed to mitigate impacts where possible. Indirectly, planned zoning density concerns the infrastructure of streets and land use compatibility. Unnecessary congestion occurs if existing streets cannot handle the additional traffic

© Chris Stewart/The San Francisco Chronicle

Heavy commute traffic north of Marin County could jeopardize surrounding oak woodland. Private and public interests working with developers and agricultural owners have helped to preserve a significant portion of the area's heritage oaks.

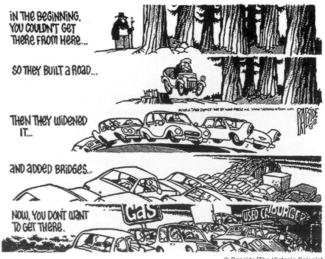

IN THE BEGINNING, YOU COULDN'T GET THERE FROM HERE...

SO THEY BUILT A ROAD...

THEN THEY WIDENED IT...

AND ADDED BRIDGES..

NOW, YOU DON'T WANT TO GET THERE.

© Raeside/The Victoria Colonist

Growing concern for planned growth has resulted in new approaches to development and revitalization of urban areas.

volume that new development may create. Land use compatibility and density often directly influence which areas succumb to urban blight. Reserving areas of open space and integrating an effective urban landscape is also vital in keeping areas enjoyable places in which to live. This includes the preservation of existing street trees, parks, and landscape corridors. Without substantial tree cover, large urban expanses often become inhospitable and uninviting.

The interface between the native landscape and new suburban landscapes is often tentative, due to pressures for expansion and the economics of development. As urban areas expand into surrounding open space, the integrity of natural habitats may be compromised unnecessarily. Development occurs as a result of population growth and is often considered necessary to sustain economic viability. But development becomes short-sighted when it is governed purely by economics, where quantity becomes more important than quality. Preservation of the natural beauty that attracts people to an area should not be pushed aside in the name of progress. Many developers have begun to realize this and are making concerted efforts to incorporate environmental sensitivity into their development projects. Much progress has also been made recently in regard to public awareness. Preservation and development interests often work together toward a common goal of integrating the native landscape into new developments.

Environmentally sensitive developments are becoming more popular, as evidenced by a growing number of advertisements depicting an unspoiled natural setting for prospective homebuyers. The presence of nearby open space corridors promotes a feeling of living closer to nature in a semi-rural environment. But surrounding habitats can be negatively affected if not carefully planned and respectfully maintained.

To lessen the impact of development, generous landscape corridors are often integrated along streets and roads, using drought-tolerant plantings as well as native species reflective of the local area as a transition to the natural setting. Reclaimed water may be used for irrigation to help offset the demand of new homes.

Sufficiently large open spaces must be preserved for the native habitat to remain viable in maintaining species diversity and natural terrain. Creeks, streams, vernal pools, and wetlands are especially well suited as open space preserves. Open spaces are most effective when connected to each other, as a continuous habitat or corridor for vegetation and wildlife to remain healthy and regenerate. When such areas are isolated or cut off within vast areas of urban or suburban development, they slowly recede and disappear.

Greenbelt connections are an effective means of linking various types of open spaces. Rather than existing as isolated pockets of preserved greenery, the habitats are connected to form a continuous corridor. In this form they also become an important part of our living environment. Creeks, streams, woodlands, and meadows that weave through suburban areas convey the feeling of a more rural setting. These open space areas need to be set aside in sufficient size to make them environmentally self-sustaining, both for the vegetation and the wildlife associated with it, yet leaving enough area to make development economically feasible.

Development implies change, specifically modifying and adapting the landscape for human purposes, and often is associated with resource depletion, environmental degradation, and the replacement of natural processes with artificial ones.

Valley Oak Woodland provides an appealing setting for housing developments. Preserving large areas as open space maintains natural habitat while providing visual and passive recreational value.

With sensitive grading, many native oaks can be preserved in new housing developments. However, homeowners who landscape beneath them face losing these trees by overwatering. Many builders provide instructions to new homeowners about the care of mature oaks, which are extremely sensitive to disturbance and summer watering.

A small community park skillfully transitioned into the adjacent open space of *Quercus lobata* and *Q. douglasii*, connecting with trails along creeks to parks throughout the city. Few ornamental trees are planted, which maintains the integrity, health, and presence of the native oaks.

This kind of development is sustainable only until renewable resources are depleted or pollution, congestion, and urban blight make an area unlivable. *Regenerative Design for Sustainable Development* (John Lyle, 1994) introduced a consolidated environmental design approach relating to land planning, site engineering, architecture, and landscape design. "Sustainable development" has since become a well-recognized concept. To be truly sustainable, development must meet the needs of the present without compromising the future. This involves an ecological understanding of the needs of humans in an environment that maintains the integrity of nature's life support systems without displacement of countless native species. In this case, maintaining does not mean simply preserving what exists.

Sensitive development requires paying closer attention to existing natural vegetation and habitat patterns and the "lay of the land" or geographical, topographical, and climatic conditions existing on site. Observing them as interconnected and interdependent bioregions involves closer analysis from the ground up, rather than taking a successful design from somewhere else and applying it arbitrarily, expecting the same results. Environmentally responsive landscapes recognize the

Parklike settings along creeks and rivers provide access and recreation along connecting trails and bike paths that may extend for miles through urban areas.

Open space corridors in floodplain areas often are used as parks, with open grassy meadows for sports and paved pathways for bicycle and pedestrian thoroughfare. These areas may be submerged under water for brief periods in winter without incurring serious damage.

important roles of native plants and preserved habitats. Such landscapes are often designed using principles of xeriscaping and incorporate the use of recycled or locally occurring materials. Landscapes can be designed to reflect and incorporate natural systems, habitats, and species of the surrounding bioregion. The concept of bioregional landscapes establishes a deeper connection to natural processes and provides a context for developing regenerative landscapes that are more ecologically sound. These types of landscapes fit with or enhance the natural setting rather than replacing it and require less maintenance, fertilization, energy, and water.

Natural processes of seasonal change become more evident in landscapes functioning as sustainable ecosystems. These natural processes link the landscape to a bioregion rather than being an arbitrarily designed form. Landscapes that allow for spontaneous evolutionary development invite biological diversity. We may design the garden but can relinquish some control, allowing natural phenomena to grow and regenerate, rather than remain static year-round. Designed landscapes that build on and blend with the larger regional landscape also connect comfortably with vistas outward toward native trees and other vegetation.

In creating an environment attuned to local conditions

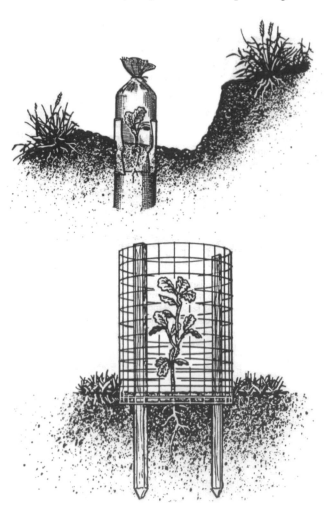

Reprinted from *Oaks on Home Grounds*, Cooperative Extension University of California–Division of Agriculture and Natural Resources–Leaflet 2783. Graphic by Carl Dennis Buehller.

Two frequently used enclosures designed to protect seedlings from damage by squirrels, birds, and deer, allowing new plantings to become established. Additional protection from gophers and other burrowing animals can be added as well.

and features, we may realize that we are building in flood-plains or areas with a high water table, on unstable soils where slides may occur, in clayey, alkaline, or expansive soils, or where recurring wildfires may present a real danger. A growing number of natural catastrophes can be avoided by looking around a bit more closely. We sometimes have trouble taking environmental issues seriously until we are affected directly or economically. Sustainable development is not governed by arbitrary rules of politics or economics, but rather by the biocratic processes of the natural world, as aptly stated in *The California Landscape Garden—Ecology, Culture and Design* by Mark Francis and Andreas Reimann (1999).

Bioregional landscape design conserves water, energy, and other resources by utilizing regional construction materials, rather than those imported from around the world. Recycled materials are utilized to reduce waste buildup. Passive design principles are utilized, such as trees placed for shading and solar radiation control. Bioregional designs ensure that materials have a benign impact on the environment, depend minimally on pesticides, artificial fertilizers, and energy consumption, and

Typical valley creekbed vegetation of *Salix* and *Populus* surrounded by *Quercus lobata*. These passageways often are cleared to allow floodwaters to flow unimpeded or development occurs too close to these zones for safe habitation. Instead, we should recognize their presence as indicators of a naturally occurring flood zone and design our developments accordingly.

Winter flooding occurs frequently in valley riparian zones. Deciduous river-bank trees are typically submerged, with their roots stabilizing the soil, along with thickets of smaller trees and shrubs slowing the water's force along the banks.

reduce the need for artificial support systems. Groundwater overdrafts and watershed draw are minimized by the use of reclaimed water for landscape irrigation or storing water from downspouts in reservoirs for later use and promoting conservation through hydrozoning and xeriscaping. Bioregional landscape design represents a conscious effort to minimize environmental degradation and loss of natural habitats.

The goal is not to exclude the use of everything except native plants, but to recognize where their use is appropriate and where they can be used in context. Native plant materials have specific cultural requirements, especially for California's seasonal dryness, and they do not always adapt well to ornamental settings that differ from those in which they occur naturally. Some also may appear too casual in stylized or formal landscapes. However, many ornamental plant species from around the world are adaptable to local climate and soils and enrich native plantings when used appropriately. Since the early Spanish missions, which introduced the use of exotic species, Californians have created urban oases in a style that has become synonymous with our state. In our Mediterranean climate these domesticated ornamental landscapes can highlight the surrounding natural setting without overwhelming or replacing them.

We are coming to the realization that available resources and natural open spaces are not unlimited. Population growth is placing higher demands on existing water supplies, and sustainability has become a real issue. Without planning accordingly, many of our ornamental landscapes could wither and die due to water shortages during droughts, which can last for a number of years. We need to plan our landscapes for the long term, including at least some species able to withstand these conditions, and utilize the concepts of hydrozoning and xeriscaping in water-conserving landscapes that will endure. Plants with similar water use requirements should be grouped together. Maintaining a thick layer of mulch in planting beds is one of the most effective means of retaining soil moisture. Mulch helps to reduce heat buildup in the soil, thereby reducing moisture loss through transpiration, and provides an organic means of inhibiting weed growth as well.

Ornamental plantings in suburban landscapes can be effectively placed near buildings to concentrate water use in small areas while creating a pleasant setting that can be more easily maintained. Vistas from the buildings then can stretch outward over a water-conserving landscape beyond, which gradually transitions to surrounding unwatered open space or native landscape with native trees. Mature native oak species are often endangered unnecessarily by creating gardens beneath them without respecting the water requirements or periods of dryness on which they have always depended. Conversely, many existing ornamental trees suffer when they no longer receive the watering to which they have become accustomed.

Native *Quercus lobata* and *Q. douglasii*, for example, are adapted to California's seasonally dry Mediterranean climate and their roots are especially sensitive to disruption of any kind. In the mid-1970s it became apparent that more attention was required for the preservation of California's native oaks. As aggressive development was moving from flat valley grassland into low-lying foothills, oak woodland was disappearing at an alarming rate. A standardized commonsense approach was drafted to promote integration of woodland oaks into new developments. Oak tree ordinances have not changed significantly since they were first introduced. They involve

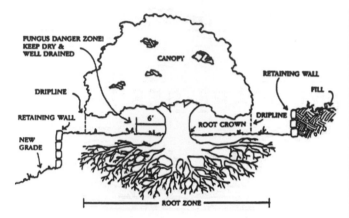

FUNGUS DANGER ZONE!
KEEP DRY &
WELL DRAINED

CANOPY

RETAINING WALL

DRIPLINE

FILL

RETAINING WALL

6'

ROOT CROWN

DRIPLINE

NEW
GRADE

ROOT ZONE

Reprinted from *Native Oaks, Our Valley Heritage*–Heritage Oaks Committee (1976).

Avoiding disturbances within the driplines of native oaks can keep them healthy. This includes grade alterations, drainage, and irrigated plantings. Soil moisture levels and water seepage cannot be changed significantly without affecting these sensitive trees.

three main elements: preservation and replacement; protection of roots within the dripline; and restraint on the use of surrounding plantings that require irrigation.

Before any new development begins, existing oaks are inventoried by a certified arborist. Each tree is tagged and numbered, and a list is prepared that indicates the species, location, health, and desirability of each tree. Trees are identified that must be removed to accommodate streets, buildings, and other necessary site improvements. Graded parcels often contain many trees that can be preserved by minor grading adjustments to maintain existing grades around them. The number and size of trees removed governs the number to be replaced in mitigation measures to offset the impact of the development. Each caliper inch of trees removed is assigned a monetary value per inch. New 15-gallon size replacement trees are usually valued at one replacement inch each, and if the number of inches replaced equals the number of inches removed, no penalty is assessed. Trees are planted at least 30 feet apart to allow for their mature size, and only the native species stipulated can be used for this purpose. Plantings beneath, if any, must be appropriate, and irrigation is provided to sustain new trees through their establishment period. Monitoring programs to insure tree survival may extend for up to five years. Any caliper inches removed but not replaced must be paid for in dollar value to the city or county involved. This fee is usually used to fund tree foundations or tree plantings in other areas, to cover the cost of purchase, planting, staking, irrigation, and subsequent maintenance of each tree planted elsewhere.

In large-scale projects, a significant portion of the site may be preserved as open space. Replacement often involves smaller-sized oaks or acorn plantings, in sizable numbers, with reduced replacement value due to their small size. These plantings are made in fall or early winter, when seasonal rainfall will help newly planted seedlings to survive. Usually sufficient moisture is provided by winter rains for them to become established under favorable conditions. While these trees may take 30-60 years to reach appreciable size, this approach provides a means to preserve and replenish valuable oak woodland stands in significantly large numbers and in large enough areas for woodland habitats to remain viable and self-sustaining.

Attention to the root zone around existing native oaks in-

volves maintaining the existing grade within this zone, for minimal root disturbance, and maintaining existing soil moisture level. Extensive retaining walls are not recommended around existing oaks, as they change the soil moisture dynamics. The retained soil becomes drier than usual, as it is higher than the soil level around it, causing root dieback. Surrounding paving affects soil porosity and moisture absorption from natural rainfall. Low walls using unmortared interlocking precast masonry units can be used instead, and any paving within the dripline is discouraged. The dripline of existing trees is usually encircled with the familiar orange plastic fencing during any type of construction, to designate areas to be kept free of any grading or other intrusion. In the past, these areas under existing trees were often used to park vehicles and store or dump materials, or limbs were trimmed, with detrimental results. With proper preservation efforts, existing native species and habitats can remain intact in a transition that enriches urban development and reflects and blends with the surrounding natural environment.

Interlocking concrete retaining walls have less effect on root systems of existing oaks. The porous joints are not grouted, allowing soil behind to breathe, and these walls do not require the extensive footings of poured concrete walls.

Existing valley and blue oaks can often be preserved in parkway medians where the street grade is even with existing grade. Existing native grasses remain in the median without irrigation, and are mowed down when they dry out. Oaks remain healthy when seasonal dryness is maintained.

TREES IN URBAN LANDSCAPE DESIGN

The Use of Trees in Urban Design

The urban landscape provides a setting for the architectural features of cities and towns, creating a pleasing environment in which to live and work. Trees in designed landscapes are often selected and placed to frame and highlight views or to reflect or evoke the natural landscape.

The urban landscape refers to landscapes designed to suit human needs in cities, towns, suburbs, and rural or agricultural areas. Ian McHarg aptly refers to this as the "domesticated landscape" in *Design with Nature* (1971). An important part of our daily lives, "man's design on the land" beautifies our world, and trees are one of the most dominant features.

In much of California, urban landscapes are able to incorporate the stunning beauty of trees from around the world. Trees from other summer-dry climates, such as Africa, Australia, and the Mediterranean, blend well with California's drought-tolerant native plants. Trees from Europe, southeast Asia, eastern and northwestern North America, Central America, and South America often require supplemental water during the dry season, and some of California's native plants fall into this category as well. Subtropical palms, cycads, and tree ferns from other parts of the world also grace irrigated landscapes in more temperate regions of California. It is im-

portant to be familiar with the cultural requirements of both native and ornamental trees and to group them in appropriate landscape situations.

Trees are of primary importance in creating an attractive and pleasing urban environment. They impart an organic element to soften the impact of buildings, paving, and streets. The vitality and pleasing nature of urban areas is often directly attributable to a balance between organic and inorganic materials, between landscaping and architectural features. Trees play a dominant role in the effectiveness of landscape settings for the urban environment.

Trees also serve biological and environmental purposes. Leaves and branches dissipate the force of falling rain, helping to control the effects of splash, runoff, and erosion. Root masses hold soil in place. Oxygen is replenished as a byproduct of photosynthesis. Other air-cleansing functions of trees include abating gaseous, particulate, and odoriferous air pollution. Twigs and foliage collect and filter air-borne particulates, which are washed away by rain. Trees transpire large amounts of water into the air, helping to settle out wind-borne pollutants. Gaseous pollutants may be directly absorbed and assimilated by foliage or masked by the fragrance of leaves and flowers.

Trees and other vegetation also help to suppress or mask noise, reduce glare, filter wind, and moderate temperatures. Sound waves are absorbed or attenuated by the soft tissues of leaves, branches, twigs, and fallen leaf mulch. Trees and shrubs can help control night or daytime glare and reflection, ranging from complete blocking to minor filtering and diffusion. As windscreens, they offer shelter and protection at ground level for inhabitants, structures, and the soil surface itself. Trees and other vegetation absorb, reflect, intercept, or filter solar radiation. Deciduous trees allow sunlight to penetrate for winter light and warmth while providing welcome shade in summer. Trees are essential to a comfortable, healthful, pleasing urban environment.

Golden Gate Park in San Francisco offers 1,200 acres of recreation and relaxation for city dwellers and remains a premier attraction for residents and for visitors to the area.

THE URBAN FOREST

Around the end of the 19th century, the combined effects of dense urban populations and the choking pollution of concentrated industrialization were coming to be recognized as causing many urban areas to be unpleasant and unhealthy places to live. By the early 20th century "new towns" in suburban settings emphasized the importance of open spaces and vegetation. In urban areas, large areas were set aside as parks, such as Central Park in New York, Golden Gate Park in San Francisco, and Elysian Park in Los Angeles. Efforts at city beautification included street trees planted in sidewalks to soften the effects of densely packed high-rise buildings and pavement.

This classic application of 17th century English-style landscape architecture is typical of many parks that were an integral part of the urban design of older cities. These parks remain a source of vitality in many cities today.

Much of this enlightened urban planning drew from the experience of European cities, where beautification with landscaped streets and large urban parks was relatively well established. Early efforts in the U.S. were experimental at best, but they were a step in the right direction, as evidenced by the parks and streetscapes in cities and towns we see today.

Urban tree planting now involves a more experienced approach based on what has worked well in the past and tempered by knowledge of the shortcomings created by haphazard or insufficient plantings. The restraints of how trees fit into surrounding urban uses, as well as the effects of the urban environment on the health of trees, directly determine which species will best serve various urban purposes. Maintaining and enhancing the "urban forest" requires continued cooperation of local and national tree foundations and local planning departments.

San Diego's Balboa Park, popular for many public uses and activities, features varied settings of ornamental, subtropical, and native oak woodland trees and other plantings.

TYPES OF URBAN LANDSCAPES

Public Parks

Parks and open spaces play a crucial role in urban landscapes, providing relief within the confining spaces of urban development. Tree-lined streetscapes provide a pleasing setting for the enjoyment of daily activities. In the first half of the 20th century, large urban parks were typically planned into most large cities, often as a means of applying the English-style landscape, where a piece of the countryside is brought into the city for all to enjoy. The arboretum approach was often used to offer a more visually interesting and unusual selection of trees and to provide an exciting ornamental setting. These urban parks continue to remain historic treasures in our cities today.

A grid of London plane trees in Golden Gate Park creates a dramatic, grandiose effect reminiscent of 17th century French-style landscape architecture.

Parks and other innovations in land use planning have improved the quality of life in urban areas. Currently, however, it is often difficult for cities to bear the enormous cost of land, equipment, and labor associated with development and maintenance of traditional parks. Experimentation with unusual trees occurs less often in newer parks in favor of more adaptable species that require less maintenance. However, preservation of older urban trees, many of which are threatened by disease and old age, is a priority for most cities, and, for now, some magnificent park specimens of the past still remain.

A high priority in current park design emphasizes activity areas, play areas, and sports fields. Cost effectiveness, durability, and low maintenance are also emphasized. Activity areas usually surround a central plaza or pavilion that offers seating and tables for individual use or group gatherings, often accented by flowering trees. Large shade trees otherwise predominate, with conifers or other evergreens providing a perimeter buffer. Trees that have objectionable litter from leaves, seeds, or fruit are discouraged around playgrounds for children, especially those that may have toxic plant parts.

Public Plazas

Public open spaces include plazas where people can gather as well as entries to large office buildings visible if not acces-

A grove of coast redwood trees may eventually provide a "forest" setting for this plaza surrounded by high-rise office buildings. Evergreen conifers are effective here in enclosing the space and buffering the urban environment.

An architectural grid of closely spaced London plane trees provides welcome shade on a hot summer day.

sible to passerby. Trees in such settings are often placed in formal arrangements in tree wells or other planters within a large expanse of paving where they serve as an integral component of the built environment. Trees may be arranged in an allée or grid, forming an architecturally designed exterior space as an extension of the building and its uses. For continuity and simplicity, a single species of tree is often used throughout the design. Trees may be closely spaced for continuous overhead shading.

Trees selected for these purposes must be able tolerate the confined root space and be relatively pest free. Trees that drop large amounts of fruit or litter are not suitable for such areas. Texture is also important in selecting trees for public plazas. Fine-textured trees such as *Albizia julibrissin, Gleditsia triacanthos,* and *Ulmus parvifolia,* despite their leaf drop, have a softer, more inviting nature than heavier textured broadleaf evergreens such as *Ceratonia siliqua* or *Magnolia grandiflora,* which can be placed further apart to allow distance between the dense canopies. *Platanus x acerifolia* is a common favorite

in plazas, often pollarded for an "old world" effect and to control canopy size. However, the practice of yearly pollarding is labor intensive and costly.

Public plazas, sometimes called mini-parks or pocket parks, provide opportunities to sit and relax in a densely urbanized environment. These spaces usually include dense groves of trees to enclose a soft landscape setting and seating areas where people can congregate, enjoy a private lunch, or watch passersby. Decorative fountains often enrich these spaces with the dynamic sights and sounds of moving water. Trees such weeping willow, white birch, and Lombardy poplar are often associated with water features.

Freeways and Circulation Corridors

With the advent of the automobile early in the 20th century, roads throughout the countryside were improved and many were paved. Soon highways were developed as major vehicular circulation routes between cities, and the first freeway was constructed in southern California. The interstate freeway system as we now know it began in the 1960s, featuring extensive beautification measures, including substantial frontages with irrigated landscapes and an abundance of trees to make the drive more pleasant. Freeways served their purpose well, encouraging people to travel and to commute to work, promoting the proliferation of development, and bringing more people to California.

In the past twenty years, budget constraints, greatly increased population, and the experience of two recent droughts have resulted in significant changes in the design and maintenance of landscapes along major highways and freeways. Population growth has required the expansion of circulation corridors at the expense of previously landscaped areas. The lush, irrigated landscapes we used to enjoy are now often considered extravagant and difficult to maintain. In many areas, only drought-tolerant trees and large shrubs remain. Groups of trees now receive a thick layer of bark mulch in a curvilinear form around the dripline, giving the visual effect of groundcover while helping to preserve soil moisture around the trees and also helping to control weeds.

A "garden type" plaza with a "soft" landscape offers a relaxed setting within an otherwise harsh urban environment.

Native trees and shrubs are increasingly planted in circulation corridors, often supplemented with faster-growing drought-tolerant ornamental trees. Even drought-tolerant plantings generally require drip irrigation for the first few years to become established. In landscapes viewed from passing vehicles at highway speeds, trees are the dominant presence, with shrubs and groundcovers having only incidental effect. Tree plantings at bridge overcrossings may take advantage of the raised grades to make the trees seem taller, while visually designating a destination in the distance over an otherwise flat expanse.

Some tree species once commonly used along freeway corridors are no longer planted today. Large swaths of *Eucalyptus, Acacia,* and *Myoporum,* especially in northern California, were devastated in the record freeze of the early 1990s. In southern California, eucalypts have been ravaged by psyllids. *Nerium oleander* was once commonly planted along freeway center

A simple planting of *Cedrus deodara* and *Pistacia chinensis* in groves near a freeway overcrossing is aesthetically pleasing and environmentally sound. These trees are drought-tolerant, and moisture-retentive mulch is used instead of groundcover. Existing grasses are left in place, usually requiring only a single mowing after they turn brown in summer.

median strips, and some remaining plantings still thrive in hot, dry conditions with little or no irrigation or maintenance. Today, many of these shrub dividers are being replaced by more efficient, though esthetically less pleasing concrete walls.

Streetscapes

Generous landscape frontages along suburban arterial streets are often referred to as parkways. Plantings here often incorporate a single dominant tree spaced uniformly along the street edge and a mixture of subordinate trees in irregular plantings behind. Streetscapes without regularly spaced trees often include random, separately massed evergreen and deciduous species. The arrangement of tree groupings may accentuate each other, with taller groves of trees and an understory of flowering accent trees.

Extensive streetscape frontages of newer housing developments typically use separate bubbler irrigation for trees. This supplements standard surface spray irrigation for lawns and shrubs by increasing moisture to the deeper level of tree roots. Providing a separate irrigation line for trees also allows trees to be watered when drought restrictions limit or eliminate irrigation of lawns and groundcovers.

A street lined with elms (*Ulmus rubra*) in a neighborhood appropriately named Elmhurst. Residential streetscapes often utilize a single large species of shade tree along a particular street. This provides both continuity and a recognizable identity that differentiates it from streets planted with other species. Trees planted in individual yards along residential streets provide secondary accent value and visual interest reflecting each residence, as an understory to the predominant street tree canopy.

Residential Landscapes

Trees in residential landscapes usually reflect individual tastes and offer an opportunity to experiment. Shade trees, flowering accent trees, evergreen screen trees, and solitary specimen trees are the most common types of residential landscape trees. Dwarf and semi-dwarf fruit tree varieties are good choices for confined spaces in residential landscapes.

ARRANGEMENT OF TREES

The gardens and landscapes that surround us are usually derived from one of two approaches: informal and natural looking or a stylized artistic arrangement. Trees form the dominant framework or structure for other plantings. The tree species selected and the manner in which they are arranged create the setting or style of the overall landscape design. Placement of trees as dominant, subordinate, understory, or single specimens, and their combined use in these ways also directly influences the plantings beneath.

Arrangement of trees may be stylized in symmetrical alignments to form geometrical patterns or randomly arranged within loose, informal groves. While a single species of street tree has often been used for a more uniform effect, use of mixed species has become more prevalent. Use of a single species throughout any formal landscape design is always subject to the loss of some trees, which disrupts the ordered appearance.

After determining a basic approach, formal or informal, the design process usually involves selection and layout of the dominant tree, followed by selection and layout of the plantings beneath. Trees planted in groves or distinguishable masses more effectively define a space. Depending on the intended use of the area and the intended effect, trees may be either randomly spaced for a more informal setting or placed in regimented rows or lines to accentuate a formalized setting. The predominant trees form the skeleton, holding the other design elements together. Subordinate trees supplement the body of the landscape in a lesser manner, with ground plane elements of shrubs, groundcovers, and lawns. Trees spaced too far apart or in insufficient quantity will "read" weakly to the eye and impart a spotty, formless quality.

Allées are formal rows of trees forming a visual corridor that directs our view toward a destination or vista beyond. Trees in an allée may be closely spaced columnar varieties, heightening the vertical accent qualities, or they may be canopy trees with the arching canopies closing together overhead. These effects may be strengthened by using double rows, or the rows may be interrupted by gaps at regular or planned intervals, providing glimpses, or windows, along the corridor to sideward vistas or elements of interest.

Agricultural themes, such as an orchard grid, are often incorporated into landscapes in an architectural manner, as in the case of trees in large public plazas and parking lots for shading purposes. The ordered arrangement of tree trunks can act as a unifying element of great visual interest, much like the diagonal rows of trees in agricultural orchards as seen from a moving vehicle.

A grid of English walnuts is a familiar scene in agricultural settings. Here, a similar grid adds visual interest along a central California freeway.

Street trees in dense urban settings are often placed in symmetrical fashion, along each side of the street, creating a linear effect. These zelkovas, with their broad canopies arching across the street, create an unusual tunnel effect, casting dense shade for residential areas in hot summer climates.

An adapted grid of Bradford pears creates a pastoral effect in an entry to a business park, directing views to the buildings beyond while allowing drivers to access a shaded parking spot with full views to each side. The bermed lawn imitates a springtime orchard of mowed grasses.

Parkway corridors provide ample room for mixed tree groupings. Mixed groves of evergreen conifers, deciduous shade trees, and flowering accent understory trees screen parking areas and provide a parklike setting along avenues and major circulation corridors in a business park.

The use of palm trees as a vertical accent is reinforced by linear alignment, directing views toward the destination. The close proximity of trunks in the driver's line of sight blocks side views toward distractions along the way, focusing attention straight ahead.

ARRANGEMENT
Geometric

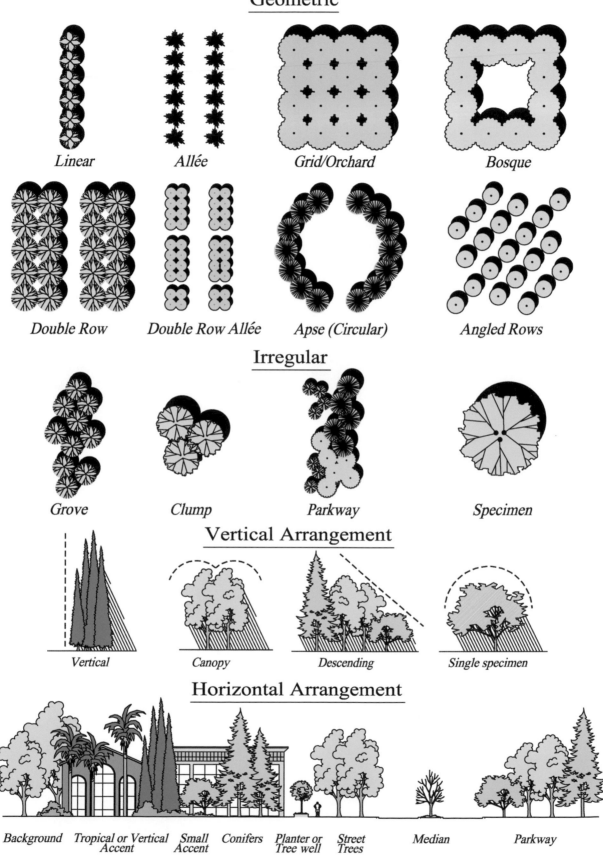

Linear   Allée   Grid/Orchard   Bosque

Double Row   Double Row Allée   Apse (Circular)   Angled Rows

Irregular

Grove   Clump   Parkway   Specimen

Vertical Arrangement

Vertical   Canopy   Descending   Single specimen

Horizontal Arrangement

Background   Tropical or Vertical Accent   Small Accent   Conifers   Planter or Tree well   Street Trees   Median   Parkway

A few of the more common layout arrangements of trees in urban settings.

TREES IN VARIOUS CONTEXTS

Landscape and architectural styles or themes can be reflected in the selection and placement of trees and other plant materials in the landscape. Trees may evoke different natural settings such as desert, subtropical, or redwood forest. They may also be used as purely architectural elements when they are arranged in geometrical or symmetrical patterns. They are often used in a cultural context, as in Spanish, Italian, or Japa-

A Spanish-style architectural theme is accentuated by a pair of windmill palms, with citrus, bougainvillea, and a specimen agave accenting the shrub plantings in a distinctive residential landscape.

This landscape imitates a mountainous or subalpine setting, utilizing evergreen conifers, rock walls, and few deciduous trees. The large conifers, including spruce, pine, and sequoia, were imported as fairly mature trees, many of them at least 25' tall. Trees in residential landscapes often are removed when they become too large for the space. The owner of this project contacted a tree mover, which utilized a "Tree Spade" to move them to their new location. Although success cannot be guaranteed, this approach provides an immediate established appearance, and the new owner typically pays equipment and labor costs.

This minimalist design exhibits quiet simplicity. The expansive lawns are a continuous ground plane element interrupted by few shrubs, allowing the architectural elements to stand out cleanly in the landscape. Trees are placed in complementary axial alignment or in groupings that frame or balance the architectural elements.

The landscape at this office complex is reminiscent of a moist springtime meadow in a redwood forest. Coast redwoods are used throughout the complex, with an understory of deciduous Japanese maples in lieu of native vine maples and bigleaf maples among ferns, azaleas, and other low shrubs.

nese garden themes. Uses such as these provide individualized landscapes with a distinctive quality in their attention to detail.

Artistic license is used to create landscapes with an a recognizable feeling or identity. Themes may reflect an artistic placement or alignment of trees that complements vistas or the composition of an overall landscape in which species are used in a specific context. Other settings may remotely integrate commonly associated species in the context of a distant native landscape setting. Such domesticated representations can add variety and interest to an otherwise ordinary or uniform landscape.

CODES AND ORDINANCES AFFECTING TREE PLANTING

Local codes, ordinances, and guidelines have been established in conjunction with the planning process governing all types of development. They are a means of establishing minimum standards for site planning and typically include landscape components to ensure that landscaping is addressed along with architectural and engineering portions of the work. In this way, the design and layout of buildings, roads, and other infrastructure must consider and provide for a meaningful landscape setting, rather than treating landscaping as an afterthought. Master plans, specific plans, and local ordinances all incorporate landscaping issues as conditions of approval. These must be reflected in a final landscape design before a development is approved. City ordinances may reflect a regional standard but are determined by the particular locale and may vary from one jurisdiction to another.

The Water Use Classification of Landscape Species (WUCOLS) developed by UC Cooperative Extension, San Mateo-San Francisco (L. R. Costello) and the California Department of Water Resources, 1992, has had a significant effect on the design of landscapes for water use efficiency. WUCOLS is used to achieve "water averaging", where portions of high water use are offset by a proportional amount of low water use. The goal is a landscape with a net medium water use that will not excessively impact local water supply. Trees, shrubs, and groundcovers are designated as high, medium, or low in their water use requirements. Plants in each category are intended to be grouped, or hydrozoned, so that species of similar water use can be sustained by the amount and frequency of irrigation provided. Lawn areas are not prohibited, but must be balanced by other areas of low water use. Landscapes designed using this approach are less likely to suffer if water rationing should occur during extended and/or severe periods of drought. Using the transpiration rate for a specific area, calculations can be made to compare the estimated water use (EWU) against the maximum allowable use (MAWA). Warmer, drier areas have a higher transpiration rate (evaporative loss) than areas with cooler, moister climates. The state is divided into six regions: North Central Coastal, Central Valley, South Coastal, South Inland Valley, High and Intermediate Desert, and Low Desert. Trees and other plants may have different moisture requirements depending the region in which they are used, indicated by a code such as: L/M/M/H/H/H. A hyphen indicates that a species is not used in that region, and a question mark indicates that information is not available.

Ordinances governing trees in parking lots came into common use in the late 1970s. They were originally developed by cities with hot summers, both to provide shading for parked cars and to reduce the impacts of large expanses of paved surfaces that absorb, retain, and reflect heat. Temperatures can be as much as 15 degrees cooler in the shade. Large expanses of paving heat up the microclimate and retain heat longer than landscaped areas. Shade trees significantly reduce heat buildup. Ordinances may apply a value for each tree indicating the canopy of shade cover over paved surfaces. This value is based on the estimated canopy size that a particular 15-gallon tree can be expected to produce in 15 years. The total value of trees proposed is compared with the total paving area in the design, with the goal of providing at least 50% shading to be effective.

Because a proposed building or complex must have suffi-cient square footage to make it economically feasible, and a certain number of parking spaces are required for each square foot of building, landscape space often is at a premium. Architects and engineers who involve landscape architects during early planning stages facilitate plan approval regarding landscape requirements of shading, screening, and minimum landscape area. Layouts often must provide at least 10% of the "envelope" area in planting, and planter widths must be wide enough to allow trees large enough to provide adequate shading. Generally planters must be 8' wide between curbs (inside face to inside face) to accommodate larger trees used in meeting minimum shading requirements. Some cities require "finger planters" for a certain number of parking spaces in linear arrangements, as well as the familiar "islands," which are effective in shading broad expanses. These islands and planters must also be spaced to accommodate light fixtures for proper distribution of light. Coordinating tree locations, planters, conduits, and light fixture locations while providing sufficient parking requires carerful planning in the initial layout.

In general, the shade values for various trees is as follows, with larger trees requiring a minimum 8' wide inside planter width and a minimum of 6' for medium-sized and smaller trees.

Large shade trees (30-35' diameter canopy)
  F: 962 sf TQ: 721 sf H: 481 sf Q: 240 sf
Medium shade trees (25-30' diameter canopy)
  F: 707 sf TQ: 531 sf H: 354 sf Q: 177 sf
Small shade trees (20-25' diameter canopy)
  F: 491 sf TQ: 369 sf H: 246 sf Q: 123 sf
Very small shade trees (15-20' diameter canopy)
  F: 314 sf TQ: 236 sf H: 157 sf Q: 79 sf

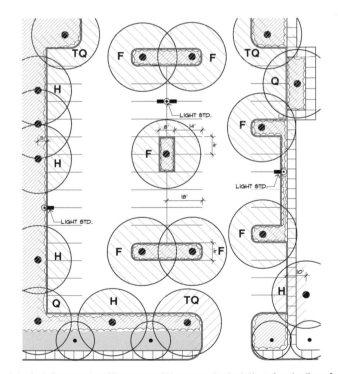

A typical diagram simplifies some of the general calculations for shading of parking areas. The area of tree shading must equal at least 50% of the overall paving area, excluding overlapping tree canopies. Trees cannot be simply placed closer together, and planters must be large enough to accommodate large trees. Coordination of tree placement and light fixture locations is essential. All of this must occur while accommodating the number of parking spaces required for the project.

An overhead canopy of shade trees is welcome in hot summer regions, providing an appealing oasis where cars can be parked for extended periods during the day. Shading can reduce temperatures 10 to 15 degrees by reducing heat absorption and reflection from large expanses of paving.

Trees used for parking lot shading are chosen from a list developed for each particular area. Generally, no more than 75% of the total number of trees can be the same species, and there may also be requirements that at least 25% of the trees be evergreen.

Problems often occur when trees are planted underneath overhead utility lines. While most new developments now place utility lines underground, trees must be kept at least 5' (to as much as 10') away from such lines, and from storm drains and sewer lines as well. Utility lines are generally engineered for placement within paved street improvements to allow for tree plantings within designated landscaped areas. However, they cannot be kept entirely separate. Existing street trees in older neighborhoods often require frequent trimming to prevent damage to overhead utility lines. Tree height is limited to 25' under high-voltage transmission lines, which require considerable clearance. Otherwise, electricity can arc to the ground through a nearby tree. The Western Area Power Authority (WAPA) has issued a list of shrublike trees for use in these areas.

Trees planted near intersections and in median plantings must be placed so as not to interfere with drivers' sight lines. Drivers must be able to clearly see oncoming traffic at intersections that do not have traffic signals. Ordinances may require as much as a 450' line of sight from the position of the driver at a stop sign to the position of an oncoming car traveling 45 miles per hour on major streets. Trees also cannot be planted too close to street lights on major streets or they will obscure the cone of light onto the street, creating a potentially hazardous situation.

DESIGN ELEMENTS IN TREE SELECTION

Space Requirements

It is difficult to imagine the mature size of an 8-foot tall, 15-gallon tree with a 1-inch caliper when it is planted. Zelkovas planted at 25-30' apart will seem to have little immediate visual impact, but in 15 years the canopies may be touching, and any closer spacing might create problems. In small residential landscapes, a shade tree may be planted that ultimately will overtake the entire yard and may require removal. A common example is Japanese black pine, which is well behaved for a few years, but needs a lot of room for its eventual size. Dynamic spatial design is required to plan for the variable and ultimate growth of trees selected to fit within the mature landscape. This involves the following elements, as well as growth rate, in proportion to a tree's mature dimensions.

Height	Spread
10-20' (small)	10-15' (small canopy)
20-30' (medium small)	20-25' (medium small)
30-50' (medium)	30-35' (medium canopy)
50-75' (medium tall)	40-50' (medium large)
75-100' (tall)	50'+ (large canopy)
100'+ (very tall)	

Growth Rate and Longevity

Trees are sometimes chosen for their fast growth rate to achieve a full-sized tree more quickly. However, most slow-growing trees are longer-lived and may be worth the wait. Flowering accent trees are usually fast growing but have a shorter life span. Trees recognized for life spans of at least 100 years generally are used as street trees, which are more permanent fixtures of the landscape. For comparative purposes, the following examples are given.

Growth Rate
Slow: under 1' per year (e.g., valley oak)
Medium: 1-2' per year (e.g., Bradford pear)
Fast: 2' or more per year (e.g., Lombardy poplar)

Longevity
0-75 year: short-lived
100-200 years: long-lived
200-500 years: very long-lived
500-2,000 years: extremely long-lived

Density

Density of the foliage mass generally determines the textural quality of a tree and, more important, the amount of shade or screening that trees will provide. Fine-textured trees planted on the shady north side of a tall structure will allow filtered sunlight to pass through but may not provide enough shade in a hot sunny exposure, where a tree with a denser canopy would be more effective. Densely foliaged trees generally require sunnier locations, and may become less dense if heavily shaded. In shade leaves may be spaced further apart and branches often become spindly or leggy.

Density is closely related to texture, and trees with fine density usually have an equally fine texture. In the case of trees with pinnately compound leaves, with many small leaflets, the effect is one of fine texture. Densely textured trees are most often used for screening. Evergreens and conifers are commonly used for screening or as a solid background for deciduous trees, and their foliage mass is especially effective in winter. A mixture of contrasting or complementary densities and textures is integral to a well designed landscape. Some examples follow.

Dense: opaque, casting heavy shade (e.g., southern magnolia)
Medium: semi-dense, medium shade (e.g., hackberry)
Fine: light, airy appearance, small leaves (e.g., honey locust)

Texture

Foliage and general appearance can determine the appropriateness of certain types of trees for the setting in which they are intended to be used. Weeping forms generally have a softer texture than trees with a dense, rounded form. Texture is a relative term determined by leaf size and density of the foliage mass. Depending on the intended effect, the textural qualities of a particular tree can be used to complement or contrast with the setting, buildings, or other trees nearby. Some examples follow.

Coarse: large, thick leaves, heavy branches (e.g., southern magnolia)
Medium: medium-sized leaves (e.g., hackberry, liquidambar)
Fine (light): visible through, casting light shade: (e.g., honey locust)

Evergreen and Deciduous

Evergreen trees are somewhat static in appearance, other than occasional flowering. Deciduous trees have more pronounced seasonal qualities and change throughout the year. Deciduous trees may exhibit profuse flowering in spring or

The soft texture of *Parkinsonia aculeata* is a pleasing characteristic where other evergreen trees might overpower the space or obscure the building.

A grouping of white birch accentuates white window frames, and the reflection of vertical white trunks on the glass makes the building facade seem semi-transparent.

summer, form a green canopy in summer, follow with a striking display of fall color, and feature a silhouette of bare branches in winter. The quality, proportions, and placement of evergreens become important in planning for the winter landscape, providing foliage mass and structure while sunlight penetrates through deciduous trees. Fine-textured or bare-branched deciduous trees are highlighted and accentuated when used with a background of dense evergreen trees.

Color and Other Special Features

Foliage colors range from light to dark green, blue, gray, yellow, and purple, as well as their changing seasonal color. Trees with showy flowering displays are highly desirable, as well as those that provide fragrance. Flowering can be timed throughout the year by planting species that bloom in succession. Certain flower colors complement each other, as well as the surrounding landscape or special features. Often flowering trees are planted as multi-trunk specimens for a more dramatic effect, emphasizing the character of their trunks and branching patterns.

Trees with showy fall color can enhance the landscape, and knowing which species and varieties can maximize that effect is worth researching in order to obtain desired color combinations. Red maple, liquidambar, tulip tree, and pistache are some of the more popular species planted for fall color, and different varieties may produce different colors or hues. Some trees, such as purple-leaf plum and Schwedler maple, are planted for their leaf color throughout the growing season as well as their spring flowering qualities.

Tree trunks can be especially appealing, especially those with attractively fissured, smooth, white, or peeling bark. White bark is more visible at night, especially in front of a dark background. Lighting can accentuate these qualities, also producing interesting silhouettes against adjacent walls.

Shape

Shape is one of the main considerations in choosing a particular tree. The choice may be for a vertical accent tree, shade canopy tree, or a tall, stately street tree. The form of a young tree can be dramatically different from that of the same tree in age, and it may take years to achieve the mature shape. As the tree gains height, the trunk thickens and the overhead canopy broadens as main branches develop, revealing its mature form. Shape and eventual size should be considered together in making a choice, allowing adequate distance from buildings and other trees. Trees planted close together may grow tall faster, but foliage is usually less dense, and the trees may be less shapely in appearance. Generally speaking, healthy, vigorous nursery tree stock should require little or no pruning to maintain the desired shape. Selective light thinning may be done to remove some of the lower branches to a desired height as the tree grows. Severe pruning of older trees is generally not desirable, and when necessary should be done by a certified arborist to maintain the health, vigor, and shape of the tree.

Tree shapes within a landscape can be chosen to provide continuity or contrast or to accent certain features. A mix of evergreen and deciduous trees provides visual interest in contrasting textures and shapes. Evergreen trees will contrast with bare branches of deciduous trees, either as a background or, when used in groves, to provide interest in the winter landscape. Tall, vertical trees may be planted in parallel rows to form allées, hedgerows, or windbreaks. These create a strong line through the landscape, whereas rounded canopy trees placed in informal groves create a pastoral or parklike effect. A landscape becomes visually integrated and balanced when trees have been carefully selected and placed for their use and desired effect. Smaller accent trees and specimens can then be added within that framework.

The vertical growth of Canary Island pine makes it popular in spots where other pines may take up too much room. This pine is especially effective planted in groups of three or four, rather than as a single specimen.

The rounded, oval shape of a red oak makes it suitable as a park, shade, or street tree if it has ample room to spread. Red oak has a heavy root system to support its mature size, and while it does well in lawns or groundcover areas, small planters should be avoided.

The classic vertical roman column shape of Italian cypress is used to create a symmetrical geometric pattern in many urban landscapes. Planted closely, they form a visual wall; further apart, they appear individually as pointed columns.

TREE SHAPES

No trunk, or short one, branching to ground; height equal to width
Mound/Shrub form
(Sambucus mexicana)

Freestanding single trunk, round headed; height equal to spread
Dome
(Prunus blirieana)

Freestanding single trunk, round headed; width 2-3 times height
Broad Dome
(Zelkova serrata)

Freestanding long single trunk, rounded canopy
Tall Dome
(Washingtonia robusta)

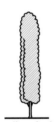

Freestanding single trunk; height 3-10 times width
Columnar
(Populus nigra 'Italica')

Freestanding single trunk; height 2-5 times width
Cylindrical
(Carpinus betulus 'Fastigiata')

Short trunk, elliptical outline; height 2-5 times width
Ellipsoidal
(Liquidambar styraciflua)

Short trunk, rounded base tapers to pointed tip; height 2-5 times width
Fastigiate
(Quercus robur 'Fastigiata')

Short trunk, triangular outline; height 3-5 times width
Conical
(Sequoia sempervirens)

Freestanding short trunk; height 3-10 times width
Broad conical
(Cedrus deodara)

Freestanding short trunk; height 3-10 times width
Tall conical
(Cupressus sempervirens 'Stricta')

Short trunk, inverted triangular shape; height 2-5 times width
Obconical
(Prunus serrulata 'Kwanzan')

Round headed, branches arch upward and outward; width 1-3 times height
Arching Vase
(Albizia julibrissin)

Short trunk, round canopy widest at base; height 1-2 times width
Ovoid
(Pyrus taiwanensis)

Short trunk, inverted oval shape; height 1-2 times width
Obovoid
(Ulmus americana)

Freestanding single trunk, with lower branches trimmed away
Standard
(Zelkova serrata)

2 or more basal trunks, radiating vertically from common base
Multi-trunk
(Arbutus unedo)

Foliage canopy with many open voids
Open-headed
(Eucalyptus cladocalyx)

Branches arching, with weeping ends
Weeping
(Salix babylonica)

Foliage canopy with irregular rounded tufts, cloudlike
Billowy
(Melaleuca linariifolia)

PLANTING AND IRRIGATION

As important as design to a successful landscape are the selection of healthy stock for planting and the use of procedures and materials that promote the vitality of newly planted trees. There are various means of propagating nursery material, including from seed, cuttings, or by grafting onto established rootstock. Many hard-to-find species are available in liners or 4" pots through specialty nurseries, but these require a much longer period to reach substantial size and are impractical for most commercial landscape installations.

With the tremendous surge in demand for plant material in recent years, availability and conformance to trade standards are a real concern. While locating the desired tree from another source may be preferable to substituting another species or variety, using substandard quality nursery stock will result in a tree of inferior quality. General nursery standards and those of the American Society of Nurserymen apply to the size and quality of materials available through nursery sources. Generally, minimum sizes for the most commonly used sizes of trees can be expected to be as follows, but may vary according to the particular species. Sizes shown below are for trees trained as single-trunk standards. Multi-trunk specimens and compact varieties are generally somewhat shorter and wider than shown.

Container Size	Trunk Caliper	Height	Canopy Spread
1-gallon	1/4-1/2"	2-4'	6"-1'
5-gallon	1/2-3/4"	4-6'	1-1/2 to 2'
15-gallon	3/4 to 1"	7-8'	2-3'
24" box	1 to 1-1/2"	9-10'	3-4'
36" box	1-1/2 to 1-3/4"	12-14'	6-7'
48" box	1-3/4" to 2"	14-16'	7-8'

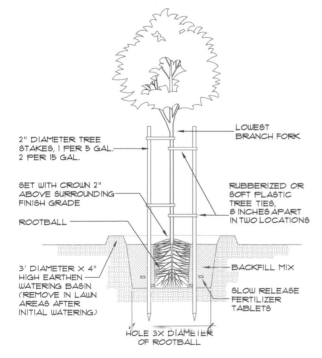

2" DIAMETER TREE STAKES, 1 PER 5 GAL. 2 PER 15 GAL.

SET WITH CROWN 2" ABOVE SURROUNDING FINISH GRADE

ROOTBALL

3' DIAMETER X 4" HIGH EARTHEN WATERING BASIN (REMOVE IN LAWN AREAS AFTER INITIAL WATERING)

LOWEST BRANCH FORK

RUBBERIZED OR SOFT PLASTIC TREE TIES, 8 INCHES APART IN TWO LOCATIONS

BACKFILL MIX

SLOW RELEASE FERTILIZER TABLETS

HOLE 3X DIAMETER OF ROOTBALL

A typical tree planting detail showing the basic elements of tree planting and staking, including planting hole, backfill mix, fertilizer tablets, and watering basin, with tree stakes and ties. Proper planting techniques and thorough watering to remove air pockets are recommended.

For commercial growing in nursery containers, young seedlings are started from seed or in flats, moved into 2" nursery pots, then 4" pots, and eventually into 1-gallon and then 5-gallon plastic containers as they mature in size and the rootball begins to fill each container size. Cuttings or grafting stock are used for varietal clones rather than seedlings, which may not be genetically pure strains. If young trees are not transplanted to larger containers when they should, they may become rootbound. This makes it difficult for roots to spread evenly after planting to form a wide-spreading or deep-reaching root structure required for health and stability when the tree matures. Visibly knotted or tangled roots on the soil surface of the container often indicate improper handling. Bare root and balled-in-burlap nursery material is occasionally offered in early spring, but is difficult to keep moist for extended periods of time prior to planting.

Single-trunk trees need a straight trunk of sufficient size and strength to support the weight of the upper branches and an upright leader that will continue to extend the height of the maturing tree. Trees that require additional staking, or those for which the nursery stake must be left in place to provide sufficient trunk strength, are susceptible to breakage. Topping or reducing the canopy mass to compensate for insufficient trunk strength creates other problems, including permanent disfiguring of the normal branching pattern. Cutting the top leader often produces weaker multiple leaders that are more likely to break in the future. Side shoots appearing on the trunks after final transplanting should be cleanly removed after the initial shock of transplanting to prevent their future regrowth. Removing them encourages new canopy growth instead.

Grafted varietal stock usually has an elbow at the graft point, but trunks with excessive elbows are undesirable, as the tree will not have its weight distributed evenly over the supporting root structure and may eventually topple or lean. Grafted varieties must be planted with the graft union fully exposed above the soil surface, as it is the stronger rootstock that supports and sustains the weaker varietal grafted portion of the tree. Sucker growth sprouting from the grafting stock should be removed cleanly at the surface.

Proper planting methods are important to the survival and

In a typical commercial nursery growing grounds, young trees have been recently planted into 1-gallon plastic containers from cutting stock in liners. The containers are placed nearly touching, for efficiency of irrigation and space. Close spacing also forces consistent upright growth, to nursery standards. Those not consistent with the rest are not suitable for commercial use.

establishment of newly planted trees. Planting holes should be at least 1-1/2 times the width of the container or rootball, with amended backfill mix placed and firmed around the rootball. The top of the rootball should rest about 2" above the surrounding finish grade, allowing for slight settling after the initial thorough hose watering to remove any remaining air pockets. The soil level should never be higher than it was for the tree in the container in which it was grown. This is especially important for varieties grafted onto rootstock. If the graft is not kept exposed, the tree may begin weaker rooting above the graft and may discard the stronger rootstock below. Soil should slope away slightly, so water does not pond at the base of the crown. This keeps the crown free of mildew and allows air to reach the rootball, which might otherwise become overly wet. Fertilizer tablets placed around the base and amended soil backfill around the rootball encourage roots to grow outward. The sides of the rootball are gently loosened, or sometimes cut in four vertical locations, to allow encircled roots to grow outward. The sides of the plant hole are scarified to break and loosen it, providing a better transition for the roots to grow into surrounding native soil. This is especially important in heavy or clayey soils.

Roots often suffocate in standing water in areas with hardpan subsurface soil or rock. Holes can be drilled with an auger to breach through to porous soil beneath, allowing water to percolate. This also allows tree roots to grow downward into the holes, where they expand and crack the hard soil around them, and smaller roots can penetrate even farther. Hardpan soil must be otherwise be fractured by heavy equipment to a sufficient depth before trees are planted.

After planting, a temporary earthen watering basin, approximately 3' in diameter, is formed around the base of the tree. This allows for thorough hose watering to soak out any air pockets left behind and ensures that soil and rootball are evenly moist. A layer of bark mulch is placed within the circle to keep the top of the exposed rootball from drying out. Watering basins are later removed, and in areas of turf the sod is held a foot away from tree trunks to minimize damage by trimmers and mowers.

Tree stakes prevent movement at the rootball, allowing roots to eventually secure the tree. By securing the trunk in an upright position, the uppermost branches are left free for movement in the wind while the base of the trunk remains steadily in place. The trunk must be of sufficient size and strength to support the weight of the foliage canopy. Single stake materials use a single metal pole, and a metal ring fits around the trunk with no anchoring involved. Standard staking in commercial landscapes often utilizes two 2" diameter pressure-treated wood stakes for each tree, driven into the ground below and outside the rootball in a plumb, vertical position. Tree ties are used to secure the trunk to the stakes on each side, just below the first branching point. The ties are pulled snug but not overly tight on the trunk and fastened to the stakes on each side. Flexible plastic ties, first developed by Cinch Tie (TM), prevent damage to the trunk while keeping it securely in place. Green plastic nursery tape or slightly stronger vinyl tape is used by nurseries on nursery stakes. This allows for rapid trunk growth without constriction, but is not strong enough for commercial landscape staking. Stakes and ties should be removed after one year.

Guying of larger trees utilizes vinyl-coated guy wires or cables, with a rubberized loop at one end to fit around a branch,

Graphic courtesy of Deep Root, Inc.

Deep Root (TM) barriers direct roots downward, helping to prevent damage to sidewalks, curbs, and paving or uplifting of lawns and adjacent planted areas.

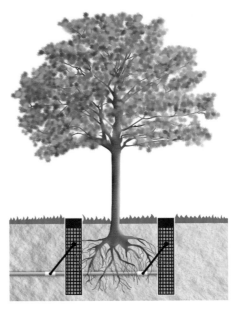

Graphic courtesy of Rain Bird Corporation

Deep watering assemblies such as this pre-assembled Rainbird RWS-BCG package encourage roots to grow downward, promoting better tree health and minimizing root damage to nearby sidewalks, curbs, paving, and planting beds. With conventional spray irrigation tree roots tend to remain along the surface where constant moisture is found.

with the other end anchored in the ground by an angled stake or underground "duckbill." Turnbuckles are located midway along each securing cable. Three cables are placed equidistant around the tree to the ground. Turnbuckles allow the cables to be evenly tightened. These cables may present a hazard to pedestrians, and they should be located out of direct paths of travel and clearly marked by flags.

Height and Space Requirements

It is important to allow adequate room for trunks and roots, which enlarge and spread as trees mature. Parklike settings with large expanses of lawn or groundcovers allow roots to spread uninhibited along or near the surface. When trees are planted adjacent to curbs, paving, or walkways or confined within tree wells or parking lot islands, the result is often cracked or buckled paving surfaces as trees mature. For confined situations, deep-rooting trees should be selected and planters should be sized to accommodate the species selected. Small trees such as crape myrtles may require only a 4-6' planter width. Medium-sized trees such as 'Aristocrat' pear or tupelo require a minimum 6-8' wide planter. Large shade trees such as London plane generally require at least a 10' wide planter.

Providing sufficient soil volume to accommodate tree roots can be a major dilemma in confined urban streetscapes, as soils under pavement must be compacted to meet engineering standards for load-bearing surfaces. The use of CU-Structural Soil (TM), developed at Cornell University in the mid-1990s, provides an integrated, continuous, root penetrable, stable pavement system that also allows for root growth, aeration, and drainage necessary for trees. This product is available through Amereq, Inc., and a network of licensed companies. A patented blend of coarse aggregate, amended clay loam, and hydrogel tackifier is proportioned so that when it is compacted in place, the aggregate forms a rigid, load-bearing stone lattice, leaving the soil within the aggregate void matrix less compacted. A geotextile is used to separate the engineered aggregate base layer of the pavement from the structural soil. Long-term benefits include significantly improved tree growth, vigor, and life expectancy and reduced movement, failures, and maintenance costs for repair of adjacent pavements and flatwork.

This elm tree was planted without sufficient room for the enormous trunk, which has grown over the curb and into the paving area.

Without mulching, bare soil around the base of this tree allows moisture to evaporate quickly from the soil below, and the roots seek moisture along the surface toward the surrounding lawn.

The most commonly used tool for reducing the likelihood of root damage to paving is a "Root Barrier," first patented by Deep Root, Inc., in 1976, and since offered by other manufacturers as well. This system utilizes interlocking plastic panels extending 18 to 24" into the ground to direct root growth downward. Roots are uninhibited below the barrier, allowing them to spread horizontally well below surface pavements and structures. Vertical "ribs" on the plastic walls also direct roots downward, preventing them from growing in circles around the enclosed area and becoming rootbound. Deep watering with bubblers or drip irrigation also helps to draw roots downward.

Tree trunks in lawns are often protected from lawn mowers and trimmers by keeping a bare soil circle around them, but without mulching this area of bare soil may heat up and dissipate moisture needed by roots. Water that ponds in the basin temporarily after watering encourages surface rooting. As these roots concentrate on the surface, a larger and larger area becomes filled with knobby surface roots. Any surface mulch to retain soil moisture is beneficial, especially in tree wells and planter islands. Bark, cobbles, or groundcovers will help hold in moisture and promote deeper root growth. Ground-up walnut shells also provides an attractive, long-lasting and economical surface mulch. The use of weedeaters often girdles the trunks of thin-barked trees planted in lawn areas. Sometimes plastic collars are positioned around the base of the trunks to protect newly planted trees.

Tree Wells and Parkway Strips

Planting strips and tree wells are a means of integrating trees into urban streetscapes and paved pedestrian plazas. Parkway strips allow for tree planting between detached sidewalks and curbs if they are sufficiently wide. These usually are planted with turf to allow unimpeded passage for pedestrians crossing from street to sidewalk. Other materials often used to provide a porous surface around tree roots in planting strips include groundcovers, interlocking pavers, unmortared brick-on-sand, decomposed granite, or small cobbles. These materials can be used in interesting combinations in parkway strips or arranged in pleasing patterns to add interest to flat, paved plazas. Flat surface tree wells create a cleaner architectural effect where trees appear to grow directly out of the pavement. Trees surrounded by shrubs or groundcovers provide a softening touch.

Tree wells can be covered with cast metal grates, which offer a pedestrian-friendly surface. A layer of mulch can be placed underneath the grates, and the grate openings allow air to penetrate to soil and roots. The heavy metal tree grates are assembled in two halves, usually fitting within a metal I-frame to keep them level and secured within the tree well opening. Some are designed to allow the circular opening left for the tree to be enlarged as the tree trunk grows. Cutouts are sometimes provided for uplights or access to capped irrigation openings. Decorative metal tree guards are also available, which attach directly to the grates. These offer additional tree protection, an attractive landscape feature, and a pleasing alternative to wood or metal stakes. Metal tree guards provide extra protection for newly planted trees in heavily traveled areas. They are especially effective in preventing damage by vehicles.

Irrigating trees in tree wells involves surface bubblers, drip emitters, or underground deep root watering assemblies, which are either prefabricated or made up on the site. These formerly utilized a corrugated flexible 3 or 4" diameter perforated flexible pipe, joined into a continuous circle near the base of the rootball. This has been simplified recently, using only a single vertical plastic net-walled pipe assembly, placed on either side of the tree. A bubbler head is located near the top to facilitate examination and adjustment, with a plastic cap covering the open end of the pipe. Subsurface drip systems such as Netafim (TM) are also effective, as below-ground irrigation loses little to evaporation and there are no exposed sprinkler heads or bubblers on the surface.

This large sidewalk tree well with irrigated turf benefits the tree and cools surface roots. The small cobble square is decorative, and also saves trimming turf around the base of the tree.

Decorative tree grates enrich the textural quality of an urban plaza, further enhanced by marble tile edging and surrounded by exposed aggregate paving.

A simple, but effective tree well treatment using a double brick soldier course edging with decomposed granite infill, provides tree roots with a porous surface that does not impede pedestrian travel.

An effective parkway treatment with lawn and a meandering sidewalk creates a more residential feel bordering an urban condominium complex.

A tree well surfaced with decomposed granite, showing irrigation bubblers flush with the surface at the base of the tree, as well as uplighting that does not pose a tripping problem for passersby. A root barrier to contain surface roots is concealed beneath the surface. There is often little available space for the tree itself. A planter of at least 8 to 10 feet is required for such tree wells, otherwise little room is left to accommodate the tree.

Tree wells are often incorporated in urban renewal projects to provide visual interest and shade streetside parking. Curbed wheel stops and metal tree guards provide additional protection for this tree in a diagonal parking situation. Where parking is parallel to the curb, trees are more subject to damage from cars without a substantially larger planter.

CREATIVE APPLICATIONS AND SPECIAL EFFECTS

Trees are often used as architectural elements, and rooftop planters can integrate a building visually with the surrounding landscape. However, tree weight and stability are major concerns in rooftop plantings, and special lightweight soil mix is used in waterproof planters. In large or narrow planters, the soil is placed over filter fabric with a hollow plastic grid underneath to facilitate subsurface drainage. Minimal space, often only two to four feet in depth, is available for root systems in such situations. Tree species selected must be durable and tolerant of these conditions. Since standard guy wires or stakes usually cannot be used to support large trees, which may be subject to toppling on windy rooftops, smaller multitrunk trees are found to be more stable.

The concrete planters in this plaza landscape become a visual extension of the building's cantilevered concrete rooftop planters, providing a seamless transition from plaza to building.

Large pots or decorative containers provide planting areas large enough for trees in difficult situations. Containers can be placed in creative arrangements, with trees providing an overhead foliage canopy. Lightweight concrete materials are available, but stability is a concern for large canopy trees in windy conditions. Surface tree wells or larger raised planters offer less restricted root space for tree stability.

Drainage and irrigation in large containers may be handled by various methods. Often a drain that connects to the storm drain system can be centered underneath the container. A drain line extends upward into the container, with either a strainer-type inlet or an atrium drain added at the surface to remove excess water. Otherwise water will drain through the hole in the bottom of the container onto surrounding paving. This often leaves a permanent stain or presents a hazardous and slippery surface for pedestrians. Irrigation can similarly be provided by lines extended through the base of the container and sealed. Xerispray emitter heads are placed with flexible tubing to irrigate without heavy sprays oversaturating the soil. Occasionally, hollow-walled planters are used, which hydroscopically transfer water from the reservoir to the planting medium, keeping it evenly moist. However, these require constant monitoring, and rainwater must be dissipated if they are used outdoors.

Trees can be used in unexpected places to create a dramatic effect, including indoors. These are usually large specimens of subtropical species that we are accustomed to seeing as house plants. Rhaphis, kentia, king palm, schefflera, ficus, tupidanthus, and cordyline or other imported species often are used as trees in enclosed malls. Exacting requirements of ambient natural sunlight and constantly maintained temperature and humidity must be met in these situations. Sunlight through skylights must be filtered, and air conditioning cannot produce cold air drafts or unnecessary air movement. In essence, conservatory conditions must be replicated without the high humidity. Only species tolerant of such conditions can be expected to do well on a permanent basis.

An attractive daytime treescape can take on an entirely different effect at night with the use of various types of lighting. Lighting can accentuate trunks, form, texture, and canopies of the surrounding trees. Uplights or downlights highlight trunks and branching structure while creating dramatic shadowing and silhouettes upon building walls and paving. Twinkle lights wrapped around trunks and branches are sometimes used not only during the holiday season but throughout the year to provide a festive atmosphere.

A colonnade of *Washingtonia filifera* accentuates a building facade, creating a visually interesting streetscape for a building and parking structure. The diagonal grid of Asian jasmine in the tree wells creates an almost lawnlike appearance when viewed from the pedestrian level. Except for the regular blowing of leaves and trimming the grids, this attractive manicured landscape is essentially low maintenance.

CULTURE AND MAINTENANCE

Besides soil, moisture, light, and temperature requirements for each tree, consideration must be given to the amount of care it will receive and where it is planted. While trees with more exacting cultural requirements may be suitable for residential use, more durable and hardy species are desirable in urban streetscapes where maintenance becomes important. In warmer climates, choosing trees that fit the hydrozone becomes critical. Trees with high water requirements cannot remain vigorous when planted in shrub and groundcover areas that receive infrequent irrigation and vice versa. Trees that require loose, fertile soil generally lose vigor and become more susceptible to disease when planted in clayey, alkaline, or infertile rocky soils, regardless of the amount of moisture applied.

One of the most common maintenance considerations concerns the litter from leaves, flowers, and fruit. While no tree is completely litter free, some are messier than others. Seedballs or other litter may pose a hazard in some urban settings or in the tight quarters of a manicured residential garden. Street trees can pose a litter problem for homeowners and city maintenance crews in keeping streets and gutters cleared. Many cities have initiated green waste programs to dispose of curbside garden materials, which are handled separately from other landfill materials and often recycled for other

By bringing an exterior landscape into an enclosed interior mall, these palms (*Washingtonia robusta*) create a dramatic vertical effect. Here, they have been placed irregularly in a circular grid, where they can be viewed from an upper floor.

A row of Lombardy poplars forms a distinctive vertical background element, contrasting with smaller purple-leaf plum trees and low shrubs to create an attractive parking lot landscape in a business park.

The classic shape of these poplars has been destroyed by topping. This practice is sometimes thought to provide better visibility into a building complex, or possibly create a more manicured effect. Instead, it detracts from the overall appearance and the health of the trees.

Trees are headed-up rather than topped to allow visibility of signage while presenting an attractive, shaded place to park.

Businesses such as automobile dealerships, which require high visibility, can become eyesores without some landscape treatment. This frontage uses privet standards, which are suited to clipping, and low shrubs to soften the otherwise stark appearance.

uses. Cities and utility companies maintain crews to undertake routine tree trimming in their jurisdictions, using chippers to grind up trimmed material into usable mulch that is sometimes available to the public at low cost.

The annual wintertime practice of pollarding fruitless mulberry continues despite the seemingly unnecessary annual expense. Unfortunately, this practice has spread to include many other unlikely subjects such as crape myrtles that are "lollipopped" in winter into neat balls and vertical trees such as 'Aristocrat' pear, white birch, Lombardy poplar, and Italian cypress, which are flat-topped and sheared.

Though tree pruning may fill the time during the winter dormant season when lawn mowing and leaf blowing are no longer a priority, overzealous pruning consumes time that could be better used to produce more beneficial results. Winter fertilization and replenishing mulch are commonly overlooked, even though they are important in maintaining soil chemistry and moisture retention. Tree stakes, which should be removed after one season, are sometimes left in place for years, allowing the tree tie to girdle the trunk. Maintenance companies with certified arborists or foremen with horticultural expertise can direct attention to details that are otherwise neglected. A properly maintained landscape is less vulnerable to disease and enhances property values.

Trees are sometimes severely clipped, sheared, or eliminated entirely to increase visibility by businesses in order to attract customers. This often results in a commercial strip environment that does not add to the enjoyment of the customer. Improperly maintained areas can easily succumb to urban blight, and the customer base may move to other more desirable areas. Selection of trees that allow visibility can often prevent this situation. In established landscapes, removing a few lower limbs on larger trees can provide clear visibility of storefront signage while attracting customers to a shaded place to park while they shop. Some trees can be grown as small standards that require less pruning.

Maintaining the health and vitality of urban trees is a major concern. Besides having a full-time city parks and maintenance staff, many cities also have established tree foundation organizations. These include knowledgeable staff, volunteers, and professionals with expertise and interest in maintaining an inventory of trees within a city, monitoring their health, and encouraging the planting of new trees for optimum effect. As age, stress, and disease take their toll on older urban trees, these organizations often collaborate with other cities throughout the country to research methods that have had the best results and to monitor the spread of disease in other areas.

Some of the many reasons for excessive shearing or pruning trees include removing lower limbs for sight line visibility, maintaining limb clearance for overhead utility lines, or removing limbs that pose a danger to people or to structures. In the case of Modesto ash, which has severe mistletoe infestations, nearly all of the limbs are stripped, resulting in a flush of subsequent growth that is pest-free. However, timely maintenance can avoid the need for such severe measures. Though these trees may eventually regain vigor, replacement with a different species may be more desirable if the infestation is likely to recur.

The most prevalent pests and diseases affecting urban trees today include anthracnose (in *Platanus*, also affecting *Fraxinus*), Dutch elm disease (in what remaining elms there are today), mistletoe (affecting many trees, mostly *Fraxinus* and *Pyrus*), fireblight (affecting *Pyrus* and *Malus* notably), aphids (produce honeydew drip and resulting sooty mildew on leaves, especially in *Tilia, Liriodendron, Quercus rubra,* and *Ulmus*), tent caterpillar (often affecting *Gleditsia* and sometimes *Cer-*

While heavy trimming and shearing of trees creates the need for constant maintenance, topiary and oversize imitations of bonsai can be used with dramatic results.

cis, as well as *Arbutus menziesii* in native forests). Pests and diseases are usually monitored quite closely. There are many programs to assist homeowners who have diseased trees in their yards. University extension services can also provide information about what can be done. Certified arborists continue to be the best sources of information and assistance for homeowners in caring for valued trees.

Trees are an important part of our urban environment. Choosing the right tree for the right place and providing proper maintenance once they have been planted are critical to their success. There is no perfect tree for every situation, but we can select one where the advantages outweigh the disadvantages. Ornamental trees commonly used in urban landscapes often require special care to keep them healthy and free of disease. Native trees are better suited to survive on their own but may be subject to disease in ornamental landscapes. Exceptionally large or very old trees, both native and ornamental, are considered irreplaceable and may be designated heritage trees. Preservation ordinances have been established to protect them.

THE SPECIAL USE AND APPLICATIONS LISTS

The lists that follow are provided as a general reference. The lists are not intended to be all-inclusive, and the trees are not necessarily approved for the suggested uses. The lists serve best as a starting point, for comparative purposes, to give readers an idea of where to begin in selecting the right tree for the intended location and use.

Mistletoe affects many species of trees, especially ash, oak, and locust. Though it can be removed during winter by regular maintenance, it becomes difficult to eradicate once it fully infests a tree.

EVERGREEN TREES

Abies spp.	fir
Acacia spp.	acacia
Acer paxii	Paxii evergreen maple
Agathis robusta	dammar pine
Agonis flexuosa	peppermint tree
Araucaria araucana	monkey puzzle tree
Araucaria bidwillii	bunya-bunya
Araucaria heterophylla	Norfolk Island pine
Arbutus menziesii	madrone
Arbutus unedo	strawberry tree
Brachychiton acerifolius	flame tree
Brachychiton populneus	bottle tree
Broussonetia papyrifera	paper mulberry
Callistemon citrinus	lemon bottlebrush
Callistemon viminalis	weeping bottlebrush
Calocedrus decurrens	incense cedar
Casuarina equisetifolia	horsetail tree
Cedrus atlantica 'Glauca'	blue Atlas cedar
Cedrus deodara	deodar cedar
Ceratonia siliqua	carob
Cercocarpus betuloides	mountain ironwood
Chamaecyparis lawsoniana	Port Orford-cedar
Chamaecyparis nootkatensis	Nootka cypress
Chrysolepis chrysophylla	golden chinquapin
Cinnamomum camphora	camphor tree
Cordyline australis	green dracaena
Cornus capitata	evergreen dogwood
Crinodendron patagua	lily-of-the-valley tree
Cryptomeria japonica	Japanese cryptomeria
Cryptomeria j. 'Elegans'	plume cedar
Cunninghamia lanceolata	China fir
x *Cupressocyparis leylandii*	Leyland cypress
Cupressus funebris	mourning cypress
Cupressus spp.	cypress
Cupaniopsis anacardioides	carrot wood
Cyathea cooperi	Australian tree fern
Cycas revoluta	cycad
Dicksonia antarctica	Tasmanian tree fern
Eriobotrya deflexa	bronze loquat
Eriobotrya japonica	loquat
Eucalyptus spp.	eucalyptus
Ficus spp.	ficus
Geijera parviflora	Australian willow
Grevillea robusta	silk oak
Hymenosporum flavum	sweetshade
Juniperus spp.	juniper
Lagunaria patersonii	primrose tree
Laurus nobilis	Grecian laurel
Ligustrum lucidum	glossy privet
Lithocarpus densiflorus	tanbark oak
Lophostemon confertus	Brisbane box
Lyonothamnus floribundus	Catalina ironwood
Magnolia grandiflora	southern magnolia
Maytenus boaria	mayten
Melaleuca linariifolia	flaxleaf paperbark
Melaleuca quinquenervia	cajeput tree
Metrosideros excelsus	New Zealand Christmas tree
Musa x paradisiaca	banana
Myoporum laetum	myoporum
Neolitsea sericea	Japanese silver tree
Nerium oleander	oleander
Olea europaea	olive
palms	palms
Phyllostachys bambusoides	timber bamboo
Picea spp.	spruce
Pinus spp.	pine
Pittosporum rhombifolium	Queensland pittosporum

An attractive evergreen planting with a curved massing of horizontal juniper shrubs contrasting with a background of vertical sequoias. Light green pistache and camphor trees add color behind.

Pittosporum undulatum	Victorian box
Podocarpus gracilior	fern pine
Podocarpus macrophyllus	yew pine
Prunus caroliniana	Carolina cherry laurel
Prunus ilicifolia	hollyleaf cherry
Prunus lyonii	Catalina cherry
Pseudotsuga macrocarpa	bigcone Douglas-fir
Pseudotsuga menziesii	Douglas-fir
Quercus agrifolia	coast live oak
Quercus arizonica	Arizona white oak
Quercus chrysolepis	canyon live oak
Quercus dumosa	Nuttall's scrub oak
Quercus emoryi	Emory oak
Quercus ilex	holly oak
Quercus myrsinifolia	Japanese live oak
Quercus palmeri	Palmer's oak
Quercus phillyreoides	ubame oak
Quercus rugosa	netleaf oak
Quercus suber	cork oak
Quercus tomentella	island oak
Quercus turbinella	shrub live oak
Quercus virginiana	southern live oak
Quercus wislizeni	interior live oak
Quillaja saponaria	soapbark tree
Rhus lancea	African sumac
Schinus molle	pepper tree
Schinus peruviana	Peruvian pepper
Schinus terebinthifolius	Brazilian pepper
Sciadopitys verticillata	Japanese umbrella pine
Sequoia sempervirens	coast redwood
Sequoiadendron giganteum	giant sequoia
Stenocarpus sinuatus	firewheel tree
Syzygium myrtifolia	brush cherry
Tamarix aphylla	athel tree
Tetrapanax papyriferum	ricepaper plant
Thuja occidentalis	American arborvitae
Thuja plicata	western red cedar
Thujopsis dolobrata	deerhorn cedar
Torreya californica	California nutmeg
Tristaniopsis laurina	laurel leaf box
Tsuga heterophylla	western hemlock
Tsuga mertensiana	mountain hemlock
Umbellularia californica	California bay
Yucca brevifolia	Joshua tree
Yucca gloriosa	Spanish dagger

DECIDUOUS TREES

Acer spp.	maple
Aesculus spp.	horsechestnut
Ailanthus altissima	tree-of-heaven
Albizia distachya	plume albizia
Albizia julibrissin	silk tree
Alnus spp.	alder
Bauhinia variegata	orchid tree
Betula spp.	birch
Carpinus betulus	European hornbeam
Carya illinoiensis	pecan
Castanea sativa	Spanish chestnut
Catalpa speciosa	western catalpa
Ceiba speciosa	floss silk tree
Celtis spp.	hackberry
Cercidiphyllum japonicum	katsura tree
Cercidium floridum	palo verde
Cercis spp.	redbud
Chilopsis linearis	desert-willow
x *Chitalpa tashkentensis*	chitalpa
Cladrastis kentukea	yellowwood
Cornus florida	flowering dogwood
Cornus mas	cornelian cherry
Cornus nuttallii	Pacific dogwood
Crataegus spp.	hawthorn
Erythrina spp.	coral tree
Fagus spp.	beech
Firmiana simplex	Chinese parasol tree
Fraxinus spp.	ash
Ginkgo biloba	maidenhair tree
Gleditsia spp.	locust
Jacaranda mimosifolia	jacaranda
Juglans spp.	walnut
Koelreuteria bipinnata	Chinese flame tree
Koelreuteria paniculata	goldenrain tree
Laburnum x *watereri*	goldenchain tree
Lagerstroemia indica	crape myrtle
Larix kaempferi	Japanese larch
Liquidambar spp.	sweet gum
Liriodendron tulipifera	tulip tree
Magnolia x *soulangeana*	saucer magnolia
Magnolia stellata	star magnolia
Malus spp.	crabapple
Melia azedarach	chinaberry
Metasequoia glyptostroboides	dawn redwood
Morus spp.	mulberry
Nyssa sylvatica	tupelo
Oxydendron arborea	sorrel tree
Paulownia tomentosa	empress tree
Pistacia chinensis	Chinese pistache
Platanus x *acerifolia*	London plane
Platanus occidentalis	American sycamore
Platanus racemosa	California sycamore
Populus spp.	cottonwood
Prosopis spp.	mesquite
Prunus spp.	flowering plum, cherry
Pterocarya stenoptera	Chinese wingnut
Punica granatum	pomegranate
Pyrus spp.	flowering pear
Quercus acutissima	Chinese oak
Quercus alba	white oak
Quercus coccinea	scarlet oak
Quercus douglasii	blue oak
Quercus engelmannii	Engelmann oak
Quercus gambelii	Rocky Mountain white oak
Quercus garryana	Oregon white oak
Quercus kelloggii	California black oak
Quercus laurifolia	laurel oak

The irregular twisted trunks of *Platanus racemosa*, with colorful mottled bark and orange fall foliage, highlight a riparian habitat in Laguna Canyon in southern California.

Quercus lobata	valley oak
Quercus macrocarpa	mossycup oak
Quercus morehus	oracle oak
Quercus muehlenbergii	chinquapin oak
Quercus palustris	pin oak
Quercus phellos	willow oak
Quercus robur	English oak
Quercus rubra	red oak
Rhus glabra	smooth sumac
Rhus typhina	staghorn sumac
Robinia spp.	locust
Salix spp.	willow
Sambucus mexicana	blue elderberry
Sapium sebiferum	Chinese tallow
Sophora japonica	Japanese pagoda tree
Sorbus aucuparia	European mountain ash
Tamarix parviflora	tamarisk
Taxodium distichum	bald cypress
Taxodium mucronatum	Montezuma cypress
Tilia spp.	linden
Ulmus spp.	elm
Zelkova serrata	sawtooth zelkova
Zizyphus jujuba	Chinese jujube

LONG-LIVED TREES

Abies spp.	fir
Acer tataricum ssp. *ginnala*	amur maple
Acer palmatum	Japanese maple
Acer platanoides	Norway maple
Acer rubrum	red maple
Acer saccharum	sugar maple
Aesculus x *carnea*	red horsechestnut
Aesculus hippocastanum	horsechestnut
Agathis robusta	dammar pine
Araucaria araucana	monkey puzzle tree
Araucaria bidwillii	bunya-bunya
Araucaria heterophylla	Norfolk Island pine
Arbutus menziesii	madrone
Arbutus unedo	strawberry tree
Bauhinia variegata	orchid tree
Brachychiton acerifolius	flame tree
Brachychiton populneus	bottle tree
Callistemon citrinus	lemon bottlebrush
Callistemon viminalis	weeping bottlebrush
Calocedrus decurrens	incense cedar
Carpinus betulus	European hornbeam
Carya illinoiensis	pecan
Castanea sativa	Spanish chestnut
Catalpa speciosa	western catalpa
Cedrus atlantica 'Glauca'	blue Atlas cedar
Cedrus deodara	deodar cedar
Ceiba speciosa	floss silk tree
Celtis occidentalis	common hackberry
Celtis sinensis	Chinese hackberry
Ceratonia siliqua	carob
Chamaecyparis lawsoniana	Port Orford-cedar
Chamaecyparis nootkatensis	Nootka cypress
Cinnamomum camphora	camphor
Cladrastis lutea	yellow wood
Cordyline australis	green dracaena
Cryptomeria japonica	Japanese cryptomeria
Cryptomeria j. 'Elegans'	plume cedar
Cupressus macrocarpa	Monterey cypress
Cupressus sempervirens	Italian cypress
Erythrina caffra	kaffirboom coral tree
Erythrina crista-galli	cockspur coral tree
Eucalyptus globulus	blue gum
Eucalyptus viminalis	manna gum
Fagus sylvatica	beech
Ficus microcarpa 'Nitida'	little-leaf fig
Ficus rubiginosa	rustyleaf fig
Ginkgo biloba	maidenhair tree
Juglans hindsii	black walnut
Juglans regia	English walnut
Juniperus californica	California juniper
Lagerstroemia indica	crape myrtle
Laurus nobilis	Grecian laurel
Liquidambar formosana	Chinese sweet gum
Liquidambar styraciflua	sweet gum
Liriodendron tulipifera	tulip tree
Lithocarpus densiflorus	tanbark oak
Magnolia grandiflora	southern magnolia
Magnolia x *soulangeana*	saucer magnolia
Melaleuca linariifolia	flaxleaf paperbark
Melaleuca quinquenervia	cajeput tree
Metasequoia glyptostroboides	dawn redwood
Metrosideros excelsus	New Zealand Christmas tree
Nyssa sylvatica	tupelo
Olea europaea	olive
Oxydendrum arboreum	sorrel tree
palms	palms
Picea spp.	spruce

A stately specimen of *Juglans californica* becomes a landmark feature in a residential frontage, providing welcome summer shade in a warm climate.

Pinus spp.	pine
Pistacia chinensis	Chinese pistache
Platanus x *acerifolia*	London plane
Platanus occidentalis	American sycamore
Platanus racemosa	California sycamore
Pseudotsuga macrocarpa	bigcone Douglas-fir
Pseudotsuga menziesii	Douglas-fir
Quercus spp.	oak
Schinus molle	pepper tree
Schinus terebinthifolius	Brazilian pepper tree
Sciadopitys verticillata	Japanese umbrella pine
Sequoia sempervirens	coast redwood
Sequoiadendron giganteum	giant sequoia
Taxodium distichum	bald cypress
Thuja occidentalis	American arborvitae
Thuja plicata	western red cedar
Tilia spp.	linden
Tsuga heterophylla	western hemlock
Tsuga mertensiana	mountain hemlock
Ulmus americana	American elm
Ulmus parvifolia	evergreen elm
Umbellularia californica	California bay
Yucca gloriosa	Spanish dagger
Yucca brevifolia	Joshua tree
Zelkova serrata	sawtooth zelkova

SHORT-LIVED TREES

Long or short-lived is a relative term. Trees that are relatively pest-free, tolerate general cultural conditions, and do not require constant maintenance to keep them healthy, usually for more than 100 years, are considered long-lived. Short-lived trees are generally less hardy, less sturdy, and generally live less than 100 years. Short-lived trees are more suited for residential garden use, and they do not make reliable street trees. Trees listed here are considered relatively short-lived.

Acacia baileyana	Bailey acacia
Acacia dealbata	green acacia
Acacia longifolia	Sidney golden wattle
Acacia melanoxylon	black acacia
Acer macrophyllum	bigleaf maple
Acer negundo	box elder
Acer n. 'Variegatum'	variegated box elder
Acer saccharinum	silver maple
Aesculus californica	California buckeye
Agonis flexuosa	peppermint tree
Ailanthus altissima	tree-of-heaven
Albizia distachya	plume albizia
Alnus cordata	Italian alder
Alnus rhombifolia	white alder
Betula nigra	red birch
Betula papyrifera	paper birch
Betula pendula 'Crispa'	cutleaf weeping birch
Chilopsis linearis	desert-willow
x *Chitalpa tashkentensis*	chitalpa
Cornus capitata	evergreen dogwood
Cornus florida	flowering dogwood
Cornus mas	cornelian cherry
Cornus nuttallii	Pacific dogwood
Crataegus laevigata	hawthorn
Crinodendron patagua	lily-of-the-valley tree
Cunninghamia lanceolata	China fir
x *Cupressocyparis leylandii*	Leyland cypress
Cupressus arizonica	Arizona cypress
Eriobotrya deflexa	bronze loquat
Eriobotrya japonica	loquat
Firmiana simplex	Chinese parasol tree
Geijera parviflora	Australian willow
Malus spp.	crabapple
Populus alba	white poplar
Populus fremontii	Fremont cottonwood
Populus nigra 'Italica'	Lombardy poplar
Populus trichocarpa	black cottonwood
Prunus x *blirieana*	flowering plum
Prunus caroliniana	Carolina cherry laurel
Prunus cerasifera	cherry plum
Prunus c. 'Atropurpurea'	purple-leaf plum
Prunus c. 'Krauter Vesuvius'	pink-flowering purple-leaf plum
Prunus serrulata 'Kwanzan'	Kwanzan flowering cherry
Prunus x *yedoensis*	Yoshino flowering cherry
Rhus glabra	smooth sumac
Rhus lancea	African sumac
Rhus typhina	staghorn sumac
Robinia pseudoacacia	black locust
Robinia x *ambigua* 'Idahoensis'	Idaho pink locust
Robinia x *a.* 'Royal Robe'	purple flowering locust
Salix spp.	willow
Sambucus mexicana	blue elderberry
Sapium sebiferum	Chinese tallow tree
Tamarix parviflora	tamarisk
Zizyphus jujuba	Chinese jujube

While trees such as *Alnus cordata* are sometimes used as street trees because of their initial fast growth rate, they are comparatively short-lived, to about 30-40 years. Longer-lived species are better used as primary street trees, with shorter-lived species used as subordinate trees.

STREET TREES

Trees for urban streets require careful selection. Residential street and parkway use allows more flexibility in tree selection.

Large

Celtis occidentalis	common hackberry
Celtis sinensis	Chinese hackberry
Ceratonia siliqua	carob
Cinnamomum camphora	camphor
Cladrastis kentukea	yellowwood
Eucalyptus spp.	eucalyptus
Fagus sylvatica	European beech
Ficus rubiginosa	rustyleaf fig
Fraxinus pennsylvanica 'Marshall'	Marshall green ash
Fraxinus uhdei	evergreen ash
Ginkgo biloba	maidenhair tree
Jacaranda mimosifolia	jacaranda
Liquidambar styraciflua	sweet gum
Liriodendron tulipifera	tulip tree
Magnolia grandiflora	southern magnolia
Olea europaea	olive
Pinus canariensis	Canary Island pine
Platanus x *acerifolia*	London plane
Platanus occidentalis	American sycamore
Platanus racemosa	California sycamore
Pterocarya stenoptera	Chinese wingnut
Quercus agrifolia	coast live oak
Quercus alba	white oak
Quercus coccinea	scarlet oak
Quercus ilex	holly oak
Quercus lobata	valley oak
Quercus rubra	red oak
Quercus suber	cork oak
Schinus molle	pepper tree
Sequoia sempervirens	coast redwood
Tilia tomentosa	silver linden
Ulmus parvifolia	evergreen elm
Ulmus rubra	slippery elm
Zelkova serrata	sawtooth zelkova

Medium

Acacia melanoxylon	black acacia
Acer tataricum ssp. *ginnala*	amur maple
Acer platanoides	Norway maple
Acer p. 'Schwedleri'	Schwedler maple
Acer rubrum	red maple
Acer r. 'Columnare'	columnar red maple
Agonis flexuosa	peppermint tree
Albizia julibrissin	silk tree
Alnus cordata	Italian alder
Alnus rhombifolia	white alder
Brachychiton populneus	bottle tree
Carpinus betulus	European hornbeam
Cedrus deodara	deodar cedar
Cupaniopsis anacardioides	carrot wood
Fraxinus angustifolia 'Raywood'	Raywood ash
Fraxinus holotricha 'Moraine'	Moraine ash
Koelreuteria bipinnata	Chinese flame tree
Koelreuteria paniculata	goldenrain tree
Lagunaria patersonii	primrose tree
Lophostemon confertus	Brisbane box
Nyssa sylvatica	tupelo
Pistacia chinensis	Chinese pistache
Prunus cerasifera 'Atropurpurea'	purple-leaf plum
Prunus c. 'Krauter Vesuvius'	pink-flowering purple-leaf plum
Pyrus calleryana 'Aristocrat'	Aristocrat pear
Pyrus c. 'Bradford'	Bradford pear

An unusual linear planting of Modesto ash provides a spectacular display of yellow fall color in a residential neighborhood. Street tree plantings often include a mixture of species to provide diversity and reduce spread of disease.

Quercus myrsinifolia	Japanese live oak
Quercus phellos	willow oak
Quercus robur 'Fastigiata'	upright English oak
Schinus terebinthifolius	Brazilian pepper
Sophora japonica	Japanese pagoda tree
Tilia cordata	little-leaf linden

Small

Acer buergerianum	trident maple
Acer campestre	hedge maple
Aesculus x *carnea*	red horsechestnut
Callistemon citrinus	lemon bottlebrush
Carpinus betulus	European hornbeam
Cercis canadensis	eastern redbud
Crataegus laevigata	hawthorn
Crataegus x *lavallei*	Carriere hawthorn
Eriobotrya deflexa	bronze loquat
Eriobotrya japonica	loquat
Eucalyptus ficifolia	red-flowering gum
Eucalyptus lehmannii	bushy yate
Eucalyptus nicholii	willow-leafed gum
Ficus microcarpa 'Nitida'	little-leaf fig
Geijera parviflora	Australian willow
Hymenosporum flavum	sweetshade
Laburnum x *watereri*	goldenchain tree
Lagerstroemia indica	crape myrtle
Laurus nobilis	Grecian laurel
Ligustrum lucidum	glossy privet
Maytenus boaria	mayten
Melaleuca linariifolia	flaxleaf paperbark
Melaleuca quinquenervia	cajeput tree
Metrosideros excelsus	New Zealand Christmas tree
Myoporum laetum	myoporum
Nerium oleander	oleander
Pittosporum rhombifolium	Queensland pittosporum
Pittosporum undulatum	Victorian box
Prunus x *blirieana*	flowering plum
Rhus lancea	African sumac
Sorbus aucuparia	European mountain ash

PARKING SHADE TREES

Most cities have adopted ordinances governing tree planting in parking areas to guide implementation of a minimum acceptable standard based on 15 years growth of listed trees. The species listed by each city varies, usually with local climate and soils.

30-35' Canopy

Acer platanoides	Norway maple
Alnus rhombifolia	white alder
Calocedrus decurrens	incense cedar
Cedrus deodara	deodar cedar
Celtis australis	European hackberry
Celtis sinensis	Chinese hackberry
Ceratonia siliqua	carob
Cinnamomum camphora	camphor
Cupaniopsis anacardioides	carrot wood
Fagus sylvatica	European beech
Ficus rubiginosa	rustyleaf fig
Fraxinus angustifolia 'Raywood'	Raywood ash
Fraxinus holotricha 'Moraine'	Moraine ash
Fraxinus pennsylvanica 'Marshall'	Marshall green ash
Fraxinus uhdei	evergreen ash
Ginkgo biloba	maidenhair tree
Gleditsia triacanthos	honey locust
Liriodendron tulipifera	tulip tree
Jacaranda mimosifolia	jacaranda
Magnolia grandiflora	southern magnolia
Pistacia chinensis	Chinese pistache
Pinus halepensis	Aleppo pine
Pinus pinea	Italian stone pine
Platanus x *acerifolia*	London plane
Platanus occidentalis	American sycamore
Platanus racemosa	California sycamore
Quercus agrifolia	coast live oak
Quercus coccinea	scarlet oak
Quercus douglasii	blue oak
Quercus ilex	holly oak
Quercus kelloggii	California black oak
Quercus lobata	valley oak
Quercus rubra	red oak
Quercus suber	cork oak
Quercus virginiana	southern live oak
Quercus wislizeni	interior live oak
Tilia tomentosa	silver linden
Ulmus parvifolia	evergreen elm
Zelkova serrata	sawtooth zelkova

25-30' Canopy

Acacia melanoxylon	black acacia
Acer rubrum	red maple
Alnus cordata	Italian alder
Liquidambar styraciflua	sweet gum
Lophostemon confertus	Brisbane box
Nyssa sylvatica	tupelo
Olea europaea	olive
Pinus canariensis	Canary Island pine
Pinus radiata	Monterey pine
Pinus sylvestris	Scotch pine
Pinus thunbergii	Japanese black pine
Pterocarya stenoptera	Chinese wingnut
Pyrus calleryana 'Aristocrat'	Aristocrat pear
Pyrus calleryana 'Bradford'	Bradford pear
Quercus palustris	pin oak
Quercus phellos	willow oak
Sapium sebiferum	Chinese tallow tree
Schinus molle	pepper tree
Schinus terebinthifolius	Brazilian pepper
Sequoia sempervirens	coast redwood

Zelkova serrata is an excellent shade tree and also provides dramatic fall color.

Sophora japonica	Japanese pagoda tree
Tilia cordata	little-leaf linden

20-25' Canopy

Acer tataricum ssp. *ginnala*	amur maple
Aesculus x *carnea*	red horsechestnut
Carpinus betulus	European hornbeam
Cladrastis kentukea	yellowwood
Eucalyptus nicholii	willow-leafed peppermint
Eucalyptus sideroxylon 'Rosea'	red ironbark
Gleditsia triacanthos 'Sunburst'	golden honey locust
Grevillea robusta	silk oak
Koelreuteria bipinnata	Chinese flame tree
Koelreuteria paniculata	goldenrain tree
Laurus nobilis	Grecian laurel
Ligustrum lucidum	glossy privet
Melaleuca linariifolia	flaxleaf paperbark
Melaleuca quinquenervia	cajeput tree
Prunus cerasifera 'Atropurpurea'	purple-leaf plum
Prunus c. 'Krauter Vesuvius'	pink-flowering purple-leaf plum
Robinia x *ambigua* 'Royal Robe'	purple-flowering locust
Tristaniopsis laurina	laurel leaf box
Umbellularia californica	California bay
Ziziphus jujuba	Chinese jujube

15-20' Canopy

Acer buergerianum	trident maple
Agonis flexuosa	peppermint tree
Betula nigra	river birch
Betula papyrifera	paper birch
Cercis canadensis	eastern redbud
Crataegus x *lavallei*	Carriere hawthorn
Eriobotrya deflexa	bronze loquat
Eucalyptus ficifolia	red-flowering gum
Ficus microcarpa 'Nitida'	little-leaf fig
Geijera parviflora	Australian willow
Lagerstroemia indica	crape myrtle
Malus floribunda	flowering crabapple
Maytenus boaria	mayten
Melia azedarach	chinaberry
Metrosideros excelsus	New Zealand Christmas tree
Myoporum laetum	myoporum
Prunus x *blirieana*	flowering plum
Pyrus taiwanensis	evergreen pear
Rhus lancea	African sumac
Sorbus aucuparia	European mountain ash

FRUIT AND SEEDS

Deciduous trees can have wintertime accent value by virtue of their seeds or fruit, although these may litter paving. Evergreen trees often have showy berries.

Acer spp.	maple
Aesculus californica	California buckeye
Albizia distachya	plume albizia
Albizia julibrissin	silk tree
Alnus cordata	Italian alder
Alnus rhombifolia	white alder
Arbutus menziesii	madrone
Betula nigra	river birch
Betula papyrifera	paper birch
Betula pendula 'Crispa'	cutleaf weeping birch
Catalpa speciosa	western catalpa
Celtis sinensis	Chinese hackberry
Cercis canadensis	eastern redbud
Cercis occidentalis	western redbud
Chilopsis linearis	desert-willow
Cladrastis kentukea	yellowwood
Cornus mas	cornelian cherry
Crataegus laevigata	hawthorn
Crataegus x *lavallei*	Carriere hawthorn
Eucalyptus spp.	eucalyptus
Fraxinus latifolia	Oregon ash
Fraxinus holotricha 'Moraine'	Moraine ash
Fraxinus velutina	Arizona ash
Fraxinus v. 'Modesto'	Modesto ash
Gleditsia triacanthos	honey locust
Jacaranda mimosifolia	jacaranda
Koelreuteria bipinnata	Chinese flame tree
Koelreuteria paniculata	goldenrain tree
Laburnum x *watereri*	goldenchain tree
Lagerstroemia indica	crape myrtle
Ligustrum lucidum	glossy privet
Liquidambar formosana	Chinese sweet gum
Liquidambar styraciflua	sweet gum
Liriodendron tulipifera	tulip tree
Melia azedarach	chinaberry
Nyssa sylvatica	tupelo
Oxydendrum arboreum	sorrel tree
Paulownia tomentosa	empress tree
Pistacia chinensis	Chinese pistache
Platanus x *acerifolia*	London plane
Platanus racemosa	California sycamore
Pterocarya stenoptera	Chinese wingnut
Punica granatum	pomegranate
Pyrus taiwanensis	evergreen pear
Quercus spp.	oak
Rhus glabra	smooth sumac
Rhus lancea	African sumac
Rhus typhina	staghorn sumac
Robinia pseudoacacia	black locust
Robinia x *ambigua* 'Idahoensis'	Idaho pink locust
Sapium sebiferum	Chinese tallow
Schinus molle	pepper tree
Sophora japonica	Japanese pagoda tree
Sorbus aucuparia	European mountain ash
Taxodium distichum	bald cypress
Tilia cordata	little leaf linden
Tilia europaea	European linden
Tilia tomentosa	silver linden

The flowering catkins on a grove of *Alnus rhombifolia* provide interest in an otherwise barren winter landscape and also create a weeping effect

The bright red berries of *Crataegus phaenopyrum*, Washington thorn, brighten the winter landscape and are favored by birds.

FLOWERING ACCENT

Flowering trees provide seasonal interest through the year, some showy, others noticeable only closeup. Approximate sequence of flowering is shown in parentheses. * Indicates extended or sporadic blooming periods throughout the year. (Based on Calif. Central Valley)

Albizia julibrissin **provides welcome shade and midsummer flowering color.**

Early Spring (January-March) (1-24)

Acacia baileyana (2)	Bailey acacia
Acacia dealbata (3)	green acacia
Acacia longifolia (8A)	Sidney golden wattle
Acacia melanoxylon (12)	black acacia
Acacia retinoides (18)	water wattle
Acer platanoides (male) (19B)	Norway maple
Acer truncatum (8B)	purpleblow maple
Arbutus menziesii (24A)	madrone
Cercis spp. (16)	redbud
Cornus florida (19)	flowering dogwood
Cornus mas (7)	cornelian cherry
Cornus nuttallii (21)	western dogwood
Crataegus laevigata (20A)	hawthorn
Crataegus x *lavallei* (20B)	Carriere hawthorn
Erythrina caffra	kaffirboom coral tree
Fraxinus ornus (24)	flowering ash
Laurus nobilis (8A)	Grecian laurel
Magnolia x *soulangeana* (5)	saucer magnolia
Magnolia stellata (4)	star magnolia
Malus floribunda (17)	flowering crabapple
Paulownia tomentosa (20)	empress tree
Prunus x *blirieana* (6)	pink-flowering plum
Prunus caroliniana (12A)	Carolina cherry laurel
Prunus cerasifera 'Atropurpurea' (9)	purple-leaf plum
Prunus c. 'Krauter Vesuvius' (10)	pink-flowering purple-leaf plum
Prunus serrulata (11)	flowering cherry
Pyrus taiwanensis (1)	evergreen pear
Pyrus cerasifera 'Aristocrat' (15)	Aristocrat pear
Pyrus c. 'Bradford' (14)	Bradford pear
Robinia pseudoacacia (22)	black locust
Robinia x *ambigua* 'Idahoensis' (23)	Idaho pink locust
Robinia x *a.* 'Royal Robe' (23A)	purple-flowering locust
Tamarix parviflora (19)	tamarisk
Umbellularia californica (7A)	California bay
Yucca brevifolia (25)	Joshua tree

Late Spring (April-May) (25-47)

Acer macrophyllum (25)	bigleaf maple
Aesculus californica (40)	California buckeye
Aesculus x *carnea* (27)	red horsechestnut
Agonis flexuosa (46B)	peppermint tree
Bauhinia variegata (39)	orchid tree
Brachychiton populneus (40B)	bottle tree
Callistemon citrinus (29)*	lemon bottlebrush
Callistemon viminalis (29A)*	weeping bottlebrush
Castanea sativa (26)	Spanish chestnut
Catalpa speciosa (38)*	western catalpa
Chilopsis linearis (46)	desert-willow
x *Chitalpa tashkentensis* (45)	chitalpa
Cladrastis kentukea (37)	yellowwood
Cordyline australis (44B)	green dracaena
Cornus capitata (42A)	evergreen dogwood
Geijera parviflora (47C)	Australian willow
Grevillea robusta (46A)	silk oak
Jacaranda mimosifolia (44)	jacaranda
Laburnum x *watereri* (43)	goldenchain tree
Lagunaria patersonii (47C)	primrose tree
Magnolia grandiflora (42)	southern magnolia
Melaleuca linariifolia (44)	flaxleaf paperbark

Melia azedarach (36)	chinaberry
Metrosideros excelsus (47C)*	New Zealand Christmas tree
Myoporum laetum (29B)*	myoporum
Nerium oleander (41)*	oleander
Pterocarya stenoptera (30)	Chinese wingnut
Punica granatum (35)	pomegranate
Sambucus mexicana (27B)	blue elderberry
Sorbus aucuparia (31)	European mountain ash
Tilia cordata (33)	little-leaf linden
Tilia europaea (32)	European linden
Tilia tomentosa (34)	silver linden
Yucca gloriosa (44A)*	Spanish dagger

Summer (June-July) (48-55)

Albizia distachya (53A)*	plume albizia
Albizia julibrissin (54)	silk tree
Arbutus unedo (53)*	strawberry tree
*Callistemon citrinus**	lemon bottlebrush
*Callistemon viminalis**	weeping bottlebrush
Cercidium floridum (49)*	palo verde
Cercocarpus betuloides (53C)	mountain ironwood
*Chilopsis linearis**	desert-willow
x *Chitalpa tashkentensis**	chitalpa
*Cornus capitata**	evergreen dogwood
Erythrina crista-galli (48)*	cockspur coral tree
Firmiana simplex (53)	Chinese parasol tree
Koelreuteria bipinnata (50B)*	Chinese flame tree
Koelreuteria paniculata (50A)	goldenrain tree
*Ligustrum lucidum**	glossy privet
Lophostemon confertus (51)	Brisbane box
Lyonothamnus floribundus (49)	Catalina ironwood
*Magnolia grandiflora**	southern magnolia
Melaleuca quinquenervia (52A)	cajeput tree
*Metrosideros excelsus**	New Zealand Christmas tree
*Nerium oleander**	oleander
Oxydendrum arboreum (50)	sorrel tree
*Punica granatum**	pomegranate
*Sambucus mexicana**	blue elderberry
Sapium sebiferum (52)	Chinese tallow
Sophora japonica (53)	Japanese pagoda tree
Tristaniopsis laurina (55)	laurel leaf box
*Yucca gloriosa**	Spanish dagger

Late Summer (August-September) (56-57)

*Chilopsis linearis**	desert-willow
x *Chitalpa tashkentensis**	chitalpa
Ceiba speciosa (57)	floss silk tree
*Erythrina crista-galli**	cockspur coral tree
*Koelreuteria paniculata**	goldenrain tree
Lagerstroemia indica (56)	crape myrtle
*Nerium oleander**	oleander
*Yucca gloriosa**	Spanish dagger

FLOWER COLOR

Red

Aesculus x *carnea* 'Briotii'	red horsechestnut
Callistemon citrinus	lemon bottlebrush
Callistemon viminalis	weeping bottlebrush
Cornus florida 'Cherokee Chief'	red eastern dogwood
Crataegus laevigata	hawthorn
Erythrina caffra	kaffirboom coral tree
Erythrina crista-galli	cockspur coral tree
Lagerstroemia indica vars.	crape myrtle
Malus x *floribunda* vars.	Japanese flowering crabapple
Metrosideros excelsus	New Zealand Christmas tree
Nerium oleander	oleander
Punica granatum	pomegranate

Orange

Eucalyptus ficifolia	red-flowering gum
Punica granatum	pomegranate

Erythrina caffra, kaffirboom coral tree, offers a spectacular display of bright coral red flowers in early spring in temperate southern California landscapes.

There are many cultivars of *Malus* x *floribunda*, flowering crabapple, offering many flower colors to choose from.

Pink

Albizia julibrissin	silk tree
Ceiba speciosa	floss silk tree
Cercis canadensis	eastern redbud
Cercis canadensis 'Forest Pansy'	purple-leaf redbud
Cercis occidentalis	western redbud
x *Chitalpa tashkentensis*	chitalpa
Cornus florida 'Rubra'	pink-flowering dogwood
Crataegus laevigata	hawthorn
Eriobotrya deflexa	bronze loquat
Eucalyptus sideroxylon 'Rosea'	red ironbark
Lagerstroemia indica	crape myrtle
Lagunaria patersonii	primrose tree
Magnolia x *soulangeana*	saucer magnolia
Magnolia stellata	star magnolia
Malus x *floribunda*	Japanese flowering crabapple
Nerium oleander	oleander
Paulownia tomentosa	empress tree
Prunus x *blirieana*	flowering plum
Prunus cerasifera 'Krauter Vesuvius'	pink-flowering purple-leaf plum
Prunus serrulata 'Kwanzan'	Kwanzan flowering cherry
Prunus x *yedoensis*	Yoshino flowering cherry
Robinia x *ambigua* 'Idahoensis'	Idaho pink locust
Tamarix parviflora	tamarisk

Purple

Bauhinia variegata	orchid tree
Ceiba speciosa	floss silk tree
Chilopsis linearis	desert-willow
Melia azedarach	chinaberry
Paulownia tomentosa	empress tree
Robinia x *ambigua* 'Royal Robe'	purple-flowering locust

FLOWER COLOR

Acacia longifolia, Sydney golden wattle, and other acacias are among the first to bloom in spring with a profusion of showy yellow flowers.

Yellow

Acacia baileyana	Bailey acacia
Acacia dealbata	green acacia
Acacia longifolia	Sidney golden wattle
Acacia melanoxylon	black acacia
Acer macrophyllum	bigleaf maple
Agonis flexuosa	peppermint tree
Albizia distachya	plume albizia
Alnus spp.	alder
Brachychiton populneus	bottle tree
Callistemon citrinus vars.	lemon bottlebrush
Castanea sativa	Spanish chestnut
Cercidium floridum	palo verde
Firmiana simplex	Chinese parasol tree
Geijera parviflora	Australian willow
Grevillea robusta	silk oak
Koelreuteria bipinnata	Chinese flame tree
Koelreuteria paniculata	goldenrain tree
Laburnum x *watereri*	goldenchain tree
Liriodendron tulipifera	tulip tree
Lithocarpus densiflorus	tanbark oak
Metrosideros excelsus vars.	New Zealand Christmas tree
Nerium oleander	oleander
Prunus caroliniana	Carolina cherry laurel
Pterocarya stenoptera	Chinese wingnut
Punica granatum vars.	pomegranate
Sapium sebiferum	Chinese tallow
Schinus molle	pepper tree
Sorbus aucuparia	European mountain ash
Tristaniopsis laurina	laurel leaf box
Umbellularia californica	California bay

White

Aesculus californica	California buckeye
Aesculus hippocastanum	horsechestnut
Arbutus menziesii	madrone
Arbutus unedo	strawberry tree
Bauhinia variegata 'Alba'	white-flowering orchid tree
Catalpa speciosa	western catalpa
Cercis canadensis 'Alba'	white-flowering eastern redbud
Chilopsis linearis	desert-willow
Cladrastis kentukea	yellowwood
Cordyline australis	green dracaena
Cornus capitata	evergreen dogwood
Cornus florida	flowering dogwood
Cornus nuttallii	western dogwood
Crataegus laevigata	hawthorn
Crataegus x *lavallei*	Carriere hawthorn
Crinodendron patagua	lily-of-the-valley tree
Eriobotrya deflexa	bronze loquat
Lagerstroemia indica	crape myrtle
Ligustrum lucidum	glossy privet
Lophostemon confertus	Brisbane box
Lyonothamnus floribundus	Catalina ironwood
Magnolia grandiflora	southern magnolia
Magnolia x *soulangeana*	saucer magnolia
Magnolia stellata	star magnolia
Malus floribunda vars.	Japanese flowering crabapple
Melaleuca linariifolia	flaxleaf paperbark
Melaleuca quinquenervia	cajeput tree
Metrosideros excelsus 'Alba'	New Zealand Christmas tree
Nerium oleander vars.	oleander
Myoporum laetum	myoporum
Oxydendrum arboreum	sorrel tree
Paulownia tomentosa 'Alba'	empress tree
Pittosporum rhombifolium	Queensland pittosporum
Pittosporum undulatum	Victorian box
Prunus cerasifera	cherry plum
Prunus c. 'Atropurpurea'	purple-leaf plum
Prunus serrulata 'Kwanzan'	Kwanzan flowering cherry
Punica granatum 'Alba'	pomegranate
Pyrus taiwanensis	evergreen pear
Pyrus calleryana 'Aristocrat'	Aristocrat pear
Pyrus c. 'Bradford'	Bradford pear
Robinia pseudoacacia	black locust
Sophora japonica	Japanese pagoda tree
Tilia cordata	little-leaf linden
Tilia europaea	European linden
Tilia tomentosa	silver linden
Yucca brevifolia	Joshua tree
Yucca gloriosa	Spanish dagger

Magnolia stellata, star magnolia, has sweetly fragrant blossoms of white, pink, or purple, making it a popular choice for small gardens.

FALL COLOR

Yellow

Acer circinatum	vine maple
Acer japonicum 'Aurea'	fullmoon maple
Acer platanoides 'Schwedleri'	Schwedler maple
Acer macrophyllum	bigleaf maple
Acer negundo	box elder
Acer rubrum 'Columnare'	columnar red maple
Acer saccharinum	silver maple
Acer saccharum	sugar maple
Aesculus x carnea	red horsechestnut
Aesculus hippocastanum	horsechestnut
Albizia julibrissin	silk tree
Betula nigra	river birch
Betula papyrifera	paper birch
Betula pendula 'Crispa'	cutleaf weeping birch
Carpinus betulus	European hornbeam
Carya illinoiensis	pecan
Castanea sativa	Spanish chestnut
Catalpa speciosa	western catalpa
Celtis occidentalis	common hackberry
Celtis sinensis	Chinese hackberry
Cercis spp.	redbud
Cladrastis kentukea	yellowwood
Cornus spp.	dogwood
Fagus spp.	beech
Firmiana simplex	Chinese parasol tree
Fraxinus spp.	ash
Ginkgo biloba	maidenhair tree
Gleditsia spp.	locust
Juglans spp.	walnut
Koelreuteria bipinnata	Chinese flame tree
Koelreuteria paniculata	goldenrain tree
Laburnum x watereri	goldenchain tree
Liquidambar styraciflua vars.	sweet gum
Liriodendron tulipifera	tulip tree
Magnolia x soulangeana	saucer magnolia
Magnolia stellata	star magnolia
Malus x floribunda	Japanese flowering crabapple
Melia azedarach	chinaberry
Morus alba	white mulberry
Paulownia tomentosa	empress tree
Populus spp.	poplar
Pterocarya stenoptera	Chinese wingnut
Punica granatum	pomegranate
Quercus phellos	willow oak
Robinia spp.	flowering locust
Salix spp.	willow
Sambucus mexicana	blue elderberry
Sophora japonica	Japanese pagoda tree
Tilia spp.	linden
Ulmus spp.	elm

Orange

Acer circinatum	vine maple
Acer palmatum	Japanese maple
Acer saccharinum	silver maple
Acer saccharum	sugar maple
Acer tataricum ssp. ginnala	amur maple
Betula nigra	river birch
Carpinus betulus	European hornbeam
Cercidiphyllum japonicum	katsura tree
Crataegus laevigata	hawthorn
Koelreuteria bipinnata	Chinese flame tree
Koelreuteria paniculata	goldenrain tree
Lagerstroemia indica	crape myrtle
Liquidambar styraciflua vars.	sweet gum

Pyrus calleryana 'Bradford', Bradford pear, is stunning with deep red fall color and a profusion of white flowers in early spring.

Nyssa sylvatica	tupelo
Pistacia chinensis	Chinese pistache
Platanus x acerifolia	London plane
Platanus racemosa	California sycamore
Populus fremontii	Fremont cottonwood
Quercus kelloggii	California black oak
Rhus glabra	smooth sumac
Rhus typhina	staghorn sumac
Sorbus aucuparia	European mountain ash

Red

Acer palmatum	Japanese maple
Acer rubrum	red maple
Acer tataricum ssp. ginnala	amur maple
Cornus florida	eastern dogwood
Crataegus laevigata	English hawthorn
Koelreuteria bipinnata	Chinese flame tree
Koelreuteria paniculata	goldenrain tree
Lagerstroemia indica	crape myrtle
Liquidambar styraciflua vars.	sweet gum
Malus x floribunda	Japanese flowering crabapple
Nyssa sylvatica	tupelo
Oxydendrum arboreum	sorrel tree
Pistacia chinensis	Chinese pistache
Pyrus calleryana 'Aristocrat'	Aristocrat pear
Pyrus c. 'Bradford'	Bradford pear
Quercus coccinea	scarlet oak
Quercus rubra	red oak
Rhus glabra	smooth sumac
Rhus typhina	staghorn sumac
Sapium sebiferum	Chinese tallow
Zelkova serrata	sawtooth zelkova

FOLIAGE ACCENT

Variously colored or variegated foliage can accent the green landscape in summer and add interest to spring flowers and fall foliage color.

Purplish Leaves (reddish to bronze)

Acacia baileyana 'Purpurea'	Bailey acacia
Acer palmatum 'Atropurpureum'	red Japanese maple
Acer p. 'Bloodgood'	red-leaved Japanese maple
Acer p. 'Burgundy Lace'	Burgundy Lace Japanese maple
Acer p. 'Ornatum'	red laceleaf Japanese maple
Acer platanoides 'Schwedleri'	Schwedler maple
Betula pendula 'Purpurea'	purple-leaved birch
Cercis canadensis 'Forest Pansy'	purple-leaved redbud
Cryptomeria japonica 'Elegans' (winter)	plume cedar
Eriobotrya deflexa	bronze loquat
Fagus sylvatica 'Purpurea'	purple beech
Fagus s. 'Riversii'	Rivers' purple beech
Fagus s. 'Tricolor'	tricolor beech
Prunus x *blirieana*	purple-leaf plum
Punus cerasifera 'Krauter Vesuvius'	pink-flowering purple-leaf plum
Quercus robur 'Atropurpurea'	purple English oak

Bluish Green Leaves (bluish or grayish)

Abies magnifica 'Glauca'	azure fir
Abies pinsapo 'Glauca'	blue Spanish fir
Abies procera	noble fir
Acacia baileyana	Bailey acacia
Acacia dealbata	green acacia
Agonis flexuosa	peppermint tree
Brahea edulis	Guadalupe fan palm
Butia capitata	pindo palm
Cedrus atlantica 'Glauca'	blue Atlas cedar
Cedrus deodara	deodar cedar
Chamaecyparis nootkatensis	Nootka cypress
Cunninghamia lanceolata 'Glauca'	blue China fir
x *Cupressocyparis leylandii*	Leyland cypress
Cupressus arizonica	Arizona cypress
Cupressus guadalupensis	Guadalupe Island cypress
Cupressus macnabiana	McNab cypress
Eucalyptus spp.	eucalyptus
Juniperus deppeana	alligator juniper
Juniperus scopulorum 'Tolleson's Blue'	blue weeping juniper
Picea pungens 'Glauca'	Colorado blue spruce
Pinus sabiniana	gray pine
Salix hindsii	sandbar willow
Pinus sylvestris	Scotch pine
Quercus arizonica	Arizona white oak
Quercus douglasii	blue oak
Quercus engelmannii	Engelman oak
Sequoia sempervirens 'Aptos Blue'	Aptos Blue redwood
Sequoiadendron giganteum	giant sequoia

Yellowish Green Leaves (*variegated)

Acer japonicum 'Aurea'	golden fullmoon maple
Acer negundo 'Elegans'*	gold-variegated box elder
Cupressus funebris	mourning cypress
Fagus sylvatica 'Zlatia'	golden beech
Gleditsia triacanthos 'Sunburst'	golden honey locust
Liriodendron 'Aureomarginata'*	variegated tulip tree
Robinia pseudoacacia 'Frisia'	golden locust

Acer platanoides 'Drummondii' has striking white-variegated foliage to accent any garden setting and is especially effective in brightening a densely shaded garden.

Variegated Leaves (white or yellow)

Acer negundo 'Elegans'	gold-variegated box elder
Acer n. 'Flamingo'	tricolor box elder
Acer n. 'Variegata'	variegated box elder
Acer palmatum 'Butterfly'	variegated Japanese maple
Acer platanoides 'Drummondii'	variegated Norway maple
Cercis canadensis 'Silver Cloud'	variegated eastern redbud
Cornus florida 'Rainbow'	variegated dogwood
Cornus f. 'Cherokee Sunset'	pink-variegated dogwood
Populus alba (underside)	white poplar
Quercus robur 'Variegata'	variegated English oak
Yucca gloriosa 'Variegata'	variegated Spanish dagger

The unusual variegated foliage of *Fagus sylvatica* 'Tricolor' is highly desirable in garden and park settings.

WHITE BARK

Trees with whitish bark or twigs can be a landscape accent either at night with lighting or in winter without foliage, featuring a dramatic branching silhouette against a backdrop of evergreens. The trees listed have white or gray bark that shows up well against a dark background, especially with multi-trunk, arching vase-shaped trees, or those with tall exposed trunks.

Acer circinatum	vine maple
Acer palmatum vars.	Japanese maple
Acer macrophyllum	bigleaf maple
Acer negundo	box elder
Acer n. 'Variegatum'	variegated box elder
Acer rubrum	red maple
Acer r. 'Columnare'	columnar red maple
Acer saccharinum	silver maple
Acer saccharum	sugar maple
Aesculus californica	California buckeye
Aesculus x *carnea*	red horsechestnut
Aesculus hippocastanum	horsechestnut
Albizia julibrissin	silk tree
Alnus rhombifolia	white alder
Betula pendula	European white birch
Betula p. 'Crispa'	cutleaf weeping birch
Brachychiton populneus	bottle tree
Carpinus betulus	European hornbeam
Catalpa speciosa	western catalpa
Erythrina caffra	kaffirboom coral tree
Eucalyptus citriodora	lemon-scented gum
Eucalyptus leucoxylon	white ironbark
Eucalyptus viminalis	manna gum
Fagus sylvatica 'Riversii'	Rivers' purple beech
Ficus microcarpa 'Nitida'	little-leaf fig
Firmiana simplex	Chinese parasol tree
Ginkgo biloba	maidenhair tree
Juglans regia	English walnut
Liquidambar styraciflua	sweet gum
Malus spp.	flowering crabapple
Melaleuca linariifolia	flaxleaf paperbark
Melaleuca quinquenervia	cajeput tree
Morus alba	white mulberry
Myoporum laetum	myoporum
Nerium oleander	oleander
Olea europaea	olive
Paulownia tomentosa	empress tree
Pittosporum rhombifolium	Queensland pittosporum
Pittosporum undulatum	Victorian box
Platanus x *acerifolia*	London plane
Platanus racemosa	California sycamore
Populus alba	white poplar
Populus a. 'Pyramidalis'	Bolleana poplar
Populus fremontii	Fremont cottonwood
Populus nigra 'Italica'	Lombardy poplar
Populus tremuloides	quaking aspen
Populus trichocarpa	black cottonwood
Quercus agrifolia	coast live oak

Trees with white trunks can be placed in front of a dark background of buildings or dark evergreens for maximum effect, as are these newly planted *Liquidambar styraciflua* trees, which will eventually become more shapely. The combination of *Albizia julibrissin* placed in front of *Pinus pinea* is also dramatic in winter.

Lighting can highlight trunks and branches of nearby trees, like these young *Platanus* x *acerifolia* trees. Wintertime silhouettes are especially dramatic.

MULTI-TRUNKED

Trees with multiple trunks take on a more shrublike appearance, having more character when branches are pruned up to accentuate irregular or contorted trunks with the foliage canopy suspended above. The broader the canopy, the more dramatic the effect. This is especially true of weeping trees, but pruning may weaken the structural strength of the branches.

Acacia baileyana	Bailey acacia
Acacia dealbata	green acacia
Acer circinatum	vine maple
Acer macrophyllum	bigleaf maple
Acer negundo	box elder
Acer palmatum	Japanese maple
Aesculus californica	California buckeye
Agonis flexuosa	peppermint tree
Albizia distachya	plume albizia
Albizia julibrissin	silk tree
Arbutus menziesii	madrone
Betula nigra	river birch
Betula papyrifera	paper birch
Ceratonia siliqua	carob
Cercidium floridum	palo verde
Cercis occidentalis	western redbud
Cercocarpus betuloides	mountain ironwood
Chamaerops humilis	Mediterranean fan palm
Chilopsis linearis	desert-willow
x *Chitalpa tashkentensis*	chitalpa
Cinnamomum camphora	camphor tree
Cordyline australis	green dracaena
Cornus florida	eastern dogwood
Cornus mas	cornelian cherry
Cupaniopsis anacardioides	carrot wood
Eriobotrya deflexa	bronze loquat
Eriobotrya japonica	loquat
Erythrina caffra	kaffirboom coral tree
Erythrina crista-galli	cockspur coral tree
Eucalyptus ficifolia	red-flowering gum
Eucalyptus lehmannii	bushy yate
Eucalyptus leucoxylon	white ironbark
Eucalyptus nicholii	willow-leafed peppermint
Eucalyptus polyanthemos	silver dollar gum
Eucalyptus pulverulenta	silver mountain gum
Eucalyptus sideroxylon 'Rosea'	red ironbark
Ficus microcarpa 'Nitida'	little-leaf fig
Ficus rubiginosa	rustyleaf fig
Jacaranda mimosifolia	jacaranda
Juglans hindsii	black walnut
Lagerstroemia indica	crape myrtle
Laurus nobilis	Grecian laurel
Lithocarpus densiflorus	tanbark oak
Lyonothamnus floribundus	Catalina ironwood
Magnolia x *soulangeana*	saucer magnolia
Magnolia stellata	star magnolia
Maytenus boaria	mayten
Melaleuca linariifolia	flaxleaf paperbark
Melaleuca quinquenervia	cajeput tree
Myoporum laetum	myoporum
Olea europaea	olive
Parkinsonia aculeata	Mexican palo verde
Phoenix reclinata	Senegal date palm
Phoenix roebelenii	pygmy date palm
Pinus contorta ssp. *contorta*	shore pine
Pinus halepensis	Aleppo pine
Pinus mugo	mugho pine
Pinus thunbergii	Japanese black pine
Punica granatum	pomegranate
Quercus agrifolia	coast live oak

Leptospermum scoparium 'Ruby Glow' is usually grown as a shrub. This large specimen has been "headed-up" with lower branches thinned and foliage removed to expose a multi-trunked, rather treelike character. This is effective with many large shrubs, such as *Xylosma*, *Leptospermum laevigatum*, *Eucalyptus lehmannii*, and *Camellia japonica*. Multi-trunk trees make excellent garden specimens. Special training and pruning may be necessary, but some species grow this way naturally, losing lower foliage with age. Trees that develop multiple low-branching leaders at an early age can be effectively used in this manner.

Quercus chrysolepis	canyon live oak
Quercus ilex	holly oak
Quercus virginiana	southern live oak
Quercus wislizeni	interior live oak
Rhus glabra	smooth sumac
Rhus lancea	African sumac
Rhus typhina	staghorn sumac
Salix spp.	willow
Sapium sebiferum	Chinese tallow
Schinus molle	pepper tree
Umbellularia californica	California bay

TROPICAL ACCENT

Though what generally comes to mind are exotic subtropical species, many common hardy trees that display unusual or exotic shape, texture, leaves, flowers, or fruit, strong fragrance, or succulent growth can be used as tropical accents. Whether as a single large specimen or grouped with others, compounding the effect, lush foliage or exotic-looking flowers can be quite dramatic.

Acacia dealbata	green acacia
Agathis robusta	dammar pine
Agonis flexuosa	peppermint tree
Albizia distachya	plume albizia
Albizia julibrissin	silk tree
Araucaria araucana	monkey puzzle tree
Araucaria bidwillii	bunya-bunya
Araucaria heterophylla	Norfolk Island pine
Bauhinia variegata	orchid tree
Brachychiton acerifolius	flame tree
Brachychiton populneus	bottle tree
Callistemon citrinus	lemon bottlebrush
Callistemon viminalis	weeping bottlebrush
Catalpa speciosa	western catalpa
x *Chitalpa tashkentensis*	chitalpa
Ceiba speciosa	floss silk tree
Cordyline australis	green dracaena
Cryptomeria japonica	Japanese cryptomeria
Cryptomeria j. 'Elegans'	plume cedar
Cunninghamia lanceolata 'Glauca'	blue China fir
Cupaniopsis anacardioides	carrot wood
Cyathea cooperi	Australian tree fern
Cycas revoluta	cycad
Dicksonia antarctica	tree fern
Eriobotrya deflexa	bronze loquat
Eriobotrya japonica	loquat
Erythrina caffra	kaffirboom coral tree
Erythrina crista-galli	cockspur coral tree
Eucalyptus citriodora	lemon-scented gum
Eucalyptus ficifolia	red-flowering gum
Eucalyptus lehmannii	bushy yate
Eucalyptus leucoxylon	white ironbark
Eucalyptus nicholii	willow-leafed peppermint
Eucalyptus polyanthemos	silver dollar gum
Eucalyptus pulverulenta	silver mountain gum
Eucalyptus sideroxylon 'Rosea'	red ironbark
Eucalyptus viminalis	manna gum
Ficus microcarpa 'Nitida'	little-leaf fig
Ficus rubiginosa	rustyleaf fig
Geijera parviflora	Australian willow
Ginkgo biloba	maidenhair tree
Gleditsia triacanthos 'Sunburst'	golden honey locust
Grevillea robusta	silk oak
Jacaranda mimosifolia	jacaranda
Koelreuteria bipinnata	Chinese flame tree
Koelreuteria paniculata	goldenrain tree
Lagerstroemia indica	crape myrtle
Lophostemon confertus	Brisbane box
Lyonothamnus floribundus	Catalina ironwood
Magnolia grandiflora	southern magnolia
Magnolia x *soulangeana*	saucer magnolia
Magnolia stellata	star magnolia
Maytenus boaria	mayten
Melia azedarach	chinaberry
Melaleuca linariifolia	flaxleaf paperbark
Melaleuca quinquenervia	cajeput tree
Metasequoia glyptostroboides	dawn redwood
Metrosideros excelsus	New Zealand Christmas tree
Musa x *paradisiaca*	banana
Myoporum laetum	myoporum

Schefflera actinophylla, a familiar house plant in California, is used as a street tree in the Hawaiian Islands. Tupidanthus, or *Schefflera pueckleri*, is sometimes used as a courtyard tree in mild southern California climates.

Nerium oleander	oleander
Olea europaea	olive
Oxydendrum arboreum	sorrel tree
palms	palms
Paulownia tomentosa	empress tree
Phyllostachys bambusoides	timber bamboo
Pinus patula	Jelecote pine
Pinus pinea	Italian stone pine
Pinus thunbergii	Japanese black pine
Pittosporum rhombifolium	Queensland pittosporum
Pittosporum undulatum	Victorian box
Podocarpus gracilior	fern pine
Podocarpus macrophyllus	yew pine
Prunus caroliniana	Carolina cherry laurel
Pterocarya stenoptera	Chinese wingnut
Punica granatum	pomegranate
Rhus glabra	smooth sumac
Rhus lancea	African sumac
Rhus typhina	staghorn sumac
Salix matsudana 'Tortuosa'	corkscrew willow
Sambucus mexicana	blue elderberry
Sapium sebiferum	Chinese tallow
Schinus molle	pepper tree
Sciadopitys verticillata	Japanese umbrella pine
Sophora japonica	Japanese pagoda tree
Stenocarpus sinuatus	firewheel tree
Syzygium paniculatum	brush cherry
Tamarix parviflora	tamarisk
Taxodium distichum	bald cypress
Tetrapanax papyriferus	rice paper plant
Tristaniopsis laurina	laurel leaf box
Yucca gloriosa	Spanish dagger
Ziziphus jujuba	Chinese jujube

DESERT ACCENT

A more accurate term for this landscape category might be Mediterranean, or dry subtropical, utilizing drought-tolerant plants from South Africa and Australia, along with cacti and succulents from the deserts, and trees that complement the character of the plantings. Blue and gray foliage can brighten and accent other colors and textures. While the trees listed here may not be true desert species, they blend well with other vegetation in a dry context.

Aesculus californica	California buckeye
Agonis flexuosa	peppermint tree
Albizia julibrissin	silk tree
Arbutus unedo	strawberry tree
Callistemon citrinus	lemon bottlebrush
Callistemon viminalis	weeping bottlebrush
Casuarina equisetifolia	horsetail tree
Cedrus atlantica 'Glauca'	blue Atlas cedar
Cedrus deodara	deodar cedar
Cercidium floridum	palo verde
Cercis occidentalis	western redbud
Cercocarpus betuloides	mountain ironwood
Chilopsis linearis	desert-willow
x *Chitalpa tashkentensis*	chitalpa
Cordyline australis	green dracaena
Cupressus arizonica	Arizona cypress
Cupressus sempervirens	Italian cypress
Eucalyptus nicholii	willow-leafed peppermint
Eucalyptus polyanthemos	silver dollar gum
Eucalyptus pulverulenta	silver mountain gum
Eucalyptus sideroxylon 'Rosea'	red ironbark
Eucalyptus viminalis	manna gum
Geijera parviflora	Australian willow
Grevillea robusta	silk oak
Jacaranda mimosifolia	jacaranda
Melaleuca linariifolia	flaxleaf paperbark
Melaleuca quinquenervia	cajeput tree
Olea europaea	olive
palms	palms
Parkinsonia aculeata	Mexican palo verde
Pinus patula	Jelecote pine
Pinus pinea	Italian stone pine
Pinus sabiniana	gray pine
Pinus torreyana	Torrey pine
Platanus racemosa	California sycamore
Prosopis glandulosa	honey mesquite
Prosopis velutina	Arizona mesquite
Punica granatum	pomegranate
Rhus glabra	smooth sumac
Schinus molle	pepper tree
Schinus peruviana	Peruvian pepper
Sophora japonica	Japanese pagoda tree
Yucca brevifolia	Joshua tree
Yucca gloriosa	Spanish dagger

Native pines provide an evergreen backdrop for a dry hillside planting of desert species, highlighted by the magnificent and colorful agave in the center.

Many plants from South Africa, Australia, and New Zealand are useful in dry gardens, often with decomposed granite pathways. Sunny areas must be maintained for these plants to thrive, with trees used sparingly. Multi-trunk *Cercidium floridum*, *Parkinsonia aculeata*, or *Schinus molle* cast light shade with their fine foliage and are adaptable to this setting.

WEEPING ACCENT

Flowers

Acacia baileyana	Bailey acacia
Acer macrophyllum	bigleaf maple
Agonis flexuosa	peppermint tree
Alnus rhombifolia	white alder
Callistemon viminalis	weeping bottlebrush
Carpinus betulus	European hornbeam
Carya illinoiensis	pecan
Castanea sativa	Spanish chestnut
Cladrastis kentukea	yellowwood
Erythrina crista-galli	cockspur coral tree
Firmiana simplex	Chinese parasol tree
Geijera parviflora	Australian willow
Koelreuteria paniculata	goldenrain tree
Laburnum x *watereri*	goldenchain tree
Oxydendrum arboreum	sorrel tree
Pterocarya stenoptera	Chinese wingnut
Robinia pseudoacacia	black locust
Schinus molle	pepper tree
Tamarix parviflora	tamarisk
Tilia cordata	little-leaf linden
Tilia europaea	European linden
Tilia tomentosa	silver linden

Seeds and Fruit

Acer tataricum ssp. *ginnala*	amur maple
Acer macrophyllum	bigleaf maple
Acer negundo	box elder
Acer n. 'Variegatum'	variegated box elder
Ailanthus altissima	tree-of-heaven
Albizia distachya	plume albizia
Albizia julibrissin	silk tree
Alnus rhombifolia	white alder
Betula papyrifera	paper birch
Betula pendula 'Crispa'	cutleaf weeping birch
Cercis canadensis	eastern redbud
Cercis occidentalis	western redbud
Cladrastis kentukea	yellowwood
Fraxinus latifolia	Oregon ash
Gleditsia triacanthos	honey locust
Laburnum x *watereri*	goldenchain tree
Liquidambar formosana	Chinese sweet gum
Liquidambar styraciflua	American sweet gum
Platanus x *acerifolia*	London plane
Platanus racemosa	California sycamore
Pterocarya stenoptera	Chinese wingnut
Robinia pseudoacacia	black locust
Sophora japonica	Japanese pagoda tree
Taxodium distichum	bald cypress
Tilia cordata	little-leaf linden
Tilia europaea	European linden
Tilia tomentosa	silver linden
Zizyphus jujuba	Chinese jujube

Leaves and Branches

Agonis flexuosa	peppermint tree
Betula papyrifera	paper birch
Betula pendula 'Crispa'	cutleaf weeping birch
Callistemon viminalis	weeping bottlebrush
Calocedrus decurrens	incense cedar
Casuarina equisetifolia	horsetail tree
Cedrus deodara	deodar cedar
Cercidium floridum	palo verde

The dramatic color and weeping foliage of *Cedrus atlantica* 'Glauca Pendula' creates a spectacular point of interest, contrasting with the foliage of the surrounding landscape. Weeping varieties for many species exist, though they are often not commonly planted or available.

Cupressus funebris	mourning cypress
Erythrina crista-galli	cockspur coral tree
Eucalyptus camaldulensis	red gum
Eucalyptus leucoxylon	white ironbark
Eucalyptus polyanthemos	silver dollar gum
Eucalyptus pulverulenta	silver mountain gum
Eucalyptus sideroxylon 'Rosea'	red ironbark
Eucalyptus viminalis	manna gum
Geijera parviflora	Australian willow
Juniperus scopulorum 'Tolleson's Weeping'	blue weeping juniper
Maytenus boaria	mayten
Metasequoia glyptostroboides	dawn redwood
Morus alba 'Pendula'	weeping white mulberry
Picea glauca 'Pendula'	weeping white spruce
Pinus patula	Jelecote pine
Quercus robur 'Pendula'	weeping English oak
Rhus lancea	African sumac
Robinia pseudoacacia 'Tortuosa'	twisted locust
Salix babylonica	weeping willow
Salix matsudana 'Tortuosa'	corkscrew willow
Schinus molle	pepper tree
Tamarix parviflora	tamarisk
Taxodium distichum	bald cypress

COURTYARD TREES

Acacia baileyana	Bailey acacia
Acer tataricum ssp. *ginnala*	amur maple
Acer palmatum	Japanese maple
Aesculus x *carnea*	red horsechestnut
Agonis flexuosa	peppermint tree
Arbutus unedo	strawberry tree
Bauhinia variegata	orchid tree
Betula jacquemontii	Jacquemont birch
Betula nigra	river birch
Callistemon citrinus	lemon bottlebrush
Callistemon viminalis	weeping bottlebrush
Carpinus betulus	European hornbeam
Celtis occidentalis	common hackberry
Cercidiphyllum japonicum	katsura tree
Cercidium floridum	palo verde
Cercis canadensis	eastern redbud
Cornus florida	eastern dogwood
Cornus mas	cornelian cherry
Crataegus laevigata	English hawthorn
Eriobotrya deflexa	bronze loquat
Eucalyptus ficifolia	red-flowering gum
Eucalyptus lehmannii	bushy yate
Eucalyptus nicholii	willow-leafed peppermint
Ficus microcarpa 'Nitida'	little-leaf fig
Fraxinus angustifolia 'Raywood'	Raywood ash
Geijera parviflora	Australian willow
Koelreuteria bipinnata	Chinese flame tree
Koelreuteria paniculata	goldenrain tree
Laburnum x *watereri*	goldenchain tree
Lagerstroemia indica	crape myrtle
Laurus nobilis	Grecian laurel
Ligustrum lucidum	glossy privet
Lophostemon confertus	Brisbane box
Magnolia grandiflora 'Samuel Sommer'	southern magnolia
Magnolia x *soulangeana*	saucer magnolia
Magnolia stellata	star magnolia
Malus floribunda	Japanese flowering crabapple
Maytenus boaria	mayten
Melaleuca linariifolia	flaxleaf paperbark
Melaleuca quinquenervia	cajeput tree
Metrosideros excelsus	New Zealand Christmas tree
Nyssa sylvatica	tupelo
Olea europaea	olive
Oxydendrum arboreum	sorrel tree
palms	palms
Pittosporum rhombifolium	Queensland pittosporum
Pittosporum undulatum	Victorian box
Pistacia chinensis	Chinese pistache
Podocarpus gracilior	fern pine
Podocarpus macrophyllus	yew pine
Prunus caroliniana	Carolina cherry laurel
Prunus x *blirieana*	flowering plum
Prunus cerasifera 'Krauter Vesuvius'	pink-flowering purple-leaf plum
Prunus serrulata 'Kwanzan'	Kwanzan flowering cherry
Prunus x *yedoensis*	Yoshino flowering cherry
Punica granatum	pomegranate
Pyrus taiwanensis	evergreen pear
Pyrus calleryana 'Aristocrat'	Aristocrat pear
Pyrus c. 'Bradford'	Bradford pear
Quercus ilex	holly oak
Quercus myrsinifolia	Japanese live oak
Salix matsudana 'Tortuosa'	corkscrew willow
Sapium sebiferum	Chinese tallow
Schinus terebinthifolius	Brazilian pepper
Sciadopitys verticillata	Japanese umbrella pine

The bright, colorful foliage of a purple-leaf plum draws attention as the focal point of an expansive terrace. Tight patio spaces often incorporate Japanese maples as small accent specimens, as an understory to larger shade trees, or to create heightened spatial effects.

Sorbus aucuparia	European mountain ash
Tristaniopsis laurina	laurel leaf box
Zizyphus jujuba	Chinese jujube

SMALL STANDARDS

Some shrubby, multi-trunk trees can be pruned and trained into a tree-form "standard." These are used in restricted planter areas, where a larger tree would interfere with overhead obstructions, or for a manicured look. Many large shrubs are suitable for this purpose.

Trees

Acacia longifolia	Sidney golden wattle
Acacia melanoxylon	black acacia
Agonis flexuosa	peppermint tree
Callistemon citrinus	lemon bottlebrush
Cercis canadensis	eastern redbud
Crataegus laevigata	English hawthorn
Eriobotrya deflexa	bronze loquat
Eriobotrya japonica	loquat
Eucalyptus ficifolia	red-flowering gum
Ficus microcarpa 'Nitida'	little-leaf fig
Geijera parviflora	Australian willow
Lagerstroemia indica	crape myrtle
Laurus nobilis	Grecian laurel
Ligustrum lucidum	glossy privet
Lophostemon confertus	Brisbane box
Melaleuca linariifolia	flaxleaf paperbark
Melaleuca quinquenervia	cajeput tree
Metrosideros excelsus	New Zealand Christmas tree
Myoporum laetum	myoporum
Nerium oleander	oleander
palms	palms
Pittosporum rhombifolium	Queensland pittosporum
Pittosporum undulatum	Victorian box
Podocarpus gracilior	fern pine
Podocarpus macrophyllus	yew pine
Sorbus aucuparia	European mountain ash
Tristaniopsis laurina	laurel leaf box
Umbellularia californica	California bay

Shrubs

Camellia japonica	camellia
* *Cassia excelsa*	crown of gold tree
* *Cassia leptophylla*	gold medallion tree
Dodonaea viscosa 'Atropurpurea'	purple hop bush
Ilex altaclarensis 'Wilsonii'	Wilson holly
* *Leptospermum laevigatum*	Australian tea tree
* *Leptospermum scoparium*	New Zealand tea tree
Photinia x *fraseri*	Fraser photinia
* *Prunus lusitanica*	Portugal laurel
* *Vitex agnus-casti*	chaste tree
* *Xylosma senticosa*	shiny xylosma

*usually multi-trunked

Small flowering tree standards are a popular means of adding accent color in residential yards, either under larger shade trees, or in a sunny location. The springtime blooming of a Chinese fringe tree provides a showy display.

The bright red flowers of crape myrtle brighten a sunny spot in late summer when few other trees are in bloom along a residential street lined with large shade trees.

POLLARDING

Trees are sometimes severely pruned in winter, leaving a massive trunk and strategically spaced heavy lateral branching with all smaller branches and twigs removed. In the case of mulberries, pollarding reduces the size and spread of the tree for a season or two, but results in a tremendous flush of new growth directly following. Older declining ash trees are often pruned the same way in an attempt to rejuvenate them with new fresh growth, where mistletoe and anthracnose have caused loss of vigor. Erythrinas normally look this way, especially with pruning after a heavy freeze.

London plane trees are the most often pollarded trees in urban plazas. Pollarding produces horizontal branching rather than upright growth, creating an architectural feature, but the annual procedure is labor-intensive and costly and is not otherwise recommended.

Erythrina crista-galli	cockspur coral tree
Fraxinus velutina 'Modesto'	Modesto ash
Ginkgo biloba	maidenhair tree
Lagerstroemia indica	crape myrtle
Malus x *floribunda*	flowering crabapple
Morus alba	white mulberry
Morus a. 'Fruitless'	fruitless white mulberry
Olea europaea	olive
Platanus x *acerifolia*	London plane
Punica granatum	pomegranate
Pyrus taiwanensis	evergreen pear

PLEACHING/ESPALIERS

Some trees can be espaliered on walls with careful pruning and training. The results can be an effective alternative to a standard foundation shrub planting.

Acer circinatum	vine maple
Arbutus unedo	strawberry tree
Callistemon citrinus	lemon bottlebrush
Callistemon viminalis	weeping bottlebrush
Cedrus atlantica 'Glauca'	blue Atlas cedar
Eriobotrya deflexa	bronze loquat
Grevillea robusta	silk oak
Magnolia grandiflora	southern magnolia
Malus spp.	flowering crabapple
Olea europaea	olive
Podocarpus spp.	podocarpus

A well-maintained grid of pollarded *Platanus* x *acerifolia* accentuates the wintertime landscape of an urban plaza, creating an overhead umbrella of branching. This horizontal branching pattern becomes an architectural design element.

Tree species, such as this southern magnolia, are often espaliered in a flat, vertical fashion against a wall instead of using foundation shrubs. This requires special pruning and training but provides an interesting effect.

MOST COMMONLY USED TREES

Trees most commonly seen in the California landscape include perennial favorites familiar to most people and those that are most commonly available. Most are easy to grow, have desirable characteristics, and are durable and long-lived. However, some have been found to be less desirable over time, and it is worthwhile investigating the wealth of other species, including cultivars and varieties of commonly used trees.

Acacia longifolia	Sidney golden wattle
Acacia melanoxylon	black acacia
Acer palmatum	Japanese maple
Acer rubrum	red maple
Agonis flexuosa	peppermint tree
Albizia julibrissin	silk tree
Alnus cordata	Italian alder
Alnus rhombifolia	white alder
Arbutus unedo	strawberry tree
Betula papyrifera	paper birch
Callistemon citrinus	lemon bottlebrush
Catalpa speciosa	western catalpa
Cedrus deodara	deodar cedar
Celtis sinensis	Chinese hackberry
Cercis canadensis	eastern redbud
Cinnamomum camphora	camphor tree
Cornus florida	flowering dogwood
Crataegus laevigata	English hawthorn
Cupaniopsis anacardioides	carrot wood
Cupressus sempervirens	Italian cypress
Eriobotrya deflexa	bronze loquat
Eriobotrya japonica	loquat
Eucalyptus spp.	eucalyptus
Ficus microcarpa 'Nitida'	little-leaf fig
Fraxinus spp.	ash
Geijera parviflora	Australian willow
Ginkgo biloba	maidenhair tree
Gleditsia spp.	honey locust
Jacaranda mimosifolia	jacaranda
Lagerstroemia indica	crape myrtle
Laurus nobilis	Grecian laurel
Ligustrum lucidum	glossy privet
Liquidambar styraciflua	American sweet gum
Lophostemon confertus	Brisbane box
Magnolia grandiflora	southern magnolia
Maytenus boaria	mayten
Melaleuca quinquenervia	cajeput tree
Metrosideros excelsus	New Zealand Christmas tree
Morus alba 'Fruitless'	fruitless mulberry
Myoporum laetum	myoporum
Nerium oleander	oleander
palms	palms
Pinus canariensis	Canary Island pine
Pinus halepensis	Aleppo pine
Pinus pinea	Italian stone pine
Pinus thunbergii	Japanese black pine
Pistacia chinensis	Chinese pistache
Pittosporum rhombifolium	Queensland pittosporum
Pittosporum undulatum	Victorian box
Platanus x *acerifolia*	London plane
Platanus racemosa	California sycamore
Podocarpus gracilior	fern pine
Podocarpus macrophyllus	yew pine
Prunus spp.	flowering cherry, plum
Pyrus calleryana 'Aristocrat'	Aristocrat pear
Pyrus c. 'Bradford'	Bradford pear
Pyrus taiwanensis	evergreen pear
Quercus agrifolia	coast live oak
Quercus coccinea	scarlet oak

Acacia dealbata, once commonly planted for its spectacular spring flowering, is now much less often planted or seen.

Quercus ilex	holly oak
Quercus lobata	valley oak
Quercus rubra	red oak
Quercus suber	cork oak
Rhus lancea	African sumac
Robinia spp.	flowering locust
Salix babylonica	weeping willow
Sapium sebiferum	Chinese tallow
Schinus molle	pepper tree
Schinus terebinthifolius	Brazilian pepper tree
Sequoia sempervirens	coast redwood
Sophora japonica	Japanese pagoda tree
Tilia cordata	little-leaf linden
Tristaniopsis laurina	laurel leaf box
Ulmus parvifolia	evergreen elm
Zelkova serrata	sawtooth zelkova

TREES THAT MAY BE INVASIVE

Some trees tend to become unwanted pests by virtue of their prolific reseeding or invasive roots, often escaping cultivation and invading native habitats. In some cases, native trees can become weedy in irrigated landscapes.

Ailanthus altissima	tree-of-heaven
Albizia distachya	plume albizia
Cordyline australis	green dracaena
Morus microphylla	Texas mulberry
Myoporum laetum	myoporum
Populus spp.	cottonwood, poplar
Prunus caroliniana	Carolina cherry laurel
Robinia spp.	flowering locust
Salix spp.	willow
Sapium sebiferum	Chinese tallow
Schinus terebinthifolius	Brazilian pepper
Tamarix aphylla	athel tree
Tamarix parviflora	tamarisk
Ulmus spp.	elm

TREES OF THE CALIFORNIA LANDSCAPE

Compendium

Coded abbreviations

Continents & Regions: c (central), **e** (eastern), **s** (southern), **n** (northern), and **w** (western) **Africa, Asia, Asia Minor, Canary Islands, Australia, Europe, N.A.** (North America), **N.Z.** (New Zealand), **Medit.** (Mediterranean), **Pacific Is.** (Pacific Islands), **South Africa, South America**

Native trees: California Native; (CA only) for native endemic species

Native Vegetation Zones: CO (Coastal Strand), **CM** (Coastal Marsh), **CP** (Coastal Prairie), **VG** (Valley Grassland), **AL** (Alpine), **CS** (Coastal Scrub), **MD** (Mojave Desert Scrub), **SD** (Sonoran Desert Scrub), **SS** (Sagebrush Scrub), **CH** (Chaparral), **OW** (Oak Woodland), **FW** (Foothill Woodland), **JP** (Juniper-Pine Woodland), **CC** (Closed Cone Forest), **ME** (Mixed Evergreen Forest), **RW** (Redwood Forest), **CF** (Klamath Mixed Conifer Forest), **PF** (Montane Pine Forest), **FF** (Montane & Subalpine Fir Forest)

Jepson Zones (*see map on page 15*): **CA-FP** (California Floristic Province); **NW** (Northwestern California); **NCo** (North Coast); **KR** (Klamath Ranges); **NCoR** (North Coast Ranges); **NCoRO** (Outer North Coast Ranges); **NCoRH** (High North Coast Ranges); **NCoRI** (Inner North Coast Ranges); **CaR** (Cascade Ranges); **CaRF** (Cascade Range Foothills); **CaRH** (High Cascade Range); **SN** (Sierra Nevada); **SNF** (Sierra Nevada Foothills); **n SNF** (Northern Sierra Nevada Foothills); **c SNF** (Central Sierra Nevada Foothills); **s SNF** (Southern Sierra Nevada Foothills); **SNH** (High Sierra Nevada); **n SNH** (Northern High Sierra Nevada); **c SNH** (Central High Sierra Nevada); **s SNH** (Southern High Sierra Nevada); **Teh** (Tehachapi Mountains); **GV** (Great Central Valley); **ScV** (Sacramento Valley); **SnJV** (San Joaquin Valley); **CW** (Central Western California); **CCo** (Central Coast); **SnFrB** (San Francisco Bay Area); **SCoR** (South Coast Ranges); **SCoRO** (Outer South Coast Ranges); **SCoRI** (Inner South Coast Ranges); **SW** (Southwestern California); **SCo** (South Coast); **ChI** (Channel Islands); **n ChI** (Northern Channel Islands); **s ChI** (Southern Channel Islands); **TR** (Transverse Ranges); **WTR** (Western Transverse Ranges); **SnGb** (San Gabriel Mountains); **SnBr** (San Bernardino Mountains); **PR** (Peninsular Ranges); **SnJt** (San Jacinto Mountains); **GB** (Great Basin Province); **MP** (Modoc Plateau); **Wrn** (Warner Mountains); **SNE** (East of Sierra Nevada); **W&I** (White and Inyo Mountains); **D** (Desert Province); **DMoj** (Mojave Desert); **DMtns** (Desert Mountains); **DSon** (Sonoran Desert)

Landscape Use: ACC Accent, **CNF** Conifer, **CLR** or **COL** Color, **DES** Desert, **EVG** Evergreen, **FAL** Fall, **FLW** Flowering, **FOL** Foliage, **FRAG** Fragrance, **FRU** Fruit, **NAT** Native, **ORN** Ornamental, **SCR** Screen, **SHD** Shade, **SPC** Specimen, **STR** Street tree, **VERT** Vertical, **WPG** Weeping

WUCOLS (*Water Use Classification of Landscape Species*) codes (for six regions: North Central Coastal, Central Valley, South Coastal, South Inland Valleys, High and Intermediate Desert, and Low Desert, respectively): **H/** high, **M/** medium, **L/** low, **VL/** very low, **-/** species not used in that region, **?/** information not available

PP #: Plant Patent number (plus date in parentheses)

***Jepson Manual* Horticultural Entry Codes** (*): **(CVS)** cultivar(s) available; **(DFCLT)** difficult, requires special care; **(DRN)** requires excellent drainage; **(DRY)** summer water intolerant; **(INV)** invasive; **(IRR)** requires moderate summer watering; **(SHD)** best in full or part shade; **(STBL)** good soil stabilizer; **(SUN)** best in sun or nearly full sun; **(WET)** wet or continually moist soil required

c&se Europe. CNF, EVG; IRR: M/-/M/M/-/-

Abies alba

EUROPEAN SILVER FIR

Pinaceae. No Sunset zones. Evergreen. Native to mountain regions of central and southeastern Europe. Growth rate slow to moderate to 60-80' tall or more, with a densely branched, wide, conical to pyramidal form, lowest branches usually remaining on young trees but may lose lower branches and shapeliness in maturity. Needles are shiny, dark green, 1/2 to 1-1/4" long, flattish, with rounded, notched tips, 2 broad whitish stripes on the undersides, growing straight outward from undersides of branchlets and upward and outward from uppersides, forming a distinct V-shaped void directly above branchlets. Flowers are insignificant, males reddish at ends of lower branches and broad-scaled females with shiny purple bracts at uppermost twigs of the tree. Cones are upright, barrel-shaped, dark purplish brown, 3-5" long, with broad, thin, tightly clasped, rounded scales and extending recurved bracts, disintegrating in fall and releasing 3/4" long, light yellow-brown seeds with light brown, blunt-ended wings. Twigs are pale yellowish brown, minutely hairy, with small non-resinous buds. Bark on young trees is thin, silvery gray, smooth, and covered with balsam blisters, darkening and becoming fissured and scaly on older trees.

Similar to Pacific silver fir (*Abies amabilis*) but more adaptable to cultivation in cooler foothill and mountain conditions, like the native white fir (*A. concolor*), tolerating moderate heat of foothill climates above 2,000' elevation. Not commonly cultivated, but dense, shiny green foliage is attractive in the landscape. Columnar and weeping cultivars occasionally are propagated by specialty nurseries. Longevity estimated to be 100-175 years under favorable conditions, but shorter-lived and less disease-resistant than Nordmann fir (*A. nordmanniana*).

nw N.A. California Native: CF *KR* (*w Siskyou Co., 2 populations*); CNF, EVG; (IRR: M/-/M/M/-/-)

Abies amabilis

PACIFIC SILVER FIR

Pinaceae. Sunset zones 3-7, 15-17. Evergreen. Native to cool, wet coastal mountains and higher-elevation Cascade Ranges from British Columbia through Oregon, with 2 isolated groves in extreme northwestern California. Growth rate slow to 20-50' tall (to 150-250' in habitat) by 12-15' wide, developing a conical to narrow, pyramidal form with short horizontal branches often bowing slightly upward at the ends and lower branches remaining unless heavily shaded. Needles are shiny, dark green on the upperside, 3/4 to 1-1/4" long, flattish, with blunt-ended or notched tips, not tapering toward the tip, grooved on the upperside, with 2 broad, whitish stripes on the underside, growing straight out from the underside of branchlets and upward radially from the upperside. Insignificant reddish male flowers are borne at ends of lower branches, below the female flowers, which are broad-scaled with shiny purple bracts. Cones are at the uppermost twigs, 3-1/2 to 6" long, upright, barrel-shaped, dark purplish brown, with thin, broad, rounded, tightly clasped scales, bracts inserted or hidden, disintegrating in fall and releasing light yellow-brown seeds with 3/4", light brown, blunt, wide-ended wings. Twigs are pale yellowish brown and minutely hairy, with fat resinous buds. Bark is thin, smooth, ashy gray with chalky white blotches, often developing resin warts. Older bark at the base of the trunk darkens with age, becoming scaly.

A northwestern forest tree, rare in California. Tolerant of deeply shaded forest conditions. In shade or filtered sunlight trees remain rather short with a sparse, formless shape and broad horizontal branching. In open clearings and in cultivation growth is much faster, more vertical, with shorter branching and a more conical form on young trees, often losing shapeliness and vigor at maturity. Longevity estimated to be 150-200 years in habitat. Series associations: Subfir,

w N.A. (Ca. only) California Native: ME (*n SCoRO, Santa Lucia Range*); CNF, EVG; (IRR: M/-/M/M/-/-)

Abies bracteata

BRISTLECONE FIR

Pinaceae. Sunset zones 6-9, 14-21, Jepson *DRN:4-6,**15,16**,17& IRR:2,3,**18-21**,22-24&SHD:7,8,9,**14**. Evergreen. Native to the Santa Lucia Mountains of Monterey and San Luis Obispo counties. Growth rate moderate, slower with age, to 70' tall (to 100' in habitat) and 15-20' wide, developing a narrowly conical, steeplelike crown, with short, slightly drooping branches near the top and wide-spreading branches at the base, often to the ground, older trees often developing long, cordlike, drooping branchlets. Needles are stiff, flat, shiny dark green, 1-1/2 to 2-1/2" long by 1/16-1/8" wide, twisted near the base, with sharply pointed tips, slightly grooved uppersides, 2 whitish stripes on the undersides, sometimes occurring in flattened sprays on lower branches. Insignificant orange-red male flowers occur at ends of lower branches, with female flowers above along uppermost twigs. Cones are upright, 2-1/2 to 3-1/2" long, purplish brown, with thin, broadly rounded, tightly clasped scales, each bract extending into a 1/2 to 1-3/4" long, yellowish brown, hairlike bristle, curving outward around the cone. Cones are rarely seen, disintegrating when fully mature and releasing ovoid, reddish brown, broad-winged seeds. Twigs are glabrous, with long, narrow, sharply pointed, non-resinous buds. Bark is smooth, thin, grayish brown, thickening and becoming scaly and slightly fissured, often with pitch oozing from cracks in the bark.

An unusual fir, limited in habitat and rarely cultivated, most often seen as an arboretum specimen. Young trees are attractive, especially the soft-appearing texture of the needles. In maturity the spirelike form and semi-drooping branches are distinctive, but trees may become ungainly with age. Longevity estimated to be 100-150 years in habitat. Series associations: **SanLucfir**.

w N.A. California native: ME, PF, FF (*KR, NCoR, CaRH, SNH, Teh, TR, PR, MP, Dmtns*); CNF, EVG; (IRR: M/-/M/M/-/-)

Abies concolor

WHITE FIR

Pinaceae. Sunset zones 1-9, 14-24, Jepson *DRN:1,**4,5,6**&IRR: **2**, **3,7**,15-17,24&SHD:**10,14**,18-23;CVS. Evergreen. Native to the mountains of central Oregon, Idaho, Colorado, and California to northern Baja California at 3,000-10,000' elevation. Growth rate moderate to 80-120' tall (to 200' in habitat) by 15-20' wide, developing a dense, pyramidal shape with short horizontal branching, more vertical toward the top near the open, pointed crown, lower branches drooping with age. Needles are 1-1/2 to 2-1/2" long, blue-green, flatly 2-ranked on juvenile branches, becoming twisted, upright on upper branches of mature trees, with 2 whitish stripes on the undersides, blunt or acutely tipped, with a distinct lemonlike smell when crushed. Insignificant male flowers are reddish, and female flowers, occurring at the top, are yellow-green, in late spring. Cones are short-stalked, purplish tan to olive-green, 2-1/2 to 5" long, upright on upper branches, with thin, flaky scales, broader than long, maturing in fall of the first year. Seeds are flat, oval, 1/3-1/2" long, with broad, rose-tinted wings. Twigs are glabrous with resinous buds. Bark is thin, smooth, gray, with resin blisters, thickening and darkening to dark gray-brown, with wide, flat ridges and narrow, deep furrows.

The most common native fir, extending throughout mountainous regions of California, often hybridizing with other species, especially in the northwest. Sometimes planted at lower elevations, usually as a single specimen or in a small grove, where it often grows more slowly and with a shorter lifespan than in its native habitat. Requires less moisture than Douglas-fir (*Pseudotsuga menziesii*), and tolerates dry or poor soils in its native range, mixed with other conifers. Longevity estimated to be 200-300 years. Various cultivars with a wide range of shapes and foliage colors are available in Europe and the United States. Series associations: Breoak shrub series; Aspen, BigDoufir, Blacottonwoo, Canlivoak, Engspruce, Giasequoia, Grafir, Inccedar, Jefpine, Jefpinponpin, Limpine, Lodpine, Mixconifer, Mouhemlock, Moujuniper, Ponpine, PorOrfced, Redfir, Subfir, Waspine, Wesjuniper, Weswhipn, **Whifir**, Alayelcedsta, Bakcypsta.

nw N.A. California Native: ME, RW, CF (*NCo, NCoR*); CNF, EVG; (IRR: M/-/M/M/-/-)

Abies grandis

GRAND FIR

Pinaceae. Sunset zones 1-9, 14-17, Jepson *DRN:**4-6**&IRR: 2,3,7,**15-17**,24&SHD:14,18-23. Evergreen. Native to coastal mountains from British Columbia to Sonoma County, California, eastward into Montana. Growth rate moderate to 80-200' tall (to 300' in habitat) by 15-25' wide, with a dense conical shape, broader with age as lower horizontal branches droop, usually to the ground, uppermost growth losing vigor and the top becoming rounded. Needles are shiny dark green, 1-1/2 to 2" long, flat, with blunt or notched ends, 2 white stomata bands below, grooved uppersides, sometimes alternating shorter and longer needles side-by-side, distinctively flat but arched and recurving, in 2 rows on lower branches, those near the crown shorter, denser, and pointing forward. Insignificant flowers occur in spring, 1" long male flowers pale yellow and female flowers yellow-green. Cones are cylindrical, yellowish green to greenish purple, 2-4" long, upright on upper branches, with thin, rounded, flaky scales, broader than long, and hidden inserted bracts, maturing in fall of the first year. Seeds are tan, 3/8", flatly oval, with broadly rounded, yellowish, 3/4" wings. Twigs are pubescent, yellow- to orange-green, with small, ovoid, resinous buds. Bark is thin, gray, with resin blisters, becoming reddish brown with age with thin, flat, or platy ridges and many small cracks and fissures.

A rather uncommon coastal fir that prefers a moist, cool climate. Quite limited in range in California, and seldom cultivated except in extreme northern parts of the state. Not tolerant of drier Sierra Nevada locations, where occasionally grown specimens are much smaller than in the northwest but attractive nonetheless. Subject to spruce budworm and stringy brown-rot fungus. Longevity estimated to be 100-200 years in habitat. Series associations: Beapine, **Grafir,** Redalder, Redwood, Sitspruce, Whifir.

w N.A. California Native: FF (*KR, w. Siskyou Co., 6 populations*); CNF, EVG; (IRR: M/-/M/M/-/-)

Abies lasiocarpa

SUBALPINE FIR

Pinaceae. Sunset zones 1-9, 14-17. Evergreen. Native to mountain areas from Alaska to Oregon, extreme northwestern California, and the Rocky Mountains, extending from Colorado and Arizona to northern New Mexico. Growth rate moderate to 60-90' tall (slightly taller in habitat) by 10' wide (15-20' in the northwest), developing a dense, narrow, pyramidal shape with a distinctly recognizable spirelike crown, though growth is usually stunted in the highest-elevation climates. Needles are deep blue-green tinged with silver, 3/4 to 1-1/2" long, flat, with blunt or notched ends, white stomata bands on both sides, densely crowded on branchlets, spread irregularly but mostly upward. Insignificant dark purple male and female flowers occur in early spring. Cones are cylindrical, 2-4" long by 1-1/4 to 1-1/2" wide, dark purple, clustered near the top of the tree, finely hairy, pointing vertically, with rounded, thin, flaky scales, broader than long, with inserted bracts, maturing in September of the first year. Seeds are flatly oval, tan, 1/4" long, with 3/4" long, broad, rounded, shiny brown wings. Twigs are stout, ashy gray, with a fine brown pubescence at first, later becoming glabrous, and small, ovoid, resinous buds. Bark is thin, gray, with resin blisters, later becoming reddish brown, fissured and scaly on older trees.

An important forest tree in higher-elevation wet meadows, where lower branches may lie flat on the ground, often rooting in moist soils of the northwest. Tolerates exposure, tending to recede in heavily shaded areas. Not usually successful when planted below 3,500' elevation, though occasionally grown as an arboretum specimen. Longevity estimated to be 100-250 years in habitat. Series associations: **Subfir**, KlaMouEnrsta.

w N.A. California Native: PF, FF, AL (*KR, NCoRH, CaRH, SNH, SNE*); CNF, EVG; (IRR: M/-/M/M/-/-)

Abies magnifica

= *Abies magnifica* var. *magnifica*

CALIFORNIA RED FIR

Pinaceae. Sunset zones 2-7, Jepson *DRN:**1**,4,5,15&IRR:2,**3**, **6**,16&SHD:7,14. Evergreen. Native to the mountains of southern Oregon and the northern Sierra Nevada and Coast Ranges of California at 4,500-9,000' elevation. Growth rate rather slow to 80-120' tall (125-175' or more in habitat) by 15-20' wide, developing a symmetrical, narrowly pyramidal to cylindrical form with uniformly spaced, short horizontal branch whorls and a fairly open, erect crown. Needles are stiff, 3/4 to 1-1/2" long, blunt-tipped, nearly 4-sided, with stoma on all sides but no groove down the center, growing from each side along stems, flatly along the bottom and curving upward toward the center, silvery gray new growth maturing to a bluish green. Insignificant flowers occur in May on mature trees, male flowers reddish in terminal clusters and longer female flowers solitary on upper branches. Cones are sparse, upright, cylindrical, to 8" long by 1-1/2" wide, greenish to purplish brown, ripening to tan. Cone scales have bracts, which are hidden or extend only slightly and disintegrate with the cone stalk still attached to the tree, dispersing 1/2-3/4" flat, light tan to dark brown seeds with broad, rose-colored wings. Bark is thin, smooth, chalky gray, often with resin blisters, thickening and turning reddish brown or gray, becoming corky, with distinctive rounded, diagonal, or zigzag furrows on older trees.

One of the largest native firs and a predominant species in higher-elevation subalpine forests, often hybridizing with noble fir (*Abies procera*), which extends into its northernmost range. The "silver tip" of the Christmas tree trade is red fir. Grows more slowly in cultivation at lower elevations and longevity is greatly shortened. Lower branches may remain for many years, but massive trunks become fully exposed on older trees. Roots are deep and widespreading. Longevity estimated to be 250-350 years. Series associations: Aspen, Engspruce, Foxpine, Giasequoia, Jefpine, Lodpine, Mixsubfor, Mouhemlock, PorOrfced, **Redfir**, Subfir, Waspine, Weswhipin, Whifir, Whipine, Bakcypsta, KlaMouEnrsta, Pacsilfirsta.

Naturally occurring cultivar, w N.A. CNF, EVG, FOL ACC; (IRR: M/-/M/M/-/-)

Abies magnifica 'Glauca'

AZURE FIR

Pinaceae. Sunset zones 2-7. Evergreen. Selected variety of the species, first discovered and cultivated in the Rocky Mountains. Growth rate slow to 125-175' tall or more by 15-20' wide, developing a dense, upright, pyramidal to cylindrical form with uniformly spaced, short, horizontal branch whorls and an erect crown. Needles are stiff, 3/4 to 1-1/2" long, blunt-tipped, nearly 4-sided, with stoma on all sides, curving upward toward the center, new growth stunning silver-gray, retaining its color indefinitely. Insignificant flowers occur in May on mature trees, males reddish in terminal clusters and longer females solitary on upper branches. Cones are upright, cylindrical, rather large but sparse, to 8" long by 1-1/2" wide, greenish to purplish brown, ripening to tan, disintegrating from the central stalk, and dispersing flat, light tan to dark brown, 1/2-3/4" seeds with broad, rose-colored wings. Bark is thin, smooth, chalky gray, often with resin blisters, thickening and turning reddish or grayish brown, becoming corky, with rounded, diagonal, or zigzag furrows on older trees.

A rather rare variety of the species that has been cultivated since the early 1900s, seldom seen in California, occurring more often in the Rocky Mountains where it originated. Noticeably silvery bluish foliage stands out among other firs. Longevity estimated to be 250-350 years.

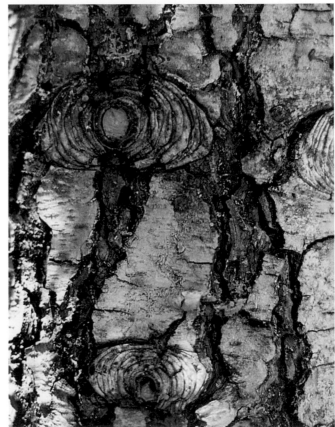

w N.A. California Native: PF, FF, AL (*KR, CaRH, s SNH*) CNF, EVG, FOL ACC; (IRR: M/-/M/M/-/-)

Abies magnifica var. shastensis

SHASTA RED FIR

Pinaceae. Sunset zones 2-7, Jepson *DRN:**1**,4,5,15&IRR:**2**, **3**,**6**,16&SHD:7,14. Evergreen. Variety native to the mountains of southern Oregon and the northern Sierra Nevada and Coast Ranges of California alongside but not nearly as common as the species. Growth rate rather slow to 125-175' tall or more by 15-20' wide, developing a symmetrical, narrowly pyramidal to cylindrical form with uniformly spaced, short, horizontal branch whorls and a fairly open erect crown, similar to the species. Needles are stiff, 3/4 to 1-1/2" long, blunt-tipped, nearly 4-sided, with stoma on all sides, growing from each side along stems, flat along the bottom and curving upward toward the center, new growth silver-gray, retaining much of its color but may fade to bluish green after the first year. Cones are upright, cylindrical, rather large but sparse, to 8" long by 1-1/2" wide, greenish to purplish brown, ripening to tan, maturing in September of the first year, the same as the species. Bracts extend beyond the cone scales to a varying extent, and buds are also slightly less resinous than the species. Bark is thin, smooth, chalky gray, identical to the species, often with resin blisters, thickening and turning reddish brown to gray, becoming corky, with rounded, diagonal, or zigzag furrows on older trees.

Less common than the species but often cultivated, especially for Christmas trees, and offered as seedling stock. Variable foliage color, from bluish to silvery green, often makes it next to impossible to differentiate from the species. Longevity estimated to be 250-350 years. Series associations: Engspruce, Foxpine, Lodpine, Mixsubfor, Mouhemlock, PorOrfced, Redfir, Subfir, Weswhipine, Whipine, KlaMouEnrsta, Pacsilfirsta.

Refer to *Abies magnifica* map, page 100.

Asia Minor: CNF, EVG; IRR: M/-/M/M/-/-

Abies nordmanniana

NORDMANN FIR

Pinaceae. Sunset zones 1-11, 14-24. Evergreen. Native to Asia Minor and Greece. Growth rate slow to 30-50' tall (taller in the northwest) and 20' wide, with a dense, pyramidal shape. Needles are glossy, dark bluish green, 3/4 to 1-1/2" long, 4-angled, blunt-ended with flexible tips, attractive double white stripes on the undersides, no resin ducts on the surface, curving upward on branchlets, completely around the topside, becoming flattened at the sides. Small, insignificant flowers cover the crowns of mature trees in April, male flowers dark purple and females, occurring only at the top, bright scarlet. On older trees cones are upright, oblong-cylindrical, vertical, 5-6" long, resinous, with variable-edged scales with a tailed, elongated, recurving bract flap, ripening to light reddish brown and releasing 1/8" black seeds with a broad, oblique, 1/2" long wing. Twigs are glaucous brown or grayish on the upperside, and buds are only slightly resinous, if at all. Bark is smooth, gray, rather thin and hard, on older trees gradually becoming mottled, with small, thin, loosely attached scales, before turning reddish to gray-brown, often scaly throughout.

An uncommon, attractive, dense park specimen tree with a clean appearance. Casts dense shade on plantings beneath, but adapts well to lawns and gardens in moderately moist, well-drained soils. Prefers half-shade inland. May develop some spreading feeder roots, but otherwise fairly deep rooted. One of the more successfully cultivated imported firs, generally reserved for specimen use rather than mass plantings. Longevity estimated to be 150-250 years.

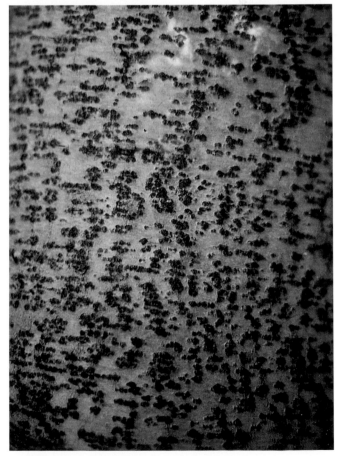

sw Europe: CNF, EVG; IRR: L/-/L/-/-/-

Abies pinsapo

SPANISH FIR

Pinaceae. Sunset zones 5-11, 14-24. Evergreen. Native to mountainous regions of southern Spain. Growth rate slow to 20' tall (older trees can reach 30-60' tall) and 15-20' wide, with a dense, pyramidal shape. Distinctive dark green needles are stiff, 1/2-3/4" long, thick, with acutely pointed or blunt-rounded tips, not notched at the ends, pale, double-striped undersides, slightly convex uppersides, whorled completely around branchlets, radially, and uniformly spaced. Small, pendulous male flowers appear in late spring, with larger, fatter, upright female flowers at ends of higher branches. Cones are erect, brown, cylindrical, purplish brown, 4-5" long, bracts hidden by 3/4" wide, smooth scales, maturing in fall of the first season, scales disintegrating from the persistent central spine the following spring, releasing small wide-winged seeds. Twigs are glabrous, brown, with resinous winter buds. Smooth young bark has balsam blisters typical of firs and ages to a smooth grayish brown, becoming cracked or fissured, with broad, flat plates.

Tolerates inland heat in moderately moist, well-drained lawn settings, as a park specimen. Rather sparse and loses symmetrical form in heavy shade. Longevity estimated to be 150-250 years in cultivation. 'Glauca' has attractive blue-gray foliage.

w N.A. California Native: CF (*n KR, CaRH*); CNF, EVG, FOL ACC; (IRR: M/-/M/M/-/-)

Abies procera

NOBLE FIR

Pinaceae. Sunset zones 2-6, 15-17, Jepson *DRN:4,5&IRR:**1-3,6,**15-17&SHD:7,14. Evergreen. Native to mountains of Washington, Oregon, and extreme northern California above 2,000' elevation. Growth rate moderate to 90-200' tall in the northwest with a 20-30' spread, developing a symmetrical, dense, pyramidal shape with rather short horizontal branching, and lower branches may droop from the weight of the foliage. Needles are pale to dark blue-green, 1 to 1-1/2" long, with a round-pointed apex, a grooved upper surface, growing outward and upward, densely covering uppersides of branches, new growth generally with a silvery tinge. Insignificant reddish purple male and female flowers occur in April. Cones are upright, flaky, 4-6" long, purplish tan to olive-green, with thin, flaky scales, broader than long, with greenish extended bracts recurving toward the base. Seeds are flat, oval, 1/2" with broad tan wings. Twigs are reddish brown, pubescent, with small, blunt, slightly resinous, conical buds. Bark is thin, smooth, gray, with resin blisters, thickening slightly with age and becoming fissured between narrow plated ridges, which peel off easily to expose reddish underbark.

An impressive large-scale, long-lived timber tree, important in the northwest and somewhat tolerant of Sierra Nevada conditions, where it is sometimes grown for Christmas trees but often loses vigor after reaching maturity. Longevity estimated to be 300-500 years in habitat, but grows more slowly when cultivated in colder regions of the eastern United States and Great Britain. Series associations: Redfir, Alayelcedsta, KlaMouEnrsta, Pacsilfirsta.

Australia: EVG, SHD, FOL, FLW ACC; IRR: L/L/L/L/-/-

Acacia baileyana

BAILEY ACACIA

Fabaceae. Sunset zones 8, 9, 13-24. Evergreen. Native to Australia. Growth rate moderate to 20-30' tall with a 20-40' spread, forming an upright to broad, billowy canopy with one or more heavy trunks and many slender horizontal branches, twiggy at the ends. Leaves are alternate, bipinnately compound, 1-3" long, blue-gray, feathery, with 16-20 pairs of lanceolate-elliptical leaflets, tightly spaced, slightly hairy, lying flatly 2-ranked, persisting into the following year. Flowers are fragrant, puffy, round, 1/4" yellow, in loose, raceme clusters, profuse and showy in early spring. Flat, tan, 1-3" long by 1/2" wide legume pods persist into fall, containing 1/8" round, flattened, shiny black seeds. Bark is thin, smooth, and grayish brown, with lighter whitish vertical lines as it expands, becoming darker and shallowly fissured.

Commonly used as a garden, patio, or small street tree and effective in groves in less manicured parkways. Often becomes shrublike if left unpruned at the base. Invasive surface roots can be a problem in gardens, but stabilize soil. Does best in a moderately dry, sunny location in well-drained soil, but benefits from occasional deep watering. Otherwise trouble-free, and drought tolerant when established. One of the hardiest acacias. Longevity estimated to be 30-50 years. 'Purpurea' has purplish to lavender foliage.

Australia: EVG, SHD, FOL, FLW ACC; IRR: VL/L/L/L/-/-

Acacia dealbata

GREEN ACACIA

Fabaceae. Sunset zones 8, 9, 14-24, borderline in zone 6. Evergreen. Native to Australia. Growth rate fast to 50' tall with a 40-50' spread, developing a tall, billowy canopy, usually branching to the ground, with a large trunk in age and branches spread horizontally to form a dense, drooping canopy. Leaves are alternate, bipinnately compound, 4-6" long, feathery and light-textured, silvery gray-green, with 20 pinnae each, composed of 30-40 paired linear leaflets, persisting into the following year. Loosely hanging clusters of up to 30 yellow, spherical, 1/4", puffy, fragrant flowers cover the tree in early spring. Smooth, bluish-tinged, light cream-colored legume seed pods are 2-3" long by 1/2" wide, remaining into summer, eventually splitting at the seams along each side, and releasing small flat, rounded, hard-shelled, black seeds. Twigs and branches are silvery gray. Bark is smooth, silvery to grayish green, with a glandular, warty, or scaly roughness, darkening to brown or nearly black with age.

Once a popular flowering accent shade tree, less used in recent years because of its allergy-producing pollen. A few large relics from the past survive, providing a brief burst of spring bloom even in abandoned landscapes. Drought tolerant. Heaviest bloom is from older wood. Severe pruning can destroy the balanced shape by producing a profusion of overly vigorous juvenile growth. Weak branches may break in strong winds. Roots are quite invasive. Reseeds readily and can become an invasive weed if not restrained. Longevity estimated to be 50-75 years.

Australia: EVG, SHD, FLW ACC; IRR: L/L/L/L/-/-

Acacia longifolia

SYDNEY GOLDEN WATTLE

Fabaceae. Sunset zones 8, 9, 14-24. Evergreen. Native to Australia and Tasmania. Growth rate fast to 20' tall with a 20' spread, usually a large, billowy, round-headed shrub unless trained to an upright oval form. Leaves are alternate, simple, bright green, linear to oblong-lanceolate, 3-6" long by 1/4-1/2" wide, somewhat sickle-shaped, with smooth edges, a thickened, leathery phyllode leaf stalk, with small gland marks near the base, 3-5 parallel longitudinal veins, smooth thickened margins, slightly hairy, with a blunt-pointed tip. Slender, erect, tassel-like, bright yellow flower spikes, 2-1/2" long in loose axillary raceme clusters, are showy in late spring, from the present year's growth. Seed pods are twisted and contorted, flat, brown, 2-6" long by 1/4-1/2" wide legumes, ripening in late summer, containing 6-10 oval, 1/4", disc-shaped, black seeds. Bark is smooth, greenish brown, thickening with age to grayish brown, with wide reddish fissures and narrow, interconnected, flat, scaly-plated ridges.

Most often used as an evergreen screen, hedge, or windbreak, less often as a standard or multi-trunk accent tree with main leaders trimmed of side branches. Relatively pest free. Tolerant of poor soils and drought. Vigorous fibrous root system binds soil in slope plantings. Longevity estimated to be 40-50 years.

Australia: EVG, SHD, FLW ACC; IRR: VL/L/L/L/-/-

Acacia melanoxylon

BLACK ACACIA

Fabaceae. Sunset zones 8, 9, 13-24. Native to southeastern Australia and Tasmania. Growth rate moderate to 40' tall with a 20' spread, developing a more upright form than *Acacia longifolia* with foliage more dense, though not as lush and shiny. Leaves are alternate, simple, oblong-lanceolate, dull dark green, 2-4" long by 1/4 to 1/2" wide, curved, somewhat sickle-shaped, with a thickened, leathery phyllode leaf stalk with small gland marks near the base, 3-5 parallel longitudinal veins, smooth thickened margins, blunt-pointed tips, slightly hairy. Occasional bipinnate juvenile leaves occur on the phyllode or rachis, with 15-20 paired leaflets on each pinna, 1/2" long by 1/8" wide, smaller toward the tip. Creamy white to yellowish, 1/4", fluffy, round flowers, noticeable but not showy, are borne in March and April in short-stemmed clustered racemes of 2-3 at leaf axils of the present year's growth. Seed pods are flat, brown, 2-6" long by 1/4-1/2" wide legumes, twisted and contorted, ripening in late summer, containing 6-10 oval, 1/4", disc-shaped, black seeds surrounded by a red appendage. Bark is smooth, greenish brown, thickening with age to grayish red-brown, with shallow reddish fissures and rough, flat, scaly-plated ridges.

Once widely used as an upright evergreen shade or screen tree. Reliable in hot dry climates. Tolerates drought and poor soils, but looks best with occasional deep watering. Wide-spreading roots are generally fibrous enough to not readily heave paving, but are invasive toward water. A good solid screen tree, but the canopy can get too big for confined areas. Longevity estimated to be 40-65 years.

Australia: EVG, SCR, FLW ACC; IRR: No WUCOLS

Acacia retinoides

WATER WATTLE

Fabaceae. No Sunset zones. Evergreen. Native to coastal Tasmania, Victoria, and southern Australia. Growth rate very fast to 20' tall with a 20' spread, usually a multi-trunked bushy shrub or small tree with brittle, slender, upward-arching branches, drooping at the ends, somewhat willowlike in texture and form. Leaves, or phyllodes, are alternate, simple, dark green, variable in form, to 6" long, mostly linear and 1/4-1/2" wide with a few more broadly lanceolate to 3/4-1" wide. A showy profusion of fluffy, round, yellow, 1/4" flowers in many short-stemmed, elongated, loosely clustered racemes of 2-3 each appear in March and April at leaf axils of the present year's growth along the ends of branches. Seed pods are reddish brown, flattened legumes, 2-5" long by 1/2" wide, containing flattened, elongated, dark brown, shiny seeds. Bark is coarse, brownish gray, becoming dark brown, often cracking with age.

Commonly used in southern California freeway and slope plantings, for its fast growth, lush appearance, and screening value. Also suitable as a semi-weeping, evergreen flowering accent tree for a protected courtyard or patio. Multi-trunk specimens are attractive trimmed up to expose dark brown trunks. Tolerates heat, drought, salt spray, and poorly drained clayey or alkaline soils, with occasional deep watering. Not reliably hardy below 25 degrees, but often quickly resprouts from the base. Longevity estimated to be 20-25 years.

e Asia: SHD, ACC, SPC, FAL COL; IRR: M/M/M/-/-/-

Acer buergerianum

TRIDENT MAPLE

Aceraceae. Sunset zones 2-9, 14-17, 20, 21. Deciduous. Native to eastern China, Korea, and Japan. Growth rate moderate to 20-25' tall and wide, forming a dense, rounded crown, often multi-trunked or low branching, with staking and pruning may be trained to a taller canopy tree form. Leaves are opposite, simple, glossy dark green, 2-3" long and nearly as wide, 3 triangular lobes near the apex pointing forward, generally smooth margins rounded at the base but may be finely serrate near the tips, pale undersides, a slender petiole to 2" long, and brilliant red fall color, especially in colder climates, with orange and yellow hues. Broad, erect, softly hairy panicles of yellowish flowers appear in early spring. Seeds are paired, yellowish to greenish, 1-3/4 to 2" long, with wings drooping to nearly parallel, persisting through winter. Twigs are smooth, shiny, reddish brown. Bark is smooth, light brown, becoming gray, fissured, and scaly-plated or shredded with age.

A clean, attractive, small park or street tree, also useful as a lawn specimen in residential settings. Introduced into cultivation near the turn of the century, but has gained popularity only since the early 1950s. Prefers moderate moisture. Tolerant of most soils. Relatively pest free. Longevity estimated to be 100-150 years or more.

Asia, Europe, N. Africa: SHD, FAL COL; IRR: No WUCOLS

Acer campestre

ENGLISH MAPLE

Aceraceae. Sunset zones 2-9, 14-17. Deciduous. Native to Asia, Europe, and North Africa. Growth rate slow to 70' tall (rarely over 30') and 30' wide, slender at first, eventually developing a tall, round-topped, oval form. Leaves are opposite, simple, palmately lobed, 1-4" long, dull green, deeply 3-5 lobed, each lobe either obtuse and entire or slightly tri-lobed at the apex, with smooth uppersides and pale pubescent undersides, petioles 1-3" long, exuding a milky sap when cut, fall color dull orange to reddish, in colder climates reddish purple to orange-yellow. Insignificant, loose, erect, pubescent corymb clusters of 10-12 hermaphroditic greenish flowers occur in April to May. Dense clusters of long-stemmed, hanging, hairy, paired seeds, with 1" long, green wings spread at 180 degrees or parallel to each other, are noticeable, even unsightly, in profusion throughout summer, turning from greenish yellow as they dry to a light tan in fall and persisting through winter. Bark is thin, brown, with vertical fissures and flat, wide, scaly plates, becoming gray-brown with age.

A useful small street, park, or patio tree, more commonly planted in the northwest. Tolerates inland heat in California with moderate moisture in fertile soils, but may not be as vigorous or shapely and fall color usually is barely noticeable. Lower branches can be pruned up as the tree becomes taller to produce more top growth, but must be cut cleanly to prevent suckering. Longevity estimated to be 100-175 years. Various cultivars exist, including dwarf, variegated, and purplish-leaved forms. 'Queen Elizabeth' is an improved variety over seedling stock.

e Asia: SHD, SPC; IRR: No WUCOLS

Acer capillipes

RED STRIPEBARK MAPLE

Aceraceae. Sunset zones 2-9, 12, 14-24. Deciduous. Native to Japan. Growth rate moderate to 30' tall and wide, developing a broad oval form, a strong trunk, upright limbs, and slender horizontal branching, becoming broad-crowned with dense foliage. Leaves are opposite, simple, light green, 2-6" long by 2-3" wide, either ovate or shallowly 3-lobed beyond the middle, smooth with sharply doubly-toothed edges, cordate at the base of the 1-2" long, reddish green petiole, bronze-tinted as they emerge, with a lighter underside, and bright yellow or reddish orange to purplish fall color, more brilliant in frosty areas. Pendulous 4-6" long raceme clusters of usually monoecious, small, long-stemmed, cup-shaped flowers appear in spring, males and females separate on the same tree, with 4 green sepals and 4 overlapping greenish yellow petals, whitish stamens protruding beyond. Hanging clusters of paired glabrous seeds, with 3/4-1" wings, in nearly 180 degree alignment, mature to reddish brown in fall. Bark is smooth, thin, bright green, with distinctive longitudinal white stripes, darkening to reddish brown with age near the base of the trunk, becoming cracked and fissured.

An unusual accent or shade tree, midsized between Japanese and red maple. Not often used, but may be available through specialty nurseries. Clean, regular branching with a neat habit. Attractive striped bark is the most notable feature, along with interesting drooping flowers. Prefers a cool, evenly moist, semi-shaded location. The most commonly available striped-bark maple and reportedly the most cold-hardy. Longevity estimated to be 50-75 years.

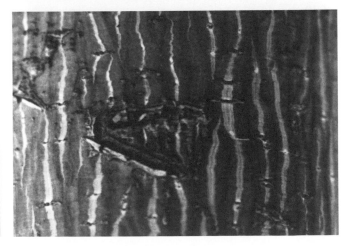

nw N.A. California Native: RW, CF (Riparian) (*n&c NW, CaRH, n SN*); SPC, FOL ACC; (IRR: H/H/-/-/-/-)

Acer circinatum

VINE MAPLE

Aceraceae. Sunset zones 2-6, 14-17, Jepson *IRR:**4-6,17**&SHD: 1-3,**7**,8,9,**14-16**,18-20. Deciduous. Native to coastal British Columbia south to northern California in moist woods and along shaded streams. Growth rate slow to 5-35' tall or more, usually shrublike or prostrate in habitat, becoming taller in a shaded setting, usually multi-trunked, developing an arching to round-topped crown, the weight of the foliage causing vigorous weak branches to droop horizontally in vinelike fashion. Leaves are opposite, simple, 2 to 4-1/2" wide, light green, oval or fanlike, with 5-11 sharply double-toothed lobes, a glabrous surface, with minute tufted hairs at vein joints on the pale underside, reddish new growth and fall color bright yellow-orange in shade, more reddish in sunnier locations, especially in colder climates. Small, loose clusters of tiny, drooping, reddish purple flowers occur in late April, nearly hidden by leaves, male and female flowers separate in the same cluster. Reddish seed capsules, joined in nearly horizontal pairing, with 1" long wings, hang among leaves through summer. Twigs are slender, smooth, green to reddish brown, with small, blunt, reddish buds. Bark is relatively thin, greenish to gray-brown, either smooth or with shallow, darker reddish-tinged fissures.

An attractive riparian native understory tree, forming thickets in its native habitat, where it tolerates deep shade. Rarely cultivated, but well suited as a small garden accent tree in moist, shaded areas within its range. Delicate form and interesting winter silhouette make it an attractive native alternative to Japanese maples. Longevity estimated to be 60-120 years in habitat. Varieties include upright and dwarf forms. 'Elegant' and 'Monroe' have lacy, deeply cut, 5-lobed leaves. Series associations: Doufir, PorOrfced, Redalder, Whifir.

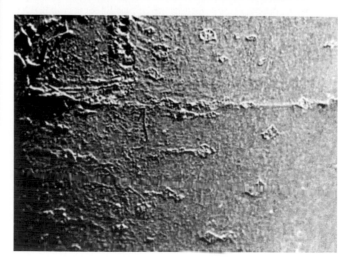

w N.A. California Native: CF, ME, PF, FF (Riparian) (*CA, to AK, c US, NM*); NAT, ORN; (IRR: No WUCOLS)

Acer glabrum

MOUNTAIN MAPLE

Aceraceae. Sunset zones 1-7, 10, Jepson *DRN,IRR:1-7. Deciduous. Native to western North America from Alaska to Arizona east to the Rocky Mountains near mountain streams at 5,000-6,000' elevation. Growth rate slow to 6' tall and wide (up to 30' in northern habitats), usually shrublike or shortly multi-trunked with slender, upright branches. Leaves are opposite, simple, palmate, 3-5" long, dark green, somewhat variable in shape but usually 3-5 lobed, rarely trifoliate, with a glabrous and shiny upperside, sharply doubly-serrate edges, very pale undersides, prominent yellowish veins, reddish-tinged petioles, and red or yellow fall color. Flowers are insignificant, mostly dioecious, male and female flowers usually on separate trees, petals yellowish green, occurring in loose, racemelike corymbs in April and May. Clusters of drooping achene seeds are glabrous, paired, with 1" long wings aligned nearly parallel, ripening to rosy red in late summer. Twigs are slender, hairless, green turning reddish brown, with pointed buds. Bark is smooth, reddish brown, becoming slightly grayish.

A hardy native understory tree in cool, rather moist canyons and on north-facing slopes. Six varieties occur in various ranges. In cultivation prefers a semi-shaded location in moist, well-drained soils and tolerates part-day full sun if well-mulched, with occasional regular deep watering. Longevity estimated to be 35-50 years or more. The *Jepson Manual* identifies *Acer glabrum* var. *glabrum* as Rocky Mountain maple. Series associations: Sitalder shrub series; Moualder.

cw N.A. NAT, ORN; IRR: M/-/-/-/-/-

Acer saccharum ssp. *grandidentatum*

BIGTOOTH MAPLE

Aceraceae. Sunset zones 1-3. Deciduous. Subspecies native to mountain rivers in Utah, Arizona, and New Mexico into northern Mexico. Growth rate moderate to 25-35' tall, forming a narrow, compact crown. Leaves are opposite, simple, palmate, orbicular, 2-5" in diameter, dark green and lustrous, shallowly 3-lobed or rarely 5-lobed above the middle, with coarsely toothed margins, pale downy undersides, and yellow to red fall color. Interesting but insignificant polygamous, drooping flowers, apetalous with a yellow calyx, appear in clusters in May, males and females on separate trees, before leaves in early spring. Clusters of paired, drooping achene seeds are reddish, with 1/2-3/4" long wings aligned at 90 degrees or slightly less. Seeds are smooth or slightly hairy, maturing to light brown in fall. Twigs are smooth, slender, reddish. Bark is gray to brown with platelike scales.

A smaller western relative of the eastern sugar maple, occurring in moist mountain and canyon woodlands. Shallow-rooted, and in cultivation prefers moist soils, well-mulched with an occasional deep watering. Longevity estimated to be 40-50 years, possibly more.

e Asia. ORN, ACC; IRR: M/M/?/?/?/?

Acer griseum

PAPERBARK MAPLE

Aceraceae. Sunset zones 2-9, 14-21. Deciduous. Native to western China. Growth rate slow to 25' tall or more and half to equally as wide, usually developing multiple upright trunks, the narrow crown becoming more spreading, with a fuller canopy, as it matures. Leaves are opposite, palmately compound,1-1/2 to 2-1/2" long, dark green, trifoliate, with coarsely toothed edges, very pale or silvery undersides, often bronzy, densely hairy, and somewhat drooping as they unfold in late spring, 1" long, reddish brown petioles also covered with dense, dark brown hairs that later recede, and orange to scarlet or crimson fall color in colder climates, elsewhere faint reddish brown to dull pink. Insignificant flowers are mostly dioecious, male and female flowers usually on separate trees, in loose racemelike corymbs with reddish petals in April and May. Showy clusters of drooping, paired achene seeds, with 1" long wings nearly parallel, ripen to rosy red in late summer. Stout twigs are reddish to grayish brown and hairy. Bark is thin, reddish brown, peeling in papery sheaths, curling back from splits.

An uncommon but attractive small accent tree with a striking winter silhouette, thick branches angling outward from multiple trunks, much like Japanese maples but less twiggy. Prefers a semi-shaded location in well-drained, moist soils. Tolerates partial to full sun, if well-mulched, with occasional deep watering in loose soils. Longevity estimated to be 50-80 years.

e Asia. ORN, ACC; IRR: No WUCOLS

Acer japonicum

FULLMOON MAPLE

Aceraceae. Sunset zones 2-6, 14-16. Deciduous. Native to Japan. Growth rate very slow to 20-30' tall with equal or greater spread, with upright trunks usually low-forked, often developing a broad, horizontal-branching canopy. Leaves are opposite, simple, palmate, cordate, 2-5" in diameter, larger than those of *Acer palmatum* and with more lobes, 7-11, which are pointed at the ends and cut less than halfway to the center, with doubly serrate edges, new leaves with silky hairs on both sides and downy stalks, later becoming smooth, and bright yellow fall color. Small, insignificant, drooping clusters of reddish purple, yellow-stamened flowers occur in mid April. Yellowish winged, 3/4-1" long seeds are slightly downy at first, with wings paired at nearly 180 degrees. Bark is smooth, greenish brown, becoming brown with age.

A distinctive, uncommon specimen tree for lawn or patio. Requires moderately moist, fertile soil in semi-shaded conditions. Otherwise more suited to the cooler northwest. Longevity estimated to be 100-150 years. 'Aureum', golden fullmoon maple, has pale yellow new leaves in spring that remain bright, light yellowish green through summer, turning yellow to red in fall.

Cultivar. ORN, FOL ACC; IRR: No WUCOLS

Acer japonicum 'Aconitifolium'

LACELEAF FULLMOON MAPLE

Aceraceae. Sunset zones 2-6, 14-16. Deciduous. Variety of the species. Growth rate slow to 15-20' tall or more with upright trunks, often multi-trunked, eventually developing a broad, horizontal to arching canopy. Leaves are opposite, simple, palmate, 2-5" in diameter, lobes cut nearly to the stalk, heavily toothed along the margins, giving them a fernlike texture, new leaves tending to droop as they emerge in spring, slowly unfurling, with silky hairs on both sides and on leaf stalks, later becoming glabrous, with deep red fall color. Small, insignificant, drooping clusters of reddish purple, yellow-stamened flowers occur in mid April. Yellowish winged seeds, 3/4-1" long, slightly downy at first, are paired at nearly 180 degrees. Bark is smooth, thin, greenish brown, becoming brown with age.

A striking lawn or garden accent or specimen tree. Requires moderately moist, fertile soils in semi-shaded conditions. The most cold-hardy Japanese maple, more so than *Acer palmatum*. Longevity estimated to be 100-150 years.

w N.A. California Native: CF, RW, ME, PF (Riparian), *CA-FP* (*exc GV*); NAT, ORN, FAL COL; (IRR: M/H/M/H/-/-)

Acer macrophyllum

BIGLEAF MAPLE

Aceraceae. Sunset zones 2-9, 14-24, Jepson *DRN:**4-6**&IRR:1-3,**7,8,9,14-24**. Deciduous. Native to moist canyons and streambanks in California foothills, extending to Alaska. Growth rate fast to 30-75' tall with a 30-50' spread, developing a tall, narrowly oval canopy, usually with multiple trunks, broader in full sun, with clear, straight trunks in heavy shade. Leaves are simple, opposite, 6-12" long, smooth light green, palmate, and 3-5 lobed, with entire or sparingly wavy-toothed edges, indented at the ends, pale undersides, and deep yellow fall color in areas of frost. Noticeable, 10" long, pendulous racemes of slightly fragrant, 1/4", bright yellow, polygamous flowers occur in April. Clustered pairs of light brown, 2" long, smooth-winged key seeds are on long slender stems, wings aligned at roughly 90 degrees. Small, sharp, bristlelike hairs on the enlarged end of the sheath covering the seed may irritate skin. Twigs are green to reddish brown, and the large, green to brown buds have hairy-margined scales. Smooth, gray-brown bark becomes reddish brown, developing deep fissures, or breaking into square plates.

An important deciduous understory canopy tree in mountain canyons, often mixed with Pacific dogwood and madrone. Not well-suited for urban use, as weak limbs break easily, and rather short-lived, but occasionally used in coastal and mountain garden settings and in riparian plantings. Grayish bark provides a striking silhouette in winter, along with persistent seeds, which resemble butterflies hanging from branches after leaves have fallen. Seeds heavily and suckers or resprouts from the base if the trunk is cut down. Longevity estimated to be 50-80 years in habitat. Series associations: Sanwillow shrub series; Arrwillow, Blacottonwoo, Blaoak, Canlivoak, Coalivoak, Doufir, Doufirponpin, Mixwillow, Pacwillow, Redwood, Sitwillow, Whialder, Whifir.

N.A. California Native: CP, VG, OW, FW (Riparian), *CA-FP*; NAT, ORN, SHD; (IRR: M/M/M/M/-/-)

Acer negundo

BOX ELDER

Aceraceae. Sunset zones 1-10, 12-24, Jepson *4-6;IRR:1-10,12-24; INV; also STBL. Deciduous. Native to California and most of the U.S. in riparian and flat river plain habitats. Growth rate initially fast to 60' tall (usually 20-50') with an equal or wider spread, from short, stocky trunks, usually multi-trunked with an arching or broad, round-topped canopy. Leaves are opposite, palmately compound, dull yellowish green, with 3-5 leaflets, 2-5" long, the terminal leaflet largest, toothed to doubly-toothed edges, hairy pale undersides, and dull yellow to brown fall color. Small, yellowish green flowers droop in racemes in spring, male flowers on unbranched threadlike stems, females on branched stems, on separate trees, from branch buds of the previous year. Noticeable clusters of paired, 1-1/4" long, winged seeds are joined at roughly 60 degrees and hang by long, slender-branched stalks, persisting into winter. Mature twigs of each season are covered with downy hairs. Bark is thin, pale grayish brown, becoming darker, with furrows and narrow ridges on older trees.

An important habitat tree, not well suited for urban use. Readily self-seeds, to the point of becoming a weed, suckers at the base, and often suffers from breakage, since the wood is weak. Sometimes pruned into a small multi-trunk canopy tree in residential, school, or park settings, or in naturalized plantings, where it grows without any special care. Longevity estimated to be 30-50 years. *The Jepson Manual* identifies var. *californicum* as having more densely tomentose leaves and stems. Series associations: Blacottonwoo, Coalivoak Frecottonwoo.

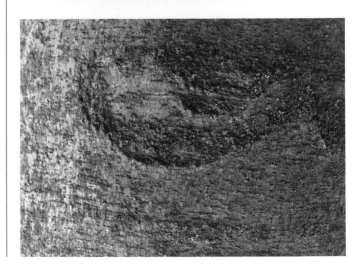

Cultivar. ORN, ACC, SHD; IRR: M/M/M/M/-/-

Acer negundo 'Variegatum'
VARIEGATED BOX ELDER

Aceraceae. Sunset zones 1-10, 14-24. Deciduous. Sport of the species. Growth rate fast to 60' tall (usually 20-50') with an equal or wider spread, developing irregular, loosely spaced branching in heavy shade, more dense and not as tall in sun. Leaves are opposite, palmately compound, bright green, with attractive white edge variegation, 3-5 leaflets, each 2-5" long, the terminal leaflet largest, edges often less heavily serrate than the species, with hairy pale undersides, and only slight yellow fall color. Small, yellowish green flowers in racemes occur in spring, males and females on separate trees, from branch buds of the previous year. Showy clusters of infertile, 1-1/4" long, white-winged seeds, paired and joined at roughly 60 degrees, hang by long, slender branched stalks until they fall in late summer. Downy hairs cover mature twigs of the current season. Bark is thin, pale greenish brown, darkening to gray-brown, with a rather rough texture, with shallow seams developing on older trees.

Somewhat less weedy than the species, and can be used as an accent tree to brighten a deeply shaded garden spot. Prefers a location that is neither too hot nor too dry. Leaves may scorch and drop in summer heat. Longevity estimated to be about 30-50 years, older trees often slowly losing vigor and becoming rather awkwardly shaped. 'Sensation' is slower growing, to 40' tall, with better branching structure and pink fall color, 'Elegans' has yellow-edged variegation, and 'Flamingo' is a tricolor form with pinkish cast leaves and pinkish leaf stalks.

e Asia. ORN, FOL ACC, SHD, SPC; IRR: M/M/H/H/-/-

Acer palmatum

JAPANESE MAPLE

Aceraceae. Sunset zones 2-10, 12, 14-24. Deciduous. Native to Japan, China, and Korea. Growth rate slow to moderate to 20' tall or more with an equal or greater spread, often with multiple trunks, which accentuate its graceful form, and usually developing a broad, rounded canopy. Leaves are oval, opposite, simple, 1-2" long, palmate, with 5-7 pointed, double-serrated lobe fingers, new growth with a reddish tinge, fall color a lasting brilliant red, orange, yellow, or a combination, depending on sun exposure and night temperatures. Insignificant, small, loose, semi-erect clusters of tiny, reddish green flowers, with many long stamens, are attractive up close in spring. Hanging pairs of 1" winged seeds, joined in a slightly less than 180 degree alignment, or slightly bowed, turn reddish in fall. Twigs are slender, green, and smooth. Bark is thin, greenish, becoming grayish brown with age, shallowly fissured or furrowed.

The most widely used maple for landscape use as a small accent tree in a semi-shaded, moderately moist location. Good bonsai plant. Sometimes tolerates moderate full sun, but leaves may scorch in hot, dry locations without adequate moisture and partial shade much of the day. Prefers moist, rich, well-drained soils. Relatively free of pests and diseases. Resistant to oak root fungus. Suffers in drought, where salt buildup in the soil may cause leaf ends to burn. Longevity estimated to be 150-200 years. Many cultivars available through specialty nurseries. Clonal cultivars are grafted onto rootstock, as seedling stock is variable.

Cultivar. ORN, FOL ACC, SHD, SPC; IRR:M/M/H/H/-/-

Acer palmatum 'Atropurpureum'

RED JAPANESE MAPLE

Aceraceae. Sunset zones 2-10, 12, 14-24. Deciduous. Cultivar of the species. Growth rate slow to moderate to 20' tall, usually with multiple trunks, forming an upright, spreading canopy. Leaves are opposite, simple, palmate, purplish to bronzy green, may fade to green or a golden-bronzy color through summer, with reddish new growth throughout the growing season and bright red fall color. Loose, drooping panicle clusters of long-stemmed, reddish flowers occur in spring with new leaves. Twigs and branchlets are dark purplish red, forming an attractive silhouette in winter. Bark of older trees darkens to grayish brown with age, shallowly fissured or furrowed.

'Bloodgood' has deeper, dark reddish purple foliage, which does not fade to green, and fall color is often darker bright red.

Cultivar. ORN, FOL ACC, SPC; IRR: M/M/H/H/-/-

Acer palmatum 'Burgundy Lace'

BURGUNDY LACE JAPANESE MAPLE

Aceraceae. Sunset zones 2-10, 12, 14-24. Deciduous. Cultivar of the species. Growth rate slow to moderate to 12 tall' and 15' wide, usually with multiple trunks, slowly forming an upright, spreading canopy. Leaves are more deeply cut than 'Atropurpureum', with reddish orange fall color. Loose, drooping, panicle clusters of long-stemmed, reddish flowers occur in spring with new leaves. Twigs and branchlets are dark purplish red. Bark of older trees darkens to grayish brown, becoming shallowly fissured or slightly furrowed. Tolerates sunny locations, but favors part shade.

Cultivar. ORN, FOL ACC, SPC; IRR: M/M/H/H/-/-

Acer palmatum 'Butterfly'

VARIEGATED JAPANESE MAPLE

Aceraceae. Sunset zones 2-10, 12, 14-24. Deciduous. Cultivar of the species. Growth rate very slow to 7' tall, with an upright form, usually from multiple upright trunks, and vertical branching, resulting in slightly irregular, dense foliage tufts. Tightly bunched leaves are opposite, simple, and irregularly variegated, 1 to 1-1/2" long, green with pinkish red tricolor tones as they emerge, later becoming dark green with white, deeply lobed and finely serrate, with little or no fall color. Loose, drooping, panicle clusters of long-stemmed, reddish flowers occur in spring, somewhat hidden by new leaves. Smooth yellowish green bark and twigs slowly turn brown.

Not as vigorous as other Japanese maples. Foliage stems with leaves that revert to all green are usually removed. Side branching may become more horizontal in moist, shady locations.

Cultivar. ORN, FOL ACC, SPC; IRR: M/M/H/H/-/-

Acer palmatum 'Dissectum'

LACELEAF JAPANESE MAPLE

Aceraceae. Sunset zones 2-10, 12, 14-24. Deciduous. Cultivar of the species. Growth rate very slow to 6' tall by 12' wide, usually with a short, stout trunk, forked or multi-branching close to the ground, a rounded crown, and irregular, broadly spreading or drooping branches, often touching the ground. Leaves are opposite, simple, green, 2-4" long, palmately lobed to the central base, appearing compound, with long, narrow fingers, which are finely dissected or serrate, with cleft lobing, and golden fall color. Loose, drooping clusters of reddish flowers occur in spring with new leaves. Thin green bark and twigs age to reddish brown.

A prized small specimen tree, used as a border foreground or in lawns as a single accent tree. Tolerates filtered or full sun with adequate moisture, though the deep green color fades in full sun. Longevity estimated to be 80-100 years or more.

Cultivar. ORN, FOL ACC, SPC; IRR: M/M/H/H/-/-

Acer palmatum 'Ornatum'

RED LACELEAF JAPANESE MAPLE

Aceraceae. Sunset zones 2-10, 12, 14-24. Deciduous. Cultivar of the species. Growth rate very slow to 6' tall by 12' wide, usually with a short, stout trunk, low-forked or multi-branching close to the ground, a rounded crown, and irregular, broadly spreading or drooping branches, often touching the ground. Leaves are opposite, simple, 2-4" long, purplish red, palmately lobed to the central base, appearing compound, with long, narrow fingers, finely dissected or serrate, and bright red fall color. Loose, drooping clusters of reddish flowers occur in spring with new leaves. Thin, purplish red bark and twigs age to reddish brown.

A small specimen tree, favored for reddish foliage, often used as a border foreground in gardens or in lawns as a single accent tree. Tolerates filtered or full sun with adequate moisture. Longevity estimated to be 80-100 years or more.

Cultivar. ORN, FOL ACC, SPC; IRR: M/M/H/H/-/-

Acer palmatum 'Sango Kaku'

REDTWIG JAPANESE MAPLE

Aceraceae. Sunset zones 2-10, 12, 14-24. Deciduous. Cultivar of the species. Growth rate slow to moderate to 10' tall or more with a nearly equal spread, forming a vigorous, upright, multi-trunk tree. Leaves are opposite, simple, palmate, 2-3" long, shiny green, with yellow fall color with a reddish tint. Branches and trunk are bright coral red.

A small specimen tree with a striking winter silhouette. Tolerates a sunny location with adequate moisture. Longevity estimated to be 80-100 years.

c&s Europe. SHD, STR, FAL COL; IRR: M/M/-/H/-/-

Acer platanoides

NORWAY MAPLE

Aceraceae. Sunset zones 1-9, 14-17. Deciduous. Native to central and southern Europe. Growth rate moderate to 50-60' tall by two-thirds to equally as wide, developing a broad, oval form, with a strong trunk, upright limbs, and horizontal branching, becoming broadly-crowned with dense foliage. Leaves are opposite, simple, yellow-green, 3-5" wide, with 5 shallow, thinly pointed lobes, sparsely long-toothed along the edges, with long, slender petioles exuding a milky sap when cut, lighter undersides hairy at vein angles, and clear deep yellow fall color. Small clusters of tiny, yellowish green flowers occur in spring, male and female flowers usually on different trees. Hanging pairs of 1-1/2 to 2" long, green-winged seeds, in clusters, with wings bow-shaped, in roughly 180 degree alignment, mature to brown in summer. Twigs are glabrous, smooth, reddish brown. Bark is thin, brown, smooth at first, soon becoming shallowly furrowed.

A hardy shade tree most commonly used in the cooler northwest as a street tree or in parks. Tolerates warmer climates with moderate moisture, as a lawn specimen tree, though shorter lived in these circumstances. Aphid infestations often produce honeydew drip, and sooty mold can discolor leaves. Deep tap root, but heavy surface roots seek water, which may heave paving. Longevity estimated to be 100-125 years. 'Almira' has a small rounded form, to 20', 'Cleveland' is an improved form, to 50', 'Emerald Queen' has fast growth, with good form, 'Millers Superform' has tall, fast growth, and 'Summer Shade' is fast growing and tolerates heat well.

Cultivar. SHD, FOL ACC; IRR: M/M/-/H/-/-

Acer platanoides 'Schwedleri'

SCHWEDLER MAPLE

Aceraceae. Sunset zones 1-9, 14-17. Deciduous. Cultivar of the species, dating to the mid-1800s. Growth rate moderate to 50-60' tall by two-thirds to equally as wide, becoming broad-crowned, with dense foliage. Leaves are opposite, simple, 3-5" wide, purplish red in spring, deepening to dark bronzy green, with 5 shallow, thinly pointed, tail-like lobes, sparsely long-toothed along the edges, with lighter undersides, hairy at vein angles, a long, slender petiole that exudes milky sap when cut, and golden yellow fall color. Small clusters of tiny, yellowish green flowers occur in spring among emerging reddish leaves, male and female flowers usually on different trees. Clusters of hanging, 1-1/2 to 2" long, paired, green-winged seeds mature to reddish brown in summer, wings bow-shaped, in slightly less than 180 degree alignment. Twigs are smooth, glabrous, reddish brown. Bark is thin, brown, and smooth or shallowly furrowed.

A rather uncommon accent tree, sometimes seen in California as a park or residential lawn specimen, though more commonly planted in the cooler northwest. Attractive purple foliage a useful alternative to widely planted purple-leaf plums. Effective in semi-shade among taller canopy trees. Longevity estimated to be 100-125 years in favorable conditions. 'Crimson King' and 'Faassens Black' are similar varieties with improved form and hold their purple foliage color better.

Cultivar. SHD, FAL COL; IRR: M/M/-/H/-/-

Acer platanoides 'Deborah'

PP #4944 (1982)

DEBORAH'S MAPLE

Aceraceae. Sunset zones 1-9, 14-17. Deciduous. Cultivar of the species. Growth rate moderately fast to 50-60' tall, with a dense, symmetrical, upright form. Leaves are opposite, simple, 3-5" wide, with 5 shallow thinly pointed tail-like lobes, sparsely toothed edges, and lighter undersides, new growth bright red becoming dark green, somewhat bronzy, with golden yellow to orange fall color.

A good shade tree, tolerating moderate heat with adequate moisture. 'Emerald Queen' is similar. Longevity estimated to be 100-125 years.

e Asia. SHD; IRR: M/M/M/M/-/-

Acer paxii

PAXII EVERGREEN MAPLE

Aceraceae. No Sunset zones. Semi-evergreen to deciduous. Native to Japan and Korea. Growth rate slow to 30-45' tall and two-thirds as wide, forming a dense, broad, oval crown, a fairly heavy, upright trunk, and casting deep shade. Leaves are opposite, simple, 1-1/2 to 2-1/2" long, leathery green, predominantly 3-lobed in the juvenile form, oblong-elliptical in the adult form, with a short tail-like tip, slightly pendent, with bronzy-pink new growth. Insignificant short panicles of tiny, greenish flowers appear in spring, followed by pendent samara seeds, paired, with wings at right angles. Bark is smooth and gray on young trees, thickening and darkening with age, becoming rough and furrowed.

An uncommon maple with an oaklike appearance. Only half-hardy, but tolerant of warm climates with moderate mois ture. Longevity estimated to be 100-125 years.

Cultivar. SHD, FOL ACC; IRR: M/M/-/H/-/-

Acer platanoides 'Drummondii'

VARIEGATED NORWAY MAPLE

Aceraceae. Sunset zones 1-9, 14-17. Deciduous. Cultivar of the species. Growth rate moderately fast to 50-60' tall with a broad crown and dense foliage. Leaves are opposite, simple, 3-5" wide, with a wide, white marginate variegation, 5 shallow, thinly pointed, tail-like lobes, sparsely toothed edges, lighter undersides, and golden yellow fall color.

A striking and unusual broad canopy accent tree, best in cooler climates, tolerating moderate heat, and requiring constant moderate moisture. Longevity estimated to be 100-125 years.

Cultivar. SHD, FOL ACC; IRR: No WUCOLS

Acer pseudoplatanus 'Atropurpureum'

WINELEAF MAPLE

Aceraceae. Sunset zones 1-9, 14-20. Deciduous. Cultivar of the species native to Europe and western Asia. Growth rate moderate to 40' tall and two-thirds to equally as wide, with an upright or broad oval form. Leaves are opposite, simple, 3-6" long, with 5 ovate lobes, short-pointed, with wavy, sawtooth margins, 5 main veins from the cordate base, veins sunken on the glabrous, dark green upperside, raised and sometimes slightly hairy on the rich, dark purple underside, new leaves bronzy green fading to dull dark green, with a long, slender, yellowish petiole, and dull yellow fall color. Flowers are yellow-green, 1/8-1/4" wide, with 5 tiny petals, in pendulous, narrow-branched, 3-6" long panicles, occurring in spring among leaves. Paired, 1" long, drooping, wide-winged, yellow-green key seeds are joined at roughly 60-90 degrees, each pair short-stalked from a thickened main yellow stalk, maturing to reddish tan in late summer. Twigs are smooth, hairless, gray, with bright green winter buds. Bark is smooth, gray, darkening to gray-brown and developing broad, flaky scales with age.

A desirable shade tree, most commonly grown along the north coast and in the northwest. Well suited as a park tree or among other shade trees, where it becomes taller if semi-shaded, though not as full and shapely. Best with moderate water in well-drained, fertile soils. Longevity estimated to be 100-125 years or more.

c&e N.A. SHD, STR, FAL COL; IRR: M/H/H/H/-/-

Acer rubrum

RED MAPLE

Aceraceae. Sunset zones 1-9, 14-17. Deciduous. Native to the central and eastern U.S. Growth rate moderate to 60' tall or more with a 40' spread or greater, developing an upright to oval symmetrical form. Leaves are opposite, simple, 2-4" long, shiny green, glabrous, with 3 short broad lobes, toothed margins, pale green undersides, and brilliant red fall color in frosty areas. Small, polygamous flowers with reddish petals and long yellow stamens occur in small, dense fascicles in late winter before leaves emerge. Attractive reddish green, wide-winged key seeds are 1/2-1" long, paired, joined at roughly 60 degrees, wings slightly divergent, hanging by slender stems from branchlets, maturing to tan in late spring. Twigs are slender, shiny red, attractive in winter, and have no fetid odor when bruised, unlike some other maples. Bark is smooth, whitish gray, darkening to gray-brown and developing shallow fissures with age.

A desirable street or shade tree, best with moderate water in well-drained soils. Effective grouped in lawns, as a foreground to an evergreen backdrop, or in small groves in park settings. Longevity estimated to be 250 years or more. Many cultivated forms of varying shape, color, and hardiness are available, and may be either male or female clones.

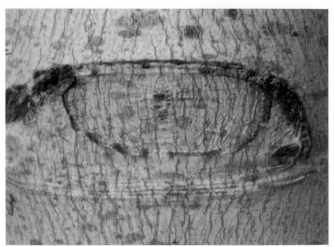

Cultivar. SHD, STR, FAL COL; IRR: M/H/H/H/-/-

Acer rubrum 'Columnare'

COLUMNAR RED MAPLE

Aceraceae. Sunset zones 1-9, 14-17. Deciduous. Cultivar of the species, introduced around 1880. Growth rate moderate to 60' tall or more by 10' wide, broadly columnar with upright branching. Leaves are opposite, simple, 2-4" long, glabrous, shiny green, with 3 short broad lobes, toothed margins, pale green undersides, and orange-yellow fall color in frosty areas. Insignificant staminate flowers do not set seeds. Attractive twigs and branchlets are red. Bark is relatively smooth and gray-brown, sometimes with a whitish cast, becoming lightly furrowed.

Clean, dramatic street, parkway, or vertical accent tree, especially effective in groves or rows that accentuate upright form. Best with regular watering, and well-suited for lawn areas, though some surface rooting develops. Attractive winter silhouette. Tolerates valley heat with moderate watering. Longevity estimated to be 200-300 years.

e N.A. SHD, FAL COL; IRR: M/M/-/M/-/-

Acer saccharinum

SILVER MAPLE

Aceraceae. Sunset zones 1-9, 12, 14-24. Deciduous. Native to eastern North America. Growth rate fast to 40-100' tall and wide, with an oval canopy and fairly open, upright branching, very twiggy at the ends. Leaves are opposite, simple, palmate, 3-6" wide, dark green, deeply 5-lobed, with pointed ends and sharply toothed margins, undersides silvery, and fall color yellow to orange to red, often in combination. Tiny, polygamous, apetalous, short-stemmed, reddish to yellow-green flowers in small, dense fascicles occur in late winter before leaves. Light brown, 1-2" long, paired, winged key seeds are joined at roughly 180 degrees, wings bowed downward, often with one malformed, hanging by long, slender stems from branchlets and maturing in late spring. Bark is smooth, silver-gray, turning darker gray on older trees, with narrow fissures and peeling scaly plates.

Fast-growing shade tree with attractive foliage and fall color. Least desirable of all maples for urban use because of invasive surface roots, weak branching structure, and susceptibility to aphids and cottony scale. Longevity estimated to be 50-80 years, then slowly losing vigor. Many cultivars, including weeping, gold-leaved, variegated, finely dissected or fern-leaved, upright, oval, and seedless forms, more often used in the eastern U.S.

c&e N.A. SHD, FAL COL; IRR: M/-/-/-/-/-

Acer saccharum

SUGAR MAPLE

Aceraceae. Sunset zones 1-10, 14-20. Deciduous. Native to central and eastern North America, with subspecies in Mexico and Guatemala. Growth rate moderate to 60' tall or more and up to 40' wide, with a stout trunk and upward-sweeping main branches, forming a tall oval crown. Leaves are opposite, simple, palmate, 3-6" long, bright green, usually 5-lobed, rarely 3-lobed, glabrous, with smooth or sparingly wavy-edged margins, pale green undersides, and brilliant yellow, orange, scarlet, or deep red fall color with an almost electric glow in areas of heavy frosts. Insignificant, tiny, yellow-green, apetalous flowers are polygamous, in crowded corymb clusters, hanging on long, hairy, pendulous stems among leaves in spring. Paired, 1" long, light brown, winged key seeds are joined with wings nearly parallel, and hang by long, slender stems from branchlets. Twigs are slender, shiny, smooth, reddish brown. Bark is thin, grayish to greenish, smooth, thickening and becoming dark grayish brown, with deep furrows and long, scaly plates.

An important forest tree on the east coast, prized for its sap and hard wood. Wood is useful in woodworking, whereas *Acer saccharinum* has soft, brittle wood. Attractive as a street, park, or lawn tree, but in California limited to foothill areas, where brilliant fall color is an attraction. Prefers deep fertile soil with moderate watering to become established. Longevity estimated to be 150-250 years. The leaf is Canada's national symbol. 'Cutleaf' has finely cut leaves and casts light shade, and 'Green Mountain' has the best shade qualities and drought tolerance.

e Asia, SHD, FAL COL; IRR: No WUCOLS

Acer tataricum ssp. *ginnala*

AMUR MAPLE

Aceraceae. Sunset zones 1-9, 14-16. Deciduous. Subspecies native to Japan, northern China, and Manchuria. Growth rate slow to moderate to 15-20' tall and wide, with slender, upright branches forming a round-topped oval canopy. Leaves are opposite, simple, 2-3" long by 1-2" wide, shiny dark green, glabrous, 3-lobed near the middle, middle lobe the longest, subcordate at the base, with doubly serrate edges, pale undersides, a 1 to 1-1/2" long red petiole, reddish main leaf vein, and bright orange-red fall color. Small, fragrant, drooping, long-stemmed, 1-1/2" yellowish flower clusters occur in April to May. Dense clusters of paired yellow-winged samara seeds, paired oppositely, hang on trees throughout summer, 1" long wings drooping to nearly parallel, turning reddish in fall. Bark is grayish brown, peeling in thin, vertical strips.

An uncommon but attractive small street or lawn tree, not particularly noticeable until the brilliant fall color, especially in colder areas. Very hardy. Drought tolerant when established. Relatively pest free. Multi-trunk specimens have more character in a garden setting, but usually do not become as tall. This subspecies has sharper, shinier, narrower leaves than the species, which is less often grown. Longevity estimated to be 100-150 years.

se Europe, w Asia. SHD, FOL COL; IRR: No WUCOLS

Acer tataricum

TATARIAN MAPLE

Aceraceae. Sunset zones 1-6, 14-16. Deciduous. Native to
southeastern Europe and western Asia. Growth rate slow to
moderate to 20-25' tall and wide, with a bushy, oval canopy.
Leaves are opposite, simple, 2-3" long by 1-2" wide, shiny dark
green, with 3 or 5 shallow lobes cut nearly to the middle of the
leaf, somewhat star-shaped, or without lobes on older trees,
with doubly-serrate wavy edges, downy at the veins on pale
undersides, subcordate at the base, from 1 to 1-1/2" long red
petioles, new leaves with a reddish tinge, turning green in summer, with brief bright yellow to orange-red fall color, especially in colder areas. Small, loose, 1-1/2" clusters of drooping,
long-stemmed, fragrant, yellowish flowers occur in April to
May. Dense clusters of yellow-winged samara seeds, paired
opposite, with 1" long wings, turn reddish and hang through
summer, wings drooping to nearly parallel. Bark is grayish
brown, peeling in thin, vertical strips.

   An attractive small street or lawn tree, most striking in fall
color. Very hardy. Drought tolerant when established. Relatively pest free. Multi-trunk specimens have more character
in garden settings. Longevity estimated to be 100-150 years.

Cultivar. SHD, FAL COL; IRR: M/M/-/H/-/-

Acer truncatum 'Pacific Sunset'

= *Acer truncatum* x *Acer platanoides* 'Pacific
Sunset'
PP #7433 (1991)

PACIFIC SUNSET MAPLE

Aceraceae. Sunset zones 1-9, 14-23. Deciduous. Hybrid selection, developed in Oregon. Growth rate moderate to 25-40'
tall, forming a tall upright crown, more spreading with age.
Leaves are opposite, simple, palmately lobed, dark green, 2-3"
long, with 5 broadly triangular pointed lobes and small-
pointed secondary lobing, smooth edges slightly recurving or
wavy, new foliage with a reddish tinge along the edges, and
yellowish orange to bright red fall color. Clusters of tiny, long-
stemmed, reddish flowers with many yellow stamens occur in
early spring. Paired winged seeds are infertile. Twigs are smooth
and greenish brown. Bark is smooth, reddish gray, becoming
fissured and slightly scaly with age.

   An attractive variety with leaf size and height midway between each parent, selected for excellent fall color. Usually has
an upright form, unless multi-trunked, which gives a more
spreading shape. Prefers moderately moist, well-drained, fertile soils. Does well in lawns, especially in warmer climates.
Longevity estimated to be 100-150 years.

e Asia. SHD, ORN, SPC, FAL COL; IRR: M/M/-/H/-/-

Acer truncatum

PURPLEBLOW MAPLE

Aceraceae. Sunset zones 1-9, 14-23. Deciduous. Native to northern China. Growth rate moderate to fast to 25' tall with an equal or nearly equal spread, developing a broad oval crown, more spreading with age, often multi-trunked or low-branched. Leaves are opposite, simple, palmately lobed, 1-1/2 to 2" long, shiny dark green, with 5 triangular pointed lobes, smooth edges, glabrous except for the slightly hairy petiole, which exudes a milky sap when cut, new growth tinged purplish along the sometimes wavy edges, later turning green, with yellow-orange to purplish red fall color. Loose, long, green-stemmed panicle clusters of bright yellow, buttercuplike flowers occur in spring with or before new leaves, male and female flowers opening at different times, rarely fertilizing. Clinging clusters of paired, winged seeds, with wings set at 90 degrees or slightly more, are not especially noticeable but flutter in breezes among the foliage. Slender twigs are smooth and reddish brown. Bark is rough, grayish brown, becoming broadly fissured with reddish highlights and broad, flat, cracked ridges.

A rather uncommon but attractive small park, lawn, or patio tree, most effective trained as a multi-trunk specimen. Often shorter and more broadly round-headed than most upright Japanese maples, and the yellow-petaled spring flowers are most unusual for a maple. Prefers moderately moist, fertile soils, especially in lawns, either semi-shaded or in part sun among taller trees. Longevity estimated to be 150 years or more.

w N.A. (Ca only) California Native: CP, VG, OW, FW, CH (Semi-Riparian) (*c&s NW, n&c CW, s CaR, SNF, Teh, sw Dmoj, GV foothills*); NAT, SHD; (IRR: VL/VL/VL/L/-/-)

Aesculus californica

CALIFORNIA BUCKEYE

Hippocastanaceae. Sunset zones 3-10, 14-24, Jepson *4-6,**7,14-17,19-24**;IRR:3,**8,9**,10,**18**;CVS. Deciduous. Native and endemic to north-facing river canyon slopes in California at 2,000 to 4,000' elevation. Growth rate slow to 10-20' tall or taller and 30' wide or wider, developing a broad oval, spreading canopy, usually stout or multiple-trunked and shrublike in its natural habitat, with leaves turning brown and withering on branches in summer dryness, beginning in June and remaining through February. Leaves are opposite, palmately compound, with long petioles and 5-7 oblong-lanceolate leaflets, 3-6" long by 1-1/2 to 2" wide, light green and short-petioled, nearly glabrous, with finely serrate edges. Showy, 12" long, raceme plumes of delicately fragrant, 1/2" white to light pink flowers with protruding stamens occur in May and June from buds of the previous year, only basal flowers fertile. Green, hanging, thin-skinned, 2-1/2 to 3" long, pear-shaped fruits split when ripe, releasing 1-2 shiny dark brown chestnut seeds. Bark is thin, smooth, gray, developing tiny raised scales.

Despite early leaf drop, can be desirable as an accent tree for showy, candelabralike flower display and striking winter silhouette. All plant parts are toxic, though squirrels seem to be immune. Seeds are poisonous if eaten raw. Pollen is toxic to honeybees. Disease-free, drought tolerant, and well suited for riparian plantings and native garden settings. Becomes leggy in shaded, moist conditions. Longevity estimated to be 50 to less than 100 years. Series associations: Intlivoakshr, Scroak & Scroakbirmou shrub series, Foopine, Mixoak.

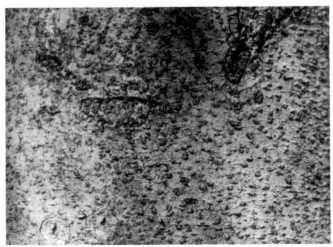

Cultivar. SHD, FLW ACC; IRR: M/M/M/-/-/-

Aesculus x *carnea* 'Briotii'

BRIOTII RED HORSECHESTNUT

Hippocastanaceae. Sunset zones 1-10, 12, 14-17. Deciduous. Cultivated variety. Growth rate moderate to 20-30' tall with a 20-25' spread, forming a broad to upright oval canopy from a short, stout trunk, and casting dense shade. Leaves are opposite, palmately compound, deep dark green, usually with 5 ovate-elliptical leaflets, 4-10" long by 2-3" wide, long-petioled and nearly sessile, with a glabrous crinkled surface, serrate to doubly-serrate edges, and yellow fall color. Showy, upright, densely clustered, 5-8" long panicle plumes of red, 1/4" bell-shaped flowers with yellow throat markings occur in April to May. Fruits are interesting 2" round, green-skinned, prickly capsules. Bark is smooth, grayish brown, becoming dark brown, shallowly fissured, and scaly.

A smaller variety of red horsechestnut with brighter red flowers and darker green foliage. Effective as a flowering accent or small shade tree. Requires moderate summer water. Leaf tips may scorch in hot, dry sun. Neat, clean appearance. Does exceptionally well in lawns as a flowering specimen. Longevity estimated to be 100-150 years.

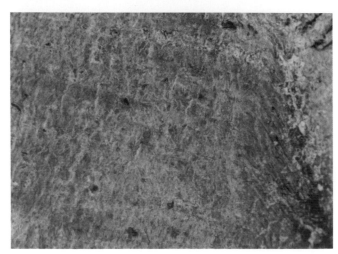

Cultivar. SHD, FLW ACC; IRR: M/M/M/-/-/-

Aesculus x *carnea*
RED HORSECHESTNUT

Hippocastanaceae. Sunset zones 1-10, 12, 14-17. Deciduous. Hybrid between *Aesculus hippocastanum* and *A. pavia*. Growth rate moderate to 40' tall with a 30' spread, forming a broad to upright oval canopy with upright branching from a straight trunk. Leaves are opposite, palmately compound, dark green, usually with 5 ovate-elliptical leaflets, 4-10" long by 2-3" wide, long-petioled and nearly sessile, with a glabrous crinkled surface, serrate to doubly serrate edges, and yellow fall color. Showy, upright, densely clustered panicle plumes of red, 1/4", bell-shaped flowers, 5-8" long, with yellow throat markings, occur in April to May after leaves. Fruits are interesting, 2" round, green-skinned, prickly capsules. Bark is smooth, grayish brown, becoming dark brown, shallowly fissured and scaly.

A desirable shade tree for small park groupings or as a residential accent or terrace tree. Requires moderate summer water. Leaf tips may scorch in hot, dry sun. Does exceptionally well in lawns, but surface roots may heave paving. Relatively pest and disease free. Longevity estimated to be 100-150 years or more.

e N.A. SHD, FLW ACC; IRR: M/M/M/-/-/-

Aesculus pavia
RED BUCKEYE

Hippocastanaceae. Sunset zones 2-9, 14-24. Deciduous. Native to the eastern U.S. from North Carolina to Illinois south to Missouri and Texas. Growth rate moderate to 12-20' tall and nearly as wide, developing a small, round-topped canopy from a rather slender trunk. Leaves are opposite, palmately compound, dark green, usually with 5 narrow, short-pointed, elliptical leaflets, 1-1/2 to 4" long, sessile or short-petioled, with a shiny glabrous surface, without hairs when mature, sparsely and finely serrate margins, and slight yellow fall color. Showy, loose, upright, 4-8" long panicle plumes of deep dark red flowers, each with a 1/2" elongated calyx tube, occur in April to May. A yellow-flowered form occurs in western Texas. Round, green-skinned, 2" fruits lack spines. Slender green twigs have bud scales with a prominent ridge and are not sticky. Bark is thin, smooth, grayish brown, often remaining so indefinitely.

An uncommon small foliage and flowering shade tree useful in small groupings or as a residential accent or terrace tree. Does well in lawns with moderate summer water. Hot sun does not scorch leaves when adequate moisture is available. Tolerates some drought if semi-shaded with occasional deep watering in well-drained soils. Produces best flowering and shapeliness in full sun, becoming leggy in heavy shade. Neat, clean appearance. Mildew resistant and relatively pest and disease free. Longevity estimated to be 50-80 years.

e Europe. SHD, FLW ACC; IRR: No WUCOLS

Aesculus hippocastanum

HORSECHESTNUT

Hippocastanaceae. Sunset zones 1-10, 12, 14-17. Deciduous. Native to eastern Europe and Mediterranean regions. Growth rate moderate to 60' tall with a 40' spread, developing a tall oval form, with dense foliage casting heavy shade. Leaves are opposite, palmately compound, light green, long-petioled, with 5-7 obovate pointed leaflets, 4-10" long by 2-3" wide, sessile or nearly so, with doubly serrate edges, pale undersides, rusty tomentose on new leaves, and yellow fall color. Showy, upright, 8-12" long panicle plumes of 3/4" bell-shaped flowers, creamy white with yellow to red throat markings, in dense terminal clusters, occur in spring after leaves. Hanging, long-stemmed clusters of conspicuous 2" round, green-skinned capsule fruit, with numerous short, sharp spines, dry to rich brown in fall and split along one side, releasing 1-2 shiny dark brown chestnut seeds. Twigs are stout, light brown, hairless, with large sticky buds. Bark is smooth, grayish brown, becoming dark brown, shallowly fissured, and scaly.

Not often used in urban situations, but makes an excellent large lawn or parkway tree, except for the prickly seed husks. Prefers well-drained, moist, fertile soil. Leaves may scorch in hot sun. Surface roots may heave paving. Longevity estimated to be 100-200 years. 'Baumannii' has double flowers and sets no fruit.

Australia. EVG, FOL ACC; IRR: M/-/M/H/-/-

Agathis robusta

DAMMAR PINE

Araucariaceae. No Sunset zones. Evergreen. Native to Australia. Growth rate slow to moderate to 40-70' tall or more with a 10-15' spread, developing a narrow columnar form from a strong central trunk with slender horizontal branching, twiggy at the ends, and dense foliage with a layered effect of distinct open spaces between tufts of clumped branches and leaves. Leaves are alternate, nearly opposite, simple, 2-4" long, thick and fleshy, glossy dark green, new growth light green to pinkish copper, ovate-lanceolate, with parallel lengthwise veining, bluntly pointed ends, smooth rounded edges, and lighter undersides, persisting 2-3 years. Male and female flowers are on separate trees, and 4" long oval cones, which contain unequally winged seeds, are rarely seen. Bark is smooth, grayish white, with bulging horizontal growth rings left from previous branch markings, covered with tiny flaky scales, and yielding a pitchy sap called dammar resin when cut.

A tender, uncommon specimen tree for parks or large gardens. Requires moderate to ample moisture. Large lawns suit it perfectly, with room for roots to spread. Retains best shape in full sun. Lower branches may die back with age, but new growth sprouts from older bark. Frosts below 25 degrees may damage foliage at the base of the tree, but trunks quickly resprout new growth on established trees. Shiny foliage sparkles after rains and flutters in breezes. Longevity estimated to be 80-100 years or more.

Australia. EVG, SHD; IRR: L/-/L/M/-/-

Agonis flexuosa
PEPPERMINT TREE

Myrtaceae. Sunset zones 15-17, 20-24. Evergreen. Native to wetter regions of southeastern Australia. Growth rate slow to moderate, eventually to 25-35' tall with a 15-30' spread, forming a broad oval canopy, more irregularly heavy-branched with age, with heavily weeping branches. Leaves are alternate, simple, 4-6" long by 1/4" wide, narrow linear-lanceolate, willowlike, dull dark green on both sides, dotted with pellucid oil glands, slightly curved or sicklelike with long, tapered, pointed ends, cuneate at the base, with smooth edges, exuding a peppermint odor when crushed, juvenile leaves broader, and new growth reddish bronze. Many tiny, white, star-shaped flowers occur in pubescent umbel heads of 10-15 in June at axils of the previous year's growth. Stemless, woody, brown fruits are aggregate clusters of 6 to 9 3-valved capsules, persistent along leaf axils into winter, containing many minute seeds. Twigs are very slender, reddish brown, and grow in a zigzag pattern. Bark is coarse and dark brown when mature, with many deep fissures, revealing reddish underbark, and flat, scaly ridges peeling in long, irregular, flat shreds.

A rather tender evergreen shade tree commonly grown in coastal areas as a specimen or in groupings or as a street or patio tree. Not a tree to walk under, since foliage weeps to the ground, but form and color are attractive from a distance. Tolerates moist, sandy soils. Best protected from high winds. Not hardy below 27 degrees. Longevity estimated to be 75-90 years.

e Asia. IRR: VL/VL/L/L/L/L Extremely invasive and weedy, its use is strongly discouraged.

Ailanthus altissima

TREE-OF-HEAVEN

Simarubaceae. Sunset zones 2-24. Deciduous. Native to northern China and naturalized throughout most of the U.S. Growth rate very fast to 50' tall and nearly as wide, developing a tall oval, round-topped canopy with a large trunk and limbs, most often multi-trunked. Leaves are alternate, pinnately compound, 1-3' long by 6-10" wide, divided into 13-25 bluntly pointed, 3-5" long, ovate-lanceolate leaflets, with smooth margins except for 1-3 glandular teeth at the base, and yellow to reddish fall color. Male and female flowers occur in panicles in early summer on separate trees. Inconspicuous, tiny, round, yellowish male flowers in large loose clusters have an objectionable odor. Female flowers produce large, noticeable clusters of reddish, oblong, semi-twisted, winged samaras, 1 to 1-1/2" long, connate-perfoliate, with a single, small, flattened seed in the center, drying to light brown and persisting into winter. Twigs are stout, reddish brown, with short velvety hairs. Bark is thin, smooth, gray, becoming shallowly fissured and furrowed with age.

An excaped exotic in urban areas, where it thrives under adverse conditions, as well as in foothill and valley riparian areas, where it forms thickets nearly impossible to eradicate. Considered a weed tree, it readily self-seeds, and numerous suckers develop as far as 100' from the trunk if invasive roots are cut. Allelopathic, sending out chemicals toxic to some other plants. Longevity estimated to be 50-75 years.

Australia. SHD, FLW ACC; IRR: L/-/L/-/-/- It is invasive, and its use is cautioned.

Albizia distachya

PLUME ALBIZIA

Fabaceae. Sunset zones 15-17, 22-24. Semi-deciduous. Native to Australia. Growth rate fast to 20' tall and equally wide or wider, developing an upright, irregular form with a rounded crown or more often rather shrublike with foliage to the ground. Leaves are fernlike, dark velvety green, alternate, bi-pinnately compound, 12-18" long by 4-6" wide, with 14-24 pinnae, each with 40-60 oblong-lanceolate leaflets, 1/4-3/8" long, persisting into the following year, if winters are not cold, with glabrous to rusty tomentose new growth. Clusters of greenish yellow, compound, central-stalked flowers are densely paired in fluffy 1-2" long spikes at branch ends sporadically throughout summer. Broadly flattened, 3" long, reddish brown legume seed pods hang into winter. Bark is smooth, light gray-ish brown, becoming shallowly fissured and rough textured.

A small flowering accent tree especially suited to coastal areas, where it has naturalized by self-seeding in sand dunes. Tolerates salt spray and coastal drought, but does not tolerate inland dryness and heat. Not hardy below 32 degrees. Longevity estimated to be 30-50 years.

Asia. SHD, FLW ACC; IRR: L/L/M/M/M/M

Albizia julibrissin

SILK TREE

Fabaceae. Sunset zones 4-23. Deciduous. Native to Asia from Iran to China and Japan and naturalized in the eastern U.S., where it is often referred to as mimosa. Growth rate moderate to fast to 40' tall with an umbrellalike canopy often twice the height and a stout, low-branched trunk with upward-arching branches, twiggy at the ends. Leaves are alternate, bipinnately compound, 6-12" long by 4-5" wide, with 8-26 pinnae, each with 30-40 leaflets, which fold closed at night, petioles with a gland 1/2-1" from the base, and yellow to reddish fall color. Fluffy, pink to reddish, many-stamened flowers occur at branch ends in clustered heads, showy in midsummer, attracting birds and bees. Flat, 6" long by 1" wide legume seed pods are pointed at the ends, maturing to golden tan, containing several flattened, hard, shiny, tan seeds, and hanging from branches into winter. Bark is smooth, grayish brown, covered with tiny glandular ridges, becoming shallowly fissured with age.

Commonly used as a showy flowering accent, though use has diminished because fallen leaves and flowers disintegrate into tiny pieces and trees leaf out quite late in spring. More attractive as a multi-trunked tree, especially against a contrasting backdrop of upright evergreens or conifers. Casts light filtered shade with a pleasant airy feeling. Does well in summer heat, and tolerates drought when established, but prefers occasional deep watering. Longevity estimated to be 75-125 years.

s Europe. SHD; IRR: M/M/M/M/-/-

Alnus cordata

ITALIAN ALDER

Betulaceae. Sunset zones 2-9, 14-24. Deciduous. Native to southern Italy and Corsica. Growth rate moderate to 40' tall with a 25' spread, developing a conical to pyramidal shape, denser, stouter, and less open than *Alnus rhombifolia.* Leaves are alternate, simple, 2-3" long by 1-2" wide, heart-shaped, glossy green, not folded in bud like other alders, smooth and glabrous, with finely serrate edges, pale undersides, brown hairs along the midrib, and dull yellow if any fall color. Insignificant, yellowish green, tassel-like male flowers, 4-5" long by 1/8" wide, occur in spring before leaves appear. Female flowers form clusters of 1" long, narrowly ovate, dark brown, conelike fruits, the ends of cone scales thick, blunt, and squarish. Cones mature in fall of the first season, persisting through winter and releasing 1/8" flat, thin-skinned, brown seeds with narrow, thin, winglike margins. Bark is thin, smooth, grayish to whitish, thickening with age and developing a rough, dark brown, warty surface.

A commonly used shade tree, effective in groupings in lawns or parkways. Relatively short deciduous period. Tolerates heat with moderate moisture in well-drained soils. Grows in poor soils, either acidic or slightly alkaline. Tolerates some drought with occasional deep watering, but growth becomes stunted. Longevity estimated to be 40-60 years in cultivation, losing vigor with age.

w N.A. California Native: CM, CP, RW, CF, ME (Riparian) (*NCo, NCoRO, CCo, SnFrB*); SHD; (IRR: H/H/-/-/-/-)

Alnus oregona

RED ALDER

Betulaceae. Sunset zones 3-7, 14-17, Jepson *WET:**4-7,16, 17**,24&SHD:2,3,14,**15**,20-23:STBL,timber. Deciduous. Native to streams and rivers of the California Coast Ranges north to Alaska. Growth rate fast to 45-50' tall with a 20-30' spread, developing a broad-based, open, oval shape, or conical to pyramidal, with a rounded crown, a fairly heavy trunk, and thin, upward-arching branches with drooping ends. Leaves are simple, alternate, 3-6" long, smooth, deep yellow-green, finely toothed in regular rounded crenate fashion, edges curving slightly under, lighter undersides with tiny rust-colored hairs, heaviest along prominent yellow midveins, and also on branchlets, and insignificant yellow fall color. Yellowish green, tassel-like male flowers, 5-6" long by 1/8"wide, hang from bare branches through winter. Female flowers form clusters of narrowly ovate, dark brown, 1" conelike fruits, with ends of cone scales thick, blunt, and squarish. Fruits mature in fall of the first season, releasing many 1/8" flattish, brown, thin-skinned seeds with narrow, winglike margins. Twigs are triangular, light green, slightly hairy at first. Bark is thin, smooth, grayish to whitish, often with mottled colorations, becoming shallowly seamed, with flat thin ridges near the base of the trunk.

A common northwestern native near streams and rivers where roots can reach the water table. Often forms tall, dense thickets in clearings between dense conifers. Quickly reseeds wherever space and moisture are available. Fairly wind tolerant along the coast, but leaves become torn along the edges. Occasionally cultivated in parks in the Central Valley. Longevity estimated to be 50-75 years. Series associations: Pacreedgrass, Narwillow, Salblahuc, Sanwillow, Yelbuslup herbaceous & shrub series; Blacottonwoo, Grafir, Hoowillow, Mixwillow, Pacwillow, **Redalder**, Sitspruce, Sitwillow.

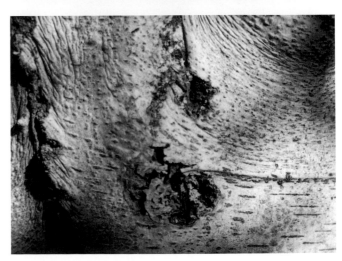

w N.A. California Native: CP, VG, OW, FW, ME, PF (Riparian) (*CA-FP, MP*); **NAT, SHD;** (IRR: H/H/H/H/H/-)

Alnus rhombifolia

WHITE ALDER

Betulaceae. Sunset zones 1-10, 14-21, Jepson *SUN:**1-3**,4-6,**7**, 8,**9**,10,**14-18**,19,**22-24**;STBL. Deciduous. Native to valley and foothill streams and rivers of California north to British Columbia. Growth rate very fast to 50-90' tall with a 40' spread, developing a broad-based, pyramidal or tall, open, broad oval shape, with a rounded crown and a heavy trunk base free of branches. Leaves are simple, alternate, 2-4" long, light green, finely toothed or rarely double-toothed, with gland-tipped variously sized teeth, lightly wavy edges curving under slightly, pale undersides minutely hairy as are prominent yellow midveins and branchlets, and barely noticeable yellow fall color. Male flowers are 4-5" long by 1/8" wide and hang in pendulous yellowish green catkin clusters, producing an almost weeping effect through winter. Female flowers develop into clusters of narrowly ovate, dark brown, 3/4" conelike fruit, the ends of the cone scales thickened, with an intended lobe, maturing in fall of the first season and releasing 1/8", flattish, brown, thin-skinned seeds. Bark is thin, grayish, becoming brown and conspicuously scaly, even on some higher branches.

A common native along foothill and mountain streams, with Fremont cottonwood, where it withstands winter flooding and reseeds in nearly any moist spot. Commonly cultivated as a shade tree for lawns in parkways or parks where it receives adequate moisture. Especially effective in informal groves with random spacing and size. Shallow watering encourages invasive surface roots. Tolerates heat and wind. Longevity estimated to be about 30-50 years. Series associations: Narwillow & Sanwillow shrub series; Blawillow, Calsycamore, Doufirtan, Hoowillow, Mixwillow, Pacwillow, Redwillow, Sitwillow, **Whialder.**

s South America. SPC, FOL ACC; IRR: L/M/-/M/-/-

Araucaria araucana

MONKEY PUZZLE TREE

Araucariaceae. Sunset zones 4-9, 14-24. Evergreen. Native to western slopes of the Andes Mountains of southern Chile and southwestern Argentina. Growth rate slow to 70-90' tall and 30' wide, with a rangy conical to oval form and heavy, brittle, spreading, ropelike branches. Leaves are simple, closely overlapping, 1-2" long needles, dark green, whorled around branchlets, ovate-lanceolate and sharply pointed, hard or leathery. Male and female flowers are on separate trees, males grouped, brownish, fuzzy-looking, conical, producing pollen in July. Larger, erect, green, 4-7", oval female cones are solitary at ends of terminal branches, ripening to brown in fall of the second year. There may be 100 or more oblong, slightly compressed, edible seeds in each cone, though they are rarely seen, except on mature trees. Bark is reddish to gray-brown, wrinkled, with shallow reddish seams marked with rings from old branch scars.

An uncommon, dramatic curiosity, often seen in older towns or parks in northern coastal areas. Also occasionally cultivated elsewhere, usually as a lawn specimen, away from pedestrian areas where sharp needles can be dangerous if left on the ground. Exceptionally tough, tolerating drought when established, and not susceptible to diseases. Hardiest of the araucarias. Longevity estimated to be 100-200 years.

Australia. SPC, FOL ACC; IRR: L/M/M/M/-/-

Araucaria bidwillii

BUNYA-BUNYA

Araucariaceae. Sunset zones 7-9, 12-24. Evergreen. Native to southeast Queensland, Australia. Growth rate slow to moderate to 80' tall or more with a 60' spread, developing a symmetrically pyramidal form, more irregular with age, dense foliage of young trees casting heavy shade and very old trees with more sparse, irregular, tightly whorled foliage on long, slender, slightly bowed branchlets. Leaves are whorled, simple, shiny, stiff, dark green, 1-2" long, appearing to lie flat along each side of the stem, or 2-ranked, due to the twisting of short stalks, with very sharp points. Catkinlike male flowers and conelike female flowers are usually borne on different trees. Female trees produce heavy, green, 10" long by 6-8" wide, ovoid, pineapplelike cones, which mature in summer of the second year and in 5-7 year cycles thereafter. Cones can weigh 10-15 pounds, with thick deciduous scales up to 3" wide, which disintegrate when dry, after the cone falls. Each scale contains one edible winged seed. Bark is light grayish brown, scaly in youth and covered with warty protuberances, becoming smoother with age and weathering to brown or gray.

As a curiosity, makes an imposing specimen. Exceptionally tough, not susceptible to diseases, and tolerates drought when established. Sharp leaves and falling cones are hazardous. Longevity estimated to be 100-200 years.

Pacific Is. SPC, FOL ACC; IRR: M/M/M/-/-/-

Araucaria heterophylla

NORFOLK ISLAND PINE

Araucariaceae. Sunset zones 17, 21-24. Evergreen. Native to Norfolk Island in the Pacific. Growth rate slow to moderate to 100' tall and 60' wide, with a symmetrical pyramidal form and distinct horizontal branching, foliage appearing layered, in flat angular sprays, in starlike arrangement from a strong central trunk. Leaves are needlelike, 1/4-1/2" long, soft to the touch, surrounding branchlets in densely regular spacing, triangular in cross-section, with a barely visible midrib, curving inward near the bluntly pointed tips. New growth on very old trees takes on a distinctly different appearance, in spiral fashion, in overlapping whorls. Globular, 4-6" long cones are occasionally seen on very old trees. Bark is thin, smooth, greenish brown, becoming whitish with horizontal lines of black protuberances and areas of thin papery skin peeling or rolling back in small sections.

Dramatic lawn or park specimen for temperate coastal climates. Commonly cultivated in groves in Hawaii, usually grown as a potted plant on the mainland, except for the ones pictured here, growing in the east San Francisco Bay Area. Requires moderate water and fertilization. Relatively free of pests and diseases. Longevity estimated to be 100-200 years. 'Albospicata' is known as silver star araucaria.

California Native: OW, CC, RW, CF, PF (*NW, CaRH, n&c SNH, CW, n ChI, WTR, SnGb, PR*); NAT, SHD; (IRR: VL/VL/-/-/-/-)

Arbutus menziesii

MADRONE

Ericaceae. Sunset zones 4-7, 14-19, Jepson *DRN,DRY&SUN: 4,**5**,6,15,17;DRN,IRR:3,7,15,24 &SHD:9,14,18,23;DFCLT. Evergreen. Native to British Columbia south to the Sierra Nevada foothill and mountain ranges from 300-5,000' elevation. Growth rate slow, with mature heights ranging from 20-100', developing either a tall oval form or a shorter, broad, rounded head in a sunny exposure, often multi-trunked. Leaves are simple, alternate, 3-6" long, leathery dark green, with shiny smooth edges, glabrous, light green undersides, persisting into the following season, yellowing and falling in late spring after new leaves appear. Occasional vigorous suckering juvenile leaves have toothed edges and a minutely hairy appearance. Showy clusters of perfect, bisexual, 1/4", bell-shaped, white to pinkish flowers occur in spring in semi-drooping terminal panicles. Semi-fleshy, rough-coated, 1/2", rounded, orange-red, drupelike berries, each with several small, angled, brown seeds, mature in fall and remain into winter. Bark is smooth, reddish brown, and peels back in paper-thin strips, though older bark at the base of the tree becomes rough, dark brown, with small, loose, peeling scales.

A fairly common though not predominant species, often mixed among conifers in semi-shaded gullies and ravines or in exposed clearings, where it has a much denser shape. Requires good drainage, acid conditions, slightly moist, well-drained soil on the heavy side, and benefits from heavy mulching. Does not tolerate wet soils. Exacting cultural requirements and a deep taproot with few fibrous roots make seedling transplantation nearly impossible. Longevity estimated to be 80-100 years in habitat. 'Marina', with rosy pink flowers, is available in nurseries and grows to 40' tall or less with an equal spread. Series associations: Deerbrush shrub series; Beapine, Bispine, Blaoak, Calbay, Canlivoak, Coalivoak, Doufir, Doufirtan, Intlivoak, Mixoak, Monpine, Orewhioak, Redwood, SanLucfir, Tanoak, Weshemlock, Whifir.

s Europe. EVG, SPC, FLW, FRU ACC; IRR: L/L/L/L/M/M

Arbutus unedo

STRAWBERRY TREE

Ericaceae. Sunset zones 4-24, often damaged in severe winters in zones 4-7, but worth the risk. Evergreen. Native to southern Europe and Ireland. Growth rate slow to 10-35' tall and wide, usually irregular multi-trunked from the base or with with lower branches trimmed up to expose somewhat twisted and gnarled trunks. Leaves are alternate, simple, 2-3" long, dark green with red stems, oblong-elliptical, with lightly serrate edges, pale undersides, persisting 1-2 years. Small, drooping, panicle clusters of whitish to pinkish, urn-shaped flowers occur in fall and winter. Plump, round, yellow, 1/2-3/4" fuzzy fruits, with fleshy pulp, edible but bland or tasteless, containing many tiny black seeds, ripen to red. Flowering continues as fruits ripen. Trunk and branches are rich reddish brown with thin, slender, peeling scales or strips.

A handsome small evergreen accent tree, especially when pruned to shape. Tolerates drought when established, as well as seashore exposure, hot, dry conditions, and pollution. Prefers acid soil. Hardy to 15 degrees. Longevity estimated to be 60-90 years. 'Compacta' and 'Elfin King' are picturesque, contorted, dwarf forms less than 5' tall, suited for containers or bonsai.

e Asia. SPC, FLW ACC; IRR: -/-/M/M/-/M

Bauhinia variegata

(*Bauhinia purpurea*)

ORCHID TREE

Fabaceae. Sunset zones 13, 18-24. Deciduous to semi-deciduous. Native to tropical Himalayan foothills, extending to the Malay Peninsula. Growth rate fast to 20-35' tall and wide, multi-trunked or staked and trained to tree form with an oval, twiggy canopy. Leaves are alternate, simple, heart-shaped, 2-4" long and wide, light to medium green, with 1" long petioles, glabrous, with an indented lobe at the end of the midrib rather than pointed, symmetrically appearing as mirrored sections, 7-11 veined from the base, with entire margins, pale undersides, and yellow fall color. Showy clusters of fragrant pink, white, or purple, broad-petaled, orchidlike flowers are 2-3" wide, occurring in sparse terminal or axillary panicles, from January to April if temperatures remain warm. Messy, 4-10" long, flattened legume pods follow, ripening to dark brown. Twigs are slender, green, with 1" long thorns concealed among foliage. Bark is smooth, greenish gray, thickening and becoming brownish gray, cracked, and fissured.

Often grown as an accent tree, attaining great size in southern California where winter temperatures remain warm. Prefers moderate water and fertilization to become established. Otherwise requires little care except occasional pruning to shape or thin dense foliage. Not reliably hardy below 22 degrees, and seldom grown in northern California, where it usually has the appearance of a redbud. Longevity estimated to be 100 years or more in frost-free areas. 'Candida', as pictured, has white flowers.

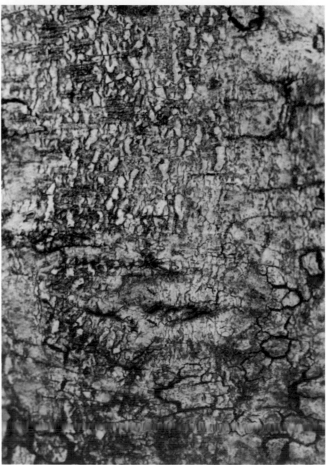

e N.A. SHD, ACC; IRR: H/H/H/H/-/-

Betula nigra

RIVER BIRCH

Betulaceae. Sunset zones 1-24. Deciduous. Native to the eastern U.S. Growth rate initially fast to 50-90' tall and 40-60' wide, developing a broadly pyramidal form, trunk often forking near the ground. Leaves are alternate, simple, 1-4" long, dull dark green, rhombic-ovate, with 5-9 paired veins, serrate edges, sometimes appearing sparsely lobed, glabrous, pale, minutely hairy undersides, wedge-shaped at the base, slightly aromatic, with orange-yellow fall color. Pendulous, monoecious, regular, apetalous, tassel-like flowers occur in early spring, with 2-3" long male flowers in clusters of 1-3 in slender pendulous aments and erect, solitary, 1/2" long females nearly invisible at ends of branch spurs below the males. Cylindrical, soft, fruiting catkins are 1 to 1-1/2" long, with rounded, overlapping, papery scale flaps, hairy, pendent from a slender central stalk, maturing to brown and opening in May or June of the first season, releasing a profusion of 1/16" seeds with thin wings on each side forming a semi-rounded outline, and a 2-tailed end. Twigs are smooth, lustrous, yellow-brown. Bark is pinkish tan, peeling in curled strips on young trees, becoming dark brown and furrowed with age.

An attractive alternative to European white-barked birches, suited to pond or terrace plantings in groves or as large multi-trunk specimens. Requires adequate moisture, and may produce surface rooting in lawns. Occasional aphid infestations may drip honeydew. Longevity estimated to be 80-100 years. 'Heritage', most commonly used, is resistant to leaf-spot and birch-borers.

w N.A. California Native: CF (Riparian) (*KR, CaRH, SNH, GB, Dmtns*); (IRR: H/-/H/H/-/-)

Betula occidentalis

WESTERN BIRCH, WATER BIRCH

Betulaceae. Sunset zones 1-7, 10, Jepson *WET:**1**-**3**,4-6,**7**,**9**, **10**,14-17&SHD:**18**,**19**,20-24;STBL. Deciduous. Native to the northwestern U.S. and the Rocky Mountains into Canada. Growth rate fast to 12-15' tall and wide, developing a broad, open crown with ascending branches, more often forming shrublike 20' high thickets in its natural habitat. Leaves are alternate, simple, 3/4 to 2" long, dull dark green, ovate, with sharply double-serrate edges or slightly lobed with acute to acuminate ends, smooth, with 3-5 paired veins, pale yellowish undersides, minutely glandular, rounded at right angles near the base to a ciliate and minutely glandular petiole, with pale yellow or orange fall color, sometimes reddish in the eastern Sierra Nevada. Pendulous, monoecious, regular, apetalous, tassel-like flowers occur in April, the 2" long yellowish males slender, pendulous aments in clusters of 1-3 and tiny, solitary, greenish, erect females at ends of branch spurs below the males. Stalkless, cylindrical, 1-1/2" long catkins mature to brown and open in fall of the first season, containing tiny 1/16" seeds with thin wings each side, forming a butterflylike outline. Twigs are clear, light yellowish brown, minutely hairy, with a lightly speckled glandular surface, and may be slightly sticky at first, later becoming shiny and glabrous. Bark is thin, shiny, dark bronze with pale horizontal lenticel markings, darkening with age and becoming shallowly furrowed, but not peeling.

A smaller western native riparian tree, not often noticed in its remote habitats, and not as tall as the eastern species, *Betula nigra*. More often cultivated in the Rocky Mountains, but adaptable to moist, well-drained soils in the San Francisco Bay Area as a small accent tree in a natural setting. Longevity estimated to be 50-90 years. Series associations: **Watbirch**.

n N.A. SHD, ACC; IRR: No WUCOLS

Betula papyrifera

PAPER BIRCH

Betulaceae. Sunset zones 1-6. Deciduous. Native to the northern U.S. into Canada and Alaska. Growth rate fast to 50-90' tall and half as wide, with an upright, open, pyramidal form, semi-weeping at the ends of rather slender branches, which are only slightly more upright than European birch. Leaves are alternate, simple, 1-3" long, dull dark green, ovate to oblong-ovate and unequally cordate at the base, with 9-11 paired veins, sharply double-serrate edges, glabrous uppersides, pale black-glandular or minutely hairy undersides, and golden yellow fall color. Pendulous, monoecious, tassel-like, apetalous flowers occur in March and April, males 4" long, in slender, pendulous aments in clusters of 1-3 and soliatary females 1 to 1-1/4" long at ends of branch spurs below the males. Pendent, cylindrical, fruiting catkins are 1-1/2" long, minutely hairy, with rounded, overlapping, papery scale flaps from a slender central stalk, maturing to brown in fall of the first season. Seeds are 1/16" long, with thin wings on the sides, forming a rounded outline. Twigs are green, hairy, becoming shiny orange-brown. Bark is creamy white, flaky, peeling in large papery sections and often covered with black markings.

Commonly used as an accent clump or irregularly spaced grove, including mixed sizes, with single and multi-trunk trees in a random informal pattern in rolling lawns with boulder outcroppings, near ponds or streams. Requires moderate moisture and fertile soils. Surface roots sucker in lawns. Suffers from occasional aphid infestations. Sometimes offered as *Betula alba*, supposedly to differentiate it from European birch. Names often used interchangeably, though species are different in character. Longevity estimated to be 50-90 years.

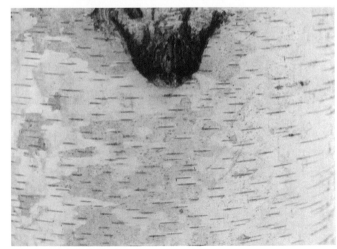

Europe, sw Asia, n Africa. SHD, ACC; IRR: H/H/H/H/-/-

Betula pendula
EUROPEAN WHITE BIRCH

Betulaceae. Sunset zones 1-12, 14-24. Deciduous. Native to Europe, southwestern Asia, and northern Africa. Growth rate moderate to fast to 30-40' tall by 15-20' wide, with an upright billowy form, branches recurving and often weeping to the ground. Leaves are alternate, simple, 1-3" long, dull dark green, rhomboid-ovate, with 9-11 paired veins, sharply double-serrate edges, smooth on both sides, with pale undersides, wedge-shaped at the base, and golden yellow fall color. Pendulous, monoecious, tassel-like flowers are apetalous, and occur in March and April, with 1-1/4" long males in clusters of 1-3 in slender, pendulous aments and 1/2-3/4" long, greenish females solitary at ends of branch spurs below the males. Pendent, cylindrical fruiting catkins are 1-1/4" long, from a slender central stalk, maturing to brown, opening in fall of the first season. Seeds are 1/16" long with thin wings that form a rounded outline. Twigs are smooth, green, glabrous, darkening to orange-brown. Bark is creamy white, flaky, peeling in thin papery flaps, with pronounced black cracks and markings.

The most widely planted birch, for its weeping accent qualities. Not considered long-lived. Needs ample moisture and does not tolerate drought. Surface roots seek water and may encroach into adjacent plantings or paved areas. Aphid infestations produce honeydew drip. Longevity estimated to be 75-90 years.

Cultivar. SHD, ACC; IRR: H/H/H/H/-/-

Betula pendula 'Crispa'

CUTLEAF WEEPING BIRCH

Betulaceae. Sunset zones 1-12, 14-24. Deciduous. Selected variety of the species. Growth rate moderate to fast to 30-40' tall and 15-20' wide, developing an upright weeping form, with branches recurving to the ground. Leaves are alternate, simple, 3-4" long, dull dark green, rhomboid-ovate, with 9-11 paired veins, sharply double-serrate edges deeply cut, and golden yellow fall color. Pendulous tassel-like flowers occur in March and April, followed by pendent cylindrical catkins. Bark is creamy white, flaky, peeling in thin papery flaps, with pronounced black cracked markings.

The most weeping variety commonly available, with deeply cut leaves. Becomes more weeping with age. Longevity estimated to be 70-90 years.

Asia. SHD, ACC; IRR: H/H/-/-/-/-

Betula jacquemontii

= *Betula utilis* var. *jacquemontii*

JACQUEMONT BIRCH

Betulaceae. Sunset zones 3-11, 14-17. Deciduous. Native to northern India. Growth rate initially fast to 40' tall (eventually to 60') with a 30' spread, developing a tall oval form. Leaves are alternate, simple, 1-4" long, dark green, ovate, with double-serrate edges, cuneate at the base, with pointed ends, glandular undersides, hairy at the veins, and yellow fall color. *Betula utilis* usually has 10-14 veins, whereas this species usually has only 7-10. Catkins are long, pubescent, pendulous, tassel-like. Bark is clear white, becoming fissured with age.

A symmetrical, dense, full-shaped birch with distinctly upright branching. First discovered in the 1880s by the French botanist it is named for, and though it has been in the U.S. for many decades, it has only recently gained popularity for its upright, symmetrical form. Longevity estimated to be 70-90 years.

Cultivar. SHD, ACC; IRR: H/H/H/H/-/-

Betula pendula 'Purpurea'

PURPLELEAF BIRCH

Betulaceae. Sunset zones 1-12, 14-24. Deciduous. Selected variety of the species. Growth rate moderate to fast to 30-40' tall by 15-20' wide, with an upright, billowy form and recurving branches. Leaves are alternate, simple, 1 to 2-1/2" long, rhomboid-ovate, new foliage bronzy to deep purplish maroon, fading to bronze-green. Fall color is orange to coppery-bronze, not as dramatic as Schwedler maple or purple-leaf plum. Longevity estimated to be 70-90 years.

Australia. EVG, SHD, FLW ACC; IRR: L/-/L/M/-/-

Brachychiton acerifolius

FLAME TREE

Sterculiaceae. Sunset zones 15-24. Evergreen, or deciduous for short periods. Native from Queensland to New South Wales, Australia. Growth rate moderate to 60' tall or more with a 30' spread, forming a tall, dense, upright canopy and a broad, heavy trunk at the base, tapering upward. Leaves are alternate, simple, 6-10" long, dark glossy green on the uppersides, with varying shapes, from entire to palmate, most often deeply 3-5 lobed, persisting 2 years, shimmering and fluttering in breezes. Showy, loose racemes or panicles of 3/4", bell-shaped red to orange-red flowers occur in late spring to early summer. Fruits are heavy, woody, 3-5" long by 1 to 1-1/4" wide, 1-celled, ovoid follicles with pointed ends, finely hairy inside, glabrous on the outside, occurring in clusters of 5 or more, palmate from a long, thick, hanging stem, maturing to dark brown in late summer and splitting open along 1 seam, exposing 10-20 shiny, light tan, 1/8-1/4" oval, pubescent seeds. Distinctive bark is smooth, green, darkening to brown, with vertical whitish lines and dark brown blotches, graying with age.

An interesting and showy half-hardy evergreen street or park tree, most commonly used in southern California, also effective in groves or as a windbreak or screen. Dense, clean, upright, drought tolerant tree for difficult situations. Litter can be a problem in paved areas. Does not tolerate overly moist conditions. Susceptible to Texas root rot. Otherwise relatively pest-free. Longevity estimated to be 100-150 years.

Australia. EVG, SHD, FLW ACC; IRR: L/M/L/L/M/M

Brachychiton populneus
BOTTLE TREE

Sterculiaceae. Sunset zones 12-24. Evergreen. Native to Australia. Growth rate moderate to 30-50' tall and 30' wide, with a tall, upright to oval canopy and a heavy trunk broad at the base, tapering quickly in bottlelike fashion. Leaves are alternate, simple, 1-1/2 to 2-1/2" long, glossy light green, ovate to ovate-lanceolate, with smooth, irregularly lobed edges and long acuminate ends, persisting 2 years, shimmering in breezes. Inconspicuous, long-stemmed clusters of small, bell-shaped, yellowish green flowers with reddish throat markings occur from May through July, noticeable only close up. Fruits are woody, 1-1/2 to 3" long, 1-celled, pointed, ovoid follicles occurring in clusters of 5 or more, palmate from a long hanging stem, maturing to dark brown in late summer, densely hairy inside and splitting open along 1 seam, exposing 10-20 light-colored, shiny, 1/8-1/4" oval seeds. Distinctive bark is smooth, green, darkening to brown, with vertical whitish lines and dark brown blotches, graying with age.

An interesting half-hardy, drought-tolerant, dense evergreen street or park tree for difficult situations, more commonly cultivated in southern California. Effective in groves as a windbreak or screen. From a distance resembles a columnar camphor tree. Litter can be a problem. Susceptible to Texas root rot. Longevity estimated to be 100-150 years.

e Asia. SHD, ACC; IRR: No WUCOLS

Broussonetia papyrifera
(*Morus papyrifera*)

PAPER MULBERRY

Moraceae. Sunset zones 3-24. Deciduous. Native to China and Japan. Growth rate moderate to 50' tall and up to 40' wide, developing a dense, broad canopy. Leaves are alternate, simple, 4-8" long, dull to grayish green, variably ovate or palmate, with 0-3 lobes, juvenile leaves intricately lobed, roughly sandpaperlike above, with velvety hairs on pale undersides, often oblique at the base, with finely toothed edges and yellow fall color. Insignificant dioecious flowers occur in late spring, with 1-3" long, yellowish green, fuzzy, drooping male catkins and oval females on separate trees. Female flowers form 3/4 to 1" round, hairy, reddish orange, fleshy, berrylike fruits, edible but tasteless, in September. Twigs are rough, hairy. Bark is thin, smooth, light tan to grayish, becoming covered with many thin ridges.

A small to medium-sized shade or accent tree. Tolerates drought, alkalinity, wind, and heat once established and is fairly hardy. Fruit and flower drop can be messy, and suckering often occurs from the base. Not often cultivated. May be considered weedy. Longevity estimated to be 50-80 years. 'Culcullata' has male flowers only, with curled leaf margins, 'Laciniata' or 'Dissecta' has finely dissected margins, and 'Variegata' is variegated, with yellow or white markings.

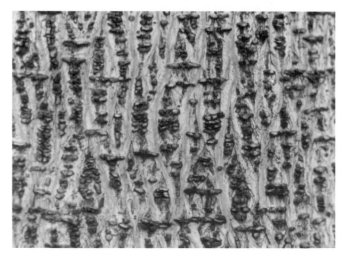

Australia. SHD, SCR, FLW ACC; IRR: L/L/L/L/-/M

Callistemon citrinus

LEMON BOTTLEBRUSH

Myrtaceae. Sunset zones 8-9, 12-24. Evergreen. Native to Australia. Growth rate fast to 20-25' tall as a trained single-trunk standard with a rounded canopy and commonly cultivated as a shrub to 10-15' tall. Leaves are alternate, simple, 2-4" long by 1/4-1/2" wide, thick, leathery, dark green, lanceolate, shiny and slightly hairy on both sides, with a prominent midrib and 2 lateral parallel veins, tapering to a slender pointed end, persisting 2-3 years, with coppery red new growth. Bright, showy red "brushes" of cylindrical 3-6" long spikes are composed mostly of 1" long stamens, grouped singly instead of in 5s like melaleuca, densely arranged completely around branchlets below new growth, and appearing in various cycles throughout the year, mainly in summer. Small, persistent, constricted woody cuplike seed capsules with contracted outer edges remain attached to branchlets for 1-3 years. Twigs are smooth, slender, and reddish brown. Bark is reddish brown, becoming gray with age and shedding in fibrous, peeling shards, revealing reddish underbark.

A tough small tree standard able to withstand adverse conditions of drought, heat, wind, and alkaline soils if drainage is good. Relatively pest-free. Flowers attract bees. Looks best with occasional trimming and deep watering. Longevity estimated to be 40-60 years.

Australia. SHD, WPG, FLW ACC; IRR: L/L/M/M/-/M

Callistemon viminalis

WEEPING BOTTLEBRUSH

Myrtaceae. Sunset zones 6-9, 12-24. Evergreen. Native to Australia. Growth rate fast to 20-30' tall and 15' wide, as a single- or multi-trunk tree with a billowy, weeping form. Leaves are alternate, simple, 1-4" long by 1/4" wide, light glossy green, lanceolate, with entire margins, rounded or sharp tips, a prominent midrib, 2 lesser lateral parallel veins, slightly hairy, persisting 2-3 years. Showy, bright red flowers are cylindrical 3-6" long spikes, mostly composed of 3/4" long stamens, grouped singly instead of in 5s like melaleuca, densely arranged completely around branchlets below new growth, which appears throughout the year, mainly in summer. Outer edges are not contracted on the small, persistent, constricted woody cuplike seed capsules, which remain attached to branchlets for 1-3 years. Twigs are smooth, slender, reddish brown. Bark is reddish brown, becoming gray with age, shallowly fissured with rounded interconnected ridges and reddish coloration at splitting seams.

Sometimes used as an evergreen weeping accent tree. Best with moderate watering and feeding, but tolerates moderate drought where roots can reach the water table. May need occasional pruning for rejuvenation, shaping, or to reduce top-heaviness. Longevity estimated to be 40-60 years.

w N.A. California Native: ME, PF (*CA exc D*); EVG, SHD, SCR; (IRR: M/M/M/M/M/-)

Calocedrus decurrens

INCENSE CEDAR

Pinaceae. Sunset zones 2-12, 14-24, Jepson *1,2,**4-7,15-17**& IRR:**3,8,9**,11,12,**14,18-24**;CVS. Evergreen. Native to the mountains of southern Oregon, California, western Nevada, and northern Baja California. Growth rate slow to moderate to 75-90' tall (up to 160' in habitat) and 10-15' wide or more, developing a tall, symmetrical pyramidal form, with a dense narrow crown, a thick buttressed trunk at the base, and drooping lower horizontal branches arching upward at the ends on older trees. Needles are rich glossy green, closely spaced, flattened, scalelike, with sharp points, occurring in flat sprays, fragrant when crushed, persisting about 2 years, as branchlets enlarge, with the main deciduous period in late summer. Inconspicuous male and female flowers occur in midwinter on separate twigs of the same branch, as yellowish thickened scaly bodies at ends of branchlets. Oblong-ovoid, 1" yellowish to reddish brown seed cones mature in September of the first season, splitting open into 5 parts, with 2 recurving away from the flat, straight center and 2 smaller scales at 90 degrees, and a sharply pointed hook at the ends. Yellowish brown, 1/4" long, linear seeds are surrounded by a 1" long, papery wing extending backward from the seed point. Bark on young trees is thin, smooth, cinnamon-red, flaking in broad, flat plates, later thickening and becoming darker brown, appearing semi-fibrous, with deep furrows and thick vertical ridges.

A common evergreen tree commonly associated with the more prolific Douglas-fir and ponderosa pine in the Sierra Nevada. Slow to establish in cultivation, but adapts to many western climates, tolerating heat, poor soils, and drought. Attractive shiny, deep green, weeping foliage and reddish bark. Older trees prefer dryness around the base of the trunk, otherwise will drop interior foliage and slowly lose vigor. Longevity estimated to be 350-500 years in habitat. Series associations: Wedceanothus shrub series; BigDoufir, Blaoak, Canlivoak, Doufir, Doufirponpin, Engspruce, Giasequoia, **Inccedar**, Jefpine, Jefpinponpin, Mixconifer, Orewhioak, Ponpine, Redfir, Whifir, Bakcypsta, KlaMouEnrsta, SanBenMousta.

Cultivar. SHD; IRR: L/M/-/-/-/-

Carpinus betulus 'Compacta'

COMPACT HORNBEAM

Betulaceae. Sunset zones 2-9, 14-17. Deciduous. Variety of the species native to Europe and Asia Minor. Growth rate moderate to 40' tall and wide, developing a dense, compact, pyramidal form, broadening with age, with twiggy outer branches drooping from a stout trunk, often fluted at the base. Leaves are alternate, simple, 2-4" long, dull dark green, ovate-oblong, with double-serrate edges, acuminate ends, pale undersides, and bright yellow fall color. Insignificant, monoecious, greenish yellow, tassel-like flowers occur in March in drooping aments, male flowers 1-1/2" long and smaller females at tips of current season twigs semi-erect at first, gradually elongating and drooping. Green fruiting catkins quickly ripen to brown, containing roughly a dozen small, flat, rounded seeds with elongated winglike bracts, and remain for a short time, hanging in loose 3" long clusters along a central stem, mostly hidden among leaves. Smooth twigs become lustrous reddish brown, with sharply pointed, scaled buds. Bark is smooth, thin, light brown to grayish with blotchy markings, becoming darker with age, with shallow furrows.

An uncommon but beautiful small, long-lived shade tree for park or streetside plantings in moderately moist, well-drained soils. Prefers colder winter climates, but tolerates northern coastal and inland conditions with adequate moisture. Occasional scale infestations can be treated. Longevity estimated to be over 200 years.

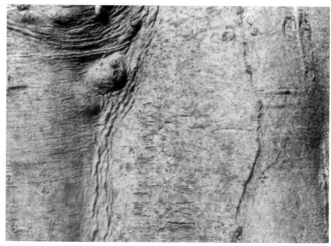

Cultivar. SHD; IRR: L/M/-/-/-/-

Carpinus betulus 'Fastigiata'
COLUMNAR HORNBEAM

Betulaceae. Sunset zones 2-9, 14-17. Deciduous. Variety of the species native to Europe and Asia Minor. Growth rate slow to moderate to 40' tall and nearly half as wide, developing a dense, upright, columnar to oval vaselike form, broadening with age, usually with a short, fluted trunk forking to multiple slender upright leaders and twiggy outer branches. Leaves are alternate, simple, 2-4" long, dull dark green, ovate-oblong, with double-serrate edges, acuminate ends, pale undersides, and bright yellow fall color. Insignificant, monoecious, greenish yellow, tassel-like flowers occur in March in drooping aments, male flowers 1-1/2" long and smaller females at tips of current season twigs semi-erect at first, gradually elongating and drooping. Green fruiting catkins quickly ripen to brown, containing roughly a dozen small, flat, rounded seeds with elongated winglike bracts, and remain for a short time, hanging in loose, 3" long clusters along a central stem, mostly hidden among leaves. Smooth twigs become lustrous reddish brown, with sharply pointed, scaled buds. Bark is smooth, thin, light brown to grayish, with blotchy markings, becoming darker with age, with shallow furrows.

An uncommon, long-lived, vertical accent tree for park or streetside planting, neater than Lombardy poplar but slower growing. The most commonly used hornbeam, requiring moderately moist, well-drained soils. Prefers colder winter climates, but tolerates coastal and inland conditions in northern California. Longevity estimated to be over 200 years.

c&s N.A. SHD, FRU; IRR: L/M/M/M/M/M

Carya illinoiensis

PECAN

Juglandaceae. Sunset zones 2&3 (warmer parts), 6-10, 12-14, 18-20. Deciduous. Native to the central and southern U.S. Growth rate fast to 70' tall and wide, with a broad open crown, a heavy stout trunk, and ascending limbs, slightly drooping at the ends. Leaves are alternate, pinnately compound, 12-18" long, medium green, glabrous, with 9-17 lanceolate leaflets, 4-7" long, with finely serrate edges, pale undersides, and yellow fall color. Monoecious flowers occur after trees are fully leafed out in spring, males greenish yellow, 3-4" long, apetalous, tassel-like, in 3-branched drooping aments, producing heavy pollen and litter as they drop, and tiny green female flowers at branch tips. Ellipsoidal 1-2" long nuts are enclosed inside a prominently 4-ribbed husk, which turns black and splits into 4 sections when ripe in late summer, sometimes persisting on bare branches into winter after nuts have fallen. Seeds have a thin, smooth, tan shell, with multiple linear black markings, enclosing the wrinkled, sweet, meaty core. Bark is smooth brown, becoming grayish, with many narrow furrows and scaly ridges.

A fast-growing tree, more graceful and lighter textured as a shade tree than English walnut. Readily self-seeds, often becoming a pest in lawns and gardens, especially where squirrels bury the nuts. Thrives in deep, fertile, well-drained soil. Requires summer watering in hot dry climates until establishment. Will not tolerate salinity. Longevity estimated to be 75-100 years.

s Europe, n Africa, sw Asia. SHD, FRU; IRR: No WUCOLS

Castanea sativa

SPANISH CHESTNUT

Fagaceae. Sunset zones 2-9, 14-17. Deciduous. Native to southern Europe, north Africa, and southwestern Asia. Growth rate moderate to 40-60' tall (up to 100') and often equally wide or wider, developing a broad, rounded crown from a strong central trunk. Leaves are alternate, simple, 5-9" long, thick, glossy dark green, oblong-lanceolate, with coarsely toothed edges, tiny sharp spikes, and deep orange to yellow fall color. Showy, golden yellow, tassel-like flowers, apetalous and monoecious, with a disagreeable odor, occur in late spring in staminate or bisexual aments. Male flowers have 10-20 stamens, in 3-7-flowered cymes, and females are solitary or 2-3-clustered. Fruits are greenish, burrlike, covered with sharp, prickly spines, maturing in fall to golden brown, and splitting along 4 seams to reveal 1-2 shiny brown, edible seeds. Occasionally, 3-4 flattened seeds occur, with little meat inside. Twigs are stout, reddish brown, and smooth. Bark on mature trees is dark grayish brown, vertically furrowed, with broad, flat ridges.

A prized shade tree, not often used, since the American chestnut (*Castanea dentata*) suffers from uncontrollable viral disease. Scattered trees are found vigorously growing as street and park trees in the Central Valley, where they are provided ample room to grow in full sun in well-drained, moderately moist soils. Resists oak root fungus. Longevity estimated to be 300-400 years.

Australia. SHD, SCR, FOL ACC; IRR: L/L/L/L/M/M

Casuarina equisetifolia

HORSETAIL TREE

Casuarinaceae. Sunset zones 8, 9, 12-24. Evergreen. Native to Australia. Growth rate fast to 40-60' tall and 20' wide, developing an irregular pyramidal shape, broadening with age, usually with a multi-forked central upright trunk and slender outward-arching branches with twiggy, pendulous ends. Delicate, lacy foliage is tightly bunched and slightly drooping. Long, slender, dark green, needlelike twigs are jointed, and leaves, reduced to tiny toothlike scales, 6-8 at each, are whorled at twig joints. Insignificant, dioecious, unisexual flowers occur on separate trees, males in slender, tan-colored terminal spikes at ends of twigs, females in small heads along axils of twigs. Fruits are oblong, 1/2", woody, conelike, comprised of many tiny-beaked capsules, slightly hairy at the valves, ripening to brown and releasing a multitude of tiny round seeds, but not tending to reseed heavily. Bark is dull dark grayish brown, becoming irregularly fissured or scaly.

Commonly used as an evergreen windbreak or buffer screen along freeways and railroad rights-of-way. Tolerates harsh conditions and requires virtually no watering after establishment. Unusually fine, lacy texture, but foliage may have a dusty appearance. Occasionally used in parks as accent specimens. Longevity estimated to be 100 years or more, losing shapeliness with age.

c&s N.A. SHD, FLW ACC; IRR: L/M/M/M/M/M

Catalpa speciosa

WESTERN CATALPA

Bignoniaceae. Sunset zones 2-24. Deciduous. Native to the central and southern U.S. from Illinois to Arkansas. Growth rate slow to moderate to 40-60' tall and 20-40' wide, developing a round-headed, broad canopy and large trunk. Leaves are large, opposite or whorled, simple, 6-12" long and nearly as wide, light green, heart-shaped or sometimes shallowly 3-lobed, with smooth margins, pale hairy undersides, and yellow fall color. Showy, loose, sparse panicles of 7-10 perfect, 2-lipped, 2-1/2" white flowers, with yellow and soft brown throat markings, occur in late spring and sporadically into summer. Seed pods are 12-18" long, slender, hanging capsules, 2-celled and thick-walled, resembling a vanilla bean, maturing to brown in fall and persisting into winter, splitting open the following spring. Seeds are numerous, flattened, 1" long, with double, elongated, round-ended wings, with ciliate ends, and a tiny seed at the center of the wings. Twigs are stout, glabrous, greenish brown. Bark is rather thin, grayish brown, becoming fissured, with flat scaly ridges.

A showy flowering shade tree with bold leaf character, well adapted to extremes of heat and cold in moderately moist and fertile soil that is neither too wet nor too dry. Pruning young trees helps to develop strong leader trunks and desired branching height. Flowers and pods are a litter problem, and leaves may be damaged in high winds. Effective in park and residential settings, but not suitable as a street tree. Longevity estimated to be 75-100 years.

Cultivar. CNF, EVG, SHD; IRR: M/M/L/M/M/M

Cedrus atlantica 'Glauca'

BLUE ATLAS CEDAR

Pinaceae. Sunset zones 3-10, 14-24. Evergreen. Variety of the species native to the Atlas Mountains of Algeria and Morocco. Growth rate slow to 60' tall or more and 30' wide, with a dense, rigid, pyramidal form, branches angled slightly upward, casting dense shade. Bluish green needles are 3/4 to 1" long, in dense tufts along the branchlets. Insignificant monoecious flowers, males sparsely grouped, greenish purple, cylindrical, vertical, in spikes, shedding pollen in early fall and fatter, greenish, solitary, vertical females at branch tips. Cones are flat-ended, 3-1/2" long by 1-2" wide, with flattened round-ended scales, maturing to tan the following summer and disintegrating on the tree, leaving a central stalk spike and releasing small, flat, wide-winged, tan seeds. Bark is grayish brown, shallowly fissured, with flat, scaly ridge plates.

A desirable accent tree or park tree in small groves, best reserved for special locations that highlight the bluish foliage. Young trees appear sparse but elegant, becoming more dense with age, and older trees develop enormous trunks. Deep rooted and drought tolerant when established. Longevity estimated to be 100-200 years. 'Aurea' has yellowish tinted foliage, 'Pendula' has long, weeping branchlets, 'Rustic' has exceptional blue color, and 'Fastigiata' is upright in habit.

Asia. CNF, EVG, SHD; IRR: L/M/L/M/M/M

Cedrus deodara

DEODAR CEDAR

Pinaceae. Sunset zones 3-10, 14-24. Evergreen. Native to the Himalayas. Growth rate fast to 80' tall with a 40' spread, developing a loose pyramidal shape, nodding at the top, with a large central trunk and strong, horizontal, bowed limbs curving upward, with smaller drooping end branches, becoming more open-trunked with age. Needles are gray-green, 1-2" long, the longest of the cedars, in tufted clusters. Insignificant flowers are monoecious, males bluish green, upright spikes in loose clusters in late summer producing a profusion of pollen and females stouter and fewer, usually high on the tree. Cones are 4", greenish, upright, maturing in 2 years to a tan-brown color, and disintegrating into scalelike pieces from the tree, releasing wide, flattened-winged seeds, with the exposed central stalk remaining attached vertically to the branch. Bark is grayish brown, shallowly fissured and furrowed, with wide, flat, scaly ridges.

A useful evergreen conifer for parks and parkways, contrasting with green-needled conifers or deciduous trees. Drought tolerant once established. Needs ample room for roots and branching. Dense shade and needle drop often kill lawns or plantings beneath. Lower branches often more effective left unpruned, sweeping to the ground, until trees become fully mature. Longevity estimated to be 100-200 years. Less commonly cultivated varieties include many with yellow or bluish foliage and weeping or upright conical forms.

Cultivar. CNF, EVG, SHD, IRR: L/M/L/M/M/M

Cedrus deodara 'Descanso Dwarf'

COMPACT DEODAR CEDAR

Pinaceae. Sunset zones 3-10, 14-24. Evergreen. Selected variety of the species. Growth rate moderate to 20-30' tall with a 15-20' spread, developing a dense, broad-based, pyramidal form with many small, slender, twiggy, upright-arching branches from the base, slightly weeping at the ends. Needles are grayish green, 1 to 1-1/2" long, in tufted buds along branches, similar to the species but slightly smaller, more densely spaced, and slightly less coarse. Flower stalks and cones are usually not present on young trees. Bark is grayish brown with flat scaly plates, peeling back to reveal reddish brown underbark.

A small, compact version of the species, rarely available commercially but occasionally seen as a small screen or accent tree, alongside or mixed with the species. Branches may be pruned up from the base in lawns, but looks best if lower branches are exposed slightly, though not removed entirely from the base, which results in a top-heavy appearance if only the main trunk remains. Longevity estimated to be 100-150 years.

South America. SHD, FLW ACC, SPC; IRR: L/-/L/L/-/M

Ceiba speciosa
(Chorisia speciosa)

FLOSS SILK TREE

Bombaceae. Sunset zones 12-24. Evergreen above 35 degrees, briefly deciduous in colder climates. Native to Brazil and Argentina. Growth rate initially slow, faster after roots develop, to 30-60' tall with a nearly equal spread, forming a broad oval canopy, uppermost foliage less during the blooming period. Leaves are alternate, palmately compound, 6-12" wide, glossy, light to medium green, on a long slender petiole, with 7 short-stemmed, lanceolate leaflets with smooth edges, pale undersides, and slight yellow if any fall color. Hibiscuslike flowers are showy, deep pink, with long ruffled petals and a white spotted throat, from a bulbous green base, covering the tree in clusters during late summer into fall. Trunk thickens at the base, slowly tapering upward and studded with stout sharp spines. Bark is smooth and green, becoming gray with age.

A spectacular accent tree commonly seen in southern California. Not reliably hardy below 22 degrees. Fast-draining soil is a prerequisite. Often drops most or all leaves before blooming, producing a more dramatic floral display. Thrives in warm inland zones, but usually cannot survive Central Valley frost. Spines are hazardous in public areas, and slowly grow back if removed, though a spineless variety is reportedly available. Surprisingly drought tolerant. Longevity in cultivation depends on temperature, but may be 50-75 years or more. *Ceiba insignis* has white flowers.

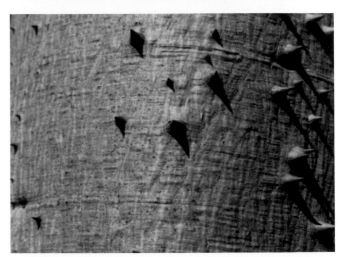

s Europe, n Africa, sw Asia. SHD; IRR: L/M/-/-/M/M

Celtis australis

EUROPEAN HACKBERRY

Ulmaceae. Sunset zones 8-16, 18-20. Deciduous. Native to southern Europe and north Africa to southwestern Asia. Growth rate moderate to 40' tall (up to 70-80' tall) with a 30-35' spread, developing an upright oval form, usually not as widespreading as other hackberries. Leaves are alternate, simple, light green, 2-5" long, elliptical-lanceolate, with coarse, sharply serrate edges, and dull yellow fall color. Leaves are longer-pointed at the ends than *Celtis occidentalis*, with a shorter deciduous period. Small, black, 1/4" round, 1-seeded fruits are noticeable through summer. Young twigs are dark brown to blackish, contrasting with light green leaves. Bark is grayish, warty, and rough, eventually becoming furrowed.

Drought tolerant, but looks best with occasional deep watering. Resists oak root fungus, but occasionally suffers from aphid infestations, which produce sticky and unsightly honeydew drip on paved surfaces. Can be invasive and weedy, as fruit is attractive to birds. Longevity estimated to be 80-100 years under favorable conditions.

w N.A. California Native: CH, FW (Riparian) (*s SNF, Teh, SnBr, PR, s SNE, e DMtns*); (IRR: L/-/-/-/L/L)

Celtis reticulata

NETLEAF HACKBERRY

Ulmaceae. Sunset zones 2-24, best in 2, 3, 7-13, 18-21, Jepson *SUN,DRN:4-16,**14-17,22-24**&IRR:**1,2,3,7-13,18-21**. Deciduous. Native to streambanks and bluffs in drier areas of southern California from San Diego to Kern and Inyo counties, south to central Mexico, north to eastern Washington, and east to Texas and central Kansas. Growth rate moderate to 25-30' tall and wide, forming a broad, rounded crown from a short trunk with widespreading branches. Leaves are alternate, simple, 2-3" long, dull dark green, ovate to oblong-lanceolate, with a raspy, rough, hairy surface above and below, lying flatly, in 2 rows, cordate to oblique-auriculate at the petiole base, with short to long-pointed ends and variable entire or sparsely toothed crenate edges, conspicuously raised netted veining on the pale undersides, and dull yellow fall color. Inconspicuous, 1/8", greenish flowers occur on axillary buds of new growth in early spring. Fruits are 1/4" round, 1-seeded, purplish or brown, smooth-skinned, from stems usually longer than those of the leaves. Twigs are stiff, dark green, slightly hairy, darkening to brown, and growing in a slightly zigzag pattern. Bark is thin, smooth, grayish, thickening and becoming rough and fissured, with prominent wartlike protuberances.

An important riparian habitat tree sometimes planted as a shade tree. Distinctly interesting foliage. Does best with regular deep watering in well-drained soils. Roots are widespreading, seeking out water. Branches occasionally have bushy deformed growths or witches-brooms caused by mites and fungi, or swollen galls caused by plant lice. Longevity estimated to be 100-120 years.

e N.A. SHD; IRR: L/L/-/M/M/M

Celtis occidentalis

COMMON HACKBERRY

Ulmaceae. Sunset zones 1-24. Deciduous. Native to eastern north America north to Quebec and south to North Carolina. Growth rate moderate to 50' tall or more with a nearly equal spread, developing a broad, rounded canopy and thick, intricately branched upward limbs, new growth drooping at the ends. Leaves are alternate, simple, 2-5" long, dull dark green, ovate to ovate-lanceolate, obliquely uneven at the base, 2 lowest veins intersecting the base of the midvein, with no gap, leaf margins coarsely serrate, with distinctly tailed acuminate ends, a rough upper surface, lighter undersides, minute hairs on veins and petioles, and dull yellow fall color. Tiny, yellowish flowers in spring are monoecious, perfect, solitary, on 3/4" long threadlike stems from leaf bases. Fruits are 1/4" round with a tiny pointed tip, ripening to sandpapery orange-brown in fall, with a thin, dry, sweet pulp covering the hard, thickshelled seed. Twigs are pale green, turning reddish brown in fall, with tiny rounded buds. Bark is smooth, grayish brown, with blackened markings and reddish-tinged fissured seams. With age, trunks become darker gray, and often fluted at the base, with irregular furrowing and shallow ridges.

A useful deep-rooted shade tree for tough situations. Tolerates hot, dry, windy conditions when established and thrives in most soils. Foliage not especially attractive, and other species are more commonly used. Longevity estimated to be 100-150 years.

e Asia. SHD; IRR: L/M/-/M/M/M

Celtis sinensis

CHINESE HACKBERRY

Ulmaceae. Sunset zones 8-16, 18-20. Deciduous. Native to China and Asia. Growth rate moderate to 40' tall and wide, with a broad oval canopy and billowy foliage tufts. Leaves are alternate, simple, 2-5" long, shiny dark green, ovate-lanceolate, uneven at the base, 2 lowest veins intersecting the base of the midvein, with crenate edges, acute ends, lighter undersides, minutely hairy on the main vein, and yellow fall color. Tiny yellowish flowers are perfect, monoecious, on 3/4" long, yellowish green, threadlike stems from leaf bases, occurring in April, males in cymes at the base of twigs and females solitary along axillary buds. Fruits are 1/4" round with a tiny blunt-pointed tip, ripening to dark orange or orange-red in fall, with a dry, sweet pulp covering the hard, thick-shelled seed. Twigs are pale green, turning reddish brown in fall, with tiny, flattened, pointed buds. Bark is smooth, grayish brown, with blackened markings and warty growths, becoming darker gray with age, often fluted at the base, with irregular shallow furrows and ridges.

A popular shade tree, commonly used in parking lots and residential lawns. Tolerates hot, dry climates if deep watered to become established but not as drought tolerant as other species. Deep-rooted and does not usually heave paving. Relatively pest free. Longevity estimated to be 100-150 years.

e Asia. EVG, SHD; IRR: L/L/L/L/-/L

Ceratonia siliqua

CAROB TREE

Fabaceae. Sunset zones 9, 13-16, 18-24. Evergreen. Native to the eastern Mediterranean. Growth rate slow to moderate to 30-40' tall and wide, forming a broad oval canopy with heavy, upward-spreading limbs and a large trunk. Leaves are alternate, pinnately compound, dark shiny green, stiff and leathery, 6-10" long, with 4-10 broadly oblanceolate leaflets, lighter undersides, and smooth, slightly recurving margins. Small, dense clusters of yellowish green, stamenlike flowers appear in spring on older branches near the base of the foliage or from trunks, with an objectionable heavy, musky odor on female trees. Thick, dark brown, flattened, leathery legume pods are 8-12" long, containing 10-16 round flattened beans, and may persist through winter. Shiny, smooth, reddish brown twigs highlight the dark green foliage. Bark is smooth, grayish brown, with small glandular warts, or slightly scaly, sometimes shallowly fissured on older trees.

A large, long-lived park or shade tree, casting dense, heavy shade, usually too dense for lawns, but shade-loving groundcovers do well beneath. Deep-rooted, but needs a wide area for surface roots, which may heave paving. Needs well-drained soil, otherwise often succumbs to crown rot. Foliage is damaged at temperatures below 20 degrees, but usually recovers the following year. Seeds are edible but not considered a delicacy. Male trees do not have the unpleasant odor or fruit. Longevity estimated to be 75-125 years.

e Asia. SHD, FOL ACC; IRR: No WUCOLS

Cercidiphyllum japonicum

KATSURA TREE

Cercidiphyllaceae. Sunset zones 2-6, and in part shade zones 14-16, 18-20. Deciduous. Native to Japan and China. Growth rate slow to 40' tall or more and as wide as tall, developing a tall oval form, low-forked or multiple-trunked, with upward-angled branches. Leaves are opposite, simple, 2-4" long, bluish green, oval or heart-shaped, with lighter undersides, lightly serrate to crenate edges, and reddish tints throughout the year, with slender reddish petioles, and brilliant red to yellow fall color. Inconspicuous, yellowish green, dioecious flowers occur on separate trees in April before leaves emerge, males 1/3" long with 15-20 red stamens and females in bundles of 3-5, 1/4" long, with twisted red styles that elongate into legume seed pods. Pods are greenish, rounded, fat, 1/2-2" long, with small seeds inside, ripening to tan in fall. Twigs are slender, smooth, and reddish. Smooth, greenish gray bark becomes grayish brown and fissured, with reddish colorations and flat, gray, scaly plates.

An unusual and handsome small accent tree with interesting foliage and brilliant fall color. Requires ample moisture to prevent leaf burn, even in shade inland, though this becomes less of a problem when roots are fully established. Tolerates a sunny exposure in moderately moist lawns. Male trees are more upright in form, females more bushy. Longevity estimated to be 50-80 years. 'Sinense' is taller, with better form, and 'Pendulum' is a grafted weeping, smaller form.

sw N.A. California Native: SD (Riparian) (*Dson*); SHD, NAT, FLW ACC; (IRR: VL/VL/VL/L/-/L)

Cercidium floridum

PALO VERDE

Fabaceae. Sunset zones 8-14, 18-20, Jepson *SUN,DRN: 7,**8,9**, 10,**14,19-23**,24&IRR:11,**12,13**;alsoSTBL. Deciduous. Native to deserts, washes, and valleys of southern California, Arizona, New Mexico, and Baja California. Growth rate slow, after initial fast growth, to 35' tall and 30' wide, usually developing a large, multi-trunked base with a broad, oval canopy, rather irregular, and stout, contorted, or zigzig branches, twiggy at the ends. Leaves are alternate, compound, evenly pinnate, 1" long, pale bluish green, with 2-3 paired, narrow-oblong, glaucous, 1/8-1/4" long pinnae, many falling as soil dries in late summer. Showy, loose, 2" raceme clusters of 4-5 bright yellow, nearly perfect, 3/4" flowers, 5-petaled and sometimes with red specks, cover the tree in spring, with sporadic reflowering. Flat, thin, or slightly bulging legume pods, 1-1/4 to 4" long by 1/4-1/2" wide, short-pointed at the ends, often constricted between 2-8 flattened, ovoid, beanlike seeds, mature to yellowish brown, falling in late summer. Twigs are smooth, stout, yellow-green to blue-green, hairless, with a straight, slender, 1/4" long spine at bud nodes. Bark is thin, smooth, olive-green, becoming reddish brown, furrowed, and scaly.

A common desert habitat tree, where it tolerates extreme drought. Also cultivated as a garden accent specimen. Prefers well-drained, sandy or gravelly soils and is intolerant of regular watering. Readily reseeds in preferred conditions. Longevity estimated to be up to 100 years in habitat. *Cercidium microphyllum* and *C. texanum* are similar. Series associations: Catacacia & Ocotillo shrub series; **Blupalveriro.**

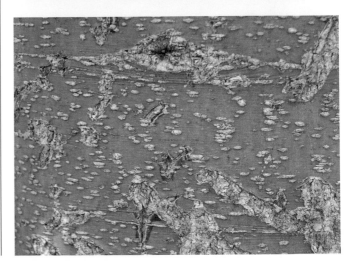

e N.A. SHD, FLW ACC, SPC; IRR: M/M/M/M/-/-

Cercis canadensis
EASTERN REDBUD

Fabaceae. Sunset zones 1-24. Deciduous. Native to the eastern U.S. Growth rate moderate to fast to 25-35' tall and wide, developing a broad, oval-arching canopy, usually with multiple slender, many-forked trunks, and branching rather irregular, horizontal, often twiggy at the ends, becoming somewhat drooping in shade. Leaves are alternate, simple, 3-4" round, deep green, smooth, broadly ovate to reniform, with acute, small-tailed, pointed ends, smooth margins, lighter undersides, new growth reddish, and brief yellow to orange fall color. Showy fascicle clusters of small, bright, rosy pinkish purple, pealike flowers occur from axillary branch buds, covering the tree in early spring just before leaves emerge. Flat, oblong legume seed pods, 2-3" long, tan with a purplish tinge, hang in fascicles along branches from summer well into winter, containing 4-6 1/4" long, reddish brown, hard seeds. Bark is thin, smooth, gray, thickening on older trees, becoming dark brown and fissured with irregular scaly plates.

A reliable, small accent tree with attractive foliage. Fastest growing, largest, and most treelike of the redbuds. Deep watering helps develop a deep root system to tolerate drier soils. Otherwise requires little care. Occasional tent caterpillar infestations may destroy foliage. Longevity estimated to be 50-75 years or more.

Cultivar. SHD, FOL, FLW ACC, SPC; IRR: M/M/M/M/-/-

Cercis canadensis 'Forest Pansy'

PP #2556 (1965)

PURPLE-LEAF EASTERN REDBUD

Fabaceae. Sunset zones 1-24. Deciduous. Variety of the species. Growth rate moderate to fast to 25-35' tall and wide, forming a broad oval canopy from slender trunks, with irregular, twiggy, horizontal branching, somewhat drooping in shade. Leaves are alternate, simple, 3-4", deep purplish green, smooth, broadly ovate to reniform, with elongated acute ends, smooth margins, lighter undersides, long, slender petioles, new growth bronzy purplish red, and early, brief, yellow-orange-red fall color. Showy fascicle clusters of small, bright rosy pinkish purple, pealike flowers from branch buds cover the tree in early spring just before leaves emerge. Flat, oblong, tan legume seed pods, 2-3" long, hang in clusters along branches well into winter, and contain 4-6 ovoid-oblong, flattened, 1/4" long, reddish brown, hard seeds. Bark is thin, smooth, gray, thickening on older trees, becoming dark brown and fissured, with irregular scaly plates.

A desirable accent tree by virtue of its purplish foliage, spring flowering, and fall color or as a patio or lawn specimen. Requires moderate moisture and well-drained, fertile soil in full sun or part shade. Vigorous growth. Foliage color fades, but new purplish leaves emerge constantly. Longevity estimated to be 50-75 years or more.

Cultivar, SHD, FOL, FLW ACC, SPC; IRR: M/M/M/M/-/-

Cercis canadensis 'Silver Cloud'

VARIEGATED EASTERN REDBUD

Fabaceae. Sunset zones 1-24. Variety of the species. Growth rate moderate to fast to 20-30' tall with 10-15' spread, forming a broad oval canopy. Leaves are alternate, simple, 3-4", dark green with white variegation, glabrous, broadly ovate to reniform, with bluntly pointed or elongated-acute ends, smooth margins, and brief yellow to orange fall color. Fascicle clusters of small, pale pinkish, pealike flowers occur in early spring as leaves emerge, showy but not as profuse as other redbuds.

An unusual form used as a flowering accent tree in garden settings, with characteristics similar to the species. Introduced in the 1960s but only occasionally available.

sw N.A. California Native: CH, OW, FW, PF (Riparian) (*NW, CaR, SN, SnJV, PR*); SHD, FOL, FLW ACC, SPC; (IRR: VL/VL/L/L/-/-)

Cercis occidentalis

WESTERN REDBUD

Fabaceae. Sunset zones 2-24, Jepson *DRN,SUN:4-6,**14-16**,17, **18,22-24**;&IRR or part SHD:1-3,**7-9**,12,**19-21**;also STBL;CVS. Deciduous. Native to the California foothills below 3,000' elevation, extending to Utah, Nevada, and Arizona. Growth rate slow to 10-18' tall and wide, usually developing a twiggy, round-headed, arching vase- to mound-shaped, multi-trunked shrub with upright, slender branches or a small tree with a low-forked, short trunk. Leaves are alternate, simple, 2-3" long, rounded, oval or heart-shaped, with an obtuse to emarginate apex and an auriculate base from which the main, straight veins arise, a slender, 1" long petiole, new growth slightly bronzy at first, becoming dark green with a bluish cast, and showy yellow to orange-red fall color. Miniature, perfect, pea-like, magenta flowers occur in simple fascicle clusters, covering branches in spring before leaves appear. A profusion of flattened, short-stalked, hanging, oblong legume seed pods, pointed at both ends, bulging around the seeds, turn purplish tan to brown as they ripen in late summer and persist into winter, releasing flattened, oblong-ovoid, 1/4" long, tan-colored seeds. Twigs are slender, smooth, and reddish green. Bark is thin, smooth, grayish, roughening slightly with age and becoming dark gray.

A small native accent or understory tree, tolerating heat and drought. Deep watering during establishment speeds growth. Rarely grown as a single-trunk standard, and most often a large shrub or multi-trunked tree. Flowering most profuse in sun where temperatures fall below 28 degrees. Resistant to oak root fungus. Longevity estimated to be 50-75 years.

sw N.A. SHD, FLW ACC, SPC; IRR: L/L/?/?/?/?

Cercis reniformis

TEXAS REDBUD

Fabaceae. Sunset zones 3-24. Deciduous. Native to mountain ranges of southern Oklahoma to northern Mexico. Growth rate slow to 10-20' tall and wide, forming a twiggy, round-headed tree or a large, multi-trunked shrub, with branching to the ground. Leaves are distinctly rounded, alternate, simple, 2-3" long, shiny, medium green, appearing waxed, with strongly undulating edges, an obtuse to emarginate apex, deeply cordate at the base, from which the main straight veins arise, with a 1" long petiole, and dull yellow fall color. Miniature, pealike, pink flowers occur in simple fascicle clusters, covering branches in spring before leaves emerge. Sets few if any seed pods. Twigs are smooth and brown. Bark is thin, smooth, grayish, roughening slightly with age and becoming dark gray, with vertical reddish seams appearing as bark cracks.

   Noticeably different in appearance and generally smaller than eastern or western redbud, with clean, shiny leaves and a dense, uniform shape. Drought and heat tolerant, but often does quite well in well-drained lawns. Longevity estimated to be 50-70 years. Pink and white varieties are available.

sw N.A. California Native: CH, OW, FW, CC, ME, PF (*NW, CaR, SN, CW, CHI, TR, PR, MP*); NAT, SHD; (IRR: VL/VL/VL/VL/VL/-)

Cercocarpus betuloides

MOUNTAIN IRONWOOD

Rosaceae. Sunset zones 3, 5, 7-10, 13-24, Jepson *SUN,DRY:1-3,5,**7**,8-10,**14-21**,22-24;also STBL. Evergreen. Native to dry slopes of southwestern Oregon and California to Baja California below 8,000' elevation. Growth rate slow to 5-12' tall and wide, usually multi-trunked, developing a wide, oval to arching, vase-shaped canopy, with upright, thin, stiff branches with slightly arching ends. Leaves are alternate, simple, 1-2" long, stiff, dark dull green with a grayish cast, wedge-shaped, elliptical to ovate, tightly bunched near the buds, sometimes hairy, with prominent, straight, feathered veining, toothed edges and ends, lighter undersides, persisting 1-2 seasons and yellowing slightly before falling. Insignificant, 1/8" long flowers with tiny cream-colored petals occur in dense clusters of 3-8 at each bud in late spring. More noticeable are the thin, contorted, hairy, white tail-like plumes, extending 1" or more from thin, cylindrical, sheathed woody seeds, which have a short stem, unlike other species, and persist through summer. Silky reddish twigs have a pleasant aroma. Bark is thin, smooth, dull gray, becoming reddish brown and flaky with age.

A tough native shrub or tree usually found in hot, dry areas with poor soil. Can become a handsome focal point in a native garden with good soil and infrequent deep watering. Blooms best in full sun. Requires only occasional minor pruning and mulching. Longevity estimated to be 50-60 years or more. Series associations: **Birmoumah**, Foopine, Bluoak.

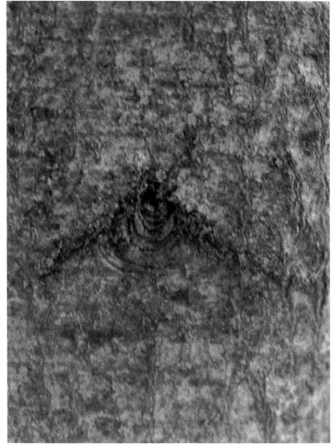

w N.A. California Native: ME, CF, PF (n Calif.) (*NCo, KR, NcoRO*);
EVG, SHD; (IRR: M/M/-/-/-/-)

Chamaecyparis lawsoniana

PORT ORFORD-CEDAR

Cupressaceae. Sunset zones 3-6, 15-17, Jepson *IRR:1,2,**4-6,15-17**,24,&SHD3,7,**14**,18-23:CVS. Evergreen. Native to coastal Oregon and extreme northern California Coastal Ranges. Growth rate moderate to fast to 60' tall (70-200' in habitat) with a 30' or greater spread, developing a dense, pyramidal or columnar shape with branching often to the ground. Lacy, drooping, flat foliage sprays are variable blue-green, with minute, flattish, scalelike leaves, soft to the touch, with short, blunt points. Insignificant reddish or brown male flowers appear as swollen bulbs at leaf tips and shed pollen in spring. Tiny green female flowers form berrylike, reddish brown, 3/8" long, oval cones among the foliage, maturing in fall of the first season, when the 8 wide, flat scales open, exposing the center portion when ripe. Seeds are brown, 1/8" long, with a common wing encircling each side. Bark is thin, brown, becoming gray with age, with irregular shallow seams and flat, shallow ridges, frequent diagonal crossings, peeling in narrow, flat, vertical strips.

A stately forest tree, common in its northwest habitat, though slowly being decimated by *Phytophthora lateralis*, a parasitic root-rotting fungus, which is easily spread by vehicle tires and hikers' boots. Valuable for timber, providing wood that has a resinous substance toxic to termites. Occasionally planted in botanical parks in central California, and commonly cultivated in western Europe. Longevity estimated to be 150-200 years, possibly to 600 years, in habitat. Many cultivars. Series associations: Darlingtonia shrub series; Doufir, Doufirtan, Jefpine, **PorOrfced**, Redfir, Weshemlock, Whialder, Whifir, Alayelcedsta, KlaMouEnrsta.

Cultivar. EVG, CNF, SHD, SPC; IRR: M/M/-/-/-/-

Chamaecyparis lawsoniana 'Stewartii'

STEWART'S GOLDEN CEDAR

Cupressaceae. Sunset zones 3-6, 15-17. Evergreen. Variety of the species. Growth rate slow to 30' tall or more, usually developing a densely pyramidal or conical shape. Flat foliage sprays of deep green, scalelike foliage with a distinct yellow variegation droop noticeably at the ends. Small reddish or brown male flowers appear as swollen bulbs at leaf tips, shedding pollen in spring. Tiny green female flowers form berry-like, bluish brown, 1/4" long, oval cones with 8 wide, flat scales, among the foliage, and mature in fall of the first season, turning brown and opening along the seams, often remaining attached through the following spring. Seeds are brown, 1/8" long, with circular wings. Brown bark thickens and grays with age, with reddish vertical seams, wide flat ridges, peeling in thin plates.

A slow growing, symmetrical, evergreen specimen tree commonly found in north coastal areas but adaptable to moderate inland or foothill heat in moderately moist soils, or lawns, in sun or semi-shade. Longevity estimated to be 150-200 years. 'Golden King' and 'Lutea' are similar.

nw N.A. California Native: CF (*KR, sparse*); EVG, SHD; (IRR: M/M/-/-/-/-)

Chamaecyparis nootkatensis

NOOTKA CYPRESS, ALASKA CEDAR

Cupressaceae. Sunset zones 2-6, 15-17. Evergreen. Native to Alaska and mountain ranges of Oregon and Washington, with a few sparse groves in the Siskyou Mountains of northwestern California. Growth rate slow to 80' tall or more with a 25' spread at the base, developing an open, dense, narrowly conical form, with weeping flat sprays, a slender leader, often drooping at the tip, and a wide-based trunk. Foliage is dense, blue-green, finely textured, harsh and prickly, scalelike, in flat, pendulous sprays. Small reddish or brown male flowers appear as swollen bulbs at leaf tips, shedding pollen in spring. Tiny green female flowers form small, 1/4" long, round, deep brown cones with a whitish cast, ripening in early fall. Cones have wide, flat scales, opening to expose the center portion, tucked among the foliage. Brown, 1/8" long seeds have a common wing encircling each side. Bark is thin, brown, becoming gray with age, irregularly and finely broken by shallow seams, with wide, flat, scaly ridges, frequent diagonal crossings, peeling in narrow, flat, vertical strips.

A valued but uncommon forest tree of the northwest, growing in canyons and on north-facing slopes. Sometimes planted in northern coastal towns of California, but older specimens are slowly disappearing in the central state. Does best with moderate moisture and shading of roots. Withstands cold and poor soils better than Lawson cypress. Longevity estimated to be 200-275 years, and up to 500 years in habitat. Series associations: Alayelcedsta. KlaMouEnrsta.

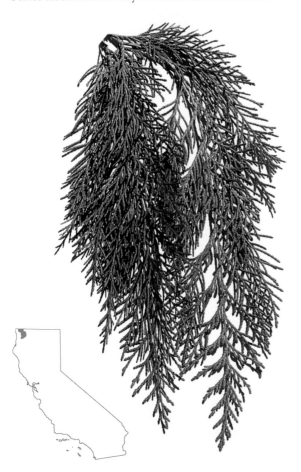

sw N.A. California Native: MD, SD (Riparian) (*D, adjacent TR, PR*); NAT, FLW ACC; (IRR: VL/VL/VL/L/M/M)

Chilopsis linearis

DESERT-WILLOW

Bignoniaceae. Sunset zones 3, 7-14, 18-23. Deciduous. Native to desert washes and streambeds below 5,000' elevation in southern California east to Nevada and Texas and south to Mexico. Growth rate fast to 15-30' tall and 10-20' wide, often multi-branching near the base, with slender, weeping branches forming an open, arching canopy. Leaves are simple, 5-10" long by 1/4" wide, alternate, drooping, opposite or whorled, grayish to bluish green, linear to linear-lanceolate, with smooth edges and blunt acuminate ends, young leaves often slightly sticky, and dull yellow fall color. Showy, loose clusters of perfect, tubular, bell-shaped, 1-2" long flowers in racemes are white or purplish pink, white at the base with a yellow spot in the throat, resembling flowers of a catalpa, with a faint violet-like fragrance, occurring intermittently throughout spring and summer. Slender, 7-12" long, vanilla-bean seed capsules are packed with many small, flat, hairy-winged seeds. Twigs are slender, yellowish to light brown, sticky at first, sometimes hairy, with tiny lateral buds, often with no winter terminal bud. Bark is thin, rough, dark brown, with many shallow interconnected furrows and ridges.

A useful but not widely cultivated flowering accent tree for native gardens. Tolerates aridity as well as poor or sandy soils. Twisted and slightly contorted trunks occur naturally. Reseeds profusely in the Mojave Desert, usually the white-flowered type. Longevity estimated to be less than 50 years, then losing vigor and shape. Series associations: Blupalveriro.

Cultivar. SHD, FLW ACC, SPC; IRR: L//M/L/L/L/M

x *Chitalpa tashkentensis*

CHITALPA

Bignoniaceae. Sunset zones 3-24. Deciduous. Hybrid between *Catalpa bignonioides* and *Chilopsis linearis* developed in the U.S.S.R., introduced to the U.S., and widely available since the early 1980s. Growth rate fast to 20-30' tall and wide, forming a broad oval canopy as a single-trunk standard or more commonly as a multi-trunk large shrub. Leaves are alternate, simple, 4-7" long by 1" wide, shiny dark green, lanceolate, with a crinkled appearance between the veins, pale undersides, minutely hairy on both sides, and an acutely pointed apex. Showy, erect, raceme clusters of short-stemmed, 1" long, white to pink, trumpet-shaped flowers, with a frilled collar, occur throughout spring into late summer. A sterile hybrid, does not produce pods. Twigs are green and smooth. Bark is smooth, gray, thickening and becoming grayish brown, with fissures and scaly ridges.

A useful accent tree for riparian settings or native gardens. Tolerates moderate aridity, but looks best with regular deep watering, otherwise leaves tend to turn brown in late summer. Requires staking and pruning to develop strong trunks. Often used like a willow, but more attractive. Longevity estimated to be 50 years or more. 'Morning Cloud' has white flowers.

w N.A. California Native: CF, CC, CH (*NW, nCaR, n SNH El Dorado County, CW exc ScoRI*); (IRR: No WUCOLS)

Chrysolepis chrysophylla

GOLDEN CHINQUAPIN

Fagaceae. Sunset zones 4-7, 14-17, Jepson *DRN:**4-6**,15-17,& IRR:1-3,7,14,18;DFCLT. Evergreen. Native to the Cascade and Coast ranges and Sierra Nevada from southern Washington to central California. Growth rate slow to 90' tall and 50' wide, forming a dense, upright, conical to rounded canopy with multiple slender, clear trunks in forest habitat, in open areas or alpine climates densely shrublike and not as tall. Leaves are alternate, simple, 2-4" long, lanceolate to oblong-ovate, dark glossy green, leathery, with smooth margins slightly recurving to acuminate ends, golden tomentose undersides, persisting 2-3 years, with a slight yellowing before falling. Monoecious, apetalous, deep golden yellow, erect, tassel-like flowers occur in early summer, staminate males in sparse clusters along the flower stem and a few similar-looking females near the base of the stem. Fruits are 1/2-3/4" long, greenish-spined, burr-like, maturing to golden brown in the second year, when the spiny skin splits along 4 seams, releasing small, edible, shiny, dark brown, chestnutlike seeds. Slender twigs are quite stiff, with golden tomentose scales, becoming smooth reddish brown. Bark is thick, dark reddish brown, becoming furrowed, with broad, scaly ridges.

An attractive native evergreen, uncommon in its widespread native habitat and adaptable to poor, dry soils. Develops a deep taproot, with widespreading lateral roots. Rarely offered for cultivation, as it usually does not transplant well, but useful as a native garden specimen or background grouping. Does best with regular deep watering in well-drained soils. Closely related to *Castanopsis*, or Japanese chinquapin, sometimes seen in parks and arboretums. Longevity estimated to be 150-200 years. Various cultivars are more readily available. Series associations: Doufir, Doufirtan, Whifir.

e Asia. EVG, SHD; IRR: M/M/M/M/-/M

Cinnamomum camphora

CAMPHOR TREE

Lauraceae. Sunset zones 8, 9, 12-24. Evergreen. Native to China, Japan, and Taiwan. Growth rate slow to moderate to 50' tall or more and up to 60' wide, forming a broad, round-headed, billowy canopy, a heavy trunk, and upright spreading branching, eventually casting deep shade. Leaves are alternate, simple, 1-3" long, glossy yellowish green, ovate, smooth and waxy, with long slender pointed tips, pale yellowish undersides, aromatic when crushed, older leaves turning reddish in spring before falling after new growth begins. Insignificant flowers appear from new growth buds in April as new leaves become fully formed. Trees are covered with lightly fragrant, loose, cloudlike, 2" long panicles of 1/8" long, whitish, star-shaped flowers, 6-petaled, from a slightly bulbous green base. Small, ovoid, pea-sized, juicy fruits with a fattened, green, cuplike base and a single tiny seed turn black in winter, and remain hanging among leaves from a 1/2-3/4" long stalk. Twigs are smooth, glabrous, yellow-green, darkening to greenish brown in fall. Young bark is thin and green, thickening and becoming grayish brown, furrowed, with scaly squarish plates on mature trees.

A commonly used street or shade or tree for large lawns with a clean, shiny appearance. Eventually becomes quite large with a massive trunk, large limbs, and competitive roots. Difficult to plant beneath, and roots may heave paving. Prefers well-drained soils, otherwise leaves droop and look chlorotic. Half-hardy, limiting its range to lower elevations. Tolerates heat, but drought slows growth. Longevity estimated to be 100-150 years.

se N.A. SHD, FLW ACC; IRR: No WUCOLS

Cladrastis kentukea

(*Cladrastis lutea*)

YELLOWWOOD

Fabaceae. Sunset zones 2-9, 14-16. Deciduous. Native to Arkansas, Kentucky, Tennessee, and North Carolina. Growth rate slow to 30-50' tall and 15-25' wide, developing a broad oval crown with slender, upright branching from a strong central trunk. Leaves are alternate, pinnately compound, 10-12" long, with 5-11 obovate pointed leaflets, 1-3" long, bright yellow-green, with smooth edges, pale undersides, and bright yellow fall color. Pendulous, 6-12" long clusters of showy white legume flowers appear from new growth buds in April in long branched panicles, resembling wisteria, with an intriguing heavy, spicy, clovelike fragrance. Seeds are in drooping clusters of 3-4" long, flat, short-stalked, legume pods, each with 4-6 hard flat brown seeds, ripening to grayish brown in late summer, and may persist on branches through winter. Twigs are slender, glabrous green at first, later turning reddish brown. Bark is thin, grayish brown, smooth, nearly shiny, with tiny glandular warts.

An uncommon, attractive street, lawn, or patio tree, excellent as a shade tree, with a neat, clean appearance. Relatively pest free. Does not bloom until at least 10 years old, and then not every year. Hardy, deep rooted, and tolerates relatively poor soils with occasional deep watering, becoming heat and drought resistant when established. Longevity estimated to be up to 200 years.

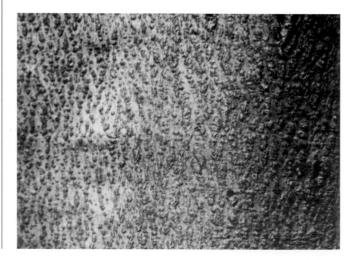

e Asia. EVG, SHD, FLW SPC; IRR: No WUCOLS

Cornus capitata

EVERGREEN DOGWOOD

Cornaceae. Sunset zones 8, 9, 14-20. Evergreen. Native to the Himalayas and western China. Growth rate moderate to 20-30' tall and wide, rather columnar in youth, broader with age, with an upright central trunk and an irregular, upright, billowy form, side branches sometimes semi-weeping. Leaves are simple, 2-4" long by 3/4 to 1-3/4" wide, green to grayish green, oval-elliptical, with pointed ends, smooth margins, pale downy undersides, some turning reddish yellow when about half the foliage drops in winter. Tiny, greenish yellow, perfect flowers are bisexual, in a tight rounded head, surrounded by 4 flower bracts, which are 1-2" long, creamy white with pointed ends, noticeable but not showy, in early summer. Fruits are curious, long-stemmed, succulent, berrylike, bright pinkish red to orange-purple, oval to flattened, tightly packed, bitter tasting, fleshy drupes, each containing tiny, hard seeds, maturing in fall, and can be messy on the ground. Twigs are greenish gray with tiny, round, buttonlike flower buds through winter. Bark is thin, smooth, reddish gray, darkening with age and becoming slightly fissured, or broken into irregularly scattered small scales.

An interesting, but rather tender evergreen dogwood for gardens, with summer flowering and unique fruit. Does best in semi-moist, well-drained soil. Prefers half shade inland. Does not bloom until a sizable trunk has developed. Not as showy as other dogwoods. Longevity estimated to be 40-60 years, gradually losing vigor, and up to 90 years as a gnarled mature tree.

e N.A. SHD, FLW ACC; IRR: H/H/H/-/-/-

Cornus florida
EASTERN DOGWOOD

Cornaceae. Sunset zones 2-9, 14-16. Deciduous. Native to the eastern U.S. Growth rate slow to moderate to 20-30' tall, taller in shade, with an equal or greater spread, developing an upright, billowy form accented by horizontal branching and foliage tufts. Leaves are opposite, simple, 2-5" long by 2-3" wide, bright green, oval-elliptical with pointed ends, arcuately veined, finely hairy, with lighter undersides, and red to yellow fall color. Tiny, greenish yellow, perfect, bisexual flowers are in a tight, rounded head, surrounded by 4-8 showy flower bracts, 2-3" long, white, pink, or red, with notched ends, appearing in early April. Fruits are bright scarlet red, 1/2", ovoid, bitter-tasting drupes in clusters of 3-4, containing 1 or 2 hard-shelled dark seeds, maturing in fall, sometimes remaining briefly on bare branches after leaves have fallen. Twigs are greenish gray and develop tiny, round, buttonlike flower buds in summer at turned-up ends of the present year's growth. Bark is thin, smooth, reddish gray, becoming darker and slightly fissured or broken into small square plates.

A prized flowering accent tree for gardens. Prefers well-drained, fertile soils with regular moisture. Tolerates some sun inland in lawns in central and northern portions of the state where soils are not alkaline. Longevity estimated to be 90-120 years. 'Cherokee Chief' has rosy red flower bracts, 'Cherokee Princess' has a profusion of white blooms, and 'Rubra' has delicate pink to rose blooms.

Cultivar. FLW, FOL ACC; IRR: H/H/H/-/-/-

Cornus florida 'Rainbow'

PP #2743 (1967)

VARIEGATED DOGWOOD

Cornaceae. Sunset zones 2-9, 14-16. Deciduous. Variety of the species. To about 15' tall or more, usually with a small broad oval canopy. Leaves are opposite, simple, oval-elliptical, green with deep yellow edging, somewhat thick but not leathery, raised at the veining, with bright red fall color, sometimes with a purplish tinge. White bracts have a U-shaped inset at the ends, brown with a reddish tinge. Fruits are bright red.

An attractive, variegated flowering accent tree, smaller than other dogwoods. Relatively recently introduced and not always available, but worth looking for at specialty nurseries. Not overly fussy, but leaves sometimes burn or fade in hot sun.

Cultivar. FLW ACC; IRR: H/H/H/-/-/-

Cornus florida 'Cherokee Sunset'

PP #6305 (1988)

PINK VARIEGATED DOGWOOD

Cornaceae. Sunset zones 2-9, 14-16. Deciduous. Variety of the species. To 20' tall or more, slightly less vigorous and smaller than the species. Leaves are opposite, simple, green, oval-elliptical, with deep yellow edging, rather thick, puckered between the raised veins, with red fall color. Bracts are deep pink to red, with a U-shaped inset at ends, with a brown tinge. Fruits are bright red.

e Asia. FLW ACC; IRR: No WUCOLS

Cornus kousa

KOUSA DOGWOOD

Cornaceae. Sunset zones 2-9, 14-17. Deciduous. Native to Japan and Korea. Growth rate moderate to 20' tall or more and equally wide, developing an irregular upright form with slender, horizontal branching. Leaves are opposite, simple, shiny green, ovate, 2-4" long, with distinctly acuminate ends, rusty-colored hairs at the veining on undersides, and yellow to reddish fall color. White 4-bracted flowers, with distinctly pointed ends, occur in late spring and early summer on long stems rising above the foliage, followed by drooping red fruits in fall.

A less disease-prone alternative to native and eastern dogwoods. Longevity estimated to be 80 years or more. 'Milky Way' has abundant, pure white flowers, 'Summer Stars' has white flowers and often reblooms in midsummer, and 'Rosabella' has pink flowers.

s Europe, Asia. FLW ACC; IRR: No WUCOLS

Cornus mas

CORNELIAN CHERRY

Cornaceae. Sunset zones 1-6. Deciduous. Native to southern Europe and Asia. Growth rate slow to 15-20' tall and wide, forming an upright, loose, twiggy, often multi-trunked tree. Leaves are opposite, simple, 2-4" long, shiny green, oval-lanceolate, with red or yellow fall color. Tiny yellow-green flowers appear in small, tightly packed clusters along bare branches in early spring. Fruits, usually concealed by leaves, are loose clusters of 3/4" long, sweet-tasting, edible drupes, slowly ripening to bright scarlet in September, sometimes hanging on bare branches into winter. Twigs are greenish, slightly hairy at first, becoming glabrous and turning tan in winter. Bark is smooth, light brown, becoming gray with age, peeling in flakes, and exposing patches of orange-red underbark.

A small garden accent tree, useful for its multi-trunk form and interesting winter silhouette. Tolerates alkaline soils. Prefers moderately moist, well-drained soils. Very hardy, withstanding the coldest temperatures. Longevity estimated to be 80-125 years.

w N.A. California Native: ME, RW, PF (Riparian) (*CA-FP less common in s. CA*); **NAT, FLW ACC**; (IRR: M/M/-/M/-/-)

Cornus nuttallii

WESTERN DOGWOOD

Cornaceae. Sunset zones 3-9, 14-20, Jepson *DRN:**4,5**&IRR:**6**,&SHD:1,2,7,14-17,22,2;DFCLT. Deciduous. Native to shaded habitats along streambanks of the Pacific Northwest and northern California foothills. Growth rate moderate to 50' tall with a 20' spread, usually with multiple or low-forked upright trunks, forming a rounded canopy in sunnier locations, more sparsely branched in heavy shade. Leaves are simple, opposite, 3-5" long, oval, light green, with entire edges, minutely hairy, pale undersides, and glowing pastel pink, red, and yellow fall color. Tiny, greenish yellow, perfect, bisexual flowers appear in April in a tight, rounded head, surrounded by 4-8 showy, 2-3" long flower bracts. They have rounded, shortly pointed ends, not notched, and are white, sometimes with a pink tinge, occasionally reblooming in early fall. Fruits are bright scarlet red, 1/2" long, ovoid, bitter-tasting drupes, appearing in clusters of 10 or so in fall and containing 1 or 2 hard-shelled dark seeds. Twigs are greenish gray with tiny, round, buttonlike flower buds, formed in summer at ends of the present year's growth. Bark is thin, smooth, gray, remaining so with age, often with a whitish or blackish lichen blotching in moist canyon locations.

A desirable native riparian understory tree for higher elevations, less adaptable to low-elevation gardens than eastern dogwood. Requires cool, moist, well-drained, acid soil. Prefers filtered shade. Longevity estimated to be 100 years or more in habitat. Series associations: Doufir, Whifir.

cw N.A. California Native: CP, SS, RW, ME, CF (*MP*); (IRR: M/M/-/M/M/-)

Crataegus douglasii
BLACK HAWTHORN

Rosaceae. Deciduous. No Sunset zones. Jepson *SUN:**4-6**,15-17&IRR;1,**2**,**7**,10,14,18. Native to southern Alaska to central California east to New Mexico and Michigan. Growth rate slow to 20-30' tall, with a compact rounded crown, usually forming thickets in habitat. Leaves are alternate, simple, 1-3" long, glossy green, obovate to ovate, with sharply serrate edges, sometimes lobed, with a short-pointed tip. Flowers are in flat clusters, white-petaled with pink stamens, appearing in spring. Fruits are 1/2" berries, ripening to shiny black, and containing 3-5 hard seeds. Twigs are slender, shiny red, usually with 1" long spines. Bark is smooth, grayish brown, becoming scaly.

An attractive, hardy tree, rarely cultivated but adaptable to garden settings. Prefers a cool, semi-moist location. Longevity estimated to be 50-75 years.

Cultivar. SHD, FLW ACC; IRR: M/M/-/M/M/-

Crataegus x *lavallei*
CARRIERE HAWTHORN

Rosaceae. Sunset zones 3-12, 14-21. Deciduous. Hybrid between *Crataegus crus-galli* and *C. pubescens*. Growth rate moderate to 25' tall with a 15-20' spread, forming a heavy, broad, round-topped canopy. Leaves are alternate, simple, ovate, 2-4" long, dark green, with lightly serrate edges and reddish orange fall color. Showy white flower clusters appear in late spring, in terminal corymbs, attractive to bees. Red fruits in showy clusters resemble pyracantha, contain 1-5 bony dark seeds, and attract birds into winter. Bark is dark gray-brown, slightly scaly or peeling in thin, narrow strips.

Slightly larger and heavier in texture than other hawthorns. Does best in moist, fertile soil in full or half sun inland. Subject to aphids and occasional fireblight. Longevity estimated to be 75-100 years.

Cultivar. SHD, FLW ACC; IRR: M/M/-/M/M/-

Crataegus phaenopyrum
WASHINGTON THORN

Rosaceae. Sunset zones 2-12, 14-17. Deciduous. Native to the southeastern U.S. Growth rate slow to moderate to 25' tall and 20' wide, with upright growth and light-textured, slender branching. Leaves are alternate, simple, 2-3" long, glossy green, ovate, with 3-5 pointed lobes, serrate edges, and reddish orange fall color. Showy 3-4" clusters of 1/2" white flowers occur in late spring, attracting bees. Shiny red fruits resemble pyracantha, each with 1-5 bony dark seeds. Young trees have smooth gray bark.

An attractive tree with a delicate branching habit. Less susceptible to fireblight than other hawthorns, but subject to aphids. Longevity estimated to be 75-100 years.

Europe. SHD, FLW ACC; IRR: M/M/-/M/M/-

Crataegus laevigata

(*Crataegus oxyacantha*)

ENGLISH HAWTHORN

Rosaceae. Sunset zones 2-12, 14-17. Deciduous. Native to Europe. Growth rate slow to moderate to 18-25' tall with a 15-20' spread, developing a tall, irregularly branching, semi-arching canopy. Leaves are alternate, simple, ovate, 2-3" long, dark green, with 3-5 lobed indentations, appearing palmate, with serrate margins, and red or orange fall color. Masses of showy white flower clusters in terminal corymbs in late spring are attractive to bees. Clusters of red, fleshy, pome berry fruits, resembling pyracantha, contain 1-5 bony dark seeds, hang into winter, showy on bare branches, and attract birds. Irregular twigs are reddish brown with short spines. Bark is dark reddish brown to gray, either slightly scaly or peeling in thin, narrow strips, may vary with grafting stock.

A showy small flowering tree, best in moist, fertile soil in full or half sun inland. Often needs light pruning to develop an attractive habit, but older trees should not be severely pruned. Fireblight causes occasional dieback. Subject to aphids. Longevity estimated to be 75-100 years. 'Paul's Scarlet' has rose-red double flowers, 'Double White' and 'Double Pink' have double flowers, 'Crimson Cloud' and 'Superba' have single red flowers with white centers, and 'Punicea Flore Pleno' has double pink flowers with mottled white centers.

TOP RIGHT: Photograph courtesy of Wayside Gardens.

South America. EVG, SHD, FLW ACC; IRR: M/-/M/?/-/-

Crinodendron patagua

LILY-OF-THE-VALLEY TREE

Eleaocarpaceae. Sunset zones 14-24. Evergreen. Native to Chile. Growth rate moderate to 25' tall with a nearly equal spread, forming a dense oval or broad-headed tree with upright branching and twiggy, drooping ends, resembling an evergreen oak. Leaves are alternate, simple, ovate-elliptical, 2-1/2" long by 1/2-1" wide, dark green, with finely toothed edges, slightly wavy and curving under, with blunt ends, light green undersides, persisting 1-2 years. Clustered masses of paired, 1/2-3/4", bell-shaped flowers appear in early summer at leaf axils of the present year's growth, somewhat hidden by foliage but noticeable up close, with multiple white, elongated, flat, overlapping petals that curve slightly outward at jagged-edged ends, hanging from 1-1/4" long petioles. Light green, elongated, bladderlike pods with swollen, crinkled side seams, outer skin turning glowing scarlet red in late summer, give the appearance of a tree in flower. Pod seams split when dry in an outward recurving fashion, pagodalike, with tiny, dark black, beanlike seeds loosely attached to the light tan inner surface. Twigs are slender, smooth, and green. Bark is reddish brown with many shallow fissures and broad, scaly ridges.

An unusual, small, evergreen shade tree with attractive foliage. Tolerates wet soils and thrives in lawns. Growth more dense in a somewhat sunny location. Deep watering reduces surface rooting. Fruit drop can be messy around paved areas. Longevity estimated to be 60-90 years.

e Asia. EVG, FOL ACC, SPC; IRR: M/H/H/H/-/-

Cryptomeria japonica
JAPANESE CRYPTOMERIA

Taxodiaceae. Sunset zones 4-9, 14-24. Evergreen. Native to Japan and China. Growth rate moderate to fast to 60-100' tall with a 15-30' spread, developing a strong, upright trunk with pyramidal growth and horizontal branching, resembling a giant sequoia in youth but thinning with age, exposing the trunk through patches of dense foliage tufts and drooping branchlets. Leaves are overlapping, linear-rectangular, needlelike, 1/2-3/4" long, dark green, with soft but sharp points curving slightly inward, turning reddish purple in fall and winter and returning to green in spring, persisting 4-5 years, with a cedarlike fragrance. Small clusters of 1/4" catkinlike male flowers release pollen in March, the scaly, globose, greenish females at branch ends enlarging and becoming rosette-shaped in summer. Small, rounded, 3/4", reddish brown cones, with 20-30 woody scales, each with 4-6 dull-pointed hooked spurs, contain tiny seeds. Reddish brown bark peels in long vertical shreds.

An unusual accent tree, dense evergreen screen, or lawn specimen, eventually becoming a towering skyline tree. Prefers deep, fertile soil in full sun, and deep watering hastens growth. Resists oak root fungus. The national tree of Japan, where it is a forest tree and also widely cultivated. Longevity estimated to be over 1,000 years.

Cultivar. EVG, CNF, SPC; IRR: M/H/H/H/-/-

Cryptomeria japonica 'Elegans'
PLUME CEDAR

Taxodiaceae. Sunset zones 4-9, 14-24. Evergreen. Variety of the species. Growth rate slow to 20-60' tall and 20' wide, developing a broad-based, compact pyramidal or obovoid form with age, either densely symmetrical or irregular and billowy, with twiggy branching often to the ground and feathery, light-textured foliage densely set on many slender branchlets. Needlelike leaves are grayish green, 1/2-3/4" long, awl-like, with soft, sharp, straight points, whorled and angled outward from branchlets, turning reddish purple in cold weather, returning to green in spring, and persisting 3-4 years. Tiny flowers are insignificant, and the small, globular, 1/8-1/4" long cones are rarely seen. Green twigs are succulent and pliable. Bark is reddish brown, fibrous, becoming shallowly fissured, with broad, thin, flaking ridges.

An accent specimen for gardens or near buildings, sometimes small groves, usually in lawns. Can be pruned to expose trunk and branches in giant bonsai fashion or clipped in symmetrical shapes. Moderately moist, well-drained soil maintains vigor. Resists oak root fungus and tolerates smog. Longevity estimated to be 500 years or more.

e Asia. EVG, CNF, FOL ACC, SPC; IRR: No WUCOLS

Cunninghamia lanceolata

CHINA FIR

Pinaceae. Sunset zones 4-6, 14-21. Evergreen. Native to China. Growth rate moderate to 30' tall or more with a 20' spread, developing a pyramidal to columnar form with a strong trunk, semi-weeping, horizontal branches, and softly textured foliage. Leaves are whorled, simple, linear-lanceolate, gray-green, needlelike, 2 to 2-1/2" long, with finely serrate edges, long-pointed but not sharp, somewhat pliable, lighter undersides with 2 broad bands of white pores, turning reddish bronze in colder weather. Monoecious flowers are fat clusters of oblong-cylindrical, greenish yellow males, which shed pollen in April, and solitary, globose, 1/2" females at branch ends. Cones are leathery, dark brown, 1-2" long, with thick-pointed scales, drooping from a 1/2" stem as they mature in late summer, and containing light brown, short-winged seeds. Bark is reddish brown aging to gray-brown, with long, shallow furrows and narrow, scaly ridges.

An interesting accent or specimen tree with an attractive soft texture. Not especially hardy. Can be rather untidy, with dead leaves and branchlets remaining indefinitely unless removed and becoming irregularly shaped with age. Needs moist, fertile soil and protection from hot, dry or cold winds. Tolerates moderate inland heat in a fairly moist, parklike setting or in semi-shaded lawns. Longevity estimated to be 75-100 years or more. 'Glauca' has stunning silvery blue foliage.

Australia. EVG, SHD, FOL ACC; IRR: M/-/M/M/-/-

Cupaniopsis anacardioides

CARROT WOOD

Sapindaceae. Sunset zones 16-24. Evergreen. Native to Australia. Growth rate slow to moderate to 40' tall with a 30' spread, developing an upright, oval form, open-branched in youth and becoming quite dense with age. Leaves are alternate, pinnately compound, 6-12" long, glossy green, with 6-10 oblanceolate blunt-ended leaflets, 4" long, with smooth, slightly recurving or wavy edges, resembling those of a carob but less rounded and leathery. Insignificant tiny yellowish flowers appear in spring in small loose panicle clusters. Fleshy yellow fruits harden into woody capsules, with 3 or 4 bulging compartments, spliting at the seams, containing black seeds the size of pine nuts. Twigs are smooth and reddish brown. Bark is chalky gray, becoming fissured and slightly scaly, darkening to brown as the trunk thickens and fissures widen.

A commonly grown street, lawn, or shade tree in coastal southern California. Often grown as a multi-trunk accent tree, most effective as a courtyard or terrace specimen. Deep rooted, with a clean appearance. Tolerates poorly drained soils, but prefers occasional deep watering in well-drained soils. Not reliably hardy inland. Longevity estimated to be 100 years or more in mild climates.

Bigeneric hybrid. EVG, FOL ACC, SCR; IRR: M/M/M/-/M/M

x *Cupressocyparis leylandii*
LEYLAND CYPRESS

Cupressaceae. Sunset zones 3-24. Evergreen. Naturally spontaneous hybrid, developed in Great Britain, between *Cupressus macrocarpa* and *Chamaecyparis nootkatensis*, but without the long-lived properties of either parent. Growth rate initially rapid to 20' tall (eventually to 60-70') by 8-15' wide, developing a narrow, vertical, pyramidal form with angular branches radiating upward from the trunk, though may become more open and floppy with age. Leaves are pointed, 1/16" long, scale-like, bluish to grayish green, in flattened foliage sprays, older needles turning brown after 2 years, persisting for another year underneath new foliage before falling. Insignificant, tiny yellow flowers occur at branch ends, oblong male flowers producing pollen in March; females more rounded. Cones are dark green, 3/4-1", globular, on short scaly stems, with blunt-knobbed scales, maturing to shiny tan in the second season, each scale containing 5 dark brown seeds, 1/16-1/8" long, oblong to triangular, with thin, narrow lateral wings. Bark is smooth, grayish brown, becoming dark and ridged, peeling in thin, stringy shreds.

A dense, fast, evergreen screen. Tolerant of wind and heat with average watering in poor soils. Prefers coastal climates. Suffers from drought or overwatering after prolonged dryness. Subject to coryneum canker virus dieback, which occurs suddenly. Longevity estimated to be 20-30 years in warm interior climates, though longer-lived in the northwest with consistent moderate moisture. 'Castlewellan', golden Leyland cypress, has gold-tipped new foliage.

w N.A. California Native: CH, OW, JP, PF (s Calif.) (*s SNH, Teh, PR; Z, n Baja CA, n&c Mex.*); EVG, SCR; (IRR: VL/VL/VL/L/L/L)

Cupressus arizonica

= Cupressus arizonica ssp. arizonica*

ARIZONA CYPRESS

Cupressaceae. Sunset zones 7-24. Evergreen. Native to transitional dry mountains and high desert plateaus in Arizona and New Mexico into Mexico and in scattered groves in southeastern California, naturalized throughout other dry zones of the state. Growth rate initially moderate to fast, slowing with age, to 40' tall with a 20' spread, developing a dense, tall, sharply conical to pyramidal form with foliage to the ground. Leaves are bluish to silvery gray-green and scalelike, each scale 1/16" long and pointed, with glandular depressions on the back, a skunklike odor if crushed, turning brown after 2 years, and persisting another 2 years underneath new foliage. Insignificant, tiny, yellow, oblong male flowers occur at terminal branch ends, producing pollen in February. Female flowers are more rounded. Dark green, 3/4-1", oval cones have short scaly stems, 6-8 peltate scales with stout, incurving, blunt spikes, maturing in 2 seasons to dark reddish brown, each with 16-20 dark brown, 1/16-1/8", oblong to triangular seeds with thin, narrow lateral wings. Bark is dark brown, loosely scaled, peeling in thin curling shreds, exposing a smooth, bright cherry-red layer underneath.

Limited in its natural range in dry rocky soils in pinyon-juniper woodland, but quite adaptable and commonly cultivated, often as a fast-growing evergreen windbreak, noise buffer, or accent tree. Tolerates drought when established. Longevity estimated to be 50-100 years, often more in habitat. 'Blue Pyramid' has blue-gray foliage and a dense pyramidal form, 'Gareei' has lush, silvery blue-green foliage, and 'Pyramidalis' is compact and symmetrical with silvery glaucous foliage.

*In nurseries more often available as *Cupressus glabra*.

e Asia. EVG, CNF, SHD, WPG, FOL ACC, SPC; IRR: No WUCOLS

Cupressus funebris
MOURNING CYPRESS

Cupressaceae. No Sunset zones. Evergreen. Native to China. Growth rate moderate to 30-60' tall with a nearly equal spread, with heavy, widespreading limbs and pendulous, weeping branch tips. Lacy, drooping, flat foliage is in sprays of minute, yellowish green, flattish scalelike leaves, with short blunt points, non-glandular on the surface, and growing in a threadlike, slightly zigzag fashion. Reddish brown male flowers appear as swollen bulbs at leaf tips, shedding pollen in spring. Tiny green female flowers form green, juniperlike, 3/8" long, oval cones among the foliage, maturing to brown in fall of the first season. The 8 wide, flat, shortly-horned scales open, exposing the center portion when fully ripe and releasing tiny brown seeds, which have a common wing encircling each side. Some cones may remain attached through the following season. Bark is thin, reddish brown, peeling in thin, hairy strips.

A seldom used but classic park tree that needs room to spread and display its attractive shape. Tolerates wet soil better then Lawson cypress, and does well in groupings around ponds. Lower branches usually trimmed up to expose colorful reddish trunks and limbs or to allow lawns to grow in the dense shade. Longevity estimated to be 150-250 years.

sw N.A. EVG, FOL ACC, SCR, SPC; IRR: L/L/VL/VL/-/-

Cupressus guadalupensis

GUADALUPE ISLAND CYPRESS

Cupressaceae. No Sunset zones. Evergreen. Native to Guadalupe Island, off the coast of southern California and Mexico. Growth rate moderate to fast to 15-30' tall with a 15-20' spread, developing a dense, rounded conical to pyramidal form, with horizontal, upward-spreading branches. Lower branches may remain to the ground, but most foliage is concentrated at twiggy branch ends, with inner limbs bare. Bright green to bluish green foliage is scalelike, each scale 1/16" long, with dull or rarely pointed ends, persisting 2 years. Tiny yellow, oblong, male flowers at branch ends produce pollen in February. Female flowers are more rounded. Dark green, 3/4-1" oval cones, on short scaly stems, with 8-10 peltate scales, mature in the second season to dark reddish brown, each scale with up to 70 short 1/16" brown seeds, oblong to triangular, with thin, narrow lateral wings. Bark is dark brown, loosely scaled, peeling in thin, curling shreds, exposing smooth, reddish underbark.

An uncommon and seldom cultivated tree attractive as an arboretum specimen. Closely resembles but is distinct from *Cupressus forbesi*, which occurs in San Diego and Orange counties. Tolerates drought when established, but does well in well-drained lawns in full sun. Longevity estimated to be 100-200 years.

California Native: CH, OW, CC, PF (n Calif.) (*NCoRI, CaRH, n SnF*); (IRR: No WUCOLS)

Cupressus macnabiana

MCNAB CYPRESS

Cupressaceae. No Sunset zones. Jepson *4-6,**7**,14,**15,16**,17,**18**, 19,20,**21**,22-24;CVS. Evergreen. Native to drier foothills of the Coast Ranges and Sierra Nevada, scattered throughout northern California, north to Oregon. Growth rate moderate to fast to 15-40' tall with a 15-20' spread, developing an open, conical or pyramidal form, usually shrubby, with foliage usually remaining low to the ground. Foliage is dark or bright green, scalelike, with a pungent odor, persisting 2 years beneath new growth. Each scale is about 1/16" long, with rounded terminal tips, conspicuously glandular-pitted on the back. Tiny, yellow, oblong male flowers, at branch ends, produce pollen in February. Female flowers are more rounded. Cones are dark green, 1/2-3/4", globose, with short scaly stems and 6-8 peltate scales, each with a protruding hook on the backside, maturing in the second season to dark reddish brown. Brown 1/16" seeds are oblong to triangular, with thin, narrow lateral wings. Bark is reddish brown to gray, becoming fissured and scaly.

An uncommon juniper in its widely distributed native habitat, which often overlaps with Sargent cypress. Sometimes grown as an arboretum specimen, but otherwise rarely cultivated outside its range. Longevity estimated to be 100-200 years. Series associations: Leaoak shrub series; **McNcypress**, Sarcypress.

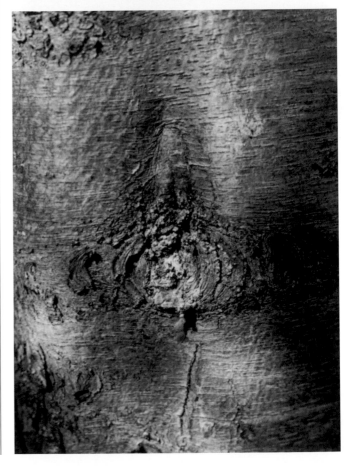

w N.A. (Ca only) California Native: CP, CC (c Calif.) (*NCo, Cco*); EVG, SCR; (IRR: M/M/M/-/-/-)

Cupressus macrocarpa

MONTEREY CYPRESS

Cupressaceae. Best in Sunset zone 17. Evergreen. Native only to the Monterey peninsula but widely planted and naturalized elsewhere along the California coast. Growth rate initially moderate to fast to 40-60' tall and 25-35' wide, with a dense pyramidal form in youth, broad at the base, with strong horizontal branching angled upward to a pointed crown. Lower branches eventually die back on older trees, as they cease vertical growth, and trees develops a wide, flat-topped crown, the massive, spreading upper branches picturesque and irregular, especially in windy coastal areas. Foliage is dark or bright green, scalelike, each scale 1/16" long or longer, with rather blunt tips, and usually without glands, persisting 2-3 years. Male flowers appear as yellowish bulbs at the ends of foliage and shed pollen in March. Greenish females form dark brown to grayish, 1" round cones, with 4-6 wide flat scales, with blunt knots, maturing in August of the second year, opening at seams between scales, and may remain attached to the tree for several years. Seeds are numerous, angular, 1/8-1/4", and shiny dark brown.. Bark is reddish brown, narrowly seamed with a network of narrow vertical ridges and smaller diagonal ones, often taking on a grayish cast as it weathers.

Famous for its stately silhouette, and cultivated in coastal areas as a dense, fast-growing windbreak, hedge, or park tree. Characteristically shorter and contorted along windswept shorelines, becoming taller with straight trunks inland. Needs a moist, cool climate, and suffers from coryneum canker fungus inland. Also may suffer from wood borers. Longevity estimated to be 100 years, up to 200 years in habitat. Series associations: **Moncypsta**.

w N.A. (Ca only) California Native: CH, CC, ME, PF (in Calif.) (*NCoR, SnFrB, ScoR*); (IRR: No WUCOLS)

Cupressus sargentii

SARGENT'S CYPRESS

Cupressaceae. No Sunset zones. Jepson *DRN:3,6,**7**,15,**16,17, 24**&IRR:14,18-23. Evergreen. Native to drier foothill slopes and valleys of northern and central California Coast Ranges from Mendocino County, where it often hybridizes with Mcnab cypress, south to Santa Barbara County. Growth rate slow to moderate to 10-40' tall or more, with narrow conical growth broadening in exposed locations, often appearing shrublike, with a broad, rounded canopy. Foliage is dull green, scalelike, with scales in 4 rows, some with a gland dot. Cones are 3/4-1", dull brown to gray, round or oblong, with 6-8 wide-rounded peltate scales, each with an inconspicuous umbo on the backside, containing many angular, dark brown seeds, and may persist unopened for years. Twigs are stiff, stout, 4-angled, branching in various directions. Bark is dark grayish brown to nearly black, thickening with age and becoming fibrous, rough, and furrowed.

One of the most widely distributed junipers in California, but seldom found in abundance. Cones open and seeds germinate after fire, establishing quickly in newly exposed areas but not on heavily thatched ground. Sometimes grown as an arboretum specimen, but otherwise not often cultivated outside its range. Longevity estimated to be 100-200 years. Series associations: Leaoak shrub series; McNcypress, **Sarcypress**.

s Europe, w Asia. EVG, FOL ACC, SCR, VERT ACC; IRR: L/L/L/L/
M/M

Cupressus sempervirens

ITALIAN CYPRESS

Cupressaceae. Sunset zones 4-24, best in zones 8-15, 18-21.
Evergreen. Native to Mediterranean regions of southern Europe and western Asia. Growth rate moderate to 60' tall or
more by 5-10' wide at the base, developing a variable, somewhat ragged, upright pyramidal form with variable rangy,
semi-twisted branching. Foliage is green to bluish green, scalelike, rhombic-shaped, in sprays, leaf scales with resinous
glands, persisting on branchlets 3-5 years. Yellowish green,
oblong-cylindrical, 1/8" male flowers, shedding pollen, occur
near branch ends in March. Greenish female flowers are conical. Dark brown, 1" ovoid cones with 8-14 woody scales, each
bristle-tipped at the center, ripen in 18 months and may persist for years, some open, some remaining closed. Bark is grayish brown, fissured, with thin, peeling ridges.

Rather common, though rarely offered in nurseries, as
upright cultivars are more popular than the species, which is
quite variable in form. Tough and durable as a background
screen or windbreak. Not especially attractive, but may be useful as an upright conifer in difficult situations. Probably one
of the oldest cultivated columnar trees, dating from Roman
times. Longevity estimated to be 500 years or more.

Cultivar. EVG, FOL ACC, VERT ACC; IRR: L/L/L/L/M/M

Cupressus sempervirens 'Stricta'

COLUMNAR ITALIAN CYPRESS

Cupressaceae. Sunset zones 4-24, best in zones 8-15, 18-21. Evergreen. Variety of the species. Growth rate fast to 60' tall or more, with a narrowly columnar form, broadening with age, with tight vertical branching. Foliage is scalelike, green to bluish green, rhombic-shaped, in dense twiggy sprays, leaf scales with resinous glands, persisting on branchlets 3-5 years. Yellowish green, oblong-cylindrical, 1/8" male flowers, shedding pollen, occur near branch ends in March. Greenish female flowers are conical. Dark brown, 1" long, ovoid cones with 8-14 woody scales, each bristle-tipped at the center, ripen in 18 months, and may persist for years, some open, some remaining closed. Bark is grayish brown, fissured, with thin, peeling ridges.

    Classic vertical accent tree, effective grouped in geometric fashion, often in rows, or as individual accents for architectural features. Sometimes requires shearing or tying if it becomes floppy from overwatering or in windy locations. Often used as a drought-tolerant windbreak or screen. Longevity estimated to be 500 years or more.

Cultivar. EVG, FOL ACC, SCR; IRR: L/L/L/L/M/M

Cupressus sempervirens 'Swane's Golden'

SWANE'S GOLDEN CYPRESS

Cupressaceae. Sunset zones 4-24, best in zones 8-15, 18-21. Evergreen. Variety of the species, introduced to the U.S. from Australia in the early 1960s. Growth rate moderate, slightly slower than the species, to 40' tall or more, developing a dense, narrow, columnar form with twiggy side branching, broadening with age. Dense, yellow foliage sprays are vertical in growth, persisting 3-5 years, fading slightly with age to green underneath. Yellowish, 1" long, ovoid cones with 8-14 woody scales, each bristle-tipped at the center, ripen to brown in 18 months, and may persist for years, tucked in the foliage. Bark is grayish brown, heavily fissured, with thin, peeling ridges.

An unusual accent or specimen tree, but can be tiresome if overused. More of a curiosity, and rarely seen, but effective as pictured here among palms and cycads in a semi-tropical landscape setting.

Cultivar. EVG, FOL ACC, SCR; IRR: M/M/M/-/-/-

Cupressus macrocarpa 'Lutea'

GOLDEN MONTEREY CYPRESS

Cupressaceae. Best in Sunset zone 17. Evergreen. Variety of the species, introduced to the U.S. from England in the early 1900s and seldom cultivated today. Growth rate moderate to slower than the species, eventually to similar height and spread, with a dense young shrubby form, broad at the base, and strong horizontal branching angled upward. New growth is bright yellow, slowly fading to green. Young cones are yellowish, 1" round, with 4-6 scales with blunt knots, ripening to dark brown, and may persist for years.

An attractive golden accent tree, May be slightly more cold tolerant than the species. Rarely seen, and not commercially available, but makes quite a splash in areas where green conifers predominate. 'Golden Pillar' is a large, slow growing, pyramidal, shrubby form.

e Asia. EVG, SHD, FOL ACC; IRR: M/M/M/M/-/M

Eriobotrya deflexa

BRONZE LOQUAT

Rosaceae. Sunset zones 8-24. Evergreen. Native to China. Growth rate moderate to 15-30' tall and may be equally wide, shrubby unless trained to single- or multi-trunked tree, with a small dense canopy, somewhat pyramidal, broadening as horizontal branching develops. Leaves are alternate, simple, dark green, 6-12" long, oblong-elliptical, pointed at the ends, with shiny, smooth, waxy, coarsely toothed, nearly spiny edges and lasting reddish bronze new growth. Attractive, 2-3", upright-pointed raceme clusters of 1/4-1/2", dark pink flowers with 5 petals, resembling *Rhaphiolepis*, are noticeable but not showy in spring. Tiny inedible orange fruits are rarely seen. Twigs are smooth, shiny, glabrous, and reddish brown. Thin, smooth, gray-brown bark becomes slightly scaly, with shallow seams.

An excellent small courtyard, patio, or lawn tree, often espaliered on walls. Most often grown as a foliage accent or small shade tree. Does well in large containers and tree wells. Does not thrive in drought. Not as hardy as Japanese loquat (*Eriobotrya japonica*). Susceptible to aphids and fireblight. Longevity estimated to be 50-75 years.

e Asia. EVG, SHD, FOL ACC, FRU; IRR: L/L/M/M/-/M

Eriobotrya japonica
LOQUAT

Rosaceae. Sunset zones 6-24. Evergreen. Native to Japan and China. Growth rate moderate to 15-30' tall and often equally wide, forming a dense, dome-shaped, rounded canopy if lower branches are trimmed up, usually taller and more slender in shade. Leaves are alternate, simple, dark green, 4-10" long by 2-4" wide, leathery, obovate to oblong-elliptical, edges with sharp, widely spaced teeth, lustrous crinkled uppersides and fuzzy brown or whitish undersides. Small, dense clusters of loosely spaced, fuzzy, dull white flower spikes at branch ends in fall have a slightly bitter, almondlike fragrance. Round, orange-yellow, 1/2-3/4" applelike pome fruits, edible and sweet tasting, contain 1-4 large, shiny black seeds. Twigs are green and covered with whitish down. Shiny, smooth, gray-brown bark, with leaf scale markings and slight fissures, may become cracked with age.

Once a popular shade tree in gardens and parks, favored for its fruit, now less often planted than bronze loquat, which has more attractive foliage and rarely sets fruit. Makes a good lawn or container tree. Needs pruning to hold a pleasing shape. Careful thinning of dense foliage promotes fresh growth and fruiting. Tolerates heat when established, but not reflected heat. Lawns suit it well. Susceptible to fireblight. Resistant to oak root fungus. Longevity estimated to be 50-80 years.

South Africa. SHD, FLW ACC, SPC; IRR: -/-/L/L/-/-

Erythrina caffra

KAFFIRBOOM CORAL TREE

Fabaceae. Sunset zones 21-24. Deciduous. Native to eastern South Africa. Growth rate slow to 24-40' tall and 40-60' wide, developing a large, thick trunk with forked limbs angular and becoming thickly angularly branched at the ends, resulting in a round-headed form with a distinct character. Leaves are alternate, palmately compound, 7" long, dark green and leathery, with trifoliate leaflets 1 to 1-1/2" long, ovate, with pointed ends, falling in January, new leaves appearing early in spring after flowers. Large, showy spikes of 1-2" long, scarlet, tubular, legumelike flowers have a large upper folded petal and two smaller lower lateral petals slightly joined at the base, with long yellow stamens protruding beyond, and often drip nectar. Hard, leathery legume seed pods, to 4-1/2" long, ripen to brown or black, containing several kidney-shaped brown beans. Twigs are smooth and yellowish green. Smooth gray bark thickens with age, developing vertical seams and shallow furrows.

A tropical flowering accent tree with picturesque branching and a thick trunk, especially attractive during a brief deciduous period. Noted for spectacular display of red flowers in early spring as new leaves emerge. Popular in coastal regions of southern California but not generally hardy north of Santa Barbara. Prefers well-drained or slightly sandy soils. Overwatering produces excessive growth, which weakens branching structure. Longevity estimated to be 80-125 years.

South America. SHD, FLW ACC, SPC; IRR: M/M/L/L/-/M

Erythrina crista-galli

COCKSPUR CORAL TREE

Fabaceae. Sunset zones 7-9, 12-17, 19-24. Deciduous. Native to tropical Brazil. Growth rate slow to 15-20' tall and wide, developing a large, thick trunk and large, forked limbs growing in an angular fashion, becoming thickly branched at the ends and forming an arching, round-headed canopy with a subtle weeping character. Leaves are alternate, palmately compound, 6" long, dark green, leathery, with trifoliate leaflets 1 to 1-1/2" long, oval-oblong pointed, with short spines on long petioles, appearing in late spring and turning deep yellow in fall. Showy, 12-18" long spikes of 1-2", deep pink to scarlet red, legumelike flowers, with a large upper folded petal and two smaller lower lateral petals slightly joined at the base, with long yellow stamens protruding beyond, droop from branch ends in late spring and intermittently throughout summer. Hard, leathery legume pods ripen to brown or black, containing several kidney-shaped brown beans. Smooth twigs are yellowish green. Bark is fairly thick and reddish brown to gray, with wide, shallow furrows and scaly, interconnected flat ridges.

A fairly common flowering accent tree in southern California landscapes. A few are seen in the Central Valley to the north, thriving in lawns with full sun. Trees that suffer frost burn usually resprout eventually from branch ends. Young trees need pruning and staking to develop a strong multi-trunk structure to support heavy limbs. Longevity estimated to be 80-125 years once established.

Australia. EVG, SHD, SCR, FOL ACC; IRR: No WUCOLS

Eucalyptus baueriana
BLUE BOX

Myrtaceae. Sunset zones 5, 6, 8-24. Evergreen. Native to Australia. Growth rate moderate to fast to 35-75' tall and 25-45' wide, developing a tall, upright, billowing canopy, often with a low-branching trunk, large upright limbs, and slender outward branches, twiggy and drooping at the ends. Leaves are alternate, simple, glaucous, bluish gray, 2-4" long, broadly oval to ovate-lanceolate, blunt-ended or slightly retusely notched. White, 1/2" flowers with many stamens occur in attractive but insignificant, dense panicles at leaf axils in spring and summer. Clusters of cylindrical, 1/2" wide, cup-shaped, gray seed capsules containing many tiny seeds persist into the following year. Twigs are reddish gray, smooth, and slender. Thin, smooth bark on young trees is whitish to gray-green, thickening with age, becoming grayish brown, furrowed and cracked, with many irregular peeling ridgeplates.

Grown for its interesting, lush foliage, often in parks, screening close to buildings, and along walkways. Leaves resemble those of *Eucalyptus polyanthemos*, though brighter blue and slightly less gray. Hardy to 10-18 degrees. Longevity estimated to be 60-90 years.

Australia. EVG, SHD, SCR; IRR: VL/L/L/L/M/M

Eucalyptus camaldulensis

RED GUM

Myrtaceae. Sunset zones 5, 6, 8-24. Evergreen. Native to Australia, where it is the most common eucalyptus, occurring along bottomland streams. Growth rate moderate to fast to 45-150' tall and 45-105' wide, developing a rather dense, upright, ellipsoidal to fastigiate form with smaller branches drooping slightly at the ends, often to the ground. Leaves are alternate, simple, 6-10" long, leathery, dull green to bluish green, ovate-lanceolate, usually glabrous, with yellowish petioles and midrib. Insignificant 1/4" white or pale yellow flowers occur in stalked umbels of 5-15, usually 7s, in summer, with sharp-beaked buds. The end of the bud drops off and stamens unfurl from the remaining cup. Pea-sized, hanging seed capsules contain tiny tan-colored seeds, and capsule valves protrude beyond the rim. Slender round twigs are reddish. Smooth, green-tan mottled bark of young trees thickens, becoming gray, with many small furrows, also cracked and scaly.

Commonly used as an evergreen screen tree, slightly more hardy and drought tolerant than blue gum (*Eucalyptus globulus*). Planted extensively in the Central Valley, along with the more appealing red ironbark (*E. sideroxylon*), though not as much as it once was. Hardy to 12-23 degrees. Longevity estimated to be 60-90 years or more.

Australia. EVG, SHD, SCR, FOL ACC; IRR: VL/L/L/L/-/-

Eucalyptus cinerea

SILVER DOLLAR GUM

Myrtaceae. Sunset zones 5, 6, 8-24. Evergreen. Native to flat-lands of southeastern Australia. Growth rate moderate to fast to 20-55' tall with a 20-45' spread, developing a rather sprawly, irregular form with twisted and recurving top branches. Leaves are simple, glaucous, silvery blue, juvenile leaves paired opposite and occurring throughout the foliage of adult trees, appearing connate-perfoliate but actually separate, 1 to 2-1/2", oval to ovate, and overlapping, and adult leaves alternate, lanceolate, 5" long by 1" wide. Axillary clusters of small white flowers occur in spring, close to each leaf node, from diamond-shaped, pointed, glaucous buds. Tan to brown, broadly cup-shaped capsules have a thick rim and 3-5 valves, which split with flaps extending slightly. Twigs are glaucous and silvery blue-green, soon turning tan to reddish brown. Thick, brownish gray bark is heavily cracked, appearing semi-fibrous within deep reddish fissures.

A commonly used drought-tolerant foliage tree. Prefers well-drained, somewhat dry soil. Pruning to correct leafless, snakelike, irregular top growth encourages new juvenile growth but may destroy natural shape. Hardy to 14-17 degrees. Longevity estimated to be 60-80 years.

ne Australia. EVG, SHD; IRR: L/-/L/M/-/M

Eucalyptus citriodora

LEMON-SCENTED GUM

Myrtaceae. Sunset zones 5, 6, 8-24. Evergreen. Native to Queensland, Australia. Growth rate moderate to fast to 45-90' tall with a 45-75' spread, developing slender, graceful trunks, sometimes slightly contorted, with rather sparse drooping foliage heaviest at the ends of long, pendulous branches, exposing attractive bark with dramatic effect. Leaves are alternate, simple, light green, 3-7" long, narrowly lanceolate, lemon-scented, with juvenile leaves rough-surfaced and hairy with slightly wavy edges and adult leaves slightly thicker and smooth. Insignificant white-stamened flowers occur in clusters of 3-5 in winter and early spring from blunt-tipped buds. Urn-shaped, 3/8" seed capsules have a single end opening, which curves outward slightly. Bark is smooth, grayish or chalky white with pinkish tones, quite elegant, with sparse areas of darker old bark peeling off in large, flat shreds.

Commonly grown in southern California as a graceful accent tree. Especially effective planted in close irregular groves to accentuate the vertical trunks and increase wind resistance. Shallow surface roots may cause trees to topple in heavy winds. Hardy to 24-28 degrees. Longevity estimated to be 80-100 years.

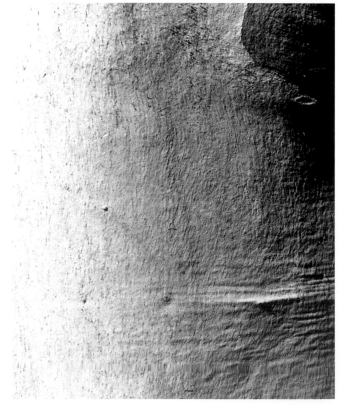

Australia. EVG, SHD, SCR; IRR: L/-/L/L/-/-

Eucalyptus cladocalyx

SUGAR GUM

Myrtaceae. Sunset zones 5, 6, 8-24. Evergreen. Native to southern Australia. Growth rate moderate to fast to 45-90' tall with a 45-75' spread, developing a large, graceful, upright form with straight trunks and a billowy, open, round-topped canopy with cloudlike puffs of denser growth at branch ends and little or no foliage in the interior. Leaves are alternate, simple, to 3-6" long, glossy to waxy, gray-green, and ovate-lanceolate to lanceolate, though juvenile leaves may appear to be opposite. Insignificant white flowers occur in dense umbels at leaf axils in July and August from cylindrical buds with a constricted midrib. Seed capsules are 3/8", short-stalked, urn-shaped, ribbed, in clusters of 5-10, with 3 or 4 valves and an open orifice at the end. Rough gray bark peels off in sheets or small plates, revealing white or pinkish red patches of underbark.

Tough, drought-tolerant shade tree. Needs plenty of room. Constantly sheds litter. Not the most attractive eucalyptus, but commonly found along freeway corridors or planted as a windbreak. Hardy to 23-28 degrees. Longevity estimated to be 80-100 years. 'Nana' grows to 25' tall.

Australia. EVG, SHD, SCR; IRR: L/L/L/L/-/-

Eucalyptus conferruminata

(*Eucalyptus lehmannii*)

BUSHY YATE

Myrtaceae. Sunset zones 5, 6, 8-24. Evergreen. Native to Australia. Growth rate moderate to fast to 12-27' tall and 15-30' wide, usually multi-trunked with a dense, oval, spreading form, usually with foliage to the ground unless pruned up to expose trunks. Finer texture than most eucalypts. Leaves are alternate, simple, 1-1/2 to 2" long, light green, elliptical to oval, with blunt ends. Flowers are light green in 4" wide clusters. Large, fused clusters of elongated, green, spidery-looking seed capsules shrink from the ends to form a tight fist as they turn brown and may remain on the branches indefinitely. Twigs are slender and reddish green. Grayish brown bark peels in long, thin shreds, exposing underbark with pinkish and cream tones.

A useful screen or attractive courtyard tree for coastal areas as far north as San Francisco Bay. Effective when multiple trunks are trimmed up, with foliage above, high enough to walk under. Hardy to 25-28 degrees. Longevity estimated to be 75-100 years. 'Max Watson' is a low, compact form.

Australia. EVG, SHD, SCR, FOL ACC; IRR: L/-/M/M/-/-

Eucalyptus ficifolia
RED-FLOWERING GUM

Myrtaceae. Sunset zones 5, 6, 8-24. Evergreen. Native to Australia. Growth rate slow to moderate to 18-45' tall and 15-60' wide as a single-trunked standard with a large, dense, rounded canopy or a low-branching, multi-trunk tree with broader form. Leaves are alternate, simple, 3-5" long, leathery, glabrous, shiny dark green, with pale undersides, a heavy, pale yellow-green center vein, younger leaves tapered ovate and adult leaves ovate-lanceolate. Spectacular, 12" round, flat-topped cyme clusters of red, orange, or pink flowers occur nearly year-round, heaviest in July and August. Caps on coral red flower buds fall off, allowing red stamens and dark red anthers to emerge. Seed capsules are heavy, woody, broadly urn-shaped, 1/3-1" wide, with many tiny, reddish brown, winged seeds in a lopsided compartment, ripening to brown and persisting into the following year. Attractive twigs are smooth and reddish green. Bark is thin, scaly gray-brown, flaking into many paper-thin strips, revealing orange underbark.

A desirable medium-sized flowering accent, street, or courtyard tree, commonly cultivated along the coast as far north as San Francisco Bay. Hardy to 25-30 degrees. Prefers somewhat dry coastal conditions, where it makes a fine specimen. Longevity estimated to be 40-60 years.

Australia. EVG, SHD, SCR; IRR: L/L/L/M/-/-

Eucalyptus globulus
BLUE GUM

Myrtaceae. Sunset zones 5, 6, 8-24. Evergreen. Native to Australia. Growth rate fast to 45-165' tall with a 30-75' spread, forming a tall, narrow, rounded crown with heavy masses of billowy foliage from a heavy, straight trunk. Leaves are alternate, simple, 6-10" long, sickle-shaped, pendulous, leathery, dark green to bluish green, margins smooth, obtusely rounded at the base and tapering gradually to a long, pointed end. Juvenile leaves are soft, oval or lanceolate, closely paired, overlapping at the base, and covered with a heavy glaucous, silver-gray, powdery film. Groupings of solitary, short-stemmed, 3/4-1", creamy white flowers with a flattened bulbous base occur in winter and spring at leaf axils on higher branches from 4-sided buds with 4 fused petals, forming a rounded cap that falls away, allowing stamens to unfurl. Large 1" round, 4-parted seed capsules are bluish gray, woody, and heavily ribbed on the flattened base, with a warty texture, maturing in fall and releasing many tiny seeds. Twigs are succulent, gray-green, slender, and pliable. Bark peels in large shreds almost constantly.

A majestic tree commonly associated with coastal windbreaks, but otherwise limited in use. Needs plenty of room. Leaf, bark, and seed drop are extremely messy. Wood is brittle, and roots are invasive. Hardy to 17-22 degrees. Longevity estimated to be 100-150 years.

Cultivar. EVG, SHD, SCR; IRR: L/L/L/M/-/-

Eucalyptus globulus 'Compacta'

DWARF BLUE GUM

Myrtaceae. Sunset zones 5, 6, 8-24. Evergreen. Selected variety of the species. Growth rate initially fast to 30-60' tall and almost as wide, multi-branched nearly to the ground unless trimmed up to a neat, dense, shrublike, oval form. Leaves, flowers, and seed pods similar to *Eucalyptus globulus*, though may be slightly smaller. Leaves are alternate, simple, 4-8" long, sickle-shaped, pendulous, leathery, dark green to bluish green, obtusely rounded at the base, margins smooth, tapering gradually to a long pointed end. Juvenile leaves are soft, oval or lanceolate, closely paired, overlapping at the base, and covered with a heavy glaucous, silver-gray powdery film. Groupings of solitary, short-stemmed, 3/4-1" creamy white flowers, with a flattened bulbous base, occur in winter and spring at leaf axils on higher branches from 4-sided buds with 4 fused petals, forming a rounded cap, which falls away, allowing stamens to unfurl. Seeds capsules are 1" round, 4-parted, bluish gray, woody, heavily ribbed on the flattened base, with a warty texture, maturing in fall and releasing many tiny seeds. Twigs are slender and soft gray-green. Reddish brown bark cracks in long seams, forming large, irregular woody plates that break loose.

Sometimes used as a windbreak or compact screen tree, with all the drawbacks for which the species is notorious, but not quite as messy. Easily sheared to form a 10' tall hedge, otherwise a 30' high evergreen tree with dense symmetrical form. Hardy to 17-22 degrees. Longevity estimated to be 100 years or more.

Australia. EVG, SHD, SCR; IRR: L/L/L/L/?/-

Eucalyptus gunnii
CIDER GUM

Myrtaceae. Sunset zones 5, 6, 8-24. Evergreen. Native to Tasmania. Growth rate moderate to fast to 30-75' tall and 18-45' wide, often upright and narrow, with irregularity, forming a tall, billowy canopy, with dense, light-textured foliage masses from slender, twiggy terminal branches. Leaves are simple, dull grayish green, juvenile leaves opposite, clasping stems in pairs, often overlapping, and nearly round or oval and adult leaves alternate, 2 to 3-1/2" long, lanceolate-elliptical, leathery, with a yellow midrib and a 1/2 to 3/4" long, yellow petiole. Clusters of fluffy white flowers, bunched in 3s at leaf axils, occur from April to July, when the reddish caps fall off the buds, allowing white, yellow-tipped stamens to unfurl. Greenish gray seed capsules in clusters of 5-12 are 1/4" round, goblet-shaped, short-stemmed, with many tiny, dark, round seeds, ripening to reddish brown and often persisting into the following year. Twigs are smooth and reddish brown. Bark is thin, smooth, or slightly mottled, light gray or greenish, becoming grayish brown and peeling to expose whitish underbark.

A small, fine-textured accent or screen tree most commonly grown for its unusual juvenile foliage, which is used in floral arrangements. Quite drought tolerant. One of the hardiest eucalypts and tolerates wetter soils than most. Hardy to 5-10 degrees. Longevity estimated to be 75-100 years. Two rarely seen cultivars are 'Undulata', with larger, wider leaves with undulating margins, and 'Acervula', with yellowish green foliage and capsules.

Australia. EVG, SHD, SCR; IRR: L/L/L/L/-/M

Eucalyptus leucoxylon

WHITE IRONBARK

Myrtaceae. Sunset zones 5, 6, 8-24. Evergreen. Native to coastal
southern Australia. Growth rate moderate to fast to 30-90'
tall and 18-60' wide, developing a slender, upright, though
variable form and an open habit, with pendulous branches.
Leaves are alternate, simple, 3-6" long, grayish to blue-green,
juvenile leaves sickle-shaped and ovate-lanceolate and adult
leaves lanceolate to broad-lanceolate. Insignificant white to
pinkish flowers with fluffy stamens of unequal length occur
in clusters of 3, intermittently in winter and spring, from
beaked buds on long, yellow pedicels, the cap falling off and
allowing stamens to unfurl. Goblet-shaped seed capsules in
clusters of 2-7 are 3/8" wide, thick-rimmed, yellowish brown
and speckled, with 4-6 valve openings, flaps extending slightly
below the rim of the capsule. Brown to gray bark peels in large,
thin strips, exposing smooth white underbark and creating a
mottled effect.

Once a commonly cultivated screen tree or semi-weeping
shade or accent tree. Leaf, bark, and seed drop messy. One of
the hardiest eucalypts, to 14-18 degrees. Tolerates drought and
adverse conditions. Longevity estimated to be 80-100 years.

Australia. EVG, SHD, SCR, FOL ACC; IRR: L/L/M/M/M/M

Eucalyptus nicholii
NICHOL'S WILLOW-LEAFED PEPPERMINT

Myrtaceae. Sunset zones 5, 6, 8-24. Evergreen. Native to southern Australia. Growth rate moderate to fast to 36-48' tall and 15-36' wide, usually developing an upright main trunk and a wide-spreading oval crown. Leaves are alternate, simple, 3-5" long by 1/8 to 1/4" wide, dull bluish green, often slightly purplish new growth in spring, narrowly lanceolate, pendulous, delicate in texture, with a peppermint fragrance when crushed. Tiny axillary clusters of inconspicuous white flowers, in 7s, emerge from yellowish buds and stems in summer. Tiny, rounded hemispherical seed capsules, clustered closely along branches, are 3-valved with 4 tiny round openings in a square arrangement, flaps extending slightly. Bark is soft gray-brown, fibrous or stringy, with coarse vertical furrowing.

A fast-growing garden or street tree, evergreen screen, graceful weeping accent tree, or courtyard specimen. Tolerates adverse conditions once established, but looks best with some watering. Requires light pruning to maintain shape if secondary branching becomes too irregular. Heavy pruning produces an abundance of juvenile growth, and overall appearance may suffer. Hardy to 12-15 degrees. Longevity estimated to be 80-100 years.

Australia. EVG, SHD, SCR, FOL ACC; IRR: L/L/L/L/M/M

Eucalyptus polyanthemos

SILVER DOLLAR GUM

Myrtaceae. Sunset zones 5, 6, 8-24. Evergreen. Native to Australia. Growth rate moderate to fast to 30-75' tall and 15-45' wide, developing an upright, billowing canopy, often with a multiple or low-branching trunk, large upright limbs, and slender outward branches twiggy and drooping at the ends. Leaves are alternate, simple, 2-4" long, dull gray or bluish green, broadly oval to ovate-lanceolate, blunt-ended, or slightly retusely notched on oval-shaped leaves, and covered with a powdery, glaucous gray film. Drooping clusters of 1" white flowers appear in spring and summer, noticeable but not showy. Clusters of cylindrical, 1/2" wide, hemispherical, cup-shaped, brownish gray seed capsules containing many tiny seeds persist into the following year. Twigs are attractive, smooth, slender, and reddish gray, accentuating the gray foliage. Bark is grayish brown, mottled with age due to shedding of variously colored large, thin, flaking shards, and may become slightly furrowed at the thick, dark base.

Commonly cultivated for its interesting foliage in parks or screening close to buildings or along walkways. Especially effective in informal groves. Hardy to 14-18 degrees. Longevity estimated to be 60-90 years. *Eucalyptus baueriana* is more full bodied, with broader, rougher leaves.

Australia. EVG, SHD, SCR; IRR: L/M/M/M/-/M

Eucalyptus pulverulenta
SILVER MOUNTAIN GUM

Myrtaceae. Sunset zones 5, 6, 8-24. Evergreen. Native to sub-alpine Australia. Growth rate moderate to 18-30' tall and 6-15' wide, forming an upright, billowy canopy from strong multiple trunks, with branches drooping at the ends, more so with vigorous heavy growth. Leaves are alternate, simple, silver-gray, and glaucous, adult leaves narrow, lanceolate, 4-6" long and stemless juvenile leaves 3/4-1" long, half-rounded and opposite, overlapping, but not connate-perfoliate, appearing as a single oval leaf with the stem protruding through the center. Clusters of white flowers, in 3-flowered umbels, from leaf axils, appear nearly year-round with new growth. Seed capsules are 1/3" across. Bark is brown to grayish and ribbony, becoming fissured and peeling in long, thick strips.

A rather common large evergreen foliage tree, usually planted as a single specimen, often in residential or park settings. Needs plenty of room. Tolerates heat, poor soils, and drought. Does best if not heavily watered. Light pruning should be done only to maintain desired shape. Heavy pruning results in a profusion of juvenile growth. Hardy to 15-21 degrees. Longevity estimated to be 50-65 years.

Cultivar. EVG, SHD, SCR; IRR: L/L/L/L/M/M

Eucalyptus sideroxylon 'Rosea'

RED IRONBARK

Myrtaceae. Sunset zones 5, 6, 8-24. Evergreen. Selection of the species native to Australia. Growth rate fast to 30-90' tall and 30-60' wide, forming an irregular, semi-weeping canopy, with multiple upright trunks and slender secondary limbs. Leaves are alternate, simple, 3-6" long, gray-green to blue-green, lanceolate, leathery, pendulous, from 1" long yellowish to reddish leaf stalks, turning purplish bronze in winter. Pendulous, fluffy, showy, light pink to crimson flower clusters usually occur in fall to late spring. Clusters of 3/8" oval, goblet-shaped seed capsules with thin, oblique discs, containing many tiny, dark, round seeds, ripen to reddish brown and persist into the following year. Twigs are smooth and reddish brown, accentuating the gray foliage. Bark is dark grayish brown with many fissures revealing reddish underbark, often marked with shiny crystals of reddish brown resinous gum.

A commonly used highway or parkway screen or a tall accent tree, also effective in small groups. Form is rather variable, always slightly irregular. Does well inland or near the coast. Needs well-drained soil, becoming chlorotic in heavy soils. Flowers produce sweet-tasting honey attractive to bees. Hardy to 20-25 degrees. Longevity estimated to be 60-75 years.

Australia. EVG, SHD, SCR, FOL ACC; IRR: L/L/L/M/-/M

Eucalyptus viminalis

MANNA GUM

Myrtaceae. Sunset zones 5, 6, 8-24. Evergreen. Native to Australia. Growth rate fast to 30-150' tall and 24-45' wide, forming a semi-open, billowing canopy, with foliage held high overhead, exposing large, erect trunks, often multi-trunked, with loose, drooping, willowlike branches to the ground. Leaves are alternate, simple, 4-6" long by 1/4 to 3/8" wide, medium green, leathery, narrowly linear-lanceolate with acute ends and a 1" long, waxy, yellow-green petiole. Insignificant white flowers, in corymb clusters of 6-8, from leaf nodes on current year's growth, occur throughout the year, usually too high to be seen. Unopened flower buds have a long pointed beak. Pea-sized, rounded, broadly flat-ended seed capsules produce year-round litter. Bark is whitish to pinkish and shedding, constantly peeling in large strips, causing considerable litter around the base of trees.

Once commonly used in parks and open spaces, many remain as stately park trees, creating airy shade. Requires ample room and well-drained soils. Single trees more subject to wind breakage than those in protected groves, and falling branches and debris undesirable for pedestrian areas. Difficult for shrubs to grow under, but lawns often do well, as the high canopy doesn't restrict sunlight at the base. Hardy to 12-15 degrees. Longevity estimated to be 80-120 years.

Europe. SHD, FOL ACC; IRR: M/H/-/-/-/-

Fagus sylvatica

EUROPEAN BEECH

Fagaceae. Sunset zones 2-9, 14-21. Deciduous. Native to Europe. Growth rate slow to 90' tall with a 60' spread, usually smaller, developing a large central trunk, with upward-ascending limbs, twiggy and slightly drooping at the ends. Leaves are alternate, simple, 2-5" long, elliptical to oblong-ovate, dark bronzy green, with 5-9 paired veins, finely serrate edges, shiny pale undersides, and deep yellow fall color. Fluffy, round, pea-sized, monoecious flowers are apetalous and dangle from 1" long stems in spring after leaves appear, many-stamened yellow males numerous and 1 or 2 female clusters occurring at branch ends. Fruits are woody, short-stemmed, with a 4-valved husk, covered with short, hairlike burrs, turning brown when mature in fall and splitting in starlike fashion from the base. Seeds are brown, 1/4-1/2", triangular. Young twigs are smooth, green, often growing in a zigzag pattern, from side-to-side between buds, turning shiny reddish brown in fall, with slender, long-pointed, scaly buds. Bark is thin, smooth, and grayish, often with mottled colorations.

A hardy, long-lived street, park, or lawn tree. Requires ample space at maturity. Young trees have surface roots that spread over a wide area. Mature trees cast dense shade, making lawn maintenance difficult. Requires moderately moist soils or regular deep watering. More shapely in full sun. Salts burn leaves and stunt growth. Occasional aphid infestations cause honeydew drip. Longevity estimated to be 200-300 years. 'Asplenifolia' has deeply cut, filigree leaf margins, 'Dawyk' has a columnar form, 'Pendula' has green leaves and a weeping form, 'Roseo-marginata' has pink-edged purple leaves, 'Spaethii' has deep purple, non-fading leaves, and 'Zlatia' has yellow new growth, fading to green.

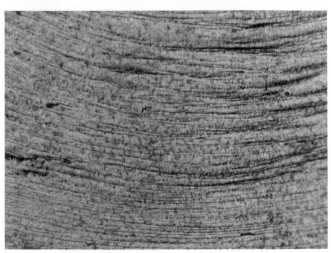

Cultivar. SHD, FOL ACC; IRR: M/H/-/-/-/-

Fagus sylvatica 'Riversii'

RIVERS' PURPLE BEECH

Fagaceae. Sunset zones 2-9, 14-21. Deciduous. Selected variety of the species. Growth rate slow to 50-60' tall with a 35-45' spread, developing an upright to pyramidal form with a slender pointed tip, becoming more oval with age, often with branches to the ground. Leaves are alternate, simple, 4" long, glossy, deep purple, elliptical to oblong-ovate, with 5-9 paired veins, margins either slightly wavy and smooth or sparsely and finely serrate, with shiny pale undersides and deep yellow fall color. Fluffy, round, pea-sized, monoecious, apetalous flowers dangle from 1" long stems in spring after leaves appear, with numerous many-stamened yellow males and 1-2 female clusters at branch ends. Woody, short-stemmed, 4-valved husks are covered with short hair-like burrs, turning brown when mature in fall of the first season and splitting in starlike fashion from the base. Seeds are brown, triangular, 1/4-1/2" long. Young twigs are smooth, shiny, reddish brown, hairy at first, often growing in a zigzag pattern between slender, long-pointed, scaly buds. Thin, smooth, grayish bark often has mottled colorations.

A desirable street, park, lawn, or accent tree. Requires ample space. In lawns, young trees develop surface roots that spread over a wide area. Requires moderately moist soils. Develops best shape in full sun. Salt buildup in dry soils will burn leaves and stunt growth. Subject to aphids, which cause honeydew drip. Longevity estimated to be 200 years or more.

Cultivar. SHD, FOL ACC; IRR: M/H/-/-/-/-

Fagus sylvatica 'Purpurea'

= *Fagus sylvatica* 'Atropurpurea' or 'Atropunicea'

PURPLE BEECH, COPPER BEECH

Fagaceae. Sunset zones 2-9, 14-21. Deciduous. Named variety of the species. Growth rate slow to 50-60' tall with a 35-45' spread, becoming more oval with age, with branches to the ground. Leaves are alternate, simple, elliptical to oblong-ovate, reddish to bronzy-purple, 4" long, with 5-9 paired veins, slightly wavy, edges smooth to sparsely and finely serrate, shiny, pale undersides, and deep yellow fall color. Pea-sized, monoecious flowers are fluffy, round, apetalous, dangling from 1" long stems in spring after leaves appear, many-stamened yellow males numerous and 1-2 female clusters at branch ends. Seeds are woody, short-stemmed, 4-valved, the husk covered with short, hair-like burrs.

A desirable street, park, lawn, or specimen tree. Requires ample space and deep watering. Salt buildup in dry soils will burn leaves and stunt growth. Prefers full sun. Subject to aphids, which cause honeydew drip. Longevity estimated to be 200 years or more.

Cultivar. SHD, FOL ACC; IRR: M/H/-/-/-/-

Fagus sylvatica 'Tricolor'

TRICOLOR BEECH

Fagaceae. Sunset zones 2-9, 14-21. Deciduous. Named variety of the species. Growth rate slow to 25' tall or more, usually less than 40' tall. Leaves are alternate, simple, elliptical to oblong-ovate, 4" long, green with white variegation, with pink, wavy, smooth edges and yellow fall color. Fluffy, round, pea-sized flowers occur in spring after leaves appear. Fruit is a woody, short-stemmed, 4-valved husk covered with short, hair-like burrs. Bark is smooth and gray.

An unusual garden or lawn accent tree occasionally grown as a large container specimen. Prefers semi-shade, under taller trees. Foliage burns in full sun or hot, dry, windy areas. Requires moderately moist soils. Salts burn leaves and stunt growth. Subject to aphids. Longevity estimated to be 200 years or more.

se Asia, Australia. EVG, SHD; IRR: -/-/M/M/-/-

Ficus macrophylla

MORETON BAY FIG

Moraceae. Sunset zones 17, 19-24. Evergreen. Native to Australia. Growth rate fast to 25-30' tall by 35-40' wide, with a broad, rounded canopy. In ideal conditions a 100 year old specimen can reach 75' in height with a 150' spread. Leaves are alternate, simple, 6-10" long by 3-4" wide, thick and leathery, shiny dark green uppersides and brownish undersides, elliptical to oval or oblong, thick, with smooth edges and blunt tips. Flowers insignificant. Fruit is 1" diameter, globular, and purplish spotted with white. Trunks usually become quite massive, often developing a wide buttressed base, with heavy exposed surface roots. Bark is smooth, whitish gray, with a rough warty texture.

Often cultivated in southern California as a large specimen tree. Does best in full or part sun. May become leggy and rangy in heavy shade. Needs plenty of room at maturity. Generally requires fertile, semi-moist soils for best growth. Young trees frost sensitive, becoming hardier with age. Foliage begins to show damage at 24-26 degrees. Trunks can withstand slightly lower temperatures and will resprout new leaves if severely damaged by frost, though may be unsightly until fully recovered. Longevity in cultivation estimated to be over 200 years.

Cultivar. EVG, SHD; IRR: M/-/M/M/-/-

Ficus microcarpa 'Nitida'

LITTLE-LEAF FIG

Moraceae. Sunset zones 9, 13, 16-24. Evergreen. The name *Ficus microcarpa* 'Nitida' has no botanical standing and may include weeping forms that resemble the species or exhibit upright branching. Growth rate fast, eventually to 20-30' tall and wide, developing a round, oval canopy, with dense foliage on upright branches. Leaves are alternate, simple, 2-4" long, shiny light green, ovate-lanceolate, leathery. with smooth edges, cuspidate ends, and pale undersides. Insignificant flowers are rarely seen. Fruit is reddish orange, 1/3" long, with blunt basal bracts. Bark is smooth and whitish with a rough, warty texture.

A widely used street tree in coastal areas with a clean, shiny appearance. Tolerates windy conditions. Can be sheared to desired form. Thrips may disfigure leaves. Not reliably hardy below 32 degrees. Longevity in cultivation estimated to be 100 years.

Australia. EVG, SHD, FOL ACC; IRR: M/-/M/M/-/-

Ficus rubiginosa
RUSTYLEAF FIG

Moraceae. Sunset zones 18-24. Evergreen. Native to Australia. Growth rate initially fast, slowing with age, to 20-50' tall and 30-50' wide, forming a dense, broad-spreading crown with a large buttressed trunk and heavy limbs, often multi-trunked and low branching. Leaves are alternate, simple, 5-6" long by 2-3" wide, shiny dark green, leathery, often with rusty tomentose undersides and stems, oval-elliptical, with thick smooth edges and blunt-tipped ends, exuding a milky sap when cut, persisting 2-3 years. New growth is noticeably reddish brown. Small flowers are insignificant. Small, short-stalked, spherical fruits are inedible. Smooth grayish brown bark has thin, flaky scales and warts.

A large canopy tree for warm climates commonly used in parks, plazas, and patio terraces, casting dense shade. Needs ample room for the broad canopy and voracious roots, which often run along the surface of the ground. Not frost hardy below 30 degrees, and not commonly used except in protected coastal areas. Longevity estimated to be 100-200 years or more.

e Asia. SHD, FOL ACC; IRR: No WUCOLS

Firmiana simplex
CHINESE PARASOL TREE

Sterculiaceae. Sunset zones 5, 6, 8, 9, 12-24. Deciduous. Native to China and Japan. Growth rate slow to 15-30' tall and nearly half as wide, forming a broad, oval, round-topped canopy usually from a single trunk with 3 or 4 main leaders and irregular twiggy branch ends, becoming more irregular and open in shade. Leaves are thick, alternate, simple, to 1' wide, dark green, with 3-5 palmate lobes, often varying slightly in form, on long stems, with smooth margins, often finely hairy, with brief pale yellow fall color. Large, showy, greenish to cream-colored flowers are 5-parted, with a bell-shaped calyx, 1/4-1/2" long, with 12-15 stamens, no petals, a single pistil, appearing in 3' long raceme clusters in May and June. Curious hanging clusters of elongated, leathery green to tan bracts split and expand into 4 or 5 individual fleshy pea pods suspended in umbrellalike fashion at the end of a long stem, with 1-3 pea-sized green seed-fruits attached along the margins. Dried fruit becomes unsightly in late summer, but after bracts detach in fall the star-like stem stubs remain, sparsely spaced from a long main stem, and are curiously interesting. Twigs are stout and green. Bark is light gray-green, often with distinctive light tan fissures.

An uncommon half-hardy shade or accent tree, sometimes seen in parks, courtyards, or terraces sheltered from strong winds. Requires regular water when young. Full sun to filtered shade among taller trees in lawns suits it well. Longevity estimated to be 50-75 years.

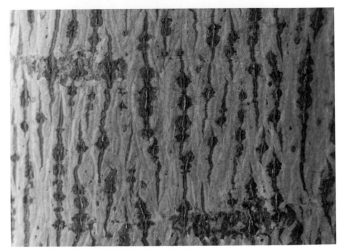

Cultivar. SHD, FAL ACC; IRR: M/M/M/M/M/M

Fraxinus angustifolia 'Raywood'

= *Fraxinus oxycarpa* 'Raywood'

RAYWOOD ASH, CLARET ASH

Oleaceae. Sunset zones 2-9, 12-24. Deciduous. Cultivar of the southwestern European and north African species, developed in Australia and introduced into North America in the mid-1950s. Growth rate fast to 25-35' tall and 25' wide, with an upright oval form. Leaves are opposite, pinnately compound, 5" long, dark shiny green, with 5 or up to 7-9 narrow lanceolate leaflets, sparsely spaced and sessile along the midrib stem with sparsely serrate edges and dark purplish red fall color. Insignificant flowers are rarely seen, occurring higher up on older trees. Dense clusters of male flowers are purplish brown, turning yellow with pollen, appearing in early spring before leaves, pale green female flowers clustered more loosely, either on separate trees from the males or on the same tree, either on the same flower or separately from the males. Tightly bunched clusters of winged achene seed capsules, rarely seen, are 1-1/2" long, maturing in late summer and persisting into winter. Twigs are reddish brown. Thin, smooth, greenish gray bark becomes grayish brown, with shallow vertical fissures and slightly scaly plates.

A widely used, attractive street, parking lot, or patio shade tree. Requires moderate watering in well-drained soils. May suffer dieback in drought. Susceptible to wind breakage. Susceptible to aphids. Longevity estimated to be 90-125 years.

w N.A. California Native: CH, OW, FW (Riparian) (*NCoR, CaRF, n SNF, c&s SN, CW, TR, PR*); SHD, FLW ACC; (IRR: No WUCOLS)

Fraxinus dipetala

FOOTHILL ASH

Oleaceae. No Sunset zones. Jepson *DRN:**7,14-16**,17,**22,23**, 24&IRR:**8,9,18-21**. Deciduous. Native to dry foothill canyons and slopes to 3,500' elevation in California, northwest Arizona, southwest Utah, and southern Nevada to Baja California. Growth rate slow to moderate to 6-18' tall. Leaves are opposite, pinnately compound, 2-5" long, dark green, usually with 3-9 elliptical to ovate, 3/4-1" long leaflets, with blunt or short-tipped ends, pale undersides, coarsely saw-toothed margins, hairless when mature, and slight yellow fall color. Long-stemmed, loose, drooping, 4" long clusters of 3/16" flowers with 2 wide white petals are rather showy, but not fragrant, appearing before leaves in April from the previous year's leafless axillary buds. Clusters of oblong-shaped key seeds, 3/4-1" long, and flat-winged nearly to the base of the flattened seed body, often notched at the tip, mature to tan in early summer. Twigs are slender, green, usually slightly 4-angled when young, usually hairless, becoming gray in winter. Bark is light gray, rough, and scaly.

   An attractive small native flowering tree, rarely cultivated but sometimes used as a habitat tree in riparian and foothill plantings. Self-seeds readily where adaptable. Tolerates moderate aridity and drought once established. Longevity estimated to be 50 years or more. Series associations: Scroakbirmou shrub series; Calbuckeye, Calwalnut.

se Europe, sw Asia. SHD, FLW ACC; IRR: No WUCOLS

Fraxinus ornus

FLOWERING ASH

Oleaceae. Sunset zones 3-9, 14-17. Deciduous. Native to southeastern Europe and southwestern Asia. Growth rate rapid to 40-50' tall and 20-30' wide, developing a broad, rounded canopy and upward- and outward-arching branches from a central trunk. Leaves are opposite, pinnately compound, 8-10" long, shiny light green, with 7-11 oval-elliptical, 2" long leaflets with serrate margins, and pale lavender-tinted yellow fall color. Showy clusters of 3-5" long, fluffy, fragrant, 4-petaled, creamy white flowers occur on terminal buds in April after new leaves. Drooping clusters of 1" long, winged key seeds follow, persisting into winter. Bark is smooth and gray.

   Sometimes used as a small to medium-sized lawn or shade tree. Benefits from occasional deep watering in summer. Loses shape in too much shade. Longevity estimated to be 60-80 years. 'Emerald Elegance' and 'Victoria', both developed in the northwest and occasionally available, are seedless, with a more compact oval form, useful as small street trees.

Cultivar. SHD, FAL ACC; IRR: M/M/-/-/M/M

Fraxinus holotricha 'Moraine'

PP #1768 (1958)

MORAINE ASH

Oleaceae. No Sunset zones. Deciduous. Cultivar of the species native to southwestern Europe and north Africa. Growth rate moderate to 30-35' tall and wide, developing a broad, oval form, with a short trunk, often low-forked, and multiple upright-arching limbs, often twiggy at the ends. Leaves are opposite, pinnately compound, 5" long, light green, with 7-9 lanceolate leaflets rather sparsely spaced and sessile along the midrib stem, and clear yellow fall color, lighter than most ashes. Flowers are polygamo-dioecious, with dense clusters of purplish brown male flowers turning yellow with pollen early in spring just before new leaves and pale green female flowers clustered more loosely, either on separate trees from the males or on the same tree, either on the same flower or separately from the males. Seed capsules are tightly bunched clusters of quadrangular, winged achenes, 1-1/2" long, with slightly notched wing tips, maturing in late summer and persisting into winter. Twigs are reddish brown. Bark is thin, smooth, greenish gray, becoming grayish brown, with shallow vertical fissures and slightly scaly plates.

A commonly used residential street tree with a very clean appearance and a light airy texture from a distance, the most delicate lacy look of all the ashes. Does best with occasional deep watering. An excellent choice for lawns or narrow parkways. Longevity estimated to be 75-125 years.

California Native: CP, VG, OW, CF, ME, PF (Riparian) (*NW, CaR, SN, GV, SnFrB, MP*); SHD, NAT; (IRR: H/H/-/-/-/-)

Fraxinus latifolia

= *Fraxinus oregona*

OREGON ASH

Oleaceae. Sunset zones 3-9, 14-24, Jepson *4-6,&IRR:1-3,7-9,14-16,17-24. Deciduous. Native to streambanks and moist valleys in the Sierra Nevada and Coast Ranges from California to British Columbia. Growth rate moderate to 40-80' tall and 30-50' wide, forming a compact, oval, semi-open crown. Leaves are opposite, light green, pinnately compound, 5-12" long, with 5 to 7 oblong-elliptical to ovate leaflets, 3-6" long by 1 to 1-1/2" wide, the terminal leaflet largest, 1/2" long petiolules or sessile on the rachis, edges either entire or finely toothed, lighter fuzzy undersides, hairy leaf stems grooved, and orange-yellow to dull orange-brown fall color. Inconspicuous panicles of yellowish, dioecious, long, droopy male flowers and shorter, compact, fluffy female flowers appear in April, on separate trees, corolla absent. Flat papery seeds are thin, 1-2" long, oblong to elliptical, narrow, with round-ended wings, thickened at the base around the seed cavity, occurring in dense, hanging clusters, and persisting into winter. Twigs are stout, round, and reddish brown with small, conical, hairy brown buds. Bark is thick, dull gray to grayish brown, becoming deeply and regularly furrowed, with wide ridges interconnected with smaller ones.

A useful riparian habitat tree, tolerating wet conditions and poor soils near streams or rivers, where it reseeds readily, becoming quickly established. Suitable as a park or parkway tree in small groves. Longevity estimated to be 150-200 years in habitat, often losing vigor with age. Series associations: Mexelderberr shrub series; Blacottonwoo, Frecottonwoo, Jefpinponpin, Valoak, Whialder.

c & ne N.A. SHD, FAL ACC; IRR: M/M/-/-/M/M

Fraxinus pennsylvanica

GREEN ASH

Oleaceae. Sunset zones 1-6. Deciduous. Native to the eastern and central northeastern U.S. Growth rate fast to 30-40' tall and wide, forming a tall, oval crown, usually with a large trunk, and a dense, twiggy structure. Leaves are opposite, pinnately compound, 10-12" long, dark green, with 7-9 oblong-lanceolate to elliptical, 3-5" long leaflets, 1/8-1/4" long petiolules, edges entire or finely toothed, silky hairy pale undersides, and yellow fall color. Inconspicuous, dioecious, yellow-green flowers in compact loose panicles, males and females on separate trees, appear in early spring with new leaves. Thin, flat, 1-2" long papery seeds are narrow-lanceolate, in dense hanging clusters, with slender, elongated wings no wider than the seed, persisting until fall. Twigs are stout, reddish gray with faintly pale hairs. Bark is thin and greenish brown, thickening and darkening to brown with a reddish tinge, with shallow fissures and scaly ridges.

A widely used, clean looking, desirable though somewhat plain shade tree for parks and residential streets. Requires moderate water and thrives in lawns. Longevity estimated to be 80-125 years. 'Marshall' is a non-fruiting male form.

sw N.A. EVG, SHD; IRR: M/M/M/M/H/H

Fraxinus uhdei

EVERGREEN ASH

Oleaceae. Sunset zones 9, 12-24. Semi-deciduous. Cultivated plants of this species are descendants of seedlings developed from a single tree selected in Mexico and introduced into California in 1938. Growth rate initially fast to 25-30' tall and 15' wide, eventually reaching 70-80' tall or more with a 60' spread, developing a broad, oval form and erect, symmetrical branching from a central trunk. Leaves are opposite, pinnately compound, 6-10" long, shiny light green, divided into 5-9 short-stalked, ovate-lanceolate leaflets, with finely serrate edges, and pale, smooth undersides, often remaining through mild winters, new leaves appearing when mild temperatures return, often in late winter. Dense clusters of small, yellowish green flowers occur in early spring, with male and female flowers on separate trees. Paired winged seeds, resembling canoe paddles, occur on female trees. Twigs are stout, greenish gray, smooth, with broad, rounded terminal buds, clearly visible in winter. Bark is thin, smooth, greenish gray, thickening slightly with age, usually remaining smooth with only slight fissuring and a slight reddish brown color until fully mature, then becoming thick and gray.

Commonly used as a shade tree with lush foliage and a short deciduous period. Invasively shallow rooted unless deeply watered. Limbs can break in strong winds. Generally hardy to 15 degrees, but late frosts below 20 degrees may temporarily damage new growth. Resists oak root fungus. Texas root rot may affect young trees. Longevity estimated to be 75-100 years.

w N.A. California Native: OW, FW, MD (s Calif.) (Riparian) (*s SN, SCo, TR, PR, s SNE, Dmoj*); SHD, FAL ACC; (IRR: M/M/M/M/M/M)

Fraxinus velutina

ARIZONA ASH

Oleaceae. Sunset zones 3-24. Deciduous. Native to dry slopes and washes of western Texas to southern Nevada and Utah, northern Mexico, New Mexico, Arizona, and rarely in the high desert mountains of southeastern California. Growth rate initially fast to 30' tall (possibly to 50') and 30' wide, with a slender, pyramidal form in youth, later developing stout trunks with multiple leaders and a broader, symmetrically rounded, dense crown. Leaves are opposite, compound pinnate, 4-6" long, dull yellowish green, somewhat leathery, with 3-5 narrow-elliptical to ovate leaflets, 1 to 1-1/2" long, with finely to sparsely crenate to serrulate edges from the middle toward the ends, lighter, softly tomentose undersides, smooth uppersides, hairy petioles, and yellow to dull brown color holding well through fall. Dioecious male and female flowers appear in April, corolla absent, from new growth buds, in pubescent panicles, on separate trees. Female trees have thick clusters of drooping, 1 to 1-1/4" long, oblong-ovate to elliptical, flat, papery thin seeds, with notched to round-ended wings wider than the seed at the end. Twigs are slender, rounded, reddish brown, and briefly covered with fine woolly hairs. Bark is thin, grayish with reddish-tinged furrows and broad ridges, often spongy, and scaly on the surface.

In cultivation since the 1900s and once a popular street and residential shade tree, now seldom planted, though many older trees remain. Suffers from mistletoe. Subject to anthracnose, which can infect any nearby plane trees and causes defoliation of new growth. Tolerates heat. Hardy to -10 degrees. Longevity estimated to be 50-75 years. Series associations: Fanpalm.

Cultivar. SHD, FAL ACC; IRR: M/M/M/M/M/M

Fraxinus velutina 'Modesto'

MODESTO ASH

Oleaceae. Sunset zones 3-24. Deciduous. Vigorous selection of the species, from Westside Park, Modesto. Growth rate fast to 50' tall with a 30' spread, forming a broad, oval canopy with upright, slender branching. Leaves are opposite, compound pinnate, 6" long, medium green, with 3-5 shiny, 1-2" wide elliptical leaflets, edges finely to sparsely crenate to serrulate toward the ends, and bright yellow, fairly long-lasting fall color. Dioecious male flowers only, in drooping panicles from new growth buds, occur in April. Does not set seeds. Twigs are slender, rounded, reddish brown, and may have minute sparse hairs for a short while. Thin, grayish bark develops shallow, reddish brown fissures, with broad, flat, scaly gray ridges.

Until recently a widely used street and shade tree, more vigorous and attractive than the species. Now less often planted because of susceptibility to diseases, including anthracnose, which causes leaf scorching and temporary leaf drop in moist spring weather. Also subject to aphids, verticillium wilt, psyllas, and spider mites. Resistant to oak root fungus. Longevity estimated to be 60-80 years.

Australia. SHD, FOL ACC; IRR: M/M/L/M/M/M

Geijera parviflora

AUSTRALIAN WILLOW, WILGA

Rutaceae. Sunset zones 8, 9, 12-24. Evergreen. Native to drier regions of southeastern Australia. Growth rate slow to moderate to 25-30' tall with a 20' spread, forming a dense, oval canopy, broadening with age, main branches sweeping upward and outward and smaller branches drooping at the ends. Leaves are alternate, simple, 3-6" long by 1/4-3/8" wide, light green to grayish green, narrowly linear, curving slightly, with a leathery, waxy feel, and aromatic when crushed. Inconspicuous, tiny, creamy white rosette flowers occur in sparse panicles at branch ends on older trees in early spring. Small globular carpel fruits contain a single hard, tiny black seed. Twigs are smooth, greenish to brown, attractive with the foliage. Bark is dark brown to grayish brown, becoming deeply fissured and cracked with age.

A graceful, fine-textured specimen or small street or patio tree. Tolerates some drought, but prefers moderate watering and well-drained soils. Does not perform well in heavy, wet soils. Roots are deep and not invasive. Casts light, airy shade, similar to a weeping willow, but becomes sparse if overly shaded by larger trees. Longevity estimated to be 50-80 years.

e Asia. SHD, FAL ACC; IRR: M/M/M/M/M/-

Ginkgo biloba

MAIDENHAIR TREE

Ginkgoaceae. Sunset zones 1-10, 12, 14-24. Deciduous. A prehistoric relic thought to have originated in China and not known to exist in the wild. Growth rate slow to 35-50' (or 70-80') tall with a 25-40' spread, developing a broad canopy with several erect leaders and outward angular branching or, less often, an upright conical form. Leaves are simple, fan-shaped, alternately spaced on longer terminal branches, closely bunched at short, stubby lateral buds, light green, 2-5" long by 1-3" wide, tapering to the petiole, with nearly parallel ribbed veins and usually a clefted lobe in the center of the wavy-edged end margin. Fall color is a spectacular bright golden yellow, but leaves often drop quickly in gusty winds. Insignificant flowers are dioecious and unisexual, on separate trees, in March. The scaly-bracted male catkinlike spikes are about 1" long and yellowish. Females are paired on single stalks, with only one producing yellow, foul-smelling, messy, 1" long, plumlike fruit covered with a whitish glaucous film in summer. Properly cleaned and roasted seed nuts are edible and were once popular at Chinese weddings, but the butric acid juice can cause skin rash. Axillary twigs are gray-green, short, and stout. Bark is thin, smooth, and grayish green, becoming gray and corky, with wide, shallow furrows and scaly ridges. Trunks often become fluted at the base.

An excellent park or street tree or single lawn specimen. Does not thrive in heavy or soggy soils. Resistant to oak root fungus and generally pest free. Longevity estimated to be 500 years or more. Grafted or cutting-grown stock from male plants, including 'Autumn Gold' and 'Fairmont', do not have fruit.

Cultivar. SHD, FOL ACC; IRR: L/L/M/L/L/L

Gleditsia triacanthos 'Inermis'

THORNLESS HONEY LOCUST

Fabaceae. Sunset zones 1-16, 18-20. Deciduous. Selected variety of the northern and central North American species. Growth rate fast to 35-70' tall with a 25-35' spread, developing an upright trunk, slender, arching branches, and a round-topped canopy, casting filtered shade. Leaves are alternate, bipinnately compound, 6-12" long, light green, with many tiny, oblong-lanceolate leaflets with finely serrated margins, pale undersides, and light yellow fall color. Pinnate juvenile leaves sometimes occur. Inconspicuous, regular, polygamous, greenish white flowers, with 3-5 petals, are in small catkinlike axillary racemes, usually hidden by the leaves, males about 2" long and females smaller and fewer. Many flattened, dark brown legume seed pods, thick and leathery, 12-18" long by 1" wide, containing 12-14 dark brown, hard, flattened seeds, mature in late summer, hanging from branches as leaves fall. Axillary twigs are stout and do not have the single or 3-branched, 2-3" long spines of the species. Bark is grayish brown, with long, shallow fissures and scaly or plated ridges.

A commonly cultivated filtered shade tree. Slow to leaf out in spring and drops leaves early in fall, litter posing a problem in paved areas. Tolerates heat, alkalinity, and moderate drought. Does well in hot climates with seasonal change. Caterpillar infestations can temporarily defoliate trees in late summer. Longevity estimated to be 50-100 years. 'Imperial' is tall and spreading, to 35', casting denser shade, 'Moraine' has fast growth but suffers from wind breakage, and 'Shademaster' is upright, with the fastest growth. All are thornless.

Cultivar. SHD, FOL ACC; IRR: L/L/M/L/L/L

Gleditsia triacanthos 'Sunburst'

= *Gleditsia triacanthos* 'Suncole'
PP #1313 (1954)

= *Gleditsia triacanthos* 'Skyline'
PP #1619 (1957)

GOLDEN HONEY LOCUST

Fabaceae. Sunset zones 1-16, 18-20. Deciduous. Selected cultivar of the northern and central North American species. Growth rate moderate to fast to 30-50' tall with a 25-35' spread, often smaller, developing an upright trunk and slender arching branches with twiggy, weeping ends, forming a round-topped canopy and casting filtered shade. Leaves are alternate, bipinnately compound, 8-12" long, and divided into many tiny oblong-lanceolate leaflets, with yellow new growth fading to light green, and yellow fall color. Sometimes pinnate juvenile leaves occur. Inconspicuous flowers, in small catkinlike axillary racemes, about 2" long, do not set seed pods. Bark is dark grayish brown with shallow furrows and flat-plated scaly ridges.

An excellent, light-textured shade tree with colorful foliage. Slow to leaf out in spring and leaves drop early in fall, littering paved areas. Good lawn tree. Does best in moderately moist, well-drained soils. Tolerates heat and alkalinity. Caterpillar infestations may temporarily defoliate trees in late summer. Longevity estimated to be 75-100 years. 'Ruby Lace' is smaller, with reddish bronze new growth, but suffers wind damage in unprotected areas.

Australia SHD, FOL, FLW ACC; IRR: L/L/L/M/-/M

Grevillea robusta

SILK OAK

Proteaceae. Sunset zones 8, 9, 12-24. Evergreen. Native to Australia, where it once occurred in large groves but is now receding. Growth rate moderate to fast to 50-60' tall (rarely 100') with a 30-35' spread, developing a conical or pyramidal form, older trees broader and more irregular. Leaves are alternate, bipinnately compound, 4-10" long, fernlike, with roughly a dozen lanceolate to deeply lobed pinnae, dark shiny green with silvery undersides, persisting 1 year, leaf drop heaviest in spring and sporadic throughout the year. Curiously showy, stiff, paired, 3-6" long, brushlike spikes of 1/4", golden yellow, slightly fragrant flowers occur in dense clusters in late spring, from axillary buds. Flowers are apetalous, only slightly fragrant, on 1" yellow stems, from a green central stalk, with 2 curled honeysuckle-like "petals," which have a dark reddish, sticky, juicy center, with a 1" long curled yellow proboscis "stamen" unfurling and protruding beyond. Boat-shaped follicle fruits, with a slender beak, dry to brown in fall, splitting along one seam, and contain 1-2 flat, oval winged seeds. Bark is brownish gray with small wartlike growths, becoming darker brown, with shallow fissures.

A fast-growing, short-lived, evergreen shade tree with attractive foliage that contrasts nicely with other broadleaved trees and conifers. Brittle wood is easily damaged by heavy winds. Leaf drop requires constant cleanup in paved areas. Flowers may attract bees. Needs ample room for its extensive root system, which can be invasive. Tolerates heat and poor soils, but grows best in well-drained soils. Heavy pruning destroys its natural shape. Longevity estimated to be 50-65 years.

Australia. SHD, FLW ACC; IRR: M/-/M/M/-/-

Hymenosporum flavum

SWEETSHADE

Pittosporaceae. Sunset zones 8, 9, 14-24. Evergreen. Native to Australia. Growth rate slow to moderate to 12-40' tall with a 9-20' spread, developing a slender, upright, dense, conical form and a variable oval canopy with age, in shade branches spaced in whorled fashion from the trunk with gaps between horizontally aligned tiers. Leaves are alternate, simple, 2-6" long by 1 to 1-1/2" wide, shiny dark green, obovate, with smooth, slightly wavy edges, constricting to a short pointed tip, clustered at ends of branches, persisting 2-3 years. Dense, showy clusters of waxy, cream-colored, 1" long, tubular flowers at branch ends are long-stemmed, with a sweet orange-blossom fragrance, becoming dark golden yellow with reddish throat markings, blooming heaviest in May and June and sporadically through September. Fleshy skin of the 1/2-1" round capsule fruit hardens, maturing to brown, containing many tightly packed, many-winged seeds. Twigs are smooth and grayish brown. Bark is thin, smooth, and grayish, browning with age and becoming slightly cracked or fissured.

An attractive, evergreen flowering tree for street, terrace, and garden use. Rather tender and does best in coastal areas protected from direct winds. Prefers well-drained, moderately moist, fertile soil. Slender upright limbs can be tip-pruned to develop a denser canopy. Weak crotches tend to break in wind, less so when trees are planted closely in small groves. Suffers from occasional aphid infestations in summer. Longevity estimated to be 60-80 years.

South America. SHD, FOL, FLW ACC; IRR: M/M/M/M/-/M

Jacaranda mimosifolia

(*Jacaranda acutifolia*)

JACARANDA

Bignoniaceae. Sunset zones 12, 13, 15-24. Semi-deciduous. Native to Brazil and Argentina. Growth rate initially fast, slowing to moderate, to 25-40' tall with a 15-30' spread, forming an oval dome canopy, open, airy, and fine-textured, casting light shade. Leaves are alternate, bipinnately compound, 10-12" long, light green, finely cut, with 14-24 pairs of tiny, oblong-lanceolate, fernlike leaflets resembling a mimosa, slight yellow fall or winter color, only briefly deciduous in warmer climates, with new growth appearing in February to March. Flowers are 2" long, drooping, tubular, bell-shaped, lavender-blue, in showy 8-12" clusters, appearing in June with sporadic flowering throughout summer. Thin, flat, 2" round, oblong seed pods turn hard and brown, splitting after drying along the entire edge and popping open from the end, releasing tiny winged seeds. Twigs are smooth and greenish brown. Bark is smooth and brownish gray, becoming thinly scaly with age.

Rarely seen in central California but commonly used in southern California as a graceful shade tree or lawn specimen and also effective in groups. Thrives in heat and may not flower well in cool, windy coastal areas. Needs well-drained soil with moderate moisture. Resists oak root fungus. Hardy to 25 degrees in youth, usually recovering from occasional tip burn and becoming more hardy with age. Older multi-trunk specimens with irregular branching are picturesque. Longevity estimated to be 80-125 years.

w N.A. California Native: VG, OW, FW (n. Calif. Riparian) (*s NCoRI, s ScV, n SnJV, SnFrB*); SHD, FRU; (IRR: M/M/-/L/-/-)

Juglans hindsii

= Juglans californica var. hindsii

NORTHERN CALIFORNIA BLACK WALNUT

Juglandaceae. Sunset zones 5-9, 14-20, Jepson *4-7,14-24,IRR: 8,9,10,11. Deciduous. Native to valley and low foothill streamside locations in northern California below 500' elevation. Hybridizes with *Juglans nigra* in most locations, and only a few pure stands of the species exist. Growth rate slow to 30-60' tall with a broad, domelike canopy, clear short trunks, usually multi-trunked, with large branches curving upward and outward, often drooping to the ground. Leaves are alternate, pinnate, 9-14" long, drooping, light yellow green, with 15-19 smooth, pointed, lanceolate leaflets, 3-5" long, with serrate edges, occasional tawny hairs underneath at vein angles, and brief yellow fall color. Thinly husked fruits are oval and green, with a finely velveted surface, maturing in late summer of the first year, the skin drying to black and becoming cracked and wrinkled. Nuts are 1-1/2 to 2" with a rather smooth but irregularly grooved shell, which is extremely hard and generally remains unopened along the single seam dividing the two halves. The tan to brown seed inside has a wrinkled surface and distinct tasting white meat. Bark on young trees is thin and ashy colored, becoming blackish brown, deeply and sharply furrowed, and ridged on older trees.

A common riparian tree, but requires well-drained soils. Can be quite drought tolerant with infrequent deep watering and plenty of mulch around roots. Resistant to oak root fungus. English walnut is often grafted onto less disease-prone black walnut rootstock. Rarely cultivated otherwise. Roots and leaves are allelopathic, producing juglone, a chemical toxic to many plants. Longevity estimated to be 50-75 years, then gradually losing vigor. Series associations: **Hinwalsta,** Frecottonwoo.

sw N.A. SHD, FOL ACC, FRU; IRR: No WUCOLS

Juglans microcarpa

LITTLE WALNUT, TEXAS WALNUT

Juglandaceae. No Sunset zones. Deciduous. Native along streams from southwest Kansas to New Mexico and northern Mexico. Growth rate moderate to 10-20' tall, forming a broad, rounded canopy and a stout, low-forked trunk. Leaves are alternate, pinnate, 8-13" long, light green, with 7-13 narrow-lanceolate, 2-3" long leaflets, short-stalked and usually curved together, with finely serrate or entire edges, and yellow fall color. Insignificant greenish flowers occur in spring, numerous male catkins with about 20 stamens, and few females with a 2-lobed style at terminal buds of the same twigs. Fruit is oval, hairy-husked, 1/2-3/4" long, drying to black, with a wrinkled surface. Nuts have a hard, grooved shell. Twigs are hairy and grayish. Bark is smooth and gray, becoming deeply furrowed.

Seldom used as a small shade tree or individual lawn specimen with attractive, unusual, fine-textured foliage. Should be planted only where it can be well maintained. Longevity estimated to be 50-90 years.

w N.A. California Native: VG, OW, FW (s Calif. Riparian) (*SCoRO Santa Lucia Mtns, SCo, s TR, n PR Santa Ana Mtns*); SHD, FRU; (IRR: L/-/L/L/-/-)

Juglans californica

Juglans californica var. *californica*

SOUTHERN CALIFORNIA BLACK WALNUT

Juglandaceae. Sunset zones 18-24, Jepson *4-6,**7,14-24**,IRR: **8,9**,10,11. Deciduous. Native to valley and low foothill streamsides of southern California. Growth rate slow to 15-30' tall and wide, with a broad, rounded canopy, or shrublike with branches drooping to the ground. Leaves are alternate, pinnate, 6-9" long, shiny green, with 11-15 oblong-lanceolate, stalkless leaflets, 1 to 1-1/2" long, with finely serrate edges and tufted hairs along the veins on the pale underside, and brief yellow fall color. Insignificant greenish flowers occur in early spring, male catkins with 30-40 stamens along axillary buds and females with a 2-lobed style at terminal buds. Fruits are oval, husked, 1 to 1-1/4" long, maturing in late summer, drying to black, with a wrinkled surface. Nuts are enclosed within a hard shell, grooved with longitudinal furrows. Bark is dark brown to black, becoming roughly furrowed with wide, flat ridges.

Smaller in its drier native habitat than *Juglans hindsii*, but becoming as large or larger in parks and lawns in the northern part of the state under favorable conditions. Resistant to oak root fungus. Longevity estimated to be 50-90 years. Series associations: Pursage, Scalebroom, Sumac shrub series; Big-Doufircan, **Calwalnut**, Engoak, Frecottonwoo, Hinwalsta.

sw Asia, se Europe. FRU, SHD, FOL ACC; IRR: M/M/M/M/-/-

Juglans regia
ENGLISH WALNUT

Juglandaceae. Sunset zones 4-9, 14-23, some varieties in zones 1-3. Deciduous. Native to southeastern Europe, with other varieties from the Himalayas and China. Growth rate initially fast to 40-70' tall and wide, with a broad, oval canopy and a heavy, short trunk with large upward limbs and slender horizontal end branching. Leaves are alternate, pinnately compound, 8-16" long, smooth and green, with 5-7 shiny, ovate-elliptical, 3-6" long leaflets with smooth edges, sessile to the central stalk, and yellow fall color. Insignificant greenish yellow tassel-like flowers occur in June, males 2-4" long and females producing oval, green-husked fruit, with skin drying to brown or black, peeling off when ripe to reveal the wrinkled-surfaced nut. Young twigs are smooth, green, and hairless. Bark is smooth, whitish gray, with wide, deep cracks, usually with a black base, which is the black walnut rootstock.

Almost exclusively grown as an orchard tree or as a shade tree in residential settings. If a new landscape is planned under an existing orchard tree, the area within the dripline is best left undisturbed and watering kept away from the trunk to prevent rot. Otherwise requires infrequent deep watering. Occasionally subject to aphids, scale, moths, and mites. Longevity estimated to be 80-125 years.

sw N.A. California Native: MD, CH, FW, JP, CC (*NCoRI, SNF, SCoRI, TR, PR, D Mtns.*); (IRR: L/L/L/L/L/L)

Juniperus californica

CALIFORNIA JUNIPER

Cupressaceae. Sunset zones 3, 6-12, 14-24, Jepson *DRN:3, 6,**7**,10,**14-16**,17,**18-23**,24&IRR:8,9,11,12. Evergreen. Native to lower-elevation dry slopes and canyons of the California Coast Ranges and Sierra Nevada, extreme southern Nevada, western Arizona, and Baja California. Growth rate moderate to 10-40' tall and wide, developing a dense, symmetrical, conical shape in youth, becoming more oval and irregular with age, with a deeply fluted, slightly tapered trunk. Leaves are minute, scaly, needlelike, light green to yellowish, with a glandular pit on the back, arranged in 3s on short rounded twigs, forming branch sprays. Inconspicuous monoecious flowers occur as yellowish brown swellings at branch ends in spring. Seed capsules are light reddish brown or bluish, 1/2-3/4" long, rounded, with thin, loose, papery skin and a whitish cast, smooth except for a short burr at the end of each scale segment, tucked among foliage along branches, maturing in September of the second year. Inside the capsules is a dry, mealy, sweet pulp, somewhat fibrous, without resin cells, and 1 or 2 seeds, angular, irregularly grooved, and ridged, but lacking the glandular pit of *Juniperus occidentalis*. Bark is reddish brown, weathering to gray, with vertical interconnected furrows and ridges.

Native to scattered locations throughout the state, usually in dry foothill and lower mountain zones among chaparral and mixed conifers, extending into semiarid desert zones. Young growth is dense, but the center thins with age, exposing the trunk and branching character. Tolerates poor, gravelly soil and hot summer temperatures. Drought tolerant once established. Reseeds readily in favorable conditions. Tough enough for cultivation in desert locations, and tolerant of well-drained soils elsewhere. Longevity estimated to be up to 200-250 years in habitat. Series associations: Blabush, Cupceafreoak, Intlivoakshr, Leaoak, Nolina, Scalebroom shrub series; Birmoumah, **Caljuniper**, Sinpinyon, Piucypsta.

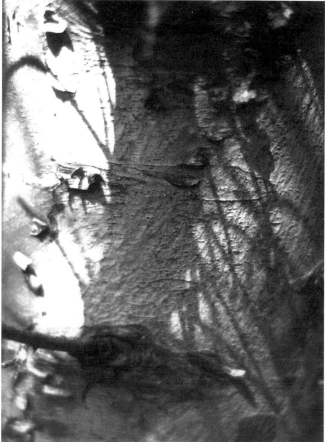

sw N.A. EVG, CNF, SHD, FOL ACC; IRR: No WUCOLS

Juniperus deppeana

ALLIGATOR JUNIPER

Cupressaceae. Sunset zones 1-3, 10-12. Evergreen. Native to upper Sonoran and transition zones in southern New Mexico and Arizona into northern Mexico. Growth rate slow to 30-60' tall and wide, developing an upright, symmetrical, dense, broad, oval shape, more sparse with age. Leaves are bluish green, scaly, needlelike, 1/8" long, minutely toothed, pointed, and conspicuously glandular pitted, persisting 2-5 years. Inconspicuous monoecious flowers occur in spring as yellowish brown swellings at branch ends. Cones are dark reddish brown, rounded, 1/3-1/2" long, tucked among foliage along branches, maturing in 2 years. There are usually 4 dark seeds per cone, distinctly grooved, and swollen on the back. Bark is reddish brown to gray-brown, becoming distinctively fissured and developing thin plates, 1-2" square, which peel up slightly at the edges, resembling an alligator's skin.

Sometimes cultivated as a garden or park specimen, mostly for its unusual bark, or as a foliage accent with pleasing bluish color contrasting with other conifers. Tolerates high desert conditions and inland summers in well-drained soils with only occasional deep watering. Longevity estimated to be 80-100 years or more.

sw N.A. California Native: SS, JP, FF, AL (*NCoRH, CaRH, SNH, SnGb, SnBr, MP, Dmtns*); (IRR: No WUCOLS)

Juniperus occidentalis
WESTERN JUNIPER

Cupressaceae. Sunset zones 1-10, 14, 18-21, Jepson *DRN:4,5, **6**,15-17,24&IRR:**1-3,7**,8-10,**14**,18-23. Evergreen. Native to arid hills and high mountain ranges from Washington to California eastward to Idaho. Growth rate slow to 50-60' tall with a 30-50' spread, developing a dense, billowy, upright form with a stocky trunk, often forked, often grooved but not fluted, tapering to the base, with multiple leaders and short, stiff, upturned branches. Leaves are short, tightly clasped, scalelike, pale grayish green, marked on the backside with a glandular pit, often whitish with resin, persisting about 2 seasons, afterward pushed off by expanding reddish brown branchlets. Leaf scales clasp twigs successively in groups of 3, forming a rounded stem, in 6 longitudinal rows. Cones are bluish black, 1/4-1/3" round, berrylike, covered with a whitish cast, maturing in September of the second season, and splitting open along various seams. Seeds are dark brown, 1/4" round, flat, bony, pitted, grooved, and bluntly pointed at one end. Older trees have light brown to reddish bark, with wide but shallow furrows and long, flat ridges connected at long intervals by narrower diagonal ridges.

Not adaptable outside its high-elevation habitat. Withstands cold, windy alpine conditions. Deep, large root system. A northern variety, var. *australis*, extends from near Susanville northward, along with ponderosa pine. A southern variety, var. *occidentalis*, extends southward, mixed with higher-elevation pines such Jeffrey pine, hemlock, singleleaf pinyon, and fir. Longevity estimated to be 125-300 years, though some may be 500-800 years old or more. Series associations: Bigsagebrush, Lowsagebrush shrub series; Blacottonwoo, Foopine, **Moujun, Wesjuniper**, Bluoak, Orewhioak, Curmoumah, Jef-Pinponpin, Bakcypsta.

Cultivar. EVG, FOL ACC; IRR: L/L/M/M/M/M

Juniperus scopulorum 'Tolleson's Blue Weeping'

TOLLESON'S BLUE WEEPING JUNIPER

Cupressaceae. Sunset zones 1-24. Evergreen. Origin unknown. Growth rate moderate to 20' tall and 10' wide, developing an irregular, upright form, usually with a slender central trunk and upward-arching lateral leaders, pendulous at the ends. Spiny, scalelike foliage is rich bluish green, in twisting, weeping branch sprays. Inconspicuous, dioecious, unisexual flowers are few and occur on separate trees, small yellowish males on branches with juvenile foliage and greenish females maturing in the second year, forming 1/4" round, purplish brown, globular, lumpy cones. Twigs are yellowish green, turning brown, later becoming scaly. Bark is thin and reddish brown, weathering to gray, with vertical interconnected furrows and thin, scaly, peeling strips.

A popular small, weeping, evergreen tree with attractive, pendulous, bluish foliage, often used as a lawn specimen or in groupings. Tolerates regular watering, but not standing water, in average well-drained soils. Lower branches are often removed to expose the trunk and allow plantings underneath, but may appear gawky if the trunk is small. A green form exists, but is seldom used. Longevity estimated to be 50-75 years.

e Asia. SHD, FLW ACC; IRR: M/M/M/M/M/M

Koelreuteria bipinnata

CHINESE FLAME TREE

Sapindaceae. Sunset zones 8-24. Deciduous. Native to western China. Growth rate slow to moderate to 20-40' tall with a 20-40' spread, developing a broad, oval form, eventually with a flattened top. Leaves are alternate, pinnately compound, 1-2' long by nearly as wide, shiny green, with 6-10 elliptical-lanceolate, smooth-edged, acutely pointed leaflets, nearly sessile on the rachis, with pale undersides, and yellow fall color, dropping slowly in fall. Showy, 24" long, loose, erect, terminal raceme clusters of bright yellow, long-stemmed, 1/2" flowers occur in June and August, followed by large clusters of inflated, triangular, thin, papery, bladderlike seed pods with 2-3 shiny, black, 3/8" round seeds. Pods are green at first, darkening to pinkish red to bronze-red in fall before drying to a warm brown, persisting briefly into winter. Bark is grayish brown with reddish fissures and broad, scaly plates.

A desirable shade or small street tree, with non-invasive roots, less commonly used than *Koelreuteria paniculata*, as it is only half-hardy. Develops best form with moderate watering and pruning to develop a strong trunk and canopy. Withstands heat, but suffers in drought and does well in lawns. Flowers attract bees. Longevity estimated to be 75-100 years.

e Asia. SHD, FLW ACC; IRR: M/M/M/M/M/M

Koelreuteria paniculata

GOLDENRAIN TREE

Sapindaceae. Sunset zones 2-24. Deciduous. Native to China and Korea and naturalized in Japan. Growth rate slow to moderate to 20-35' tall and 25-40' wide, with rather upright, open branching, becoming taller if shaded or planted closely. Handsome leaves are alternate, odd-pinnate or bipinnately compound, 10-15" long, medium green, with 7-15 ovate pointed leaflets, 1-3" long, with lobed or toothed edges, pale undersides with downy veins, and brief yellow fall color. Large, showy, loose, 24" long raceme clusters of bright yellow, long-stemmed, 1/2" flowers occur in June and August. Clusters of inflated, triangular, thin, papery, bladderlike seed pods, with 2-3 shiny, black, 3/8" round seeds turn warm brown, then reddish to purplish bronze in fall, persisting briefly into winter. Bark is grayish brown with reddish fissures and broad, scaly plates

   An excellent lawn tree with non-invasive roots, casting light shade, and effective in small groves. Somewhat more graceful and light-textured than *Koelreuteria bipinnata*. Withstands heat, cold, alkaline soils, wind, and drought, but benefits from regular deep watering. Flowers attract bees. Longevity estimated to be 75-125 years. 'Kew' and 'Fastigiata' are upright, narrow forms, not often available.

Cultivar. SHD, FLW ACC; IRR: M/M/-/-/-/-

Laburnum x watereri
GOLDENCHAIN TREE

Fabaceae. Sunset zones 1-10, 14-17. Deciduous. Cultivated hybrid between *Laburnum anagyroides* and *L. alpinum*, which are native to Europe. Growth rate moderate to 15-30' tall with a 10-20' spread, developing a tall, columnar to ovoid form, an upright trunk, and branches that may arch in a slightly irregular fashion. Leaves are alternate, trifoliately compound, larger and thicker than either species, 2-3" long, dull dark green, with 3 ovate-elliptical, sessile leaflets, a long petiole, finely hairy, with smooth edges, and yellow fall color. Yellow, wisterialike, legume flowers in clusters are showy in early spring, drooping in 6-10" long racemes. Flattened, dull brown, legume seed pods, with small, black, hard-shelled seeds, are sparsely set. Bark is smooth and greenish gray, becoming more gray with age.

An uncommon but dramatic flowering accent tree or single lawn specimen, especially with an evergreen background, or in a grove or allée. Best in semi-shade with moderate watering. Occasional heavy aphid infestations in deep shade. Needs protection from hot, dry sun. May become chlorotic in alkaline soils. Light pruning may be required to retain shape or remove unwanted suckers. All parts of the plant are poisonous. Longevity estimated to be 50-90 years.

TOP RIGHT: Photograph by Debbie Martinez.
MIDDLE RIGHT: Photograph by Kathryn Martinez.

Cultivars. SHD, FLW ACC; IRR: No WUCOLS

Lagerstroemia fauriei hybrids
JAPANESE CRAPE MYRTLE

Lythraceae. Sunset zones 7-10, 12-14, 18-21. Deciduous. Growth rate moderate to 20-30' tall and wide with vigorous, upright branches that arch outward. Leaves are simple, mostly opposite, occasionally alternate on upper branches, 1-1/2 to 3" long, bright green, ovate-lanceolate, with smooth edges, smooth uppersides, slightly hairy at veins on the undersides, with combinations of red, orange, and yellow fall color. Crinkled, crapelike flowers with 6 petals occur in showy, upright, over 1' long clusters at branch ends in June and July. Clusters of round, 2-seamed seed capsules mature to hard-shelled, shiny, light brown in fall, persisting into winter and containing many tiny dark seeds. New twigs are light green to reddish brown, appearing 4-angled, and either smooth or slightly hairy. Bark is grayish tan and peels off in thin flakes, exposing light and dark patches of shiny, cinnamon-colored underbark.

Lagerstroemia fauriei, native to Japan and seldom cultivated, has small white flowers, attractive peeling brown bark, and resistance to powdery mildew. White-flowered 'Fantasy' and 'Townhouse' are selected seedling sports with with improved flowering and form. Hybrids between *L. indica* and *L. fauriei* selected by the U.S. National Arboretum for flower quality, hardiness, and mildew resistance include: 'Acoma', 10' x 11', white; 'Biloxi', 20' x 12', pale pink; 'Comanche', 12' x 13', coral pink; 'Hopi', 8' x 10', pink; 'Lipan', 13' x 13', lavender; 'Muskogee', 25' x 12', lavender; 'Natchez', 25' x 12', white; 'Pecos', 8' x 6', pink; 'Sioux', 8' x 5', pink; 'Tonto', 15-20' x 15-20', red; 'Tuscarora', 22' x 12', pinkish red; 'Tuskegee', 15' x 18', deep pink; 'Yuma', 13' x 12', lavender; and 'Zuni', 9' x 8', lavender. All do best in well-drained fertile soils with deep watering. Longevity estimated to be 100-175 years.

e Asia. SHD, FLW ACC; IRR: L/L/M/M/M/M

Lagerstroemia indica

CRAPE MYRTLE

Lythraceae. Sunset zones 7-10, 12-14, 18-21, mildew in zones 15-17, 22-24. Deciduous. Native to China. Growth rate slow to moderate to 25' tall and wide, developing an upright, oval canopy, usually low-branched unless trained early as a standard. Leaves are simple, mostly opposite, occasionally alternate on upper branches, 1-2" long, deep glossy green, oblong-elliptical, with smooth edges, short-stalked or nearly sessile, glabrous and shiny on the upperside, pale undersides sometimes slightly hairy along the lower midvein. new leaves with a reddish bronze tint, and red, orange, or yellow fall color. Crinkled, crapelike, odorless flowers, with 6 petals, are in showy, upright, 6-12" long clusters at branch ends from July to September, colors ranging from red, pink, purple, white, to red with white edging. Clusters of round, 2-seamed seed capsules mature to hard-shelled, shiny, light brown in fall, and persist into winter, containing many tiny dark seeds. Twigs are slender, light green to reddish brown, appearing 4-angled, and either smooth or slightly hairy. Bark is smooth, shiny, tan, and peels off in thin flakes, resulting in light and dark blotches.

A popular lawn or courtyard tree that blooms best in warm, dry climates and full sun. Otherwise subject to mildew. Does best in well-drained fertile soils with deep watering, Many named varieties with various flower colors. Also compact and dwarf forms, with a wide range of applications as a small shrub, larger multi-trunk shrub or tree, or single-trunk standard. 'Catawba', 15' x 15', dark purple; 'Glendora White', 25' x 20', white; 'Near East', 15-20' x 15-20', soft pink; 'Peppermint Lace', 15-20' x 15-20', pink with white edge; and 'Watermelon Red', 20-25' x 20-25', bright red. Longevity estimated to be 100-175 years.

Australia. SHD, FLW ACC; IRR: L/-/L/L/-/-

Lagunaria patersonii

COW ITCH TREE

Malvaceae. Sunset zones 13, 15-24. Evergreen. Native to eastern Australia and the Norfolk Islands. Growth rate moderately fast to 20-50' tall, with an upright, narrow form, often somewhat pyramidal, with ascending branches in youth, broadening to an irregular, billowy form at maturity, sometimes to 40' wide, with a heavy, dense, foliage mass. Leaves are alternate, simple, 2-4" long by 3/4 to 1-1/2" wide, oblong-ovate, with smooth margins, rough-textured, dark olive-green uppersides, and pale gray, finely tomentose undersides. Pendulous, 5-petaled flowers are bell-shaped, 1-2" long, rose pink, rarely white, occurring from May to July, solitary at upper leaf axils. Fruits are light green, ovoid, olivelike, 1" long, with rough tomentose skin, forming capsules that mature to brown and, when fully ripe, split into 5 sections, exposing smooth, reddish brown, shiny seeds in fuzzy-lined compartments. Pods are used in flower arranging, but short fibers inside can cause skin irritation. Twigs are fuzzy and grayish green. Bark is dull grayish brown with many deep furrows, narrow, interconnected, scaly ridges, and occasional warty protuberances.

A fairly tough accent or flowering background tree used in groupings or as a street tree in narrow planters. Sometimes cultivated in southern California, less often in the San Francisco Bay Area, as it is not reliably hardy below 25 degrees. Tolerates heavy and poor soils, inland heat, and drought once established, but often does best in coastal conditions. Longevity estimated to be 80-100 years.

e Asia. EVG, CNF, SHD, FOL ACC; IRR: No WUCOLS

Larix kaempferi

JAPANESE LARCH

Pinaceae. Sunset zones 1-9, 14-19. Deciduous. Native to Japan. Growth rate moderately fast to 60' tall or more with a 20-30' spread, developing a tall, graceful, slender, pyramidal form with slender, bowed branches and drooping branchlets. Bluish green needles, 1/2 to 1-1/4" long, arising tuftlike from buds along slender twigs, are only slightly stiff but sharply pointed, and turn soft golden yellow in fall. Insignificant, monoecious flowers are sparsely grouped, the vertical, cylindrical, greenish purple male spikes shedding pollen in summer and fatter, greenish yellow, vertical, 1/2 to 1-1/2" long female conelets maturing to small, tan, rounded cones in late fall. Cone scale tips are turned under at the ends, and have 1/4" long, tan, winged seeds, 2 per scale. Cones remain on bare branches through winter, creating a curious effect. Bark is grayish brown, shallowly fissured, with numerous small, flat, scaly ridgeplates.

An uncommon introduced species, adaptable only to cooler, moister climates of the northwest, with a form much like *Tsuga heterophylla*, but deciduous, and with a softer texture than *Cedrus deodara*. More graceful than the western native *Larix occidentalis*, which is found only at the highest northwest elevations and has much sparser foliage. Requires cold winters for vigor. Longevity estimated to be 150-200 years.

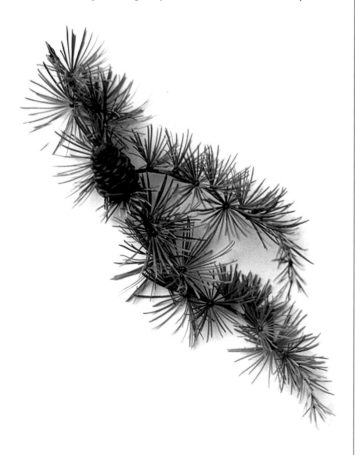

MIDDLE RIGHT: Photograph by Monty and Diane Knudsen.

Mediterranean. EVG, SHD, FLW ACC; IRR: L/L/L/L/M/M

Laurus nobilis
GRECIAN LAUREL

Lauraceae. Sunset zones 5-9, 12-24. Evergreen. Native to the Mediterranean. Growth rate slow to 12-40' tall and wide, usually multiple-trunked at the base, narrowly upright in youth, becoming wide-spreading with age, more conical if trained as a single-trunk standard. Leaves are alternate, simple, 2-4" long, dark green, shiny and leathery, lanceolate with acutely pointed ends, and aromatic. Insignificant small, yellow, dioecious, unisexual flowers occur in clusters at leaf axils in spring, with male and female flowers on separate plants. Dark purple or black pea-sized drupe-berries containing a woody-stone seed mature in fall and are favored by birds. Twigs are smooth, dark green. Bark is thin, smooth, and grayish, often developing black markings.

Commonly used as a large, dense, background screen or as a small shade tree or tree standard in tree wells. Tolerates clipping and heavy shearing as a hedge or large container plant. Tree forms often sucker heavily at the base. Drought tolerant once established, but needs good drainage. Aromatic leaves are the "bay leaves" used in cooking, the "laurels" used by Greeks and Romans, and also make good Christmas wreaths. Longevity estimated to be 200-400 years or more.

e Asia. EVG, SHD, FLW ACC; IRR: L/L/M/M/M/M

Ligustrum lucidum

GLOSSY PRIVET

Oleaceae. Sunset zones 5-24. Evergreen. Native to Japan. Growth rate initially very fast to 20-40' tall and wide, either as a large shrub or trained as a single-trunk tree with a round-headed canopy from upright limbs, which become twiggy at the ends. Leaves are opposite, simple, 4-6" long, glossy dark green, leathery, with smooth edges, acuminate ends, and dull pale undersides. Large, dense, upright, pointed panicles of white flowers occur in summer, with a heavy fragrance that some find objectionable. Clusters of 1/8 to 1/4" purple or black drupe-berries contain tiny, hard-shelled seeds, and most persist into winter until eaten by birds. Twigs are smooth and greenish yellow. Bark is smooth and gray with whitish wartlike markings, becoming grayish brown, either shallowly fissured or scaly on older trunks.

Once commonly used as a small street or courtyard tree, and many still remain. Effective as a dense screen for narrow planters or as a standard in large containers. Reseeds readily. Tolerates clipping and shearing. Fast growth and drought tolerance are its main virtues, though allergies to the flowers make it less popular than in the past. Leaves and berries are poisonous. Longevity estimated to be 50-80 years.

e Asia. SHD, FAL ACC; IRR: No WUCOLS

Liquidambar formosana

CHINESE SWEET GUM

Hamamelidaceae. Sunset zones 4-9, 14-24. Deciduous. Native to China. Growth rate moderate to 40-60' tall by 25' wide, pyramidal in youth, older trees developing more irregular, rangy branching. Leaves are alternate, simple, palmate, to 4-1/2" wide, glossy green, distinctly 3-lobed, with finely toothed edges and pointed ends, pale undersides, and orange-yellow fall color. Insignificant, round, monoecious, yellow-green, tufted flowers occur in spring as leaves emerge, males in 2-3" long terminal racemes and long-stalked females paired or solitary. Fruits are round, long-stemmed, 1", burrlike seed balls covered with stiff, dark brown, hairy spines, with 2 tiny, short-winged seeds in each capsule, maturing in late fall. Bark is smooth and grayish, darkening to grayish brown, furrowed, with narrow, somewhat scaly ridges.

Attractive grouped informally in parks and parkways, as a shade tree, or less often as a street tree. Less symmetrical, more open, and upwardly branched than *Liquidambar styraciflua*. Deep-rooted but with voracious surface roots. Prefers well-drained, moderately moist soils. Longevity estimated to be 200-250 years. 'Afterglow' has lavender-purple new growth and rose-red fall color.

e N.A. SHD, FAL ACC; IRR: M/M/M/M/M/-

Liquidambar styraciflua

AMERICAN SWEET GUM

Hamamelidaceae. Sunset zones 3-9, 14-24. Deciduous. Native to the eastern U.S. Growth rate moderate to about 60' tall (to 100' or more on the east coast) with a 20-25' spread, erect, with lateral side branches curving upward in a rather symmetrical fashion, broadening somewhat with age. Leaves are alternate, simple, 4-6" wide, glossy deep green, palmate, 5-7 lobed, with finely toothed edges, pointed ends, pale undersides, minute tufted hairs at vein axils, and brilliant red, purple, orange, or yellow fall color, often in combination. Insignificant, monoecious, round, yellow-green, tufted flowers occur in spring as leaves emerge, males in 2-3" long terminal racemes and long-stalked females paired or solitary. Fruits are round, 1", burrlike seed balls covered with stiff, blunt, woody spines, with 2 tiny short-winged seeds in each capsule, maturing in late fall and hanging like ornaments into winter. Bark is smooth and grayish, darkening to grayish brown and becoming furrowed, with narrow, somewhat scaly ridges.

Grown primarily for fall color and vertical, upright form as a street, park, or shade tree. Prefers moderately moist, well-drained soil. Deep-rooted, but voracious surface feeder roots can heave paving. Spiny seed balls are a nuisance in pedestrian areas. Resists oak root fungus. Longevity estimated to be 200-250 years. 'Burgundy' has deep wine or purple fall color, 'Festival' a combination of yellow, pink, orange, and red, and 'Palo Alto' yellow to orange-red.

Cultivar. SHD, FOL ACC; IRR: M/M/M/M/M/-

Liquidambar styraciflua 'Rotundiloba'

ROUNDLEAF SWEET GUM

Hamamelidaceae. Sunset zones 3-9, 14-24. Deciduous. Variety of the species native to the eastern U.S. Growth rate moderate to 50' tall or more with a 20-30' spread. Leaves are alternate, simple, with 5 rounded lobes, and a combination of yellow, red, and purple fall color. Insignificant, yellow-green, tufted flowers occur in spring as leaves emerge. Does not set seed balls.

Touted as a more desirable form of the species and becoming more readily available.

Cultivar. SHD, SPC ACC; IRR: M/H/M/M/M/-

Liriodendron tulipifera 'Tortuosum'

CORKSCREW TULIP TREE

Hamamelidaceae. Sunset zones 2-12, 14-24. Deciduous. Variety of the species native to the eastern U.S. Growth rate moderate to 40' tall or more with a 15-20' spread, erect with slender vertical branching, often contorted in a semi-corkscrew fashion, broadening somewhat with age. Leaves are the same as the species. Bark is smooth and grayish on young trees.

An unusual lawn or accent tree, mostly used as a curiosity, dramatic in groupings among other tall-trunked trees. Lower branches usually trimmed free of foliage to expose trunks.

Cultivar. SHD, FAL ACC; FOL ACC; IRR: M/H/M/M/M/-

Liriodendron tulipifera 'Aureomarginatum'

VARIEGATED TULIP TREE

Magnoliaceae. Sunset zones 2-12, 14-24. Deciduous. Variety of the species native to the eastern U.S. Growth rate moderate to fast to 60' tall or more with a 40' spread, developing a columnar or pyramidal crown. Leaves are alternate, simple, 4-6" long by 4-8" wide, the same as the species but with yellow-edged variegation, and solid yellow fall color. Tulip-shaped flowers occur in summer, hidden among foliage.

Seldom available, but suitable as a large, tall, street, park, or lawn shade tree with pleasing form and variegated foliage.

e N.A. SHD, FAL ACC; IRR: M/H/M/M/M/-

Liriodendron tulipifera

TULIP TREE

Magnoliaceae. Sunset zones 2-12, 14-24. Deciduous. Native to the eastern U.S. Growth rate moderate to fast to 60-80' tall with a 40' spread, developing an upright columnar or pyramidal form with branches rising from the straight trunk in a regular pattern. Leaves are alternate, simple, 4-6" long by 4-8" wide, light green, 4-lobed, with smooth margins, pale undersides, truncated at the base and truncated or notched at the apex, looking like the end has been cut off, in a tuliplike outline, with bright yellow fall color. Tulip-shaped, perfect flowers, barely noticeable among foliage in early summer, are 2-3" across, with 6 greenish yellow, overlapping petals, orange or reddish throat markings, and many large yellow stamens around a central style. Erect, short-stalked, 2-3" long, composite seed fruits are conelike samaras, spirally arranged around a central spine. Seeds are 1-1/2" long, 4-angled, with an elongated, thin, terminal wing, maturing to light tan in fall and persisting into winter. Shiny, reddish brown twigs have buds with 2 green flaps, which temporarily remain at the base of emerging leaf petioles. Bark is smooth and dark green, becoming gray-brown and furrowed, with rough, rounded ridges.

An excellent street, park, or lawn shade tree with showy fall color in colder areas. Prefers moderately moist fertile soils. Needs ample room for spreading surface roots. Occasionally suffers from aphid and scale infestations, but not affected by oak root fungus. Longevity estimated to be 200-250 years. 'Arnold' is more columnar and blooms at a young age.

w N.A. California Native: OW, CC, ME, RW, PF, FF (*NW, CaR, SN, CW, WTR*); (IRR: L/-/L/L/-/-)

Lithocarpus densiflorus

TANBARK OAK

Fagaceae. Sunset zones 4-7, 14-24, Jepson *DRN:4,**5**,6,**17**,& IRR, part SHD:1-3,**7**,14,**15**,**16**,18-24;CV. Evergreen. Native to the Coast Ranges from southern Oregon to Santa Barbara and the eastern Sierra Nevada at 3,000 to 4,000' elevation. Growth rate slow to moderate to 40-80' tall with a 30-50' spread, forming a broad, oval, round-headed canopy with heavy, drooping, lower horizontal branches, becoming taller and more sparse in shade. Leaves are alternate, simple, 2-4" long, shiny, smooth, light green, covered with a creamy to yellowish fuzz when they first appear, shallowly and rather sparsely toothed, edges slightly recurving, dense reddish brown hairs on the undersides, becoming whitish to faintly bluish later in the year, persisting 3-4 years. Pale yellow, drooping, tassel-like flowers with an unusual odor occur in late spring. Shiny, tan, 1 to 1-1/2" long acorns are slightly downy, maturing in the second year, with a finely hairy, bristly scaled cup covering 1/4 of the base of the nut. New twigs are covered with densely woolly, tiny, star-shaped hairs, becoming deep brown with a reddish tinge and a whitish bloom by the second year. Bark on young trees is smooth, grayish, and un-broken, firming and thickening to darker tan to brown tinged with red, with gray blotches, and deep, narrow seams form-ing wide, squarish plates.

A rather common evergreen riparian tree found in cool, moist canyons along mid-elevation mountain creeks and riv-ers, but rarely cultivated outside its habitat. A shrub form oc-curs in the northern ranges into Oregon. Bark has been a main source of tannin. Longevity estimated to be 150-250 years in habitat. Series associations: Caloatgrass, Deerbrush shrub series; Calbay, Doufir, Doufirtan, Grafir, Intlivoak, Knopine, Mixconifer, PorOrfced, Redwood, SanLucfir, **Tanoak**, Wes-hemlock, Weswhipin, Whialder.

e Australia. EVG, FLW ACC, FOL ACC, SHD; IRR: M/-/M/M/-/-

Lophostemon confertus
(*Tristania conferta*)

BRISBANE BOX

Myrtaceae. Sunset zones 15-17, 19-24. Evergreen. Native to eastern Australia. Growth rate slow to moderate to 30-45' tall with a 25' spread, developing an oval form with a dense rounded crown, becoming more irregular and open with age, with foliage tufted at branch ends. Leaves are opposite, simple, 4-6" long by 1-1/2-2" wide, broadly elliptical, smooth, thick, and leathery, glossy dark green, with pale undersides, entire margins, persisting 1 year. Noticeable clusters of small, delicately long-frilled, creamy white, 1-1/2" long, star-shaped flowers, with many stamens, occur in June at leaf axils. Seed capsules are short-stalked, smooth, bell-shaped capsules resembling eucalyptus, ripening to brown in late summer, 3 blunt-ended valves splitting to release a few 1/2", angular, wedge-shaped seeds. Twigs are smooth, shiny, and reddish brown. Bark is smooth and reddish, becoming grayish brown, with large, paper-thin, peeling flakes, resulting in interesting mottled colorations on trunks.

A small evergreen accent tree resembling a eucalyptus but with shiny green foliage. Usually cultivated as an upright standard, but also attractive as a multi-trunked courtyard tree or in large containers. Tolerates heat and coastal conditions. Not reliably hardy below 30 degrees. Otherwise not especially fussy. Longevity estimated to be 60-80 years. 'Elegant' has reddish new growth and 'Variegata' has yellowish edge variegation.

w. N.A. (Ca. only) California Native: s Ca Is. (OW, CH) (*CHI*); EVG,
FOL ACC, FLW ACC; (IRR: L/-/VL/L/-/-)

Lyonothamnus floribundus ssp. *aspleniifolius*

FERN-LEAVED CATALINA IRONWOOD

Rosaceae. Sunset zones 14-17, 19-24. Evergreen. Native to the
Channel Islands off the coast of southern California. Growth
rate moderate to 20-35' tall with a 15' spread, developing a
conical shape, older trees broadening with more picturesque
irregularity. Fernlike leaves are deep green, glossy, opposite,
and variable, either simple with entire to slightly indented
imperfect lobes or more commonly bipinnate, 4-6" long, divided into 3-7 deeply notched or lobed leaflets, with gray, hairy
undersides. Showy, flat-topped, 8-18" umbels of small white
blossoms occur in midsummer, accentuated by dark foliage,
but may become unsightly afterward. Fruit matures in fall of
the second season, composed of 2 very small, closely joined
capsules with a bristly glandular skin, enclosing 4 tiny elongated seeds, dispersed slowly as capsules gradually open. Twigs
are smooth, shiny, and reddish brown. Bark is deep reddish,
turning gray with age, becoming fissured, with thin, flaky strips,
and peeling in long shreds.

A highly desirable but not often cultivated tree for coastal
areas, where it tolerates moderate drought. Otherwise requires
fast-draining soils and light winter pruning to shape. Tolerates inland heat surprisingly well, with moderate deep watering, but becomes chlorotic in heavy soils. Longevity estimated
to be 50-65 years, losing vigor and shape with age. Subspecies
floribundus, which occurs only on Santa Catalina Island, has
simple, undivided, linear leaves. Series associations: **Catirosta**.

se N.A. EVG, SHD, FLW ACC; IRR: M/M/M/M/-/H

Magnolia grandiflora
SOUTHERN MAGNOLIA

Magnoliaceae. Sunset zones 4-12, 14-24. Evergreen. Native to the southeastern U.S. Growth rate slow to moderate to 80' tall with a 60' spread, developing a towering, round-topped, dense canopy. Leaves are alternate, simple, 6-10" long, leathery, dark shiny green, broadly elliptical with slightly recurving smooth edges, lighter fuzzy undersides, and acute to acuminate at the ends, persisting 2 years. Pure white, waxy flowers are perfect, 8-10" wide, with 6-12 petals in multiples of 3, powerfully fragrant, occurring throughout spring into summer. Fruits are unusual, 4-6" long, brown, ovoid, conelike aggregates of spirally arranged follicles, maturing in late summer of the first year, with small slotted compartments holding pea-sized drupelike seeds with a bright scarlet red, fleshy, outer coat. Bark is grayish brown and smooth, except for many horizontal lines of glandular warts, and occasionally slightly scaly.

Eventually becomes a magnificent street or lawn specimen. Older trees require litter cleanup throughout the year. Needs ample room. Heavy surface roots and dense shade make planting underneath difficult. Longevity estimated to be 150-200 years. 'Majestic Beauty' is a vigorous, broad pyramidal form, 'Samuel Sommer' has large flowers and leaves rusty tomentose on the undersides, and 'St. Mary' is dense though small, to 20' tall, and a heavy bloomer.

Cultivar. SHD, FLW ACC, SPC; IRR: M/M/M/M/M/-

Magnolia x *soulangeana*

SAUCER MAGNOLIA

Magnoliaceae. Sunset zones 2-10, 12-24. Deciduous. Thought to be a hybrid between the Chinese species *Magnolia denudata* and *M. liliiflora*, originally raised in France. Growth rate slow to moderate to 25' tall with equal or greater spread, often taller and more sparse in shade, usually forming a large, multi-trunked, oval canopy. Leaves are alternate, simple, 4-6" long by 2-3" wide, dull medium green, obovate to broadly oblong, with smooth edges, a narrow apex, pale downy undersides, and yellow fall color. Showy saucer-shaped, white to pink or purplish, 5-10" blossoms with 6-9 petals are lightly fragrant, covering the tree as leaves appear in early spring. Fruits are cucumberlike, cylindrical, 2-4" long, greenish brown, ovoid, conelike aggregates of spirally arranged follicles, maturing in late summer of the first year, with slotted compartments containing 1/4" drupelike seeds with a bright scarlet-red, fleshy outer coat. Twigs are brown and glabrous with grayish lenticel markings. Bark is thin, grayish brown, usually smooth, except for many glandular warts, appearing shallowly fissured.

A popular small accent or lawn tree. Prefers fertile, well-drained, moderately moist soil in part shade to full sun. Longevity estimated to be 75-125 years. Variable from seed. 'Alba' is upright and early blooming, with purple-tinged white flowers, 'Alexandrina' has deep purplish flowers, white inside, midseason, 'Linnei' has large goblet-shaped flowers, white inside, late blooming, and 'Rustica Rubra' has cup-shaped, reddish purple flowers, midseason.

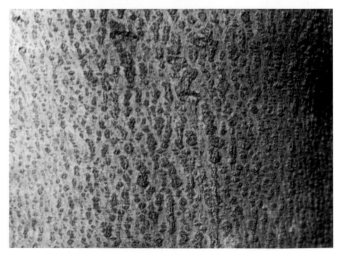

e Asia. SHD, FLW ACC, SPC; IRR: M/M/M/M/M/-

Magnolia stellata
STAR MAGNOLIA

Magnoliaceae. Sunset zones 2-9, 14-24. Deciduous. Native to Japan. Growth rate slow to 10' tall with a 20' spread, developing an oval shrublike form unless pruned up to expose multiple trunks. Leaves are alternate, simple, 1-2" long by 1/2-3/4" wide, dull light green, oblong to narrow-elliptical, with obtusely pointed ends, pale undersides with soft hairs, and yellow fall color. Showy, fragrant, white, pink, or purplish, 3" star-shaped flowers, with 9-15 narrow, elongated, 2" by 1/4" wide petals occur in early spring as leaves appear. Fruits are small, odd-looking, miniature, cucumberlike, 1-2" long, cylindrical, fleshy, conelike aggregates with 1/8-1/4" red seeds in individual compartments, drying and opening as they ripen in summer of the first year. Twigs are slender, brownish gray, glabrous, and closely set. Bark is smooth, ashy gray, and may become scaly, with black markings.

A commonly grown small accent tree for lawn, patio, or garden as a multi-trunk specimen or in groupings. Prefers fertile, well-drained, moderately moist soil. Longevity estimated to be 80-125 years. 'Centennial' has large white petals, 'Pink Star' and 'Rosea' have pink flowers fading to white, 'Rubra' has purplish rose flowers, 'Waterlily' has pink buds opening to large white flowers and an upright, bushy form.

Cultivar. SHD, FLW ACC; IRR: M/M/-/M/M/-

Malus x floribunda
JAPANESE FLOWERING CRABAPPLE

Rosaceae. Sunset zones 1-11, 14-21. Deciduous. Hybrid of Japanese origin. Growth rate moderate to fast to 25' tall and wide, developing a broad, oval canopy, usually with a heavy trunk and low branching forming an irregular umbrella shape. Leaves are alternate, simple, 1-3" long, dark green to bronzy green, ovate to ovate-oblong, edges serrate or slightly lobed, long-pointed ends, lustrous uppersides with a slightly crinkled texture, pale undersides, and red, orange, or yellow fall color. Dark pink flower buds open to a profusion of 1" white flowers, regular and perfect, in long-stemmed rounded racemes, in clusters of 4-7, the outside of the petals dark pink, in April and May. Long-stemmed, hanging clusters of 1/2" yellow, obovoid, pome fruits, edible but bland tasting, containing tiny black seeds, ripen in late summer. Twigs are reddish brown. Bark is thin, smooth, reddish gray-brown, thickening and darkening, becoming fissured, with broad scaly or thin-plated, peeling ridges.

Commonly cultivated as a small flowering accent tree in lawns or patios. Prefers fertile, well-drained, moderately moist soil, but tolerates mildly acidic or alkaline soil and heat. Longevity estimated to be 30-60 years. Many cultivars widely available, with single to double, white to red flowers, upright to arching or semi-weeping form, and varying disease resistance.

South America. EVG, SHD, FOL ACC; IRR: M/M/M/M/-/-

Maytenus boaria

MAYTEN

Celastraceae. Sunset zones 8, 9, 14-24. Evergreen. Native to Chile. Growth rate slow to moderate to 20' tall (may reach 30-50') with a 15' spread, developing an oval, billowy form with long, pendulous branches, the graceful semi-weeping appearance accentuated by dark furrowed bark on trunks, especially after rains. Leaves are small, thin, alternate, simple, 1-2" long by 1/4-1/2" wide, glabrous, light green, lanceolate to ovate-lanceolate, with finely toothed margins, tapering to tip and base, somewhat sparsely spaced on a multitude of overlapping, drooping branchlets. Insignificant clusters of tiny, 1/16" spherical, greenish white flowers, densely and closely set, occur along leaf axils in spring. Fruits are pea-sized, scarlet capsules, maturing to black in fall, barely visible among foliage. Twigs are slender and greenish brown, drooping branchlets flexible and swaying in the slightest breeze. Bark is dark brownish black with many shallow fissures and small, irregular, scaly plates.

Commonly cultivated as a small street or parkway tree and as a residential lawn or patio shade tree. Best with regular deep watering, suffers in drought, and requires well-drained soil. Longevity estimated to be 75-90 years. 'Golden Showers' has uniform, lush, deep green foliage and 'Carsonii' is fast growing with dark, broad foliage.

Australia. EVG, SHD, FLW ACC; IRR: L/L/L/L/-/-

Melaleuca linariifolia

FLAXLEAF PAPERBARK

Myrtaceae. Sunset zones 9, 13-24. Evergreen. Native to New South Wales and Queensland, Australia. Growth rate moderate to fast to 20-30' tall and 20-25' wide, forming a dense, shapely, broad oval canopy, slightly spreading, from a thickened, short trunk, with many smaller branches, densely twiggy at the ends. Often multi-trunked, or with several leader stems, and usually bushy, requiring training to become more tree-like. Leaves are alternate, simple, 1-1/4" long by 1/8" wide, bright green to bluish green, narrow-lanceolate, tapering to a long, narrow point, nearly sessile, and semi-whorled on drooping branchlets. Fluffy, dense clusters of small, cottony, yellowish white, exceptionally showy flowers cover the tree in early summer. Brown, 1/8" bell-shaped capsule fruits are sparsely spaced along branches, persisting 2-3 years. Bark is whitish brown and papery, peeling in thin shreds.

An attractive street, parkway, courtyard, or flowering shade tree or screen for adverse conditions. Not especially clean, but tolerates heat, wind, heavy or poor soils, drought, and coastal conditions. Longevity estimated to be 75-95 years. *Melaleuca nesophila*, with pink flowers, is similar, often used as a shrub, though it can become equally tall.

Australia. EVG, SHD, FLW ACC; IRR: L/L/M/M/-/M

Melaleuca quinquenervia

CAJEPUT TREE

Myrtaceae. Sunset zones 9, 12, 13, 15-17, 20-24. Evergreen. Native to Australia. Growth rate moderate to fast to 20-40' tall and 15-25' wide, developing an upright, semi-open to dense form with a rounded oval canopy and slightly weeping branches. Leaves are alternate, simple, 2-4" long by 3/4" wide, dull green or with a silvery sheen on the upperside, stiff and shiny, elliptical, tapering toward tip and base, slightly hairy, parallel-veined with entire margins, new growth tinged reddish, becoming purplish in winter cold. Showy, 2-3" long clusters of short, yellowish white flowers resembling bottlebrushes, sometimes in pink or purple shades, with many stamens, are grouped in 5s, densely arranged around branchlets below new foliage in summer and fall. Cup-shaped, brownish capsules surround stems in 2-3" long clusters, persisting 1-3 years. Bark is thick, whitish, and peels in large, spongy sheets.

Useful as a parkway or courtyard accent tree with attractive bark and grayish foliage. Effective as a multi-trunk specimen or in groupings. Tolerates heat, wind, poor soil, drought, and coastal conditions. Generally too tender for the Central Valley. Longevity estimated to be 75-100 years.

Asia. SHD, FLW ACC; IRR: VL/L/VL/L/L/L

Melia azedarach

CHINABERRY

Meliaceae. Sunset zones 6-24. Deciduous. Native to India and China. Growth rate initially fast, slowing at maturity, to 30-50' tall and wide, developing a wide-spreading oval canopy, thick trunk, heavy limbs, and dense foliage casting deep shade. Leaves are alternate, bipinnately compound, 1-3' long, with 10-30 ovate, glabrous leaflets, 1-2" long, with toothed or lobed margins, acuminate ends, and yellow fall color. Loose panicle clusters of showy, fragrant, lilac or white flowers occur in May. A profusion of round, inedible, 1/2 3/4" long, yellow, fleshy, drupe berries mature in early fall, somewhat attractive on bare branches but messy as they drop or shrivel and turn brown, persisting into winter, poisonous if eaten in quantity but attractive to birds. Hard seeds are said to have been used to make rosaries, and sometimes called bead tree. Twigs are dark brown, short and stout, with buds set closely together. Dark brown bark becomes deeply furrowed, with reddish tones as outer bark expands.

Once commonly used as a street or shade tree with lush foliage and flowers, especially in hot, dry climates where choices are few. Still commercially available, but not widely used because of fruit drop, weak wood, and frequent suckering. Occasionally grown as a residential lawn specimen. Tolerates extreme drought when established, as well as alkaline soils, wind, and moderate seacoast exposure. Longevity estimated to be 50-100 years. 'Umbraculifera' is a compact dome-shaped tree to 30' tall but less picturesque than the species.

e Asia. CNF, SHD, FOL ACC; IRR: H/H/H/H/-/-

Metasequoia glyptostroboides

DAWN REDWOOD

Pinaceae. Sunset zones 3-10, 14-24. Deciduous. Native to China, dating back millions of years. Once thought extinct, a small grove was discovered in Szechuan, China, in 1945, and it has since become widely available. Growth rate moderate to 90' tall with a 20' spread after 40 years (maximum size unknown) developing a conical form resembling *Sequoiadendron giganteum* but deciduous in winter. Leaves are pinnately compound, light green, paired opposite along branchlets, with 20-30 flat needlelike, 1/3 to 1/2" long leaflets on petioles, angling upward along each side of branchlets, in frondlike fashion, soft to the touch, turning light bronzy yellow to orange-tan in fall before dropping. Inconspicuous flowers are monoecious, unisexual, light green 1/8" females barely noticeable in early spring. Cones are ovoid, 1/2-3/4" long, woody-scaled, reddish brown, and pendulous from long slender stalks. Scales open to release many tiny brown seeds when mature in late summer of the first year, and persist into winter, hanging on long stems from bare branches. Buttressed trunk tapers to the crown. Bark is fibrous, reddish brown, with shallow fissures and stringy, peeling ridges.

An excellent lawn, shade, or accent tree, somewhat of a curiosity for its soft delicate foliage and deciduous character. Requires moderate moisture, but easy to grow and garden under despite some surface rooting. Longevity may be 1,000 years or more.

N.Z. EVG, SHD, FLW ACC; IRR: L/-/M/M/-/-

Metrosideros excelsus

NEW ZEALAND CHRISTMAS TREE

Myrtaceae. Sunset zones 16, 17, 23, 24. Evergreen. Native to New Zealand. Growth rate slow to moderate to 30' tall or more with an equal spread, developing a dense, oval form, either with upright ascending branches, becoming a spreading crown with age, or branching from the ground up, unless trained to tree form and staked. Leaves are opposite, simple, 2 to 3-1/2" long by 1 to 1-1/2" wide, leathery, dark green and shiny, oval-elliptical, with white, densely woolly undersides. Large, showy umbel clusters of dark red, long-stamened flowers occur at branch ends from May to July. Green seed capsules persist, maturing to brown in winter. Twigs are slender, green, and pubescent. Bark is dark brown, smooth at first, becoming shallowly fissured with broad, scaly, peeling plates. Older trees in humid climates may develop wiry, reddish brown aerial roots extending from lower branches to the ground.

Commonly cultivated as a small street or patio tree, most often in coastal locations. Does not do well in hotter, drier inland climates. Rather tender but tolerates wind, sandy soils, and salt spray dependably and requires little water along the coast once established. Does not tolerate wet soils. Heavy roots may heave paving. Longevity estimated to be 80-125 years. 'Aurea' has yellow flowers, and 'Alba' has white flowers.

e Asia. SHD, FAL ACC; IRR: M/M/M/M/M/M

Morus alba
WHITE MULBERRY

Moraceae. Sunset zones 2-24. Deciduous. Native to China. Growth rate fast to 30-50' tall and wide, forming a broad, rounded canopy. Leaves are alternate, simple, with serrate edges, juvenile leaves 2-3" long, thinly palmate, with distinct narrow fingers and mature leaves 4-6" long by 2-4" wide, nearly oval lanceolate or 3-lobed, with long pointed ends, pale undersides, minutely hairy, and golden yellow fall color. Greenish yellow, monoecious flowers occur in spring after new leaves, 1" long males and 1/2" long females, in drooping tassels. Fruits are 1/2", oblong, red, berrylike syncarps, edible but bland tasting, favored by birds, turning black and juicy as they ripen in late summer. Bark is thin, greenish gray, darkening with age and becoming shallowly fissured, often with a grayish cast.

Not as commonly grown as the fruitless variety, but still seen in urban residential yards, casting heavy shade. Voracious heavy surface roots. Fruits stain paving. Tolerates heat, moderate drought, coastal conditions, and alkaline soils. Longevity estimated to be 75-100 years.

s N.A. SHD; (IRR: No WUCOLS) Extremely invasive and weedy, its use is strongly discouraged.

Morus microphylla
TEXAS MULBERRY

Moraceae. No Sunset zones. Deciduous. Native to southern Oklahoma and Texas to Arizona and northern Mexico. Growth rate fast to 20' tall and wide, forming a small tree or clumping large shrub with a low, broad oval canopy. Leaves are alternate, simple, 1-3" long, thick dark green, with coarsely serrate edges, variable in shape, either ovate or 3-lobed, with pointed ends, 3 main veins from the base, which is often heart-shaped, with short, rough hairs on the upperside, lighter, downy undersides, and yellow fall color. Tiny, greenish yellow flowers occur in dense, 1/2-3/4" long clusters in spring among leaves, males and females on separate trees. Reddish to purplish black mulberries, ripening in May and June, are 1/2" long, edible, sweet and juicy. Twigs are slender and greenish brown, hairy at first, and exude a milky sap when cut. Bark is smooth and light gray, becoming furrowed and scaly.

Small tree sometimes grown for fruit or for shade. Tolerates heat and drought when established. Rarely cultivated on the west coast in favor of other more desirable species, though it has reseeded often enough to become a nuisance. Longevity estimated to be 50 years, losing vigor with age.

Cultivar(s). SHD, FAL CLR; IRR: M/M/M/M/M/M

Morus alba 'Fruitless'
FRUITLESS WHITE MULBERRY

Moraceae. Sunset zones 2-24. Deciduous. Growth rate fast to 30-50' tall and wide, forming a broad, rounded canopy. Leaves are alternate, simple, 6-8" long by 5-7" wide, medium to dark green, oval lanceolate or palmately lobed, glossy and nearly glabrous, with pale undersides, coarsely serrate edges, and deep yellow to orange fall color. Greenish yellow monoecious flowers occur in spring after leaves appear, with 1" long males and 1/2" long females in drooping tassels. Does not set fruit. Smooth bark with a whitish cast becomes furrowed, turning brown to gray-brown with age.

'Fruitless' refers to any of 3 commonly used cultivars: 'Fan-San' P.P. 2681 (1966), 'Kingan' (1932), or 'Stribling' (about 1952). Can be left upruned, or with minimal sucker shoot pruning. Pollarding when it becomes too large for most spaces produces vigorous growth the next year. 'Stribling' has larger, more deeply lobed leaves than 'Kingan', which is reportedly most commonly used. Fast, reliable, casting dense shade without the messy fruit of the species. Too large for most spaces unless pollarded, which produces vigorous growth the following year, requiring yearly maintenance. Heavy surface roots and many yellow fibrous roots invade gardens beneath, which is why they are nearly always found in lawns. Tolerates heat, dryness, and poor or alkaline soils. Longevity estimated to be 75-100 years.

Asia. SHD, FRU; IRR: No WUCOLS

Morus nigra

BLACK MULBERRY

Moraceae. Sunset zones 4-24. Deciduous. Native to Asia. Growth rate initially fast to 30' tall and 35' wide, with a dense, heavy-branched, low, broad oval canopy. Leaves are alternate, simple, 3-5" long, thick, dark green, with coarsely serrate edges, ovate or 1-3 lobed, heart-shaped at the base, with short rough hairs on the upperside, lighter downy undersides, and yellow fall color. Greenish yellow monoecious flowers occur in spring after leaves, with 1" long males and 1/2" long females in drooping tassels. Fruits are oblong, berrylike, 1", red syncarps, edible but bland tasting, favored by birds, turning black and juicy as they ripen in late summer. Bark is dark brown, scaly, and furrowed.

Widely cultivated in southwestern Asia and southern Europe, but seldom grown here, where it remains mostly a curiosity tree, for fruit or shade, in a park or residential setting. Tolerates heat and drought when established. Fruits stain paving and other surfaces. Heavy surface roots may develop without periodic deep watering. Does not appear to reseed invasively. Longevity estimated to be 75-100 years, losing vigor and shapeliness with age.

N.Z. EVG, SHD, SCR; IRR: L/M/M/M/-/-

Myoporum laetum
MYOPORUM

Myoporaceae. Sunset zones 8, 9, 14-17, 19-24. Evergreen. Native to New Zealand. Growth rate moderate to fast to 25-30' tall with a 20' spread, forming a broad, dense, billowy canopy, often multi-trunked and bushy or low-branched from a short trunk, with dense foliage casting heavy shade. Leaves are alternate, simple, 3-4" long, bright, shiny dark green with translucent spots, lanceolate to elliptic-oblong, margins finely toothed toward the tips, nearly sessile along stems, and closely whorled, persisting 2-3 years. Small umbel clusters of 1/2", white, bell-shaped flowers with tiny purple throat spots occur in summer in clusters of 2-6, followed by pea-sized, 1/8" round, reddish to purple berry fruits. Twigs are stout, glabrous green, slightly sticky, covered with a gummy film at first, turning brown with age. Bark is smooth grayish brown, becoming slightly scaly, with many shallow fissures.

Often seen as a shade or background tree, especially along the coast, where it has also naturalized, or inland with adequate moisture. Best in full sun, otherwise leggy and less shapely. Needs frequent pruning to develop desired form. Roots are somewhat invasive. Fruit drop may be messy. Not reliably hardy, and may be seriously damaged in heavy frosts. Leaves turn black in smoggy locations, but wash off in rain. Succulent growth considered somewhat fire retardant. Longevity estimated to be 70-90 years.

e Asia. EVG, SPC, FOL ACC, FRAG; IRR: M/?/M/H/-/H

Michelia figo

BANANA SHRUB

Magnoliaceae. Sunset zones 6 (borderline), 9, 14-24. Evergreen. Native to southeastern China. Growth rate slow to 8-20' tall and 6-10' wide, forming a dense oval canopy when the trunk is exposed. Leaves are alternate, simple, 1-3" long by 1/2 to 3/4" wide, light green, stiff and shiny, with a waxy appearance, smooth edges, pale undersides, and blunt pointed tips, persisting 2 years, new growth covered with soft, brown hairs. Small waxy flowers, from April through June, are creamy white or slightly yellow, with a brown to purplish coloration, 1 to 1-1/2" wide, with 6 elongated petals cupped at the edges, an elongated pistil resembling a proboscis, with many short stamens, a bananalike fragrance, at leaf axils rather than branch ends like other magnolias. Cucumberlike fruit is small, fleshy, 1" long. Twigs and bark are smooth and gray.

Generally shrublike, eventually becoming a small tree. Attractive as a courtyard standard or entry or garden specimen with lush, shiny foliage and delightful fragrance. Flowers noticeable only up close. Habit and cultural requirements similar to camellia. Longevity estimated to be 75-100 years.

e Asia. EVG, SHD, FOL ACC; IRR: No WUCOLS

Neolitsea sericea

JAPANESE SILVER TREE

Lauraceae. Sunset zones 4-9, 14-24. Evergreen. Native to eastern China, Korea, Japan, and Taiwan. Growth rate relatively fast to 30' tall and 10-20' wide, forming a dense oval canopy. Leaves are alternate, simple, 4-7" long by 1-2" wide, thick, shiny, dark green with a pale silky sheen on the undersides, 3-veined near the base, with entire margins and acuminate ends, aromatic when crushed, new growth in April bronzy to golden with shiny, silky hairs. Inconspicuous yellow flowers occur in early fall, females producing 1/2 to 5/8" long, ovoid, green berries that mature to glossy red. Twigs are greenish brown, silky, and hairy. Bark is smooth and green, maturing to rough and brown.

A useful evergreen accent or shade tree prized for its silky, golden new foliage, contrasting with glossy dark green mature leaves. Prefers moderately moist, well-drained, fertile soils in full sun or part shade, but not fussy otherwise. Only recently available in North America. Longevity estimated to be 75-100 years.

Mediterranean, e Asia. EVG, FLW ACC; IRR: L/L/L/L/M/M

Nerium oleander

OLEANDER

Apocynaceae. Sunset zones 8-16, 18-24. Evergreen. Native to the Mediterranean and eastward to Asia. Growth rate fast to 15-20' tall with a 4-12' oval crown when trained as a tree, either single- or multi-trunk. Leaves are opposite, simple, 4-6" long by 3/4-1" wide, thick, glossy, dark green, lanceolate, with a slender pointed tip, featherlike veining, smooth edges, persisting 2-3 years. Clusters of showy tubular flowers, either single or double, with 5 petals flatly recurved at the ends, occur at branch ends all summer, in colors ranging from white, pink, red, or yellow to salmon, sometimes slightly fragrant. Slender, drooping, 4-5" long, brown seed pods composed of 2 elongated follicles mature in late summer, splitting along the central seam to release many small, flat, brown seeds with hairy, bristle-winged ends. Twigs are smooth, yellow-green, sometimes covered with a sticky, gummy substance, and exude a poisonous milky white sap when cut. Bark is smooth and grayish.

A common if not ubiquitous shrub, useful as a small evergreen standard or multi-trunk flowering accent tree, in constant summer bloom after other trees have finished flowering. Tolerates heat and extended drought once established, and also poor and moderately salty soils. Not attractive to deer, but subject to aphid, gall, and scale infestations. All plant parts are poisonous. Longevity estimated to be 50-60 years.

e N.A. SHD, FAL ACC; IRR: M/M/M/H/-/-

Nyssa sylvatica
TUPELO

Nyssaceae. Sunset zones 2-10, 14-21. Deciduous. Native to eastern North America, and distantly related to dogwood (*Cornus* sp.). Growth rate slow to moderate to 30-50' tall and 15-25' wide, with an upright pyramidal shape when young, becoming more irregular and spreading. Leaves are alternate, simple, 2-5" long by 1 to 1-1/2" wide, glossy dark green, with entire or wavy margins, pale, often hairy undersides, and deep purplish fall color turning glowing reddish orange. Inconspicuous, small, regular, greenish white, long-stemmed flowers are polygamo-dioecious, and occur in small dense panicles in June. Fruit is a 1/3-2/3" fleshy, ovoid, bluish black drupe with an indistinctly ribbed seed pit, maturing in late summer, attractive to birds. Twigs are slender and greenish to reddish brown, growing in a somewhat zigzag fashion. Bark is thick and reddish brown, becoming fissured, with square, scaly ridges.

An excellent street, lawn, parkway, or patio tree with dependable fall color even in mild winter climates. Grows well in any soil. Tolerates alkalinity and poor drainage. Relatively pest free. Deep rooted, with shallow lateral roots. Larger trees difficult to transplant. Longevity estimated to be 100-150 years.

e Mediterranean, w Asia. EVG, SHD, SPC, FRU; IRR: VL/VL/L/L/M/M

Olea europaea

OLIVE

Oleaceae. Sunset zones 8, 9, 11-24. Evergreen. Native to the eastern Mediterranean and western Asia. Growth rate moderate to 20-35' tall and wide, developing a broad, oval canopy, with strong, upright, exposed branching, usually multi-trunked. Leaves are opposite, simple, 2-4" long by 1/2" wide, leathery, oblong-lanceolate, with smooth margins, dull gray-green on the upperside, with silvery pale undersides. Insignificant but fragrant, small, axillary, greenish to yellowish white flowers occur in small branching clusters in late spring on the previous year's branches. Fruits are oily, fleshy, 1/2 to 1-1/2", ovoid olive drupes with a single hard-shelled seed nut or pit inside, blackening in fall, inedible without processing but attractive to birds. Bark is smooth and gray, becoming grayish brown, slightly scaly and fissured.

A shade or accent tree often used in Spanish, Mediterranean, or adobe/desert landscapes or cultivated for fruit. Tolerates drought, and needs full sun. Picturesque when foliage is pruned up to expose gnarled trunks. Fruit drop stains pavement. Large trees often shipped and transplanted as mature specimens. Longevity estimated to be 300-500 years or more. 'Manzanillo' is the most common fruiting variety, with a small round canopy. 'Swan Hill' has a similar habit, with little pollen, and is fruitless, as is 'Majestic Beauty'. 'Skylark Dwarf' is 10-15' tall, has small leaves, and sets few fruits.

e N.A. SHD, FAL COL, FLW ACC; IRR: No WUCOLS

Oxydendrum arboreum

SORREL TREE

Ericaceae. Sunset zones 2-9, 14-17. Deciduous. Native to the eastern U.S. Growth rate slow to 15-30' tall (up to 50') with a 20' spread, developing a tall oval canopy, from a slender trunk, to a pointed top, with horizontal branches somewhat drooping. Leaves are alternate, simple, 4-6" long by 1-1/2 to 2-1/2" wide, dark shiny green, oblong to lanceolate, finely toothed, nearly glabrous, with pale undersides, and brilliant scarlet fall color. Showy, 6-10" long, drooping panicles of creamy white, regular, perfect, bell-shaped flowers occur at branch tips in late summer. Showy, dry, grayish, 1/2-1/3" long capsules, enclosing 1/8" seeds, are highlighted against fall foliage, rachii persisting on bare branches, creating a unique weeping winter silhouette. Twigs are slender, smooth, and reddish brown. Bark is thick and grayish with reddish-tinged fissures and wide scaly ridges.

A highly decorative accent tree for lawn or terrace. Requires moderately moist but not wet, well-drained soil in semi-shade. Shallow surface roots are not competitive, but may be sensitive to cultivation. Does well as a singular lawn specimen. Longevity estimated to be 50-75 years or longer where exacting cultural requirements are met.

Palms

Australia. SHD, ACC; IRR: M/M/M/M/-/-

Archontophoenix cunninghamiana

KING PALM, BUNGALOW PALM

Palmaceae. Sunset zones 21-24; or indoors. Native to Queensland and New South Wales, Australia. Growth rate slow eventually to 50' tall or taller with a 10-15' spread, forming a 12-14" diameter columnar trunk, usually not enlarged at the base and usually smooth except for leaf stub depressions. Stiff, green, feather-shaped fronds may reach 8-10' long, with numerous 1 to 1-1/2' long by 2-3" wide leaf blades, green on the upperside and gray beneath, smooth, green, 3-6' long by 1-3" wide petiole stalks clasping the trunk, and 1-3' long spineless sheaths as green as the leaves. Old fronds fall off cleanly. Attractive lilac or purplish monoecious flower clusters arise from the trunk below fronds in late summer and fall. Fruits are 1/2-3/8" long waxy, bright red datelike drupes.

A clean and handsome palm, often sold as an indoor houseplant and commonly grown outdoors as a specimen in areas that are nearly frost-free. An excellent tree for streets, medians, parks, and lawns, either as a single specimen or in groves. Tolerates heat and wind, but hardy only to about 32 degrees.

sw N.A. SHD, ACC; IRR: L/-/L/L/L/L

Brahea edulis

GUADALUPE FAN PALM

Palmaceae. Sunset zones 12-24. Native to Guadalupe Island, off the coast of Baja California. Growth rate slow, eventually to 30' tall with a 15' spread, on a short, heavy trunk, usually textured with leaf stub depressions. Leaves are silver-blue to bright green, stiff, fan-shaped, nearly circular, 3' long, with 70-80 folds, open 1/3 of the way to the center base. Leaf margins often have a few threadlike filaments along otherwise smooth edges, with split ends, and are on 3-6' long by 1-3" wide stalks, with a persistent sheath, which usually is cut off with the base of the stalk protruding from the trunk. Fragrant, showy, creamy white flower clusters are borne on a long drooping stem in summer. Fruits are 1", black, shiny, edible drupes enclosing a brown, flat, hard-shelled seed, ripening in fall.

A medium-sized palm with attractive green fronds and showy flowers, commonly used in warmer climates. Tolerates heat, drought, and wind. Hardy to 18 degrees.

Palms

South America. SHD, ACC; IRR: L/L/L/L/L/L

Butia capitata

PINDO PALM

Palmaceae. Sunset zones 8, 9, 12-24. Native to eastern Brazil. Growth rate slow to 10-20' tall with a 10-15' spread on a single heavy trunk. Fronds are bluish green, pinnate, feathery, 6-12' long, with as many as 100 paired, folded, 12-18" long, linear-lanceolate pinnae, fairly stiff and acutely pointed, with whitish undersides, a smooth sheath to the base, and persistent until cut off close to the trunk. Flowers are rather striking, pale yellow to pinkish red, monoecious, on a long fleshy spike, male and female flowers on the same tree. Fruits are 3/4", orange, ovoid, sweet-tasting and edible drupes, ripening in late summer, and containing an oily dark seed pit.

A small accent palm with stunning bluish gray foliage. Patterned stubs from old leaf sheaths on trunks can be attractive, especially if trimmed uniformly. Tolerates drought as well as moderate moisture in a sunny location. One of the hardier palms available, to about 15 degrees.

se Asian tropics. SHD, ACC; IRR: H/-/M/-/-/-

Caryota urens

FISHTAIL PALM, WINE PALM

Palmaceae. Sunset zones 23, 24; or indoors. Native to India, Ceylon, and the Malay Peninsula. Growth rate slow, eventually to 40' tall, with a 15' canopy, forming a 12-14" diameter columnar trunk, usually not enlarged at the base and usually smooth except for leaf stub depressions. Green, bipinnate, feather-shaped fronds to 10-20' long are covered with numerous overlapping wedge-shaped leaflets resembling a fishtail. Smooth, green, 3-6' long by 1-3" wide petiole stalks without spines clasp the trunk. Sheaths may remain indefinitely if not removed. Showy panicles of scented cream-colored flowers may reach 10' long. Fruits are 1/2" long, red, datelike drupes, with one or two seeds, covered with a substance that irritates skin.

A handsome and strikingly unusual accent palm, sometimes sold as an indoor house plant. Also grown outdoors in frost-free coastal areas of southern California. Freezes permanently at 32 degrees.

Palms

Mediterranean. SHD, ACC; IRR: L/L/M/M/M/M

Chamaerops humilis

MEDITERRANEAN FAN PALM

Palmaceae. Sunset zones 4-24. Native to the Mediterranean. Growth rate slow to 20' tall and wide, with clumping, 6-12" diameter trunks, forming from offshoots, covered with short, dark stumps if old leaves are removed. Leaves are dark green, fan-shaped, nearly circular, 2-3', with many narrow, stiff folds open 3/4 of the way to the center base, smooth margins, on 1' long by 1/2" wide, tooth-edged stalks, with a persistent sheath, usually cut off with the base of the stalk protruding from the trunk. Small, yellow, perfect flowers are borne on rather short panicles, which hang slightly amongst the fronds.

Somewhat shrubby in youth, and prized for its eventual multi-trunk character. Fairly drought tolerant but responds well to water and fertilization. One of the hardier palms, to 6 degrees. Magnificent specimens in southern California are usually over 100 years old.

South America. SHD, ACC; IRR: L/M/L/M/-/-

Jubaea chilensis

CHILEAN WINE PALM

Palmaceae. Sunset zones 12-24. Native to Chile. Growth rate slow to 50-60' tall and 25' wide, with a single 3-4' diameter, smooth trunk with patterned leaf stalk scars. Feathery fronds are dark, glossy green, pinnate, 6-12' long, with many paired, folded, 20-30" long, linear-lanceolate, stiff and acutely pointed pinnae, and a hairy sheath to the base, persisting about 2 years, then turning brown and hanging skirtlike, dropping from the trunk at the end of the second season. Small, cream-colored, monoecious flowers droop from the crown in long, loose clusters, males and females on the same tree. Coconutlike fruits are edible, yellow, 1 to 1-1/4" long, fibrous, ovoid, containing a dark seed pit, ripening in late summer. Fruits and sugary liquid sap have been popular for various uses in Chile.

Not common, but most often grown as a park or lawn tree, the massive trunk imposing as a specimen or in rows. Benefits from regular watering until establishment, later becoming quite heat and drought tolerant. Hardy to 20 degrees.

Palms

Canary Is. SHD, ACC; IRR: L/L/L/L/M/M

Phoenix canariensis

CANARY ISLAND DATE PALM

Palmaceae. Sunset zones 8, 9, 12-24. Native to the Canary Islands. Growth rate slow to 60' tall with a 50' spread, developing a single, heavy, 2-3' diameter trunk with long, stiff, gracefully arching fronds forming an oval head, thatched dead fronds hanging below. Fronds are dark green, pinnate, feathery, 15-20' long, with many paired, folded, stiff, and fairly sharply pointed pinnae, smooth to the sheath base, and persisting until cut off close to the trunk, stubs gradually decaying, leaving rounded shinglelike impressions on the bark. Small, whitish, dioecious flowers in large, many-branched clusters droop from the crown on a 2-4' long, thick central stem, males and females on separate trees. Fruits are 1/2-3/4" long, yellow-orange, inedible, ovoid date drupes with a dry flesh, containing a dark seed pit, and ripening late in summer.

A strong, classic, commonly cultivated palm, impressive when grouped in large-scale settings and streetscapes. Tolerates wind and heat. Hardy to 20 degrees.

w Asia. SHD, ACC; IRR: L/L/L/L/M/M

Phoenix dactylifera

DATE PALM

Palmaceae. Sunset zones 8, 9, 12-24. Native to the Middle East. Growth rate slow to 80' tall with a 20-40' spread, usually forming a single trunk and suckering from the base. Fronds are grayish green, pinnate, feathery, 15-20' long, glaucous and waxy, arching upward and drooping at the ends. Clusters of ellipsoidal, 1-3" long, edible orange dates droop in large, loose clusters from a long stem under the foliage.

Classic though not native palm of the California desert and the commercial source of dates. Smaller specimens occasionally seen in northern parts of the state. Resembles Canary Island date palm but foliage is more grayish. Adaptable to seaside locations. Hardy to 20 degrees.

Palms

Africa. SHD, ACC; IRR: -/-/M/M/-/M

Phoenix reclinata

SENEGAL DATE PALM

Palmaceae. Sunset zones 9, 13-17, 21-24. Native to tropical Africa. Growth rate slow to moderate to 20-30' tall and wide, with several upward-curving trunks and, where freezing has caused regrowth from the base, it forms a broad, short-based clump with little or no trunk and a short, bushy appearance. Fronds are dark green, pinnate, feathery, 6-10' long, with many paired, folded, stiff and sharply pointed pinnae, smooth to the sheath base and persistent until removed. Small, cream-colored, dioecious flowers droop from the crown in long, stringy clusters, males and females on separate trees. Fruits are small, 1/2-3/4" long, red or black, ovoid, inedible date drupes with dry flesh, containing a dark seed pit, ripening in late summer.

A prized palm in southern California where multiple long-curving trunks develop on specimens 50-100 years old. Northern California specimens are rarely over 15' tall, with shorter trunks or very bushy at the base, with juvenile sucker growth in areas of frost. Foliage hardy to 32 degrees, though trunks may withstand 23 degrees.

se Asia. ACC; IRR: -/-/M/M/-/M

Phoenix roebelenii

PYGMY DATE PALM

Palmaceae. Sunset zones 13, 16, 17, 22-24. Native to Laos. Growth rate very slow to 6-10' tall with a 6-8' spread, forming a slender, 4-8" diameter trunk covered with matted, stringy, dark hairs, side shoots developing a multi-trunk clump, with finely textured, delicate, arching foliage. Fronds are dark green, pinnate, feathery, 1-3' long, with many paired, folded, limp, acutely pointed pinnae with slightly stringy margins, often with a white glaucous cast, shiny and smooth to the sheath base, and persistent until removed. Small, yellow, dioecious flowers droop from the crown in short, stringy clusters, male and female flowers on separate trees. Fruits are 1/2" long, red or black, ovoid date drupes with dry flesh, containing a dark seed pit, ripening late in summer.

A fine-textured dwarf palm with somewhat exacting requirements for filtered sun and rich, well-drained but moist conditions or in a sunlit atrium with reflected sunlight. An attractive silvery gray variety exists. Hardy to 20 degrees.

Palms

South America. SHD, ACC; IRR: L/M/M/M/M/M

Syagrus romanzoffianum

(*Arecastrum romanzoffianum***)**

QUEEN PALM

Palmaceae. Sunset zones 12, 13, 15-17, 19-24. Native to Brazil. Growth rate moderate to fast to 50' tall with a 20-25' spread, forming a single straight, smooth, trunk. Fronds are glossy, bright green, pinnate, feathery, 9-15' long, with many paired and folded, linear-lanceolate, 14-18" long pinnae, soft, flexible, and subject to damage by heavy winds, with a smooth sheath to the base, persisting 1 year, then turning brown and hanging skirtlike, separating from the trunk at the end of the second season. Small, white, monoecious flowers droop from the crown in loose, 20-40" long clusters among the foliage, males and females on the same tree. Fruits are 1" long, yellow-orange, globose date drupes, containing a dark seed pit, ripening late in summer.

While not as commonly used as Canary Island date palm, the fronds have an appealing delicate, tropical appearance. Responds well to water and fertilizing. Generally hardy to 25 degrees.

e Asia. SHD, ACC; IRR: L/M/M/M/M/-

Trachycarpus fortunei

WINDMILL PALM

Palmaceae. Sunset zones 4-24. Native to China. Growth rate very slow to 30' tall with a 10' spread, forming a single slender, 3-6" diameter trunk, strangely thicker at the top than at the base, and covered with flat sheets of stiff, dark, matted, hairy fibers and the remains of cut foliage stalks. Fronds are 2', nearly round, fan-shaped, with 30-50 folds, on short 1-2' stalks with no teeth, sheaths and thatch persisting indefinitely but usually trimmed off, leaving a short base of the stalk protruding through matted fibers. Small, dense, tightly packed clusters of small, creamy flowers droop slightly on 6-12" long stems near the base of the foliage above the thatch.

A small vertical accent tree for small-scale or tight spaces. Responds well to but doesn't require extensive watering or fertilization. One of the hardier palms, to 10 degrees.

Palms

sw N.A. California Native: SD (Oases) (*Dson*); SHD, ACC; (IRR: L/M/L/L/M/M)

Washingtonia filifera

CALIFORNIA FAN PALM

Palmaceae. Sunset zones 11-24, Jepson *SUN:**14,16**,17,**22,23,24**&IRR:**8,9**,10,**11-13,18-21**. Native to the southwest, often near springs in the Sonoran Desert. Growth rate moderate to 60' tall with a 20' spread, forming a single heavy, 2-3' diameter trunk, with a thatchlike mass of hanging dead fronds, except near the base. Fronds are 3-6' long, green, fan-shaped, nearly circular, with 40-70 folds, open 1/2-1/3 from the center to the base, with threadlike filaments on the margins and split ends, stalks 3-5' long by 1-3" wide with stout, hooked spines along the edges. Sheaths persist indefinitely, but usually are cut off with the base of the stalk protruding or are ground off for a smooth trunk. Fragrant, small, greenish white flowers are borne in clusters on an 8-10' long stem, followed by 3/8" long, black, shiny, edible drupes enclosing a brown, flat, hard-shelled seed.

A large-scale accent tree often used in groves, clumps, as single specimens, or as street trees. Tolerates heat and drought, but thrives on moisture in well-drained soils. Hardy to 18 degrees. Series associations: **Fanpalm**.

se N.A. SHD, ACC; IRR: L/M/L/L/M/M

Washingtonia robusta

MEXICAN FAN PALM

Palmaceae. Sunset zones 11-24. Native to Mexico. Growth rate moderately fast to 100' tall with a 10' spread, forming a single, slender, 1-2' diameter trunk with a thatchlike mass of hanging dead fronds, except near the base. Fronds are green, fan-shaped, 2-5', nearly circular, with 70-80 folds, open 1/2-1/3 from the center to the base, with no or a few threadlike filaments on the margins and drooping ends, stalks 1-3' long by 1-2" wide with a red streak on the underside and stout, hooked spines on the edges. Sheaths persist indefinitely, but are usually cut off with the base of the stalk protruding or are ground off for a smooth trunk. Trunks may develop a curvature if planted at an angle. Small, fragrant, greenish white flower clusters on 3-5' long stems are followed by 3/8" long, black, shiny, edible drupes enclosing a brown, flat, hard-shelled seed.

A large-scale palm with a smaller head and trunk than *Washingtonia filifera* but used for the same purposes. Hardy to 20 degrees.

sw N.A. SHD, FLW, FOL ACC; IRR: VL/VL/L/L/L/L May become weedy or invasive.

Parkinsonia aculeata

MEXICAN PALO VERDE

Fabaceae. Sunset zones 8-24. Deciduous. Native to northern Mexico, southern Texas, and Arizona and naturalized in California deserts and parts of the San Joaquin Valley. Growth rate fast to 15-30' tall and wide, usually developing a large multi-trunked base with a broad oval canopy and extremely fine-textured spiny branches, branchlets, and evergreen leaf stalks, which often persist after tiny leaflet pinnae fall. Leaves are green, alternate, bipinnately compound, appearing pinnately compound, with one slender, spine-tipped axis and 1-3 pairs of thin, wiry, 8-20" long, evergreen streamers, with 25-30 pairs of tiny, 1/8-1/4", narrow, oblong leaflets. Showy clusters of upright, bright yellow, 3/4" flowers with 5 rounded petals cover the airy mistlike foliage, appearing sporadically throughout the year, heaviest in spring. Seed pods are 2-4" long, dark brown legumes, narrowly cylindrical and pointed, constricted between the 1-8 beanlike seeds, maturing in fall and hanging into winter. Slender, yellowish green twigs are finely hairy at first, with 2 short spines next to a larger brownish spine at the leaf axis. Bark is thin, smooth, and green, becoming mottled with brown and somewhat scaly at the base.

A stunning multi-trunk centerpiece in a cactus or succulent garden. Rather common in the San Joaquin Valley, the south coast, the western Transverse Ranges, the Peninsular Ranges, and various desert areas, less often seen as a curiosity tree or accent specimen in hot inland and valley locations. Tolerates extreme heat and drought. Naturalizes easily, often spreading along roadsides and invading native habitats, much like *Tamarix*. Longevity estimated to be 50-75 years, then losing vigor and shape.

e Asia. SHD, FLW ACC; IRR: No WUCOLS

Paulownia tomentosa

EMPRESS TREE

Scrophulariaceae. Sunset zones 4-9, 11-24. Deciduous. Native to China. Growth rate fast to 40-50' tall with a nearly equal spread, forming a tall oval canopy from a heavy trunk with a billowy foliage mass and many short, angular branches, resembling catalpa. Leaves are opposite, simple, 5-12" long by 4-7" wide, light green, heart-shaped, variable in form, either unlobed or 3-5 lobed on larger leaves, with toothed edges, densely hairy, pale undersides, and yellow fall color. Brown flower buds form in fall and are visible through winter. Showy, upright, 6-12" long panicles of fragrant, 2-lipped, 2" long, lilac-blue, trumpet-shaped flowers, with purple spotting and a yellow-striped throat, appear in May. Seed pods are leathery, dark brown, oval, pointed capsules, 1-2" long, splitting down the center to release many small-winged seeds, and persist into winter along with newly forming flower buds. Greenish twigs are soft and hairy. Bark is shiny gray, developing mottled fissures, with a reddish coloration.

A seldom used accent or shade tree that can be a spectacular focal point in the right garden setting. Surface roots, dense shade, and heavy leaf drop make it difficult to garden under. Generally used in lawns, in a parklike setting. Young shoots damaged by frost usually regrow, but older trees weather better as they become acclimated, with thickened bark. Longevity estimated to be 50-75 years in temperate climates.

c Europe. EVG, CNF, FOL ACC; IRR: M/M/M/-/-/-

Picea abies

NORWAY SPRUCE

Pinaceae. Sunset zones 1-6, 14-17. Evergreen. Native to central Europe, Scandinavia, and Russia. Growth rate moderate to fast to 100-150' tall with a 20' spread, developing a symmetrical pyramidal form with horizontal branching and branch ends drooping with age. Needles are 1/2-3/4" long, dark green, stiffly pointed, sometimes curved, and whorled completely around twigs. Insignificant flowers open in May, reddish 1/4-3/8" males, yellow with pollen, clustered at branch tips and fatter, erect, solitary females, pinkish at first, later turning green and drooping downward. Cones are cylindrical, tan, 4-7" long, with wide, thin, woody scales, somewhat pointed at the ends and overlapping rather tightly, maturing to shiny brown in fall of the first year. Twigs are slender, reddish brown, with dark brown, round buds. Bark is reddish brown with many thin, flaking scales, becoming grayish brown with age, broken into square plates.

One of the most common ornamental spruces, long cultivated around the world, with many cultivars. Sometimes used as grafting stock for other ornamental species. Often seen in parks or as a lawn specimen in older residential settings. The more colorful Colorado blue spruce (*Picea pungens* 'Glauca') has been planted more often in recent years. Tolerates wind and cold, as well as heat in moderately moist settings. Longevity estimated to be 150-200 years.

w N.A. California Native: (n Calif.) CF (*KR*); (IRR: No WUCOLS)

Picea breweriana
WEEPING SPRUCE

Pinaceae. Sunset zones 2, 3-7, 14-17, Jepson *DRN,IRR: 2,**3,4**,5, 6,15-17. Evergreen. Native to slopes and river canyons of the Siskiyou Mountains of southwestern Oregon and northwestern California at 3,500-9,000' elevation. Growth rate slow to moderate to 30-50' tall (80-120' in habitat) with a 10-12' spread, developing a dense conical shape with a slender pointed top, sparse, upright shoots, and lower horizontal branching, lowest branches having ropelike, drooping, 4-8' long side branchlets. Needles are 1-1/4" long, bright dark green, flattish, or slightly triangular in cross-section, angled more visibly on the undersides, blunt-ended, with whitish bands on the uppersides. Cones are pendulous, elliptical, 2-5" long, dark green to purplish, with thick woody scales, rounded rather than elongated ends, and entire edges, maturing in one season, turning dull brown when ripe, and releasing wide-winged seeds. Bark is dark reddish brown, developing long, thick, firmly attached scales and becoming gray-brown with age.

An uncommon spruce in isolated forest regions with a graceful weeping form. One of the least hardy of the spruces, limiting its distribution. Does not adapt easily to cultivation. Occasionally seen in arboretums of the northwest. Longevity estimated to be 500-900 years or more in habitat. Series associations: Breoak shrub series; Engspruce, Redfir, Subfir, Whifir, KlaMouEnrsta, Alayelcedsta, Pacsilfirsta.

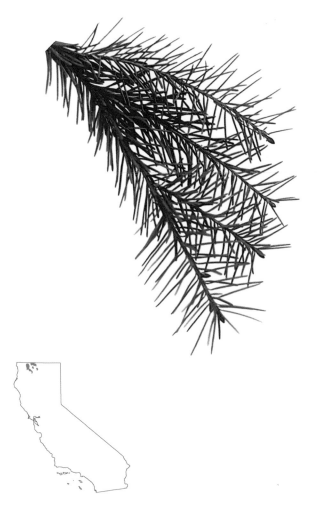

w N.A. California Native: (n Calif.) CF (*KR*); (IRR: No WUCOLS)

Picea engelmannii

ENGELMANN SPRUCE

Pinaceae. Sunset zones 1-7, 10, 14-17. Evergreen. Native to southwestern Canada, Oregon, and extreme northern California east to the Rocky Mountains. Growth rate moderate to fast to 60-130' tall and 20-25' wide, with a tall pyramidal form in youth with upward-arching horizontal branches, becoming round-topped with age, with drooping branches, from a rather large buttressed trunk. Needles are dark green to bluish green, 1 to 1-1/8" long, occasionally with glaucous white bloom, and tending to be crowded at the upper side of branches by upward-curving needles from the undersides. Bluntly pointed ends are not sharp, and needles are somewhat flexible, with no visible resin ducts on the surfaces. Insignificant dark purple male flowers and bright scarlet female flowers occur at ends of upper branches in spring. Pendulous cones are oblong to cylindrical, sessile or short-stalked, with slightly wavy and elongated, flexible, papery scales, with irregular end margins, maturing in fall of the first season to a light brown and falling shortly thereafter. Dark seeds are 1/8" long with a 1/2" long, round-ended wing. Bark is fairly thin and reddish brown, becoming grayish and broken into large, thin, loosely attached scales.

Limited in its native range in California but can be cultivated as an attractive park, lawn, or evergreen background tree. Tolerates wind and cold, as well as moderate heat with moderate moisture. Longevity estimated to be 350-500 years. 'Glauca' and 'Vanderwolf's Blue Pyramid' have steel-blue needles, 'Fendleri' is weeping, and 'Argentea' has silver-gray needles. Series associations: **Engspruce**, Subfir, KlaMouEnrsta.

Cultivar. EVG, CNF, FOL ACC, SPC; IRR: No WUCOLS

Picea glauca 'Conica'

DWARF ALBERTA SPRUCE

Pinaceae. Sunset zones 1-7, 14-17. Evergreen. Cultivar of Alberta spruce (*Picea glauca* var. *albertiana*) which is native to Canada. Growth rate very slow to 6-15' with a dense, bushy, conical form. Needles are stiff, 1/2" long, bluish green, symmetrically whorled around branchlets, slightly curved, with sharp, pointed ends.

A miniature conifer common in northern California and the northwest. Prefers full sun in moderately moist, well-drained soils. Retains its compact geometrical shape without constant shearing and grows only 1/2 to 1 inch per year. Longevity estimated to be 75 years in ideal conditions, often much less as a lawn specimen in warmer climates.

Cultivar. EVG, CNF, FOL ACC, SPC; IRR: No WUCOLS

Picea glauca 'Pendula'

WEEPING WHITE SPRUCE

Pinaceae. Sunset zones 1-7, 14-17. Evergreen. Cultivar of the species, discovered in France. Growth rate slow to 30' tall with a 10-12' spread, developing a dense, upright form, with branches weeping at the ends. Needles are stiff, green, 1/2" long, slightly curved, symmetrically whorled around branches, with sharp, pointed ends.

An unusual specimen, rarely grown, but available in northern California and the northwest. Prefers full sun in moderately moist, well-drained soils. Retains a dense, weeping form, usually without staking. Longevity estimated to be up to 100 years in the northwest, half that in residential California settings, depending on climate and culture.

e Asia. EVG, CNF, FOL ACC, SPC; IRR: No WUCOLS

Picea jezoensis

YEDDO SPRUCE

Pinaceae. No Sunset zones. Evergreen. Native to Manchuria and northern Japan. Growth rate slow to 20-30' with a nearly equal spread, developing a dense, broad-oval form, closely branched, with many twigs. Needles are stiff, bluish green, 1/2-1" long, symmetrically whorled around branches, slightly curved, with sharp, pointed ends and white bands on the upperside, lustrous below. Stout, yellow, red-tinged male flowers are insignificant, occurring along branch ends with the pale green female flowers in May. Pendulous cones are light brown, 2-3" long, oblong to cylindrical, sessile or short-stalked, with jagged scales and tiny winged seeds, occurring on older trees, maturing in fall but not falling until the second year. Bark is gray and scaly, becoming fissured and scaly-plated on older trees.

An uncommon specimen or park lawn tree, once used to a greater extent. Very adaptable. Prefers full sun in moderately moist, well-drained soils. Retains a dense, compact, geometrical shape without constant shearing. Longevity estimated to be 150-200 years.

Cultivar. EVG, CNF, FOL ACC, SPC; IRR: M/M/M/-/-/-

Picea pungens 'Glauca'

COLORADO BLUE SPRUCE

Pinaceae. Sunset zones 1-10, 14-17. Evergreen. Selection of species native to Colorado. Growth rate slow to moderate to 30-60' tall (100' in habitat) with a 10-20' spread, forming a broad, dense, oval pyramid, closely branched to the ground. Needles are very stiff, blue-gray, 1 to 1-1/4" long, 4 angled, with whitish blue resin ducts on 2 sides, sharp prickly ends, in whorled fashion, extending at nearly right angles completely around branchlets. On older trees insignificant stout, yellow, red-tinged male flowers occur in May along branches, with pale green female flowers at branch ends. Pendulous cones are light brown, 2-4" long, oblong to cylindrical, sessile or short-stalked, with flexible, narrowly elongated scales with wavy irregular margins, maturing in fall and dropping the following season. Dark 1/8" long seeds have a broad, oblique, 1/2" long wing. Bark is thin and scaly, pale to dark gray, becoming deeply furrowed, with rounded ridges.

Commonly planted in residential landscapes as a single specimen or accent tree. Also effective in small groves in parks, contrasting colorfully with green conifers and deciduous trees and shrubs. Longevity estimated to be 100-150 years or more, often losing vigor and symmetrical shape with age. 'Koster' has bluer glaucous foliage and somewhat irregular form, 'Fat Albert' is a semi-dwarf, broadly pyramidal, 'Moerheim' is compact with bluer foliage, and 'Thomsen' is symmetrical with pale blue foliage.

w N.A. California Native: (n Calif.) CF, FF (*NCo; to AK*); EVG, CNF; (IRR: No WUCOLS)

Picea sitchensis

SITKA SPRUCE

Pinaceae. Sunset zones 4-6, 14-17, Jepson *IRR:2,3,**4,6,15-17**& SHD:7,14;CVS. Evergreen. Native to coastal mountains from sea level to 3,000' elevation from British Columbia to northern California. Growth rate moderate to 80-160' tall or more with a 20-40' spread, developing a tall, open, conical crown, a broad base, upswept branches, and a narrow to tapered top, more bushy and less upright near windswept coastlines. Needles are thick, stiff, bristly, bright green, new growth distinctly more bluish to grayish green, 3/4-1" long, with prickly ends, somewhat flattened and indistinctly 4-angled, standing straight out evenly and completely around branchlets. On older trees, sparsely clustered, insignificant, reddish male flowers occur near branch tips below fattened greenish females with elongated bracts. Pendulous cones are light brown, oblong-elliptical, 2-4" long, with thin, toothed, somewhat undulating, papery scales, maturing in 1 season and falling in winter after releasing 1/8", light yellow-brown seeds with a 1/4" wing. Bark is thin, scaly, gray-brown on younger trees, becoming deep reddish brown with large, flat, easily detached scales.

Tallest of the spruces, and a valuable forest tree in the northwest, but not readily adaptable for landscape use outside its habitat in loose, acidic soils and high rainfall of temperate coastal areas. Tolerates wet soils and salt spray. Wood is highly valued for strength and is used in making violins for its resonant qualities. Longevity estimated at 400-700 years in habitat. Series associations: Pacreedgrass, Pamgrass, Salblahuc herbaceous and shrub series; Beapine, Grafir, PorOrfced, Redalder, **Sitspr**, Weshemlock.

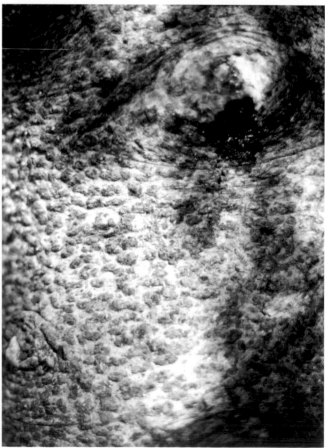

w N.A. California Native: FF, AL (*KR, CaRH, SNH, Wrn*); EVG, CNF; (IRR: No WUCOLS)

Pinus albicaulis

WHITEBARK PINE

Pinaceae. Sunset zones 1-7, 15-17, Jepson *DRN,SUN:**1**,6,15-17&IRR:**2,3**,7;DFCLT. Evergreen. Native to alpine elevations of the Sierra Nevada in central California north to British Columbia and the Cascade Range, east to Alberta, and south to western Wyoming and the Rocky Mountains, in dry gravelly soils. Growth rate very slow to 20-40' tall and half as wide, rather upright in cultivation, often with multiple twisted trunks, but shrubby or windblown in habitat, with picturesque, gnarled trunks and little foliage in harsh alpine conditions. Needles are stiff, dark green, in 5s, 1 to 2-1/2" long, tightly spaced, in tufts at ends of bare branchlets, with faint white bands on all sides, dropping after the first year. Cones are purplish to brown, ovoid, 1-3" long, nearly stalkless, and very resinous, with thickened or swollen scale ends and stout pointed tips. Cones do not fully open at maturity, and the dark brown, wingless seeds, 1/3-1/2" long, remain intact. Bark is thin, smooth, chalky white, finely fissured or with small, squarish, flaky scales.

Rarely cultivated, but an important high-elevation forest tree, occurring in loose stands. Intolerant of shade. Does well in well-drained soils or mounded in cultivation. Sometimes available through specialty sources. Longevity of some habitat specimens is estimated to be 2,000 years or more. Series associations: Rotsagebrush shrub series; Curmoumah, Foxpine, Limpine, Lodpine, Mixsubfor, Mouhemlock, Redfir, **Whipine**.

w N.A. California Native: CH, CC (*NW, CaR, SN, e SnFrB, SCoR, SnBr, PR, MP*); EVG, CNF; (IRR: L/L/L/L/-/-)

Pinus attenuata

KNOBCONE PINE

Pinaceae. Sunset zones 2-10, 14-21, Jepson *DRN,SUN:3,4-7,14-17,24&IRR:10,18-23. Evergreen. Native to rocky slopes and ridges in the Cascade, Sierra Nevada, and Coast Ranges from Oregon to Baja California. Growth rate moderate to rapid to 20-80 tall' with a 20-25' spread, forming a broad, open, pyramidal crown, with branches curving outward and upward, older trees usually having a forked trunk and an irregular form. Needles are slender, light yellowish green, 3-5" long, usually twisted, in 3s, persisting 4-5 years. Cones are light brown, 3-6" long, narrowly ovate, recurved, tapered toward the ends, thickened scale ends having a protruding hook, usually covered with pitchy sap, distinctly whorled along branch nodes and trunk, maturing in fall of the second season, and persisting indefinitely, remaining unopened unless removed or until the tree dies. Black seeds are 1/4" long with a 3/4" wing. Bark is dull brown, thin, shallowly furrowed, with ridges and large, loose scales near the base of the tree.

One of the first species to reseed after fire, closed cones opening with intense heat and often heard exploding. Grows in dry, serpentine or granitic areas that other pines find inhospitable, often forming thickets or groves. Rarely cultivated except as a specimen or in native plantings. Longevity estimated to be 100 years, relatively short for pines. Series associations: Intlivoakshr, Ionmanzanita, Leaoak, Scroakcha shrub series; BigDoufircan, Blaoak, Jefpine, **Knopine**, McNCypress, Monpine, Sarcypress, Weswhipin, Bakcypsta, KlaMouEnrsta, SanCrucypsta.

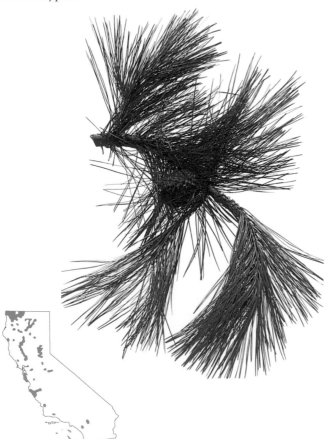

Mediterranean. EVG, CNF; IRR: L/L/L/L/M/M

Pinus brutia

CALABRIAN PINE

Pinaceae. Sunset zones 6-9, 12-24. Evergreen. Native to eastern Mediterranean regions of Cyprus and Turkey. Growth rate rapid when young, slowing with age, to 30-60' tall with a 15-25' spread, forming an irregularly rounded, oval canopy with a thick heavy trunk and branches denser and more erect than Aleppo pine (*Pinus halepensis*). Needles are dark green, 4-7" long, sheathed in 2s, somewhat more stout than those of *P. halepensis*, persisting 2-3 years. Insignificant flowers occur in May, yellowish green males, 1/2" long, and ovoid purplish females at branch ends. Cones are reddish tan, 2-3/8 to 4" long, oval to oblong, stalkless or nearly so, recurved or at right angles to branches, usually persisting many years. Bark is reddish brown, with large, flat, thin flaky scales.

A tall evergreen screen or background tree, introduced to the U.S. in the 1960s, now less often used than the more popular Aleppo pine and Italian stone pine (*Pinus pinea*). Tolerates heat, drought, wind, and poor soil. Relatively cold hardy. Longevity estimated to be 200 years or more. 'Christmas Blue' has a more symmetrical pyramidal shape and blue-green foliage.

Cultivar. EVG, CNF, FOL ACC, SPC; IRR: L/L/L/L/M/M

Pinus brutia var. *eldarica*

MONDELL PINE

Pinaceae. Sunset zones 6-9, 11-24. Evergreen. Variety of the species native to semi-arid regions of the Republic of Georgia, widely cultivated in Pakistan, Iran, and Afghanistan, and introduced to North America in the early 1900s. Growth rate moderate to fast to 30-60' tall with a 15-25' spread, developing an upright pyramidal form, fairly dense, with horizontal to upright regular branching. Needles are dark green, 2-1/2 to 6" long, sheathed in 2s or rarely in 3s. Insignificant dioecious flowers occur in May, yellowish green males and 1/2" long purplish females at branch ends. Cones are tan to reddish brown, 2 to 4-1/2" long, oblong and stalkless, at nearly right angles to branchlets or recurving slightly backwards, the same as *Pinus brutia*. Bark is rough grayish brown, deeply fissured, with orange tones, and broad, flat, scaly, cracked plates.

A useful tree in hot, dry climates, tolerating poor soils and moderate drought. Also does well near the coast. Greener and more symmetrical than *Pinus halepensis*. Longevity estimated to be 150-200 years or more.

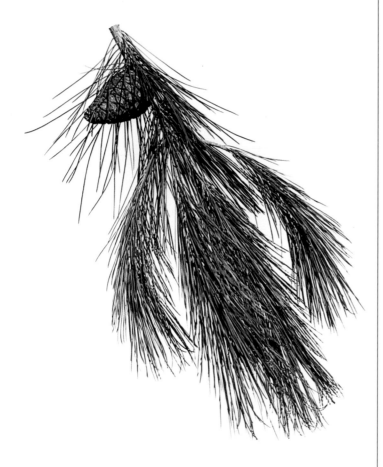

n & s China. EVG, CNF, SPC; IRR: No WUCOLS

Pinus bungeana
LACEBARK PINE

Pinaceae. Sunset zones 2-10, 14-21. Evergreen. Native to northern and central China. Growth rate very slow to 50-75' tall with a 30-50' spread, developing a rather sparse, airy foliage mass, usually multi-trunked or bushy, though erect, with an upward-arching to spreading canopy. Needles are bright green, 2 to 5-1/2" long, sheathed in 3s or sometimes 5s, sparsely spaced along branchlets, persisting 2-3 years. Insignificant dioecious flowers occur in May, males yellowish green with pollen and 1/2" long, purplish females at branch ends. Cones are light yellowish brown, 1-1/2 to 3" long, oval to oblong, stalkless or nearly so, with edible seeds. Striking bark is smooth gray, peeling in thin, curiously shaped plates, resembling a jigsaw-puzzle, with depressed mottling between raised and pointed brown ridges, becoming white on very old trees.

A rarely cultivated, picturesque accent or specimen pine, prized for its unusual bark. Tolerates inland valley heat under normal garden conditions. Hardy to subzero temperatures in mountain climates. Longevity estimated to be 100-200 years.

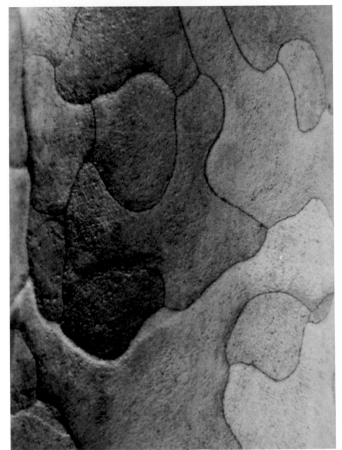

Canary Is. EVG, CNF, FOL ACC; IRR: L/L/L/M/M/M

Pinus canariensis
CANARY ISLAND PINE

Pinaceae. Sunset zones 8, 9, 12-24. Evergreen. Native to the Canary Islands. Growth rate moderate to 50-80' tall and 20-35' wide, pyramidal in form and gradually taking on a tiered, horizontal branching effect. Needles are dark green, sheathed in 3s, 9-12" long, persisting 3-4 years, with bluish green juvenile growth occurring sporadically beneath mature foliage from exposed portions of the trunk or from the base, often as unwanted suckering. When lower branches are trimmed off, more are produced. Insignificant purplish male flowers occur in dense clustered spikes, turning yellow with pollen, with 1/2" reddish brown females at branch ends, in early June. Cones are glossy brown, oblong-ovoid, 4-8" long, with scales having a dull pointed tip. Seeds are 1/4" with a 3/4" long wing, maturing in summer of the first season. Twigs are orange-brown, usually covered with thin, papery scales. Bark is dark reddish to grayish brown, irregularly fissured, with broad scaly ridges.

An attractive, soft-textured pine, commonly used as a lawn or street tree or as an evergreen screen with upright habit that does not usually interfere with other trees growing in close proximity. Young trees may appear somewhat gawky until branching pattern becomes fully established. Drought tolerant when established, but prefers deep watering for the first few years. Longevity estimated to be 250-350 years.

w N.A. California Native: (n Calif.) CO, CP (*Nco*); EVG, CNF, SHD;
(IRR: M/M/-/-/-/-)

Pinus contorta ssp. *contorta*

SHORE PINE

Pinaceae. Sunset zones 4-9, 14-24. Evergreen. Subspecies native to the Coast Ranges and coastal meadows from Mendocino County, California, to Alaska. Growth rate moderate to 20-35' tall and wide, forming an irregular, dense, rounded or pyramidal crown with many-forked, short branches often extending to the ground, more vertical if growing close together. Needles are shiny dark green, 1-1/4 to 2" long, sheathed in 2s, persisting 4-7 years. Cones are light tan, ovate, 1-1/4 to 2" long, opening in fall or remaining closed for years. Seeds are deep reddish brown, covered with black spots, and 1/8" long with a 1/2" long wing. Scaly dark brown bark becomes furrowed.

A shorter form of lodgepole pine, preferring moist, cool coastal climates of the northernmost part of the state, occurring in small scattered groves, but not as predominant as spruce, redwood, and fir. Adaptable in cultivation with moderate moisture. Similar to mugho pine, but more upright and bushy. Longevity estimated to be 100-150 years. Series associations: **Beapine**, Bispine, Mouhemlock, Pygcyp.

w N.A. California Native: ME, CF, FF (*KR, CaRH, SNH, SnGb, SnBr, SnJt, GB*); (IRR: No WUCOLS)

Pinus contorta ssp. *murrayana*

LODGEPOLE PINE

Pinaceae. Sunset zones 1-7, 14-17, Jepson *SUN:4-6,15,16& SHD:**3**,4. Evergreen. Native to the Sierra Nevada in California to Washington and the Rocky Mountains above 5,000' elevation. Growth rate slow to 50-80' tall with a 20-25' spread, developing a straight trunk, finely branched, and somewhat thin, sparse foliage, with a denser cylindrical form in open clearings. Needles are dark green, 1 to 2-1/2" long, sheathed in 2s, persisting 2-3 years. Cones are light tan, 1 to 1-3/4" long, ovate, opening in fall or remaining closed for years. Seeds are 1/8" long with a 1/2" wing, deep reddish brown with black spots. Scaly light brown bark becomes grayish and furrowed, with sharply cross-checked ridges.

A predominant forest tree in association with Jeffrey pine at higher elevations above Douglas-fir and ponderosa pine forests. Young trees often grow in thickets, with trunks twisted by heavy snowfall, but older trees have straight, narrow trunks with sparse lateral branching. Logs were often used to build mountain lodges. Not readily adaptable and rarely cultivated. Longevity estimated to be 125-175 years or more. Series associations: Blacottonwoo, Curmoumah, Engspruce, Foxpine, Giasequoia, Inccedar, Jefpine, Knopine, Limpine, **Lodpine**, Mixsubfor, Moujuniper, Ponpine, Redfir, Subfir, Waspine, Weswhipin, Whifir, Whipine, Alayelcedsta, KlaMouEnrsta, Pacsilfirsta.

sw N.A. California Native: (s Calif.) CH, FW, CC, ME, PF; (*CW, SW*); EVG, CNF, SHD; (IRR: L/L/L/L/M/-)

Pinus coulteri

COULTER PINE

Pinaceae. Sunset zones 3-10, 14-23, Jepson *DRN,SUN:**4**,5,**6**,**7**, **14-17**,24&IRR:**2**,**3**,8-10,**18-23**. Evergreen. Native to dry rocky slopes of the inland coastal ranges of southern and central California. Growth rate moderate to fast to 30-80' tall with a 20-40' spread, developing an oval form, often with distinct horizontal spreading branches with upturned ends, and a short, stout trunk. Needles are stiff, dark bluish green, 5-12" long, sheathed in 3s, persisting 3-4 years. Purplish male flowers occur in dense clusters, turning yellow with pollen, with 1/2" reddish brown females at branch ends in early June. Cones are tan, 9-14" long, weighing up to 5 pounds, elliptical to oblong-ovoid, short-stalked, with thick scales terminating in recurving horns, developing in summer, maturing in 2 years, and either partially opening or persisting unopened on trees for 5-6 years. Seeds are 1/2", hard-shelled, with 1" wings. Branchlets are very stout and rough orange-brown. Bark is dark brown to black, furrowed, with broad scaly ridges.

A rather uncommon pine, scattered in areas of chaparral, mixed coastal pine forest, and oak woodland. A full and shapely tree, more so than gray pine, which has similar, but wider cones. Sometimes cultivated in roadside and open space plantings and quite adaptable with minimum irrigation or care. Huge cones can be dangerous when used in playgrounds, parks, or around cars. Longevity estimated to be 350-500 years. Series associations: Easmanzanita, Intlivoakshr, Intlivoakcha shrub series; BigDoufir, BigDoufircan, Canlivoak, **Coupine**, Coupincanliv, Foopine, Knopine, Ponpine, SanLucfir, Tanoak, Cuycypsta, SanBenMousta.

e Asia. EVG, CNF, SPC; IRR: M/M/-/M/-/-

Pinus densiflora

JAPANESE RED PINE

Pinaceae. Sunset zones 2-9, 14-17. Evergreen. Native to Japan. Growth rate rapid when young, slowing with age, to 40-60' tall (up to 100') with a 40' spread, forming a broad, irregular canopy, often with multiple trunks at ground level unless trained to a single trunk. Needles are stiff, bright blue-green to yellowish green, 2-1/2 to 5" long, sheathed in 2s. Cones are 1-1/4 to 3" long, brown, oval to oblong, often clustered on branches. Bark is dark brown, narrowly furrowed, with wide, flat ridges.

A medium-sized pine for small, informal groves, especially in Japanese-style gardens. Tolerant of pruning, where entire branches are removed for a sparse, tiered effect, producing short, contorted horizontal branching. Forked, irregular, horizontal branching is distinctive on unpruned trees. One of the most common pines in Japan. Yellow, dwarf, or weeping cultivars sometimes available. Does not tolerate hot, dry, or cold, windy areas. Often used in coastal and inland areas where it receives moderate moisture. Longevity estimated to be 150 years or more.

Cultivar. EVG, CNF, FOL ACC, SPC; IRR: M/M/-/M/-/-

Pinus densiflora 'Umbraculifera'

TANYOSHO PINE

Pinaceae. Sunset zones 2-10, 14-17. Evergreen. Cultivar of the species native to Japan. Growth rate slow to 12-20' tall by 18' wide, forming a broad, round-topped canopy often wider than tall with many slender trunks from the base. Needles are stiff, bright blue-green to yellowish green, 1-1/2 to 3" long, sheathed in 2s, shorter and lighter than the species. Cones are brown, oval to oblong, 1/2-3/4" long, much smaller than the species, usually densely clustered close to branchlets, developing at an early age and often remaining attached for a number of years. Bark is dark brown, narrowly furrowed, with flat ridges.

A slightly larger alternative to mugho pine, eventually becoming large and bushy. The symmetrical, dense, oval canopy occurs naturally without trimming, though lower branches may need to be clipped to expose multiple trunks. Longevity estimated to be 100 years or more.

w N.A. California Native: (s Calif.) JP (*DMtns, New York Mtns*); (IRR: L/L/VL/L/L/-)

Pinus edulis

COLORADO PINYON

Pinaceae. Sunset zones 1-11, 14-21. Evergreen. Native to arid, high-elevation mountains of Colorado, Wyoming, Arizona, and New Mexico, and sparsely in the New York Mountains of the Mojave Desert in southeastern California. Growth rate slow to 10-20' tall and 8-16' wide, often with a rounded, dense, bushy form, stout forked trunks, and many rangy, upright branches. Needles are stiff, dark green, 1-2" long, sheathed in 2s, sharply pointed, with smooth edges, persisting 3-5 years or more. Cones are 1-2", yellowish tan, ovoid, appearing rather squattish, with fat, thickened scale ends. Seeds are 3/8-1/2", oval, dark brown, and thin-shelled, wings remaining attached to cone scales when seeds loosen and fall out. Bark is smooth, grayish, thickening with age and becoming brown, often with a whitish cast, furrowed or breaking into thin, scaly plates, revealing reddish underbark.

One of nearly a dozen pinyon pines endemic throughout the southwest, in rather sparse and variable populations. Rarely cultivated, though sometimes available. Usually remains compact and bushy. Intolerant of shade. Requires well-drained soils. Longevity estimated to be 100-250 years in habitat. Series associations: **Twopinsta**.

c & w N.A. California Native: PJ, FF, AL (*SNH, TR, PR, SNE*); (IRR: No WUCOLS)

Pinus flexilis

LIMBER PINE

Pinaceae. Sunset zones 1-11, 14-21, Jepson *DRN:**1**,4,6,15-17& IRR:**2**,**3**,7,10,14,**18**,19. Evergreen. Native to desert mountain slopes in southern California and the eastern Sierra Nevada, in Inyo County, at 4,000-11,000' elevation, south into Baja California, north into British Columbia, and east to the Rocky Mountains. Growth rate moderate to 30-55' tall and 15-25' wide, developing a broad, open form, taller at lower elevations, with a stout trunk and slender, flexible, horizontal branching, foliage distinctly in tufts at ends of slightly drooping or upswept branchlets. Needles are stiff, dark yellow-green, 1 to 2-1/2" long, sheathed in 5s, minutely toothed, persisting 5-6 years. Insignificant reddish male flowers turn yellow with pollen, with 1/2" reddish purple females clustered at branch ends, in early June. Cones are light yellowish brown, 5-8" long, short-stalked, ellipsoidal, with thickened, slightly recurving, unspiked scale tips, opening in fall of the second season. Seeds are 1/3-1/2" long, dark reddish brown with dark speckles, and wingless. Branchlets are smooth and gray, with broad pointed buds. Bark is thin, smooth, and gray, thickening with age and darkening to brown, becoming furrowed or breaking into thin, scaly plates.

An important high-elevation forest tree, occurring in loose stands. Similar to eastern and southwestern white pine. Numerous cultivars, not commonly available, with deeper blue foliage as well as columnar and prostrate forms. Intolerant of shade but appears adaptable to any well-drained soil. Longevity estimated to be 200-300 years. Series associations: Bripine, Foxpine, **Limpine**, Lodpine.

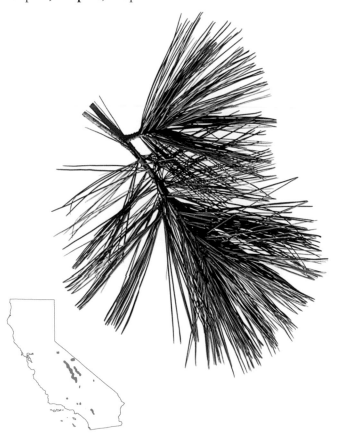

s Europe, Medit., w Asia. EVG, CNF; IRR: L/L/L/L/L/L

Pinus halepensis

ALEPPO PINE

Pinaceae. Sunset zones 7-9, 11-24. Evergreen. Native to southern Europe and Mediterranean regions to Asia Minor. Growth rate moderate to rapid to 30-60' tall and wide, developing an oval form, symmetrical in youth and irregular with age with rather rangy horizontal branching. Needles are light green, 2-1/4 to 5-1/2" long, sheathed in 2s or rarely 3s. Insignificant dioecious flowers occur in May, with yellowish green males and 1/2" long purplish females at branch ends. Cones are tan to reddish brown, 2-1/4 to 4" long, oblong and short-stalked, usually in pairs or in 3s, recurving slightly backward on branches, with thickened, slightly recurving, unspiked scale tips, and may persist for years. Branchlets are tan to brown, with non-resinous growth buds with recurving scales. Reddish brown bark grays with age, becoming fissured, with scaly plates.

Looser form, softer foliage, and less hardy than the closely related *Pinus brutia*. Often used as a parkway tree. Fast growing and fills in quickly. Best in fertile soil with moderate watering, but tolerates hot, dry, or even desert conditions. Longevity estimated to be 150-200 years.

w N.A. California Native: PF, FF (*KR, NCoR, CaR, SN, SCoRI, TR, PR, GB*); (IRR: L/L/-/-/-/-)

Pinus jeffreyi

JEFFREY PINE

Pinaceae. Sunset zones 2-9, 14-19, Jepson *DRN,SUN:1,**4-6,15-17**&IRR:**2,3,7**,8,9,**14,18,19**. Evergreen. Native to the Sierra Nevada above 5,000' elevation from Oregon to Baja California. Growth rate slow to moderate to 60-120' tall and 20-25' wide, forming a straight, heavy trunk, similar to *Pinus ponderosa* but with slightly denser foliage, thicker branchlets, and shorter branches. Symmetrical in youth, becoming more open with age. Needles are dark bluish green, 5-10" long, sheathed in 3s, persisting 5-8 years, new growth purplish with a violetlike odor if broken. Cones are oval, brown, 5-8" long, maturing in the first season, and releasing 3/8", brown, mottled seeds with a 1" long wing. Unlike *P. ponderosa*, cone scale tips are not prickly and scale tips point inward, not outward. Bark is reddish brown, becoming deeply furrowed, with narrow ridges, irregularly interconnected, and on older trees becoming deeply broken into long, wide plates, often with a yellowish coloration, and a strong vanilla-like odor.

An imposing pine with a massive trunk. Predominant in higher-elevation Sierra Nevada forests, generally with red fir, lodgepole pine, western white pine, or single-leaf pinyon. Rarely cultivated outside its range, as it prefers cold temperatures. Longevity estimated to be 300-400 years. Series associations: Bigsagebrush, Bitternrush, Blasagebrush shrubseries; Blacottonwoo, Blaoak, Curmoumah, Doufir, Doufirponpin, Giasequoia, **Jefpine**, Jefpinponpin, Limpine, Mixconifer, Moujuniper, Orewhioak, Parpinyon, Ponpine, PorOrfced, Redfir, Sinpinyon, Waspine, Wesjuniper, Weswhipin, Whifir, Bakcypsta, KlaMouEnrsta, SanBenMousta.

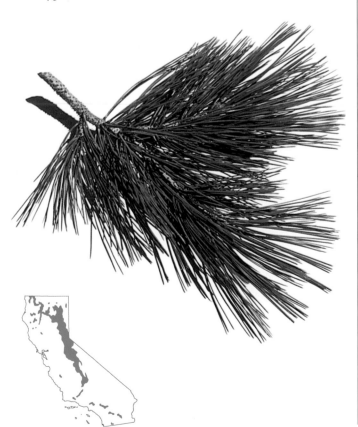

w. N.A. California Native: ME, PF (*NW, CaR, SN, SW, w GB*); (IRR: No WUCOLS)

Pinus lambertiana

SUGAR PINE

Pinaceae. Sunset zones 2-9, 15-17, Jepson *DRN:1,**4-6,15**,16, 17&IRR:2,3,7. Evergreen. Native to the Sierra Nevada and southern Cascade Ranges at 3,000 to 5,000' elevation. Growth rate initially slow, faster with age, to 200' tall or more with a 50' spread, pyramidal in youth, with slender horizontal branching and distinctly loose foliage, older trees often developing a sizable trunk, with broad sweeping limbs giving a somewhat tiered effect and long cones hanging like ornaments from branch tips. Needles are dark bluish green, 2 to 3-1/2" long, sheathed in 5s, with a whitish tinge, persisting 2-3 years, tufted at ends of slender branchlets. Cones are the largest of all pines, 10-20" long, cylindrical, 3-4" in diameter, light brown, with shiny tipped scales with a darker inner surface and sometimes a glob of pitchy sap, ripening in late summer of the second year and shedding dark brown to blackish, 1/2", flattened seeds with a 3/4-1" long, rounded wing in fall. Bark is thin, smooth, and grayish, becoming grayish brown, deeply furrowed, with long, irregular plates along the ridges.

An attractive soft-textured pine in youth, accentuated by bluish green foliage. Not readily adaptable to cultivation. One of the tallest pines, occurring naturally among other pines, firs, and incense cedar, which usually predominate. Susceptible to white pine blister rust, which limits number and size. Longevity estimated to be 300-500 years. Series associations: BigDoufir, Canlivoak, Doufir, Doufirponpin, Doufirtan, Engspruce, Giasequoia, Inccedar, Mixconifer, Ponpine, PorOrfced, Redfir, SanLucfir, Tanoak, Whifir, Bakcypsta, KlaMouEnrsta.

w N.A. California Native: (s Calif.) JP (*c&s SNH, Teh, se SCoRI, TR, PR, SNE, Dmtns*); (IRR: L/-/L/L/L/-)

Pinus monophylla
SINGLELEAF PINYON

Pinaceae. No Sunset zones. Jepson *DRN:1,**2**,**3**,4-6, **7**,10,**14-16**,**17**,**18-23**,24&IRR:8-12. Evergreen. Native to arid, high-elevation southern and eastern Sierra Nevada mountains into Utah, Nevada, and Arizona. Growth rate very slow to 10-25' tall and 10-15' wide, upright in youth, maturing to a stout-trunked, densely rounded, bushy canopy with heavily twisted branches. Needles are stiff, dark green, 1-2" long, round in cross-section, sheathed singly, sharply pointed, with smooth edges, persisting 3-5 years or more. Cones are ovoid, 2-1/2 to 3-1/2", yellowish tan, with thickened scale ends, maturing in August, and containing dark brown, thin-shelled, 3/8-1/2", oval seeds. Thin, narrow wings remain attached to cone scales, and empty cones fall in winter. Bark is smooth, dull gray, becoming darker and thicker, roughly furrowed, or breaking into thin, scaly plates.

Most common pinyon pine in high-desert regions of California, mixed with junipers and sagebrush, and an invaluable food source for wildlife. More treelike than Colorado pinyon (*Pinus edulis*). Quite adaptable, even aggressive, in habitat. Rarely cultivated. Longevity estimated to be 100-250 years in habitat. Series associations: Bigsagebrush, Bitterbrush, Balbush, blasagebrush, Cupceafreoak, Jostree shrub series; BigDoufir, BigDoufircan, Birmoumah, Caljuniper, Curmoumah, Moujuniper, Parpinyon, **Sinpinyon**, Utajuniper, Whifir, Piucypsta.

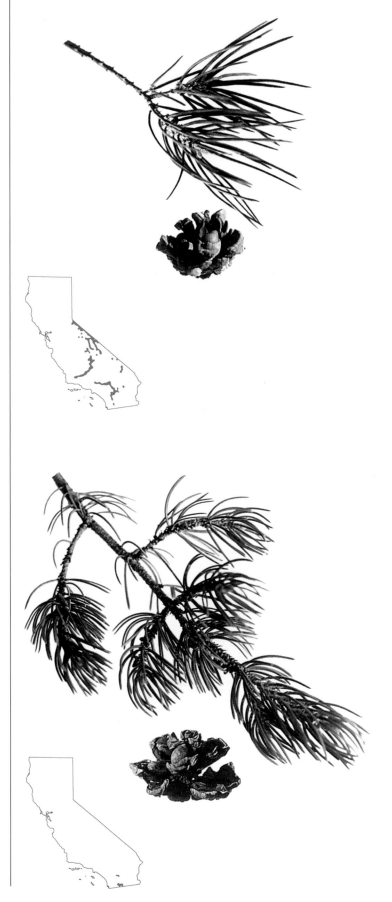

w N.A. California Native: (s Calif.) CH, PJ, PF (*SnJt, s PR*); (IRR: L/-/L/L/L/-)

Pinus quadrifolia
PARRY PINYON PINE

Pinaceae. No Sunset zones. Jepson *DRN: **3**,4-6,**7**,**15**,**16**,17,24&IRR:10,**14**, **18-23**. Evergreen. Native to arid, high-elevation southern California mountains into northern Mexico. Growth rate slow to 10-25' tall and 10-15' wide, developing a pyramidal to open, round, bushy form. Needles are stiff, dark bluish green, 3/4 to 1-1/2" long, sheathed in 4s, but may occur in 1s to 5s, curved, and whitish striped on the innerside. Cones are 1-1/4 to 2", yellowish tan, ovoid, scale ends having a ridged knob. Seeds are 5/8", dark brown, thin-shelled, with wings remaining attached to cone scales. Bark is grayish, thickening and darkening to brown, furrowed or breaking into thin, scaly plates.

Limited to a narrow strip in the Peninsular Ranges and a few locations east of San Diego, with a wider band stretching into northern Mexico. Populations are variable and considered hybridized. Rarely grown, but an important forest tree in habitat. Intolerant of shade. Requires fertile, well-drained soils. Longevity estimated to be 100-225 years in habitat. Series associations: Caljuniper, **Parpinyon**.

w N.A. California Native: FF, AL (*KR, NCoRH, CaRH, SNH, SNE, MP*); (IRR: No WUCOLS)

Pinus monticola

WESTERN WHITE PINE

Pinaceae. Sunset zones 1-7, Jepson *DRN:1,**4-6**&IRR:**2,3**,7,15, 16;DFCLT. Evergreen. Native to the California Sierra Nevada north to British Columbia and the Cascade Range and east to Montana at 7,500-10,000' elevation. Growth rate moderate, slowing with age, to 60' tall and 20' wide, with an upright, open form on young trees, a rounded, pyramidal shape at maturity, and tiered horizontal branches. Needles are bluish green, 1 to 3-1/2" long, sheathed in 5s, somewhat twisted, white banded beneath, margins with minute teeth, in tufts at ends of bare branchlets, persisting 3-4 years. Insignificant yellow male flowers and clustered reddish purple female flowers occur in late spring. Cones are long-stalked, narrowly cylindrical, 6-8" long, light brown, with shiny-tipped, smooth scales and a darker inner surface, ripening in late summer though may remain on the tree through the following year. Seeds are reddish brown, 1/4" long, with a 1" long wing, the end angled to a round-pointed, offset tip. Bark is thin, smooth, light grayish brown, becoming gray, finely fissured or with small, squarish, flaky scales.

A temperamental pine, hardy but not readily adaptable outside its natural habitat. An important native in sparse high-elevation forests, often interspersed among red fir and hemlock. In exposed areas a massive and stately tree, standing alone amidst glacial granite outcroppings where it is common. Susceptible to white pine blister rust. Intolerant of shade. Longevity estimated to be 300-500 years. Series associations: Engspruce, Foxpine, Jefpine, Knopine, Lodpine, Mixsubfor, Mouhemlock, PorOrfced, Redfir, Subfir, Waspine, **Weswhipine**, Whipine, Alayelcestad, KlaMouEnrsta, Pacsilfirsta.

Europe. EVG, CNF, FOL ACC SPC; IRR: L/L/-/M/M/-

Pinus mugo

MUGHO PINE, SWISS MOUNTAIN PINE

Pinaceae. Sunset zones 1-11, 14-24. Evergreen. Native to the eastern Alps of Europe. Growth rate slow to 4-8' tall and and 8-15' wide, variable in form from prostrate to bushy and pyramidal but usually a densely spreading, low shrub of equal height and width. Needles are dark green, 1-2" long, rigidly curved and densely crowded on slender branches, sheathed in 2s, with finely toothed edges, a short apex with a blunt, horny point, and lines of stoma on both sides. Inconspicuous, monoecious, yellow-green flowers are clustered near branch tips. Cones are light brown, 3/4 to 1-1/2" long, ovoid to conical, sessile or short-stalked, scale ends not prickly, solitary or clustered up to 4, maturing in one year, and may persist opened into the following year, releasing small winged seeds. Bark is smooth when young, with regularly spaced bumps left from fallen needles, becoming brownish gray, scaly, and peeling into irregular persistent plates.

Often used as a container, bonsai, or rock garden plant or in foundation plantings in cold-winter areas where it tolerates snow cover for extended periods. Tolerates drought when established. Susceptible to rust and scale. Longevity estimated to be 75-100 years. 'Compacta' is very dense and 'Gnome' even denser.

w N.A. California Native: CH, OW, CC, RW, CF (*NCo, CCo, n SnFB, n CHI*); (IRR: M/M/L/-/-/-)

Pinus muricata

BISHOP PINE

Pinaceae. Sunset zones 5, 14-17, 22-24, Jepson *5,15-17,24*&
IRR:**14**,22,23. Evergreen. Native to the California Coast Ranges,
Santa Cruz Island, and Baja California, usually near the coast.
Growth rate moderate to fast to 40-75 tall' and 20-40' wide,
developing a dense and rounded or slender and sparse pyramidal form, irregular with age, with bowed, upward-arching
limbs. Needles are stiff, light green, 4-6" long, sheathed in 2s,
tightly clustered at branch ends, persisting 2-3 years. Cones
are brown, 2 to 3-1/2" long, broadly oval, in groups of 3 to 5,
whorled around branches, with dark, shiny, prickly scale ends,
maturing in summer of the second season, and may persist
on limbs indefinitely but do not become embedded as do other
closed-cone pines. Seeds are 1/4" long, rough-surfaced, black
or dark brown, with a 3/4" wing, occasionally released in late
summer. Bark is dark gray to brownish black, becoming rough
and deeply furrowed.

    A variable species, with populations north of Sonoma
County having bluer foliage and rougher bark than those to
the south. Rarely cultivated but moderately adaptable away
from its preferred coastal habitat. Tolerates wind, clayey or
gravelly soils, and salt air. Longevity estimated to be 100-150
years. Series associations: Beapine, **Bispine**, Grafir, Monpine,
Pygcypress, Gowcypsta.

Santa Cruz &
Santa Rosa Is.

sw N.A. EVG, CNF, FOL ACC, SPC; IRR: M/M/M/M/M/M

Pinus patula

JELECOTE PINE

Pinaceae. Sunset zones 6-9, 12-24. Evergreen. Native to mountains of eastern Mexico. Growth rate slow to moderate to 60-80' tall and 30-40' wide, with a rather rounded form, less often with a slightly irregular trunk and branching. Needles are thin, pendulous, light green, 4-12" long, sheathed in 3s, sometimes in 4s or 5s, drooping from branches, persisting 1-2 years. Insignificant dioecious flowers occur in May, with small clusters of yellow-green cylindrical males and 1/2" long pale yellowish green females. Cones, rarely set, are shiny, pale brown, 2-1/2 to 4-3/4" long, ovoid, maturing in fall of the first season. Seeds are 1/4" long with elongated, thin wings. Bark is reddish brown and furrowed, a pleasing contrast to the soft-textured foliage.

Commonly grown as a single specimen foliage accent or silhouette feature or in small groupings, which enhances the beauty of this graceful pine. Half-hardy and suffers in drought. Does well in lawns, casting light shade. Resistant to oak root fungus. Longevity estimated to be 100-200 years.

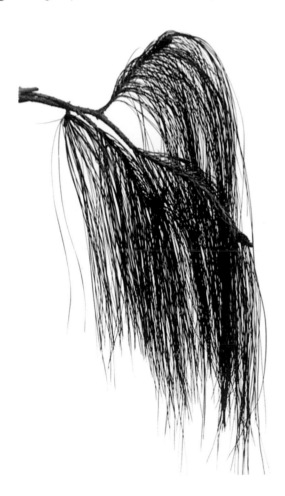

sw Europe, Medit., w Asia. EVG, CNF; IRR: L/L/L/L/M/M

Pinus pinea
ITALIAN STONE PINE

Pinaceae. Sunset zones 8, 9, 11-24. Evergreen. Native to southwestern Europe and Mediterranean regions of Greece, Turkey, and Asia Minor. Growth rate slow to moderate to 40-80' tall with a 40-60' spread, dense and rounded in youth, with age developing thick trunks with branches exposed, the canopy broadening and opening. Needles are stiff, dark green, 3-1/2 to 7" long, sheathed in 2s, rarely in 3s, becoming grayish green on older trees. Insignificant dioecious flowers occur in May, with small clusters of golden, cylindrical males and 1/2" long, pale yellowish green females. Cones are dark brown, ovate, 3-1/2 to 5" long and fairly heavy, maturing in fall of the first season and remaining closed for 3 more years. Seeds are oblong-oval, hard-shelled, 1/2-3/4" long, light brown with black colorations, and a meaty, edible fruit. Bark is dark reddish brown to black, becoming deeply furrowed, with rough plated ridges.

A useful accent, screen, or background tree, handsome in youth and picturesque at maturity, with somewhat twisting, gnarled branches spreading to a towering canopy. Susceptible to toppling. Needs ample room. Often seen near the coast. Tolerates heat and drought once established. Longevity estimated to be 300 years or more.

w N.A. California Native: ME, PJ, SS, PF (*CA-FP exc NCo, GV, SCo*); EVG, CNF, SHD; (IRR: L/L/-/L/-/-)

Pinus ponderosa

PONDEROSA PINE

Pinaceae. Sunset zones 1-10, 14-21, Jepson *DRN,SUN:1,**4-6,15,16**,17&IRR:**2,3,7**,8,9,**14,18,19**. Evergreen. Native to foothills and the Sierra Nevada from British Columbia to California and east to Nebraska and Oklahoma. Growth rate moderate, slowing with age, to 50-100' tall (200' in habitat) and 25-30' wide, straight-trunked and well-branched with an upright pyramidal form. Needles are glossy green, 5-10" long, sheathed in 3s, in heavy brushlike clusters at branch tips, persisting 2-3 years. Cones are oblong, 3-5" long by 1-2" round, with short, sharp spurs at tips of cone scales, which point inward. Cones mature in summer to light reddish brown, shedding 1/4" tan seeds with dark blotching and a 3/4" long wing in September, and falling through winter, often leaving a few basal scales on the branch. Bark is reddish brown to blackish, narrowly furrowed, becoming very thick, with reddish, broad, shield-like plates.

California's most widespread and common conifer and the predominant pine on the western slopes of the Sierra Nevada, the staple of the timber industry in that region along with Douglas-fir. Reseeds and regenerates rather quickly. Not suited to container growing, so nurseries offer smaller stock or seedlings that grow about 18" per year. Quite adaptable with minimum care once established. Longevity estimated to be 350-500 years. Series associations: BigDoufir, Blaoak, Canlivoak, Coupine, Coupincanliv, Doufir, Doufirponpin, Doufirtan, Engspruce, Giasequoia, Inccedar, Jefpinponpin, Mixconifer, Mixoak, Monpine, Orewhioak, **Ponpine**, PorOrfced, SanLucfir, Sinpinyon, Waspine, Wesjuniper, Bakcypsta, KlaMouEnrsta, SanCrucypsta.

w N.A. California Native: (Ca only) CP, OW, CC (*Cco*); EVG, CNF, SHD; (IRR: M/-/M/M/-/-)

Pinus radiata

MONTEREY PINE

Pinaceae. Sunset zones 14-24. Evergreen. Native to the central California coast. Growth rate moderate to fast to 80-100' tall by 25-35' wide, developing a shapely, broad, conical form, becoming broader with age and often picturesque. Needles are dark green, 3-1/2 to 6" long, sheathed in 3s, or rarely in 2s, persisting 3 years. Insignificant dioecious flowers occur in April, bright yellow, cylindrical males in fat clusters and dark purplish red females at branch ends. Lopsided cones are 2-5" long, with smooth, shiny, dark brown scale tips with dark inner surfaces, maturing in late summer of the second year, remaining on branches unopened for 10-30 years, opening if removed from the tree. Seeds are dark with black mottling, 1/4" long with a 3/4" long wing. Bark is reddish to blackish, deeply furrowed, with broad, flat ridges with distinct plates.

The most widely planted conifer and a significant timber tree in various parts of the world. Occurs naturally in only 3 limited coastal areas of California, and in North America rarely cultivated outside California. Exceptionally fast growing, needing ample room. Commonly found in small park groves or as specimens in residential landscapes. Requires moderate moisture, and will not tolerate extended drought. Resists oak root fungus. Shallow rooting can cause toppling in heavy winds. Subject to viral disease, suddenly turning brown, with no known remedy other than to remove affected trees. Longevity estimated to be 80-100 years. Series associations: Beapine, Bispine, Knopine, **MonPine**, Gowcypsta.

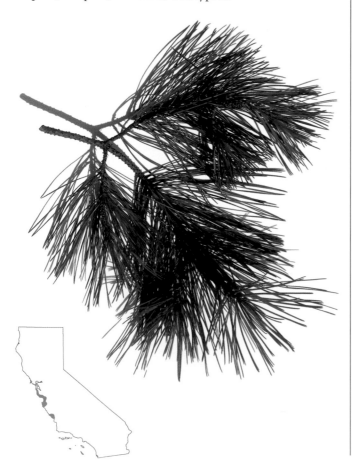

w N.A. (Ca only) California Native: CH, OW, FW, PF (*w GB, w D, C-FP exc n NW, n CaR, SnJV*); EVG, CNF, SHD; (IRR: VL/VL/VL/L/-/-)

Pinus sabiniana

GRAY PINE

Pinaceae. Sunset zones 3-10, 14-21, Jepson *DRN,SUN:3,4-7,14-24&IRR:8-11. Evergreen. Native to dry, rocky foothills and valleys in California at or below 1,000-3,000' elevation. Growth rate fast to 40-80' tall and 30-50' wide, young trees usually pyramidal, later developing multiple, irregular trunk leaders and U-shaped forks and taking on an oval form with sparse foliage on slightly drooping branches. Needles are soft, bluish gray-green, 7-13" long, sheathed in 3s. Cones are light brown, oval, 6-10" long, heavy and bulky, with tan recurving scale tips, maturing in fall of the second year, but may remain attached for years, opening gradually and releasing seeds for a long period. Seeds are edible, dark brown to black, 3/4" long with a 1/2" long wing. Bark is dull gray, darkening and thickening with age, becoming roughly furrowed with scaly ridges.

A predominant pine in foothill woodland and chaparral regions, generally below the ponderosa pine belt, in loose, scattered groves. An attractive soft-textured pine with a delicate appearance in youth, older trees developing spindly trunks with multiple forks and sparse, wispy foliage. New trees planted to replace them quickly fill in. Good value in native plantings. Does well in dry or semi-moist, well-drained soils. Tolerates inland heat. Longevity estimated to be 150 years or less. Series associations: Intlivoakshr, Ionmanzanita, Leaoak, Scroak shrub series; BigDoufircan, Birmoumah, Bluoak, Calbuckeye, Coupine, Coupincanliv, Foopine, Intlivoak, Knopine, McNcypress, Mixoak, Sarcypress, Piucypsta, SanBenMousta.

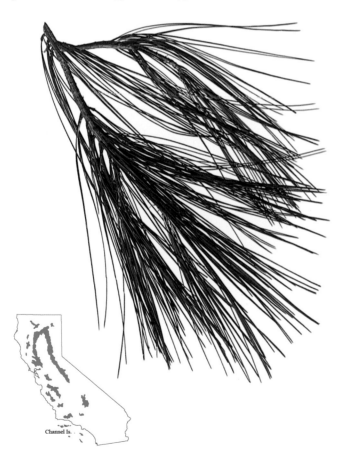

e N.A. EVG, CNF, SHD; IRR: No WUCOLS

Pinus strobus

EASTERN WHITE PINE

Pinaceae. Sunset zones 1-6. Evergreen. Native to the northeastern U.S. from Georgia and Iowa to Illinois and in Canada from Newfoundland to Manitoba. Growth rate slow, becoming faster, to 50-80' tall and 20-40' wide, dense, straight-branched, and conical in youth, broadening with age to a more open, picturesque form with irregularly drooping branches. The example shown was topped as a young tree and developed multiple leaders, with a more rounded form than single-trunk trees. Needles are soft, bluish green, 3-5" long, sheathed in 5s, twisted, with minute teeth on all sides, whorled completely around branchlets, persisting 3-4 years. Insignificant dioecious flowers occur in May, scattered clusters of yellowish males producing pollen and reddish purple females clustered at branch ends. Cones are reddish brown, 3-8" long, slender, long-stalked, with light-tipped scale ends and no spines. Seeds are reddish brown, 1/4" long, with a 1" long, angled, round-pointed, papery wing. Bark is thin, smooth, and light gray, thickening and darkening to grayish brown, with rectangular scaly plates.

A handsome, fine-textured specimen or accent tree with a bluish cast commonly cultivated around the world. Selected varieties available, though not often in quantity, including upright, weeping, compact, and contorted forms and foliage colors from bluish to pale or slightly yellowish. Prefers moderate, consistently moist, well-drained, fertile soil. Longevity estimated to be 200-250 years.

n Europe, Asia EVG, CNF, SHD; IRR: M/M/-/M/-/-

Pinus sylvestris

SCOTCH PINE

Pinaceae. Sunset zones 1-9, 14-21. Evergreen. Native to northern Europe and Asia. Growth rate moderate to 70-100' tall in habitat (usually 30-70' in cultivation) by 25-30' wide, with a dense, straight-branched, conical form, aging to more picturesque shape, with sparsely foliaged, irregular, drooping branches at maturity. Needles are stiff, bluish green, 1-1/2 to 2-3/4" long, sheathed in 2s, twisted, and whorled completely around branchlets. Insignificant dioecious flowers occur in May, yellowish males in scattered clusters, producing pollen, and reddish females at branch ends. Cones are reddish brown, broadly ovate, 2-3" long, with pyramidally thickened scale ends, pointed but not prickly, maturing in fall of the second season. Seeds are 1/16-1/8" long with 1/2" long, round-ended, papery wings. Bark is furrowed, dark brown with an orange or reddish cast, orange-tinted on young branches.

A popular Christmas tree, widely used in landscapes as a dense, evergreen screen or background tree with bluish foliage a colorful accent near green conifers. Neat in appearance with confined growth. Quite hardy. Deep, wide-spreading roots. Tolerates wind and moderate heat, but not dryness. Cultivars with very blue foliage, as well as conical and upright forms, available to a limited extent. Longevity estimated to be 100 years in cultivation at lower elevations, possibly more at higher elevations and in the northwest.

e Asia EVG, CNF, SHD; IRR: M/M/M/M/M/M

Pinus thunbergii
JAPANESE BLACK PINE

Pinaceae. Sunset zones 3-12, 14-21. Evergreen. Native to Japan. Growth rate moderate to fast to 100' tall by 30' wide in the northwest, to 20' tall by 10' wide in southern California deserts. Broadly conical with wide-spreading branches, becoming irregular unless a strong leader trunk develops. Long, light green "candles" of new growth develop at branch tips before needles form, pronounced in spring. Needles are stiff, twisted, 3 to 4-1/2" long, bright dark green, sheathed in 2s, and sharply pointed. Insignificant dioecious flowers occur in May, small reddish male clusters turning noticeably yellow with pollen and purplish red females occurring more sparsely at branch ends. Cones are dark brown, 1-1/2 to 3" long, oval, often in clusters of 6-12, with thickened scale ends, armed with a small spine, maturing in fall of the first season, sometimes persisting partially opened for another 2 years. Seeds are 1/8-1/4", dark brown with 3/4" long wings. Bark is dark brown to blackish, deeply furrowed with rough, scaly ridges.

One of the most common pines in landscape use, but may become too large for most residential gardens. Attractive but can be quite rangy. At maturity, wide-spreading horizontal branching can be picturesque as a lawn specimen with plenty of room. Tolerates drought, heat, and coastal conditions, as well as heavy pruning, or even shearing as a large bonsai or in large planters. Casts dense shade. Roots can be invasive. Longevity estimated to be 200 years.

sw N.A. (Ca only) California Native: (s Ca) CS, CH (*s SCo, n CHI Santa Rosa Is*); (IRR: L/L/L/M/-/-)

Pinus torreyana

TORREY PINE

Pinaceae. Sunset zones 8, 9, 14-24. Evergreen. Native to the San Diego coast of southern California. Growth rate moderate to 40-60' tall by 30-50' wide, developing a broad conical form with stout branches, becoming somewhat irregular or prostrate in windy coastal areas. Needles are dark bluish to gray-green, 7-11" long, sheathed in 5s, clustered in large bunches at branch tips, persisting 3-4 years. Cones are dark brown, 4-6" long, oval, with a flat base, attached strongly to branches with a thick, short stem, ripening in summer of the third year. Scales have a thickened triangular apex, with short, fat, recurving spines. A small part of the base remains after cones fall, though cones often remain attached for many years, slowly releasing seeds over a long period. Seeds are dark brown, 1/2-3/4" long, with yellow-brown blotches, and a 1/4-1/2" long wing, with a rounded tip. Bark is thick, dark brown, becoming broken into ridges, with wide, flattened, reddish brown scales.

Limited in its habitat and rarely cultivated, but adapts well to inland or high desert conditions, tolerating heat and moderate drought. In cultivation trunks are usually straight and become quite massive with a rounded canopy of light-textured foliage, whereas trees in the wild are usually smaller and more irregular. Similar in appearance to gray pine (*Pinus sabiniana*) and closely related to Coulter pine (*P. coulteri*). Longevity estimated to be 100 to 150 years. Series associations: **TorPinsta**.

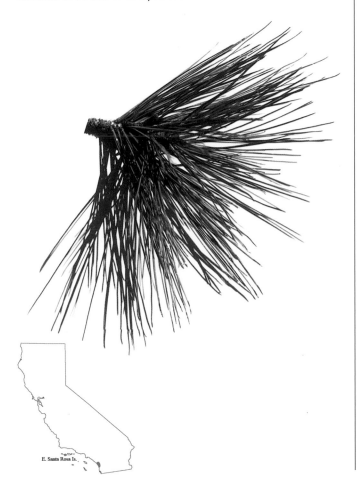

e Asia. SHD, FAL CLR; IRR: L/L/M/M/M/M Considered an invasive reseeder, but not as serious as some others.

Pistacia chinensis

CHINESE PISTACHE

Anacardiaceae. Sunset zones 4-16, 17 (warmer parts), 18-23, seldom grown in zones 4-7. Deciduous. Native to China and naturalized in eastern Texas and parts of California. Growth rate slow to moderate to 30-60' tall with a nearly equal spread, young trees developing a dense oval shape, irregular branching eventually forming a full rounded canopy with low branches sweeping to the ground if left unpruned. Leaves are alternate, pinnately compound, 4-8" long, dark to medium green, with 10-16 elliptical to ovate-lanceolate leaflets, 2-4" long by 3/4" wide, glabrous, with brilliant orange to deep red or bright yellow fall color. Insignificant, yellowish green, dioecious flowers occur in loose drooping panicles from March to May, male and female flowers on separate trees. Female trees develop clusters of small, reddish, 1/8-1/4" drupe fruits favored by birds, thin-skinned with a hard seed inside, turning bluish or black when ripe in early fall. Twigs are smooth, reddish brown, with stout, brown pointed buds. Bark is reddish brown, thickening and becoming gray-brown, finely fissured, with flat, scaly ridges.

A popular and reliable street or shade tree favored for its rounded canopy and bright fall color. Tolerates drought and alkalinity. Resists oak root fungus, and relatively pest-free. Overwatering can cause verticillium wilt. Intolerant of pruning, which often destroys natural form, though lower branches can be removed if cut closely at the trunk, leaving no stubs to sucker. Longevity estimated to be 80-100 years or more. 'Keith Davey' is a male cultivar, somewhat neater in form, more compact, with a slightly slower growth rate.

Australia. EVG, SHD; IRR: -/-/M/M/-/-

Pittosporum rhombifolium

QUEENSLAND PITTOSPORUM

Pittosporaceae. Sunset zones 12-24. Evergreen. Native to Australia. Growth rate slow to moderate to 15-35' tall with a 12-25' spread, forming an ovoid canopy as a tree standard, otherwise a large, bushy shrub. Leaves are alternate, simple, 2-1/2 to 4" long by 1-2" wide, smooth, shiny dark green and waxy, diamond-shaped, with sparsely sharp-toothed margins, teeth at the ends pointing toward the tip. Insignificant, slightly fragrant, white flowers occur in terminal clusters in late spring to early summer, followed by showy, dense clusters of bright yellowish orange, 1/4-1/2" long, round fruit, which hang from branch ends from late summer into fall. Twigs are smooth, orange-brown. Bark is smooth, grayish brown, becoming heavy and thickly scaly-plated, with many small interconnected shallow fissures.

Commonly used as a tough, reliable street tree in coastal areas or inland, where it tolerates moderate heat with moderate moisture. Fruits litter paving. Resists oak root fungus. Longevity estimated to be 50-75 years.

Australia, Africa, Asia. EVG, SHD; IRR: M/-/M/M/-/-

Pittosporum undulatum

VICTORIAN BOX

Pittosporaceae. Sunset zones 14 & 15 (with protection), 16, 17, 21-24. Evergreen. Native to Australia, New Zealand, Africa, and Asia. Growth rate fast initially to 15' tall, slowing to moderate to 30-40' high and wide, forming a dense, oval canopy as a single-trunk tree, broader as a multi-trunk specimen. Leaves are alternate, simple, 3-5" long and 1 to 1-1/2" wide, glossy light green, elliptical, with entire wavy-edged margins, older leaves darker and less wavy. Dense, 3" wide terminal clusters of 1/2" long, creamy white, 5-petaled flowers, with a heavy orange blossom fragrance, bloom heaviest in spring and intermittently into summer. Yellowish orange fruit capsules split open in fall to reveal many golden-orange seeds in a sticky orange jelly. Twigs are shiny and dark brown. Bark is grayish brown with a coarse to somewhat smooth texture.

Commonly used as a reliable, tough residential street or lawn tree, especially in coastal areas where it thrives in moist air. Specimen multi-trunked trees are often grown in large planters. Roots can become invasive. Longevity estimated to be 50-60 years.

Cultivar. SHD; IRR: L/L/M/M/H/H

Platanus x *acerifolia*

LONDON PLANE

Platanaceae. Sunset zones 2-24. Deciduous. Hybrid between *Platanus orientalis* and *P. occidentalis*. Growth rate fast to 40-80' tall with 30-40' spread, developing a broad, symmetrical, conical to oval form. Leaves are alternate, simple, palmate, 4-10" long and nearly as wide, with 3-5 lobes, coarsely toothed margins, fuzzy tomentose undersides, and fall color, if any, yellow-orange or dull orange-brown. Multiple light green flower heads are 3/4-1" round, 2 or rarely 3 to a string, hanging from and slightly covering the 5-6" long, pendulous common stem, male flowers from leaf buds of the previous year and females from present-year buds. Rough seed balls ripen to light brown in fall of the first year and drop to the ground, disintegrating when crushed, releasing many tightly packed, slender achene seeds, 1/2-3/4" long, with a fibrous bristlelike tail. Bark peels in irregular thin blotches, with upper bark and branches smoother and cream colored.

A classic shade tree for streets, parks, or large tree wells, sometimes pollarded for urban uses. Anthracnose, an airborne disease, causes temporary leaf drop and disfiguration. Tolerates heat, dust, and smog. Needs ample space. Shallow watering produces large surface roots, which can heave paving. Longevity estimated to be 150-300 years. 'Bloodgood', for inland use, is more resistant to anthracnose, and 'Yarwood' and 'Columbia' have good mildew resistance in coastal climates.

e&c N.A. SHD; IRR: No WUCOLS

Platanus occidentalis

AMERICAN SYCAMORE

Platanaceae. Sunset zones 1-24. Deciduous. Native throughout the eastern to central U.S. Growth rate fast to 40-80' tall with a 30-40' spread, developing a broad, symmetrical, conical to oval form. Leaves are alternate, simple, palmate, 5-12" long, broader than long, often droopy, with 3-5 lobes, margins with sparsely spaced, coarse teeth, tomentose undersides, and yellow-orange fall color turning to brown. Multiple light green flower heads are 3/4-1" round, often solitary, or up to 3-4, hanging from and slightly covering the 5-6" long pendulous common stem, male flowers from leaf buds of the previous year and females from present-year buds. Rough, rounded seed balls ripen to light brown in fall of the first year and drop to the ground, releasing many tightly packed, slender achene seeds, 1/2-3/4" long, with a fibrous, bristlelike tail. Bark is whitish, peeling in thin blotches, young bark and branches having a smooth, whitish cream color and mature bark becoming dark brown, with many squarish fissures and cracks and flat ridges slightly recurving along the edges.

A classic shade tree for streets or parks, large treewells or, less often, pollarded into a shaped canopy. Often holds its green color into late summer, when *P. racemosa* and *P. occidentalis* have a definite orange-yellow cast, but leafs out later in spring. Susceptible to anthracnose. Tolerates heat, dust, and smog. Longevity estimated to be 150-300 years.

sw N.A. California Native: CS, VG, OW (Riparian) (*c&s SNF, Teh, GV, CW, SW*); SHD, SPC; (IRR: M/M/M/M/H/H)

Platanus racemosa

WESTERN SYCAMORE

Platanaceae. Sunset zones 4-24, Jepson *IRR,SUN:1-3,7-9,**11**,14-17,**18-21**,22-24. Deciduous. Native to riparian streams of California foothills and Coast Ranges. Growth rate fast to 30-80' tall by 20-50' wide, developing an irregular, upright trunk and low secondary trunks or multi-trunked and low branched. Leaves are alternate, simple, light green, palmate, 5-11" long, deeply 3-5 lobed, with pointed ends, minutely to-mentose undersides, and brief orange-tan to brown early fall color. Multiple light green flower heads are 3/4-1" round, 3-7 to a string, and hang closely to the 5-10" long pendulous com-mon stem, males from leaf buds of the previous year and fe-males from present-year buds. Rough seed balls ripen to light brown in fall of the first year and either fall apart or disinte-grate when crushed, releasing many tightly packed, slender achene seeds, 1/2-3/4" long, with a fibrous, bristlelike tail. At-tractive whitish bark peels in irregular tan patches, becoming darker, ridged, and furrowed near the base of the trunk.

A common native along streambanks and adjacent flat-lands throughout the western part of the state. Less often cul-tivated than London plane (*Platanus x acerifolia*), but useful in riparian settings or as a large parkway or shade tree. Toler-ates heat and poor soils. Large, wide-spreading side roots may surface with shallow watering. Graceful twisted branching, attractive blotchy bark, and interesting winter silhouette. Tol-erates heat and wind. Susceptible to mites, leaf miner, and anthracnose, more so than London plane. Longevity estimated to be 100-200 years. Series associations: Scalebroom shrub series; Arrwillow, Balwillow, **Calsycamore**, Fanpalm, Mixwil-low, Pacwillow, Redwillow, Sitwillow, Valoak, Whialder.

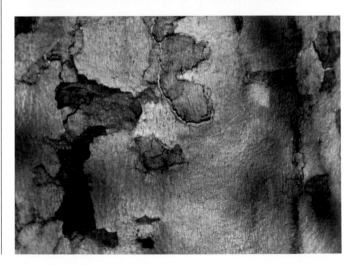

se Africa. EVG, SHD, FOL ACC; IRR: M/M/M/M/-/M

Podocarpus gracilior

FERN PINE

Taxaceae. Sunset zones 8, 9, 13-24. Evergreen. Native to southeast Africa. Growth rate moderate to fast to 20-60' tall and 10-20' wide, developing a broad, oval, upright form as a trained single-trunk tree or bushy as a large shrub. Needlelike leaves are soft, bright bluish green, narrow-pointed, 1-2" long by 1/8" wide, narrowly tapered at ends and base, spirally arranged on slender branchlets, somewhat delicate in texture though quite dense, persisting 2-3 years. Tiny flowers are inconspicuous. Fruits are purplish, cherrylike, 1/3" round drupes with a single seed, occurring only on mature female trees and only if a male pollenizer is nearby. Slender greenish twigs turn grayish. Bark is tan to grayish, fissured, and peels in small, thin, squarish gray plates, often revealing reddish underbark.

A commonly cultivated small evergreen street, courtyard, or patio tree suited to tight spaces. Tolerates heat and coastal conditions. Relatively pest free. Requires only minimal watering once established, though responds well to moderate water. Young trees may require staking until a strong leader develops. Tolerates clipping and light pruning, which is usually not required to retain dense symmetrical form. Not reliably hardy below 25 degrees. Longevity estimated to be 50-75 years.

e Asia. EVG, FOL ACC, IRR: M/M/M/M/M/M

Podocarpus macrophyllus

YEW PINE

Taxaceae. Sunset zones 4-9, 12-24, with protection in zones 3, 11. Evergreen. Native to southern China and southern Japan. Growth rate slow to moderate eventually to 15-50' tall by 6-15' wide, developing a cylindrical form, somewhat irregular with age, with single or multiple upright trunks and upward-ascending branches often drooping outward. Needlelike leaves are simple, dark green, glossy, 3-6" long by 1/8 to 1/4" wide, linear-lanceolate, loosely whorled on branchlets, tapered, with smooth edges, acute pointed ends, and yellowish green undersides. Flowers are insignificant, small, dioecious, with catkinlike yellowish males and solitary female clustered scales occurring on separate plants. Fruits are 1/4" oval, greenish purple drupes, ripening in fall, with a fleshy bluish purple receptacle and a tiny, hard-shelled seed, rarely seen. Twigs are bright yellowish green. Bark is reddish brown, becoming lightly fissured, with darker, thin, slender, peeling shreds.

A small vertical evergreen accent tree, also useful in containers or limited spaces. Tolerates heat, drought, and clipping to desired shape. Longevity estimated to be 50-90 years. 'Maki' is a slower growing, dwarf, shrub form with smaller leaves, reaching 6-8' tall in 10 years.

Europe, n Asia. IRR: M/M/M/M/H/H Extremely invasive and weedy, its use is highly discouraged.

Populus alba

WHITE POPLAR

Salicaceae. Sunset zones 1-11, 14-21. Deciduous. Native to Europe and northern Asia. Growth rate fast to 40-70' tall and wide, upright at first, though older trees become broader, usually multi-trunked with heavy, horizontal spreading limbs forming a tall oval canopy. Forms thickets, if untended, by heavy suckering at the base of trunks and along highly invasive, spreading surface roots. Leaves are alternate, simple, 2-5" long, dark green, irregularly sinuate-dentate to palmate, with 3-5 sparsely toothed lobes, a blunt-pointed apex, bases usually rounded or flat, white woolly undersides and petioles, and yellow fall color. Dioecious yellowish flowers are regular, in drooping tassel-like aments, males and females from separate buds, in early spring before leaves. Fruit capsules, 1-celled with 2-4 valves, contain many tiny seeds with long, silky hairs, and mature in profusion in early summer. Twigs and buds are pliable and white tomentose. Bark has a whitish cast, developing blackened spots and furrows.

Once planted for its attractive, fluttering, 2-toned leaves, but now considered a serious pest, as it reseeds readily and invades areas outside cultivation, especially near riparian habitats, where it can be extremely difficult to eradicate. Must be closely monitored and maintained wherever used. Longevity estimated to be 30-50 years, but thickets may persist indefinitely.

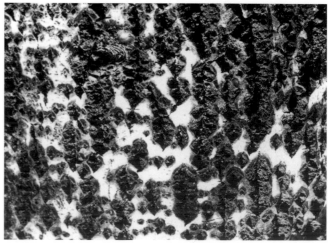

Cultivar SHD, VERT ACC; IRR: M/M/M/M/H/H

Populus alba 'Pyramidalis'

= *Populus bolleana*

BOLLEANA POPLAR

Salicaceae. Sunset zones 1-11, 14-21. Deciduous. Selection of the species. Growth rate fast to 40-70' tall with a 15' spread, developing a narrow, upright, columnar form from a thick basal trunk with multiple upright leaders and slender twigs, often contorted, in vertical alignment. Leaves are alternate, simple, 3-5" long and wide, deep glossy green, irregularly sinuate, and dentate to palmately lobed, with light green, felty petioles and undersides, and golden yellow fall color. Juvenile leaves are more deeply lobed. This variety is reproduced from cutting stock of male trees, which have yellowish tassel-like flowers but do not produce the cottony seeds typical of female trees. Twigs and buds are pliable, green, slightly hairy. Bark is greenish gray, developing brown markings and shallow furrows but not as distinctly as the species.

A vertical accent tree effective in groves or rows, but not often cultivated, with a dramatic winter silhouette, attractive bark, and upright branching, much like Lombardy poplar (*P. nigra* 'Italica'). Drawbacks inherent to the species include invasive roots and suckering and a rather short lifespan. Requires moderate moisture and does not tolerate drought. Longevity estimated to be 50-75 years.

sw N.A. California Native: VG, OW, FW, PF (Riparian) (*CA exc MP*); NAT, SHD; (IRR: M/M/M/M/H/H)

Populus fremontii

FREMONT COTTONWOOD

Salicaceae. Sunset zones 1-12, 14-21, Jepson *IRR or WET, SUN: 1,**2**,**3**,4-6,**7**-**24**;INV;STBL. Deciduous. Native to wet lowlands and rivers throughout California's Central Valley and the Sierra Nevada. Growth rate very fast to 40-60' tall or more with a 30' spread, developing an upright, tall oval form, often multi-trunked or with multiple leaders. Fairly thick limbs and drooping branchlets on older trees usually form a round-topped, open crown. Leaves are alternate, simple, 2-4" wide, thick, glossy dark green, smooth, broadly triangular, with coarsely scalloped or toothed edges, flat, yellow petioles to 1" long, and bright to dull yellow or, in colder areas, light orange fall color. Flowers are yellowish green catkins, appearing before leaves in spring, followed by cottony seeds on female trees. Twigs are smooth, pale yellow becoming yellowish gray, with shiny, yellowish green, pointed, sticky buds. Bark is pale, ashy brown with many shallow seams, thickening at the base of the trunk and becoming rough, dark grayish brown, deeply furrowed, with widely cut ridges and smaller irregular lateral ridges.

One of the most common riparian trees in California, along with alders and willows, which are close relatives. Less formal than Lombardy poplar (*Populus nigra* 'Italica'). Occasionally used in park or parkway settings to create a riparian or streamside character, or for its fast growth and fall color. Wood is relatively weak. Surface roots are invasive and may sucker vigorously if nicked. Susceptible to mistletoe. Seedless cutting-grown male stock trees are available. Longevity estimated to be 75-100 years, after which the trees lose vigor but may live to 150-200 years in habitat. Series associations: Mexelderberr, Narwillow, Sanwillow, Scalebroom shrub series; ArrWillow, Blacottonwoo, Blawillow, Calsycamore, Fanpalm, **Frecottonwoo**, Hoowillow, Mixwillow, Pacwillow, Redwillow, Sitwillow, Watbirch.

Cultivar. SHD, VERT ACC, FAL CLR; IRR: M/M/M/M/H/H

Populus nigra 'Italica'

LOMBARDY POPLAR

Salicaceae. Sunset zones 1-11, 14-24. Deciduous. Selected form of hybrid origin. Growth rate very fast to 40-100' with a 15-30' spread, developing a classic narrow, columnar form with vertical branches rising along a strong central trunk. Leaves are alternate, simple, 4" long, medium green, deltoid to broadly ovate, with a flattened to slightly rounded base, glabrous, with a shiny, waxy surface, finely serrate edges, and stunning golden-yellow fall color in colder areas. Dioecious yellowish green flowers are regular, in drooping tassel-like aments, with males and females from separate buds, in early spring before leaves. Fruit capsules are 1-celled, with 2-4 valves, containing many tiny seeds with long, silky hairs, maturing in early summer. Twigs are clear reddish brown with pointed, scaled buds covered with a sticky sap. Bark is grayish brown, quickly becoming irregularly furrowed, with narrow ridges.

A windbreak or skyline tree, often used as a vertical accent in informal groupings, or in linear or circular fashion in parks and along drives or near ponds, where there is adequate moisture. Roots are very invasive, with suckers forming thickets if not maintained. Blight can affect branchlets. Drought causes unsightly dieback from which trees never recover. Longevity estimated to be 60-80 years.

w N.A. California Native: JP, FF, AL (Riparian) (*KR, NCoRH, CaR, SNH, SnBr, GB*); FAL ACC, SHD, SPC; (IRR: No WUCOLS)

Populus tremuloides

QUAKING ASPEN

Salicaceae. Sunset zones 1-7, 14-19, Jepson *IRR,SUN,DRN:**1, 2**,3,**4-7**,14-17,**18**,19-2;INV:also STBL. Deciduous. Native to higher mountains of the western U.S. and throughout the Rocky Mountains. Growth rate slow to moderate to 20-60' tall and 15-30' wide, with an upright, loose, open habit, often forming clumps or extensive thickets in habitat, spreading by underground roots and suckering. Leaves are alternate, simple, 2-4" long, bright green, oval, pointed, with finely serrate edges, long, flattened leaf stalks, fluttering in the slightest breeze, and brilliant golden fall color in cold climates. Flowers are creamy yellow catkins, in spring, followed by small cottony seed capsules on female trees. Upright trunks are often somewhat contorted. Bark is conspicuously smooth, whitish to pale green with rounded, black, scarlike markings, with older, thickened bark at the base of trees becoming nearly black with a slight furrowing.

Famous for spectacular fall color in high-elevation landscapes. Tolerates warmer valley conditions but does not develop bright yellow fall color. Prefers moist, well-drained soil. Sometimes clumped in lawns like birches. Can be planted as a riparian grove in natural settings. Longevity estimated to be 30-50 years. Older trees lose vigor, but replacements sprout from suckers around the base. Series associations: **Aspen**, Blacottonwoo.

w. N.A. California Native: VG, SS, FW, PF, FF (Riparian) (*CA-FP*); (IRR: H/H/M/M/H/-)

Populus trichocarpa

= *Populus balsamifera* ssp. *trichocarpa*

BLACK COTTONWOOD

Salicaceae. Sunset zones 1-9, 14-24, Jepson *IRR or WET,SUN: 1,**2-7**,8,9,14,**15-18**,19-23,**24**. Deciduous. Native to streams throughout California west of the Sierra Nevada below 9,000' elevation. Growth rate fast to 30-100' tall and 25-30' wide, forming an open-crowned canopy with large, upright trunks, becoming more irregularly shaped, with short twigs giving a rounded billowy effect. Leaves are alternate, simple, 2-4" wide, finely toothed, dark shiny green, smooth and leathery, ovate, pointed, with finely crenate serrate edges, pale undersides, finely netted veining between raised midrib and secondary veins, 1" long, round, reddish green petioles, and dull yellow fall color. Midveins underneath as well as petioles and branchlets are sometimes minutely hairy. Inconspicuous 1-1/2 to 3" male and female flowers occur on separate trees in spring, with cottony seed capsules following on female trees. Yellowish to reddish, pointed, 3/4" leaf buds often appear curved and are covered with a fragrant, sticky, shiny, yellowish brown gum. New twigs are shiny, reddish yellow, with an indistinctly angled cross-section, later becoming rounded. Bark is smooth and gray with black, shallow markings and fissures, becoming dark grayish brown, deeply furrowed at the base of older trees.

One of the tallest poplars, and a massive riparian tree, growing in shallower soils than Fremont cottonwood (*Populus fremontii*). Rarely cultivated outside its native habitat but an important snowbelt tree, where it reseeds to form groves along mountain rivers. Longevity estimated to be 100-150 years. Series associations: Narwillow, Sanwillow shrub series; Arrwillow, **Blacottonwoo**, Blawillow, Hoowillow, Mixwillow, Pacwillow, Redalder, Sitwillow, Watbirch.

sw N.A. California Native: MD, SD (*SnJV, PR, D*); FOL ACC, DES ACC, SHD; (IRR: -/L/L/L/M/M)

Prosopis glandulosa

= *Prosopis glandulosa* ssp. *glandulosa*

HONEY MESQUITE

Fabaceae. Sunset zones 10-13, 18-24. Deciduous. Native to desert regions of Colorado and California and basin grasslands east to Texas and south to Mexico. Growth rate slow to 20' tall and up to 40' wide, with a short, multi-trunked base and many crooked, upward-arching to horizontal branches, weeping at the ends. Leaves are alternate, bipinnately compound, with 2 pinnae, 2-4" long, each with 20-36 elliptical, paired 1/4 to 1-1/4" long by 1/8-1/4" wide, hairless leaflets. Leaf petioles have a thickened base and are 2-1/2 to 4" long. Paired, or occasionally single, spines up to 1" long occur at the base of leaves.. Slender, cylindrical, greenish yellow flower spikes, 2 to 3-1/2" long, occur from April to June. Seed pods are flattened, sometimes twisted, 3-8" long, irregularly constricted between seeds. Twigs are smooth, brown, with noticeably knobby spur buds. Bark is dark brown, becoming deeply furrowed, with wide, flat, scaly ridges that peel back in strips, exposing reddish underbark.

Once limited to grasslands and streams of the western Great Basin but spread into extended areas by livestock and wildlife. A valuable forage tree in habitat. Seeds are eaten by animals and pollen produces a sweet-tasting honey. Cultivated as a curiosity or desert specimen tree. Tolerates extreme heat and drought, but benefits from occasional deep watering in well-drained or sandy soils. Longevity estimated to be 75 years. Series associations: Allscale, Crebush, Fousaltbrush, Mesquite, Blupalveriro shrub series; Fanpalm.

sw N.A. (*Ca non-native, SnJV, CCo, Sco*) DES, FLW ACC; (IRR: -/L/L/L/M/M)

Prosopis velutina

ARIZONA MESQUITE

Fabaceae. Sunset zones 10-13, 18-24. Deciduous. Native to desert regions of central Arizona and southwestern New Mexico into Mexico. Growth rate slow to 20-30' tall and wide, with a short forked trunk and many crooked, upward-arching to horizontal branches, weeping at the ends. Leaves are alternate, bipinnately compound, with 1 or 2 pinnae, 1-1/2 to 3" long, each with 15-20 elliptical to narrowly oblong, paired, dull green, stalkless leaflets, 1/4-3/8" long by 1/16-1/8" wide, finely hairy throughout. A few paired, or occasionally single, stout yellowish spines, 1/4-1" long, occur at leaf nodes. Hairy, slender, 2-3" long, light greenish yellow, cylindrical flower spikes occur from April to June, sometimes sporadically in fall, with a slight fragrance. Seed pods are narrow, 3-8" long, slightly flattened, finely hairy, constricted between seeds, and do not split open when mature in late summer. Twigs are short, smooth, light brown, finely hairy, and grow in a zigzag fashion, with noticeably knobby spur buds and few thorns. Bark is dark brown, becoming thick and deeply furrowed, peeling in long narrow strips, exposing reddish underbark.

Seldom cultivated, but adaptable as an attractive small desert accent tree. Occurs naturally in desert grasslands and along streambeds, with honey mesquite (*Prosopis glandulosa*), of which it is often regarded as a variety. Longevity estimated to be 75 years.

Cultivar. SHD, FLW ACC, FRU; IRR: M/M/M/M/M/-

Prunus x *blirieana*
PINK FLOWERING PLUM

Rosaceae. Sunset zones 3-22. Deciduous. Hybrid between *Prunus cerasifera* 'Atropurpurea' and *P. mume*. Growth rate initially fast but slowing, to 25' tall with a 20' spread, usually with densely twiggy growth and a round-topped canopy. Leaves are alternate, simple, 1 to 1-1/2" long, oblong-ovate, slightly crinkled, with pale glabrous undersides, finely serrate edges, bronze-green to purplish, turning dull greenish bronze in summer and yellowish orange in fall. Showy racemes of fragrant, semi-double, rose pink flowers occur in rounded clusters as reddish new growth appears in February to April. Sets few if any rounded, 1" purplish, juicy fruit. Slender twigs and buds are purplish. Bark is thin, grayish brown, becoming shallowly furrowed with reddish orange fissures and scaly plated ridges, usually with protruding knobby "knees" on the trunks.

A popular foliage and flowering accent or shade tree for lawns, terraces and patios, large containers, or as a small street or parking lot tree. Flowering appears denser and trees less vigorous than other plums. Tolerant of most soils if well-drained with moderate moisture, and benefits from occasional deep watering. Suffers from drought, which stunts growth, affecting vigor afterward. Subject to scale and sooty mold. Longevity estimated to be 50-75 years.

se N.A. EVG, SHD, SCR; IRR: L/L/M/M/M/M

Prunus caroliniana

CAROLINA CHERRY LAUREL

Rosaceae. Sunset zones 5-24. Evergreen. Native to the eastern U.S. from North Carolina to Texas. Growth rate fast to 20-30' tall and 15-25' wide if trained as a single-trunk tree, with a broader canopy and broader oval form as a multi-trunk tree. Leaves are alternate, simple, 2-4" long, dark green, smooth and shiny, oblong-lanceolate, with slightly wavy edges, pale glabrous undersides, persisting 2-3 years. Noticeable small spikes of tiny, creamy white, slightly fragrant flowers occur at branch ends in spring. Fruits are black-skinned, juicy, 1/4-1/2" round, and noticeable as they ripen. Twigs are reddish green and contrast with the shiny green foliage. Bark is grayish brown, becoming fissured, with reddish orange splitting seams.

Commonly grown as an evergreen shade tree, casting dense shade in maturity. Pruning young trees is necessary to develop a strong central trunk and desired branching structure. Often used as a large evergreen shrub or hedge. Flowers and fruit litter paving. Tolerates heat, wind, and moderate drought when established. Chlorotic in alkaline or poorly drained soils. Reseeds readily in moist settings and may be considered a pest. Longevity estimated to be 50-75 years.

se Europe, Asia. SHD, FRU; IRR: M/M/M/M/M/-

Prunus cerasifera

CHERRY PLUM

Rosaceae. Sunset zones 3-22. Deciduous. Native to southeastern Europe and Asia and widely naturalized in the U.S. Growth rate fast to 30' tall and wide, with a wide branching canopy often exceeding its height, usually forming a multi-trunked, arching vase if not maintained, and commonly self-seeding or suckering from grafted rootstock. Leaves are alternate, simple, 1 to 1-1/2" long, dark green, oblong-ovate, with finely serrate edges, pale undersides, and yellowish fall color. Showy, pure white, 3/4-1" fragrant flowers occur in racemes in spring. Fruits are juicy, sweet but bland, 1 to 1-1/4" long, spherical to oval, reddish or yellow, often in profusion. Twigs are slender and reddish with reddish buds. Bark is thin, grayish brown, becoming shallowly reddish-furrowed, with wide ridges, often scaly.

Not generally cultivated for landscape use, or as a fruiting tree, but common nonetheless and used as rootstock and in breeding for many "stone fruit" trees. One of many introduced trees naturalized throughout the U.S. Generally not invasive in large lawns, though birds carry seeds and fruit to nearby areas. Tolerant of adverse conditions when established. Longevity estimated to be 50-75 years.

Cultivar. SHD, FOL, FLW ACC; IRR: M/M/M/M/M/-

Prunus cerasifera 'Atropurpurea'

PURPLE-LEAF PLUM

Rosaceae. Sunset zones 3-22. Deciduous. Hybrid variety of the species. Growth rate fast to 25-35' tall and wide, forming a broadly rounded oval canopy. Leaves are alternate, simple, 1 to 1-1/2" long, oblong-ovate, with finely serrate edges, pale undersides, new leaves light green to coppery red, emerging as flowers begin to fade and deepening to dark purple, fading to greenish bronze in summer, and turning yellowish orange in fall. Showy clusters of lightly fragrant, single white flowers occur in racemes, a week or so after *Prunus* x *blirieana* and just before *P.* 'Krauter Vesuvius'. Abundant 1 to 1-1/2" long, oval, reddish, edible plums are sweet but bland tasting. Twigs are slender and purplish with purplish buds. Bark is thin, grayish brown, becoming shallowly furrowed, with reddish orange fissures and scaly, plated ridges.

A commonly used, perhaps overused, accent or shade tree, best in lawns or shrub areas, away from paving, where fruit drop is messy. Reliably tolerant of heat with moderate moisture in well-drained soils. Longevity estimated to be 60-80 years.

Cultivar. SHD, FOL, FLW ACC; IRR: M/M/M/M/M/-

Prunus cerasifera 'Krauter Vesuvius'

PINK-FLOWERING PURPLE-LEAF PLUM

Rosaceae. Sunset zones 3-22. Deciduous. Hybrid variety of the species. Growth rate fast to 18' tall with a 12' spread, developing upright branching and an oval to arching canopy. Leaves are alternate, simple, 1 to 1-1/2" long, shiny, dark purple to almost black, oblong-ovate, with finely serrate edges, pale undersides, and yellowish orange fall color. New leaves emerge as flowering begins and hold their color through summer. Showy, fragrant, light pink, single flowers occur in profusion in late spring in racemed clusters. Sets few if any fruits. Bark is thin, grayish brown, becoming shallowly furrowed, with reddish orange fissures and scaly, plated ridges.

A commonly used foliage and flowering accent tree without fruit. Good branching structure. Fairly disease resistant. Requires minimal care with moderate moisture in well-drained soils. Longevity estimated to be 60-80 years.

w N.A. California Native: (s Ca.) CH, OW (*s NCoR, CW, SW*); EVG, SHD, SCR; (IRR: L/L/VL/VL/-/-)

Prunus ilicifolia

HOLLYLEAF CHERRY

Rosaceae. Sunset zones 5-9, 12-24, Jepson *DRN,SUN or part SHD:5,7,14-24&IRR:8-10. Evergreen. Native to coastal areas and islands of California from Napa southward to Baja California, on dry slopes or moist soils, to 5000' elevation. Growth rate moderate to fast to 10-25' tall and wide, forming short, stout trunks with low horizontal branching and a dense, rounded canopy. Usually a large shrub in its native habitat. Leaves are alternate, simple, 1-2" long, shiny, leathery, dark green, ovate to oval, glabrous, coarsely spiny-toothed, less often with smooth edges, either short-pointed or rounded at the tips, with pale, glabrous undersides, persisting 2-3 years. Slender, 2" long spikes of 1/4" creamy white flowers with 5 rounded petals emerge from leaf axils in late spring. Clusters of oval, 1/2"-5/8" round, cherry fruits have a juicy, sweet pulp and a single hard, smooth stone seed, with branched markings, maturing in October to December. Twigs are slender and reddish brown to grayish. Bark is dark reddish brown, becoming fissured with small squarish plates.

A dense native evergreen tree or shrub, occurring more commonly in coastal areas than *Prunus lyonii*. Drought tolerant when established. Seeds germinate quickly after wildfires. Lower branches on larger specimens can be pruned up to expose trunks for a more graceful, treelike form. Longevity estimated to be 50-75 years. Series associations: BirmoumahCal, Intlivoakshr, Intlivoakscr, Sumac shrub series; Calbuckeye, Isloak, **Holchesta**.

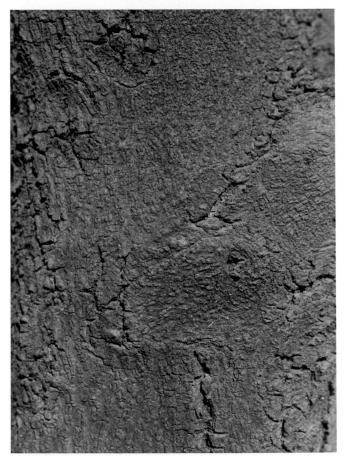

w N.A. California Native: (s Ca. Channel Islands) CH, OW (*ChI*);
(IRR: L/L/L/L/-/-)

Prunus lyonii

CATALINA CHERRY

Rosaceae. No Sunset zones. Jepson *DRN,SUN or part SHD:
14-17,18,**22-24**&IRR:**7**,8,9,**19-21**. Evergreen. Native to southern California islands south to Mexico. Growth rate moderate
to 15-20' tall and wide, forming a dense, rounded canopy as a
cultivated tree, or to 45' tall in habitat with horizontal branching to the ground and a broad, shrublike, oval form. Leaves
are alternate, simple, 2-4" long by 1/2 to 2-1/2" wide, shiny
dark green, leathery, narrowly ovate, short-pointed at the tip,
with smooth, slightly wavy edges, rarely spiny toothed, with
pale, glabrous undersides, persisting 2-3 years. Many 2 to 4-
1/2" long spikes of slender, 1/4" creamy white flowers with 5
rounded petals occur in late spring, from leaf axils. Fruits are
oval, dark purple to black, 1/2-1" long, and edible, with a thick,
sweet-tasting, juicy pulp and a single, large, hard stone pit,
maturing in late summer to fall. Twigs are stout, hairless, yellowish green to reddish brown. Bark is rough, dark reddish
brown, becoming fissured, with reddish orange seams.

A dense native evergreen, limited in habitat but often
cultivated as a screen or windbreak. Flowers and fruit litter
paving. Tolerates heat, wind, and moderate drought when
established. Chlorotic in alkaline or poorly drained soils. Longevity estimated to be 50-75 years. Series associations: Holchesta.

e Asia. SHD, FLW ACC; IRR: M/M/M/M/M/-

Prunus serrulata

JAPANESE FLOWERING CHERRY

Rosaceae. No Sunset zones. Deciduous. Native to northern Japan and Korea. Growth rate moderate to 25-30' tall and wide or wider, developing a broad, rounded crown, with upright-arching branches from a large heavy trunk, usually multiple-grafted onto rootstock. Leaves are alternate, simple, 3-5" long by 3/4 to 1-1/4" wide, glossy dark green, oblong-ovate, glabrous, with pale undersides, finely serrate edges, slender 1 to 1-1/2" long, reddish-tinged green petioles with tiny reddish warts or multiple short green stipule appendages, and yellow to orange-red fall color. Round, showy clusters of 1-1/2" broad, white-petaled, perfect, single flowers occur in profusion in loose, long-stemmed racemes in late March before leaves. Petals are dark pink-tinged at the base, and white stamens darken to a deep pinkish red. Insignificant, 1/2" round, long-stemmed, edible reddish cherries, usually unseen among the foliage, are favored by birds. Twigs are clear, shiny, reddish brown. Bark is smooth, shiny, gray to reddish brown with distinctive horizontal, swelled bumpy markings.

A medium-sized flowering accent tree. Cultivated varieties are best known and most often planted. Prefers well-drained, moderately moist, fertile soils, but not particularly fussy otherwise. Longevity estimated to be 50-85 years.

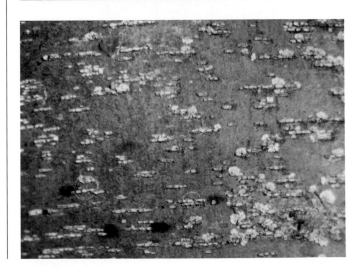

Cultivar. SHD, FLW ACC; IRR: M/M/M/M/M/-

Prunus 'Kwanzan'

(*Prunus* 'Kanzan', *Prunus serrulata* 'Sekiyama')

KWANZAN FLOWERING CHERRY

Rosaceae. Sunset zones 3-7,14-20. Deciduous. Variety of the species. Growth rate moderate, slowing with age, to 30' tall and 20' wide, with stiff, upright branches forming a distinctive inverted cone. Leaves are dark glossy green, glistening reddish bronze as they first appear, 3-5" long, oblong-ovate, with finely serrate edges, pale undersides, and yellow to reddish fall color. Showy, drooping clusters of double, 1", deep pink or white flowers appear with leaves in early spring. Bark is smooth, grayish brown with horizontal markings, becoming fissured with age.

Cultivar. SHD, WPG, FLW ACC; IRR: M/M/M/M/M/-

Prunus pendula

(*Prunus* x *subhirtella* 'Pendula')

WEEPING CHERRY

Rosaceae. Sunset zones 2-7, 14-20. Deciduous. Variety of the species. Growth rate slow to moderate to 10-12' tall and wide, with a broad, arching form, branch ends often drooping to the ground, weeping from upright grafting stock. Leaves are glossy green, oblong-ovate, 1 to 1-3/8" wide by 3-4 to 1/2" long, with finely serrate edges, pale undersides, reddish bronze in spring, and yellow to reddish fall color. Showy, double, deep pink flowers are borne in profusion March to April, as leaves emerge. Bark is smooth, grayish brown, and variable, often developing horizontal markings and becoming fissured with age. Many varieties exist. Clones of *Prunus subhirtella* (Higan cherry) often considered synonymous. *P. pendula* clones generally have longer, narrower, more finely toothed leaves. *Prunus s.* 'Autumnalis' blooms in spring and fall. *Prunus s.* 'White Fountain' is a white-flowering variety.

Cultivar. SHD, FLW ACC; IRR: M/M/M/M/M/-

Prunus x *yedoensis*

YOSHINO FLOWERING CHERRY

Rosaceae. Sunset zones 3-7, 14-20. Deciduous. Cultivated variety developed in Japan. Growth rate moderate to fast to 40' tall and 30' wide, with graceful, semi-arching branches from a short, stout trunk. Fragrant, single, white or pale pink flowers in drooping clusters of 3-5 occur in profusion in early spring. Leaves are glossy green, oblong-ovate, 2 to 2-1/2" wide by 3-1/2 to 5-1/2" long, with finely serrate edges and pale undersides. Bark is smooth, grayish brown, becoming fissured with age. This is the variety famous in Potomac Park in Washington, D.C. 'Akebono' (daybreak cherry) is found at the University of Washington in Seattle.

MIDDLE: Photograph courtesy of Wayside Gardens.

sw N.A. California Native: (s Calif.) ME, PF, FF (*s SCoRO, TR, PR*); (IRR: No WUCOLS)

Pseudotsuga macrocarpa

BIGCONE DOUGLAS-FIR

Pinaceae. Sunset zones 3-11, 14-23, Jepson *DRN:**3-7,14-17**&IRR:2,8,9,**10**,11,**18**,19,**20-23**. Evergreen. Native to southern California on dry slopes and in canyons from Kern and Santa Barbara counties south to Baja California at 1,000-7,000' elevation, outside the range of Douglas-fir to the north. Growth rate slow to 60' tall and 30' wide, developing a broad conical shape, stout trunk, long, horizontally spreading branches, irregular with age, with sparse areas between older drooping branches often sprouting juvenile growth. Needles are stout, densely set, bluish green to dark green, 3/4 to 1" long, slightly curved, whorled along branchlets, appearing 2-ranked, or semi-flattened, with a slightly grooved upperside, ends slightly more pointed than *Pseudotsuga menziesii*, and persisting about 6 years. Cones are reddish brown, 4-7" long by 2-3" wide, short-stalked, narrowly ovate, hanging from branch ends, with shortened tri-tipped, pointed bracts extending just beyond the broad, thick, rounded scales, ripening in summer, and opening in fall of the first season, but may persist into the following year. Seeds are dark brown and long-winged. Slender twigs are reddish brown, slightly hairy at first, with shiny, dark brown, pointed buds, slightly shorter than those of *P. menziesii*. Bark is dark reddish brown, thickening at an early age and becoming roughly furrowed, with wide, heavy ridges and interconnecting narrow cross-ridges.

Limited to southern California mountain habitats, along ridgetops and steep ravines, in loose scattered groves. Tolerates drier conditions than Douglas-fir, growing in chaparral and mixed conifer regions, with canyon live oak, as well as pinyon, Jeffrey, ponderosa, sugar, and gray pines. Rarely cultivated outside its native habitat. Resists oak root fungus. Longevity estimated to be 300 years or more in habitat. Series associations: **BigDoufir**, BigDoufircan, Canlivoak, Coupine, Coupincanliv, Mixconifer, Tanoak.

w & N.A. California Native: (n Calif.) RW, CF, ME, PF, FF (*KR, NCoR, n&c SNH, CCo, SnFrB, NcoRO*); EVG, CNF; (IRR: No WUCOLS)

Pseudotsuga menziesii

DOUGLAS-FIR

Pinaceae. Sunset zones 1-10, 14-17, Jepson *DRN:**4-6,15-17**& IRR:**2,3,7**&SHD:**14**;CVS. Evergreen. Native to the northwestern U.S. from northern California to Alaska east to the Rocky Mountains at 2,000 to 4,000' elevation. Growth rate fast to 80-160' tall and 20-30' wide in cultivation, up to 250' tall in habitat. Broadly pyramidal in youth, with lower branches horizontal or slightly drooping and higher branches ascending openly, older trees becoming more rounded, with an irregular, loose form. Needles are soft, dark green, new growth light green, 3/4 to 1-1/2" long, flattish, densely set, with a slightly grooved upperside, blunt to dull-pointed ends, and a distinctive fragrance, persisting 6-8 years. Cones are thin, reddish brown, 2-4" long by 1 to 1-1/2" wide, narrowly ovate with tri-tipped, pointed bracts extending among and beyond the broad, thin scales, hanging from branch ends, ripening and opening in fall of the first season. Seeds are dull brown, 1/4" long with a 5/8" long wing. Twigs are orange to brown with shiny, brown, long-pointed buds. Bark is thin, smooth, grayish brown, becoming thicker, dark brown, and roughly furrowed, with heavy ridges interconnected by narrow cross-ridges.

A common conifer in the moist northwest, as well as the drier Sierra Nevada, with a wide range throughout the northern part of the state, except for the Central Valley and sagebrush steppe, in the northeastern quarter. Tolerates lower elevations, but with slower, denser growth. Resists oak root fungus. Longevity estimated to be 150-200 years or to 375 years in habitat. 'Glauca' is a bluish form, native to the Rocky Mountains. Series associations: Caloatgrass, Pamgrass, Deerbrush herbaceous & shrub series; Beapine, Bispine, Blaoak, Canlioak, **Doufir**, Doufirponpin, Doufirtan, Engspruce, Giasequoia, Grafir, Inccedar, Jefpine, Mixconifer, Mixoak, Monpine, Orewhioak, Ponpine, PorOrfced, Redalder, Redwood, Tanoak, Weshemlock, Weswhipin, Whialder, Whifir, Alayelcedsta, Bakcypsta.

e Asia. SHD; IRR: No WUCOLS

Pterocarya stenoptera

CHINESE WINGNUT

Juglandaceae. Sunset zones 4-24. Native to China. Growth rate fast to 40-90' tall and 30-50' wide, with heavy limbs, wide-spreading, often low-branching or multi-trunked. Leaves are alternate, pinnately compound, 6-12" long, walnutlike, with 11-23 sessile or very short-stalked leaflets, 2-4" long, finely toothed, ovate-oblong to oblong-lanceolate, unequal at the base, with an acuminate apex, from a distinctive flat-winged rachis between the pinnae, shiny dark green on the upperside, pale on the underside, hairy at the midvein, remaining late into fall with only slight if any yellow coloration. Pendulous clusters of creamy, monoecious catkin flowers occur in late spring, males 5" long and females 10-12" long. Female tassels develop into drooping strands with many densely spaced, single-seeded nuts with rounded, oblique wings, hanging from branches into winter. Twigs are stout, greenish brown, growing in a somewhat zigzag manner. Bark is smooth and reddish brown, aging to grayish brown with many reddish furrows and irregular, interconnected, flat, scaly ridges.

An uncommon but attractive shade tree for streets, parks, or courtyards. Roots are invasive. Thrives in poor and even heavily compacted soils. Tolerates drought. Requires adequate room for its broad canopy. Longevity estimated to be 70-90 years or more.

e Asia. SHD, FLW ACC, FRU; IRR: L/L/M/M/M/M

Punica granatum

POMEGRANATE

Punicaceae. Sunset zones 5-24. Deciduous. Native from Persia to northwestern India and widely grown and naturalized in tropical climates. Growth rate fast to 8-10' tall or more and equally wide, usually multi-trunked with an irregularly arching, spreading canopy. Leaves are opposite, simple, 1-2" long by 3/8-3/4" wide, bright green and glossy, oblong-lanceolate, with smooth margins, reddish bronze new growth, and bright yellow fall color. Large, showy flowers are white, yellow, orange, red, or striped, in clusters of 1-5 at branch ends, with 5-7 densely crinkly, crapelike petals, many stamens, and a single pistil, from a green bulbous base, with a short stem. Most flowers set large, round, hard-rinded, juicy, pulpy fruits, 2-4" in diameter, which ripen to red when mature in late September, containing many edible, juicy red seeds. Twigs are glossy and reddish to green. Bark is smooth, gray, becoming rough, with peeling scales that often leave a mottled coloration as they fall off.

A rather common multi-trunk accent tree most often grown as a single specimen for its fruit, which does not set in coastal climates. Requires pruning at the base to control sucker growth. Flowers and fruit may be messy on paving. Blooms and sets fruit best in partial to full sun. Fairly drought tolerant, but needs regular deep watering for best fruiting. Longevity estimated to be 50-80 years. 'Alba Plena' has double creamy white flowers and rarely fruits, 'Double Red' has double orange-red flowers and no fruit, 'Legrellei' has double cream-striped red flowers and no fruit, and 'Wonderful' has single red flowers and fruit.

Cultivar. SHD, FLW ACC, FAL CLR; IRR: M/M/M/M/M/M

Pyrus calleryana 'Aristocrat'

ARISTOCRAT PEAR

Rosaceae. Sunset zones 2-9, 14-21. Deciduous. Selection of the species. Growth rate fast to moderate to 35-40' tall with a 20' spread, developing a symmetrical conical to oval shape, with multiple upright leaders from a stout trunk and shorter horizontal to slightly upward-arching branches with twiggy ends. Leaves are alternate, simple, 1-1/2 to 3" long, dark glossy green, ovate to broadly ovate, nearly glabrous, with lightly crenate edges, short acuminate ends, rounded cuneate at the base, with lighter undersides, and yellow fall color deepening to pinkish orange to scarlet red. Dense, showy, 3-4" wide, corymb clusters of heavily fragrant, white, 1/2-3/4" long, wide-petaled flowers occur in profusion in frosty areas as leaves emerge in early spring. Birds favor the small, otherwise inedible, tan, 1/2" round pear fruits, or pomes, which have a dull, rough surface. Twigs are smooth, shiny brown, with a slight white hairiness at first, becoming glabrous and white-spotted, with stout, reddish brown, hairy buds. Bark is smooth, whitish gray-brown, becoming shallowly fissured, with irregular, flat, scaly ridges.

Introduced in the mid-1970s and commonly used as an attractive street and shade tree or a flowering and fall color accent tree. Favored for its neat, clean appearance. More columnar and narrower than Bradford pear. One of the first trees to bloom in spring. Very susceptible to fireblight and mistletoe. Longevity estimated to be 70-80 years.

Cultivar. SHD, FLW ACC, FAL CLR; IRR: M/M/M/M/M/M

Pyrus calleryana 'Bradford'

BRADFORD PEAR

Rosaceae. Sunset zones 2-9, 14-21. Deciduous. Selection of the species. Growth rate fast to moderate to 50' tall with a 30' spread, developing a symmetrical, conical to oval pyramidal shape with upward-arching, erect branching from low on the stout trunk, becoming broad-based with age. Leaves are alternate, simple, 1-1/2 to 3" long, dark glossy green, ovate to broadly ovate, nearly glabrous, with lightly crenate edges and short acuminate ends, rounded cuneate at the base, with lighter undersides and deep scarlet-orange to wine red fall color. Showy, dense, 3-4" wide corymb clusters of heavily fragrant, white, 1/2-3/4" wide-petaled flowers occur in profusion in frosty areas as leaves emerge in early spring. Dull, rough tan, 1/2" round, inedible, pome fruits, or pears, are favored by birds. Twigs are smooth, shiny brown, with a slight white hairiness at first, becoming glabrous and white-spotted with stout, reddish brown, hairy buds. Bark is smooth, grayish brown, becoming shallowly fissured, with irregular flat ridges.

Introduced in the early 1950s, and once extensively used as an attractive street and shade tree, effective in small groves or in rows, but less often planted now, since the broad curving branches tend to break off at the base. Clean appearance. One of the first trees to bloom in spring. Resists fireblight. Longevity estimated to be 70-80 years.

Asia. SHD, FLW ACC; IRR: M/M/M/M/M/M

Pyrus taiwanensis

(*Pyrus kawakamii*)

EVERGREEN PEAR

Rosaceae. Sunset zones 8, 9, 12-24. Semi-deciduous. Native to Taiwan. Growth rate moderate to fast to 15-30' tall and wide, forming a broad dome canopy, taller than wide with age. Leaves are alternate, simple, 2 to 3-1/2" long, shiny dark green, ovate to obovate-elliptical, with finely serrate, wavy edges, long, slender pointed tips, lighter undersides, and orange-red fall color. Most foliage remains through winter if temperatures are not far below freezing but drop before pale yellowish green leaves occur in spring. Showy clusters of fragrant, white, 1" wide flowers cover the tree in early spring, just as or before new leaves emerge, with occasional sparse flowering in November. Fruit is small, dull, rough brown, tasteless, pea-sized, globose, with tiny seeds that generally do not germinate. Twigs are smooth and reddish brown. Bark is rough, scaly, checked, dark brown to blackish brown, contrasting with white flowers on bare branches in spring.

A small accent street or courtyard tree commonly used since the 1950s. Blooms with acacias as one of the first harbingers of spring. Tolerant of many soils and slight dryness but not drought. Benefits from occasional deep watering. Suffers from occasional aphids and susceptible to fireblight. Longevity estimated to be 60-80 years.

e Asia. SHD; IRR: No WUCOLS

Quercus acutissima

CHINESE OAK

Fagaceae. Sunset zones 2-7, 14-17. Deciduous. Native to the Himalayas, China, and Japan. Growth rate moderate to fast to 35-45' tall (or 50-70') and wide, forming an upright oval shape, more irregular and with a broad, open-headed, spreading canopy in age, with a heavy trunk and large spreading limbs. Leaves are alternate, simple, 3-5" long by 1" wide, shiny, light green, narrow-elliptical to oblong-lanceolate, with short, spiny-toothed or bristle-tipped edges, pale, glabrous undersides, resembling chestnuts, and yellow fall color. Flowers are pendulous, orange-yellow, and tassel-like, in spring. Acorns are rounded, short-pointed, 1/2-3/4" long, sessile to shortly stalked, tan, maturing in fall of the first year, with tightly scaled cups, fringed along the top edge, covering half of the nut. Bark is grayish brown, becoming fissured with age, with narrow vertical ridges, revealing reddish orange tones underneath as bark expands.

An uncommon but especially attractive park, shade, or lawn tree with shiny foliage and a clean appearance. Often develops distinctively broad, horizontal branching as it ages, with a more open, irregular form. Deep-rooted and tolerant of most soils but prefers slightly moist, consistently well-drained soils. Longevity estimated to be 100-250 years or more.

w N.A. California Native: CP, OW, ME (*NCoRO, CW, SW; Baja CA.*);
EVG, SHD; (IRR: VL/VL/L/L/-/M)

Quercus agrifolia

COAST LIVE OAK

Fagaceae. Sunset zones 7-9, 14-24, Jepson *DRN,SUN:5-7,**14-17,22-24**IRR: 8,**9**,10-13,**18-21**. Evergreen. Native to Coast
Ranges and inland foothills of California. Growth rate slow to
moderate to 20-70' tall, often with a greater spread, forming a
broad, rounded oval canopy, a short, clear heavy trunk, and
dense foliage. Leaves are alternate, simple, 1-3" long, medium
to dark green, stiff and glossy, oval to oblong, the hollylike
edging varying from densely or sparsely prickly-toothed to
nearly entire, curving under slightly at the edges, with pale
undersides, persisting 1 year and falling as new leaves appear
in early spring. Yellowish green, tassel-like flowers appear in
early spring. Acorns are 3/4-1", brown, slenderly conical, with
a thin, closely scaled cup covering 1/3 of the base of the nut,
from a sessile or short-stalked base, maturing in fall. Young
twigs are dull gray to reddish brown, with pale chestnut-colored buds, somewhat downy, with very short whitish hairs.
Bark is smooth, light grayish brown, frequently ashy white on
young trees and limbs, becoming thick and hard, dark brown
or blackish, and roughly furrowed with wide ridges on older
trees.

A native coastal oak that hybridizes with some other oaks.
Not native to interior valleys but widely planted in landscapes.
Occasionally naturalized along interior streams and rivers,
encroaching on the native *Quercus wislizeni*. Eventually becomes an impressive tree with handsome character for park
or street tree use, but requires ample room for the broad
canopy. Roots may heave paving. Newly planted trees benefit
from deep watering, but moderately drought tolerant when
established. Does well in lawns. Usually heavy leaf drop in
spring. Occasional moth larva infestations. Sudden oak death
has recently had a devastating effect on this species, raising
concern for other species that may be infected or act as carri-
ers of this viral infection. Longevity estimated to be 100-200
years. Series associations: Calbucwhisag, Pursage, Wooman-
zanita shrub series; Blaoak, Bluoak, Calbay, Calwalnut,
Coalivoak, Coupin, Coupin/Canlivoak, Engoak, Foopine,
Mixoak, Monpine, Tanoak, Valoak, Antdunsta.

e N.A. SHD, FAL CLR; IRR: No WUCOLS

Quercus alba

WHITE OAK

Fagaceae. No Sunset zones. Deciduous. Native to southeastern Canada and the eastern U.S. west to Iowa and eastern Texas. Growth rate moderate to 80-100' tall with a 30-45' spread, developing an upright oval canopy with a large trunk and heavy rising limbs. Leaves are alternate, simple, 5-9" long by 2-3" wide, shiny green, oblong to ovate, with 7-9, rarely 5, deep, rounded lobes, rounded at the apex, lighter undersides, slightly downy at first, new growth purplish-tinted, turning deep red to orange in late fall. Insignificant yellowish green, tassel-like flowers occur in early spring with new leaves. Acorns are light brown, oblong, 1/2-3/4" long, sessile or short-stalked, maturing in fall of the first season, the cup with thickened warty scales covering 1/4 of the nut. Twigs are reddish and rather stout with reddish brown, glabrous, rounded buds. Bark is light ashy gray, peeling in irregular, thin, vertical plates to expose lighter reddish brown underbark. Older trees sometimes have furrowed bark with narrow ridges near the base of the trunk.

A tall shade, street, or park lawn tree, most commonly cultivated on the east coast but occasionally found on the west coast. Distinctive foliage and a clean appearance. Deep-rooted with wide-spreading subsurface side roots. Needs adequate space. A valuable source of hard-grained wood for fine furniture and oak casks. Longevity estimated to be 500-600 years in habitat.

sw N.A. SHD; IRR: No WUCOLS

Quercus arizonica

ARIZONA WHITE OAK

Fagaceae. No Sunset zones. Evergreen. Native to dry slopes of the southwestern U.S. from central New Mexico and Arizona into central Mexico. Growth rate slow to 30-50' tall and wide, developing a broad, round-topped, oval canopy, with a fairly thick trunk, upward-arching limbs, and many slender, twiggy branchlets. Leaves are alternate, simple, 1-4" long, dark bluish green, oblong-lanceolate to broadly ovate, with entire margins, or wavy, with sparse spiny teeth, recurving along the edges, pale, densely hairy undersides, persisting until new growth appears in spring. Insignificant yellow-green, tassel-like flowers occur in spring from axils of new growth. Acorns are dark brown, oblong to ovoid, sessile or short-stalked, 3/4-1" long, maturing in fall, the shallow cup with tight, slightly woolly, red-tipped scales covering 1/4 of the nut. Twigs are stout, reddish brown, slightly hairy at first, becoming glabrous. Bark is rather thick, ashy gray, with furrows and narrow, scaly ridges.

The most common live oak of the southwest, but rarely cultivated. Requires hot, dry conditions similar to its native habitat. Effective planted in loose, informal groves in dry settings or on foothill slopes. Provides filtered shade and evergreen screening with little care. Longevity estimated to be 100-250 years in habitat.

w N.A. California Native: CS, CH, FW, CC, ME, RW, PF (*CA-FP exc GV, e Dmtns*); EVG, SHD; (IRR: VL/L/L/L/-/-)

Quercus chrysolepis

CANYON LIVE OAK

Fagaceae. Sunset zones 3-11, 14-24, Jepson *DRN,SUN:**4,7**, **14-18,22-23**,24&IRR:1-3,8,**9**,10,11,**19-21**. Evergreen. Native to upper foothill canyons, valley-facing mountain slopes, and the inland side of the Coast Ranges from 1,000-5,000' elevation in northern California to southwestern Oregon, south to the northern border of Baja California, and east into southern Arizona and New Mexico. Growth rate slow to 20-60' tall and wide, with dense, upright growth, developing a broad, spreading, heavy-branched canopy, often wider than tall, usually from a massive, multi-trunked base. Leaves are alternate, simple, 1 to 3-1/2" long by 1/2-3/4" wide, dark green, stiff, glossy and smooth, variably lanceolate, varying from completely smooth edges on older trees to distinctly saw-toothed edges on vigorous young growth, with finely woolly, yellowish to light green undersides, later becoming glabrous and more bluish, persisting for 3-4 years. Insignificant cream-colored flower tassels occur in spring. Acorns are dark brown, ellipsoidal to ovate, 1 to 1-3/4" long, covered with fine, creamy, fuzzy hairs, with a thin sessile cup, often golden, hairy to woolly, tightly scaled, covering 1/6-1/4 of the nut, maturing in 1 season. Young twigs are dark reddish brown with fine woolly hairs. Whitish young bark turns dark brown to black with squared "scales" on older trees.

A widely distributed, attractive evergreen native oak, casting dense shade, rarely cultivated but often found in low foothill canyons. Tolerates moderate heat and drought when established. Longevity estimated to be 250-300 years in habitat. Series associations: Canlivoakshr, Deerbrush, Haiceanothus, Intlivoakshr, Intlivoakcan, Intlivoakcha, Intlivoakscr, Ionmanzanita, Whimanzanita shrub series, BigDoufir, BigDoufircan, Blaoak, Calbay, Canlivoak, Coupine, Coupincanliv, Doufir, Doufirponpin, Doufirtan, Fanpalm, Incced, Jefpine, Knopine, Mixconifer, Moujuniper, Orewhioak, Ponpine, SanLucfir, Sinpinyon, Tanoak, Whifir, SanCrucypsta, Twopinsta.

e N.A. SHD, FAL CLR; IRR: M/M/-/M/-/-

Quercus coccinea

SCARLET OAK

Fagaceae. Sunset zones 2-10, 14-24. Deciduous. Native to the eastern U.S. Growth rate moderate to 60-80' tall and 40-60' wide, developing an upright, broadly oval canopy, with light, openly horizontal branching. Leaves are alternate, simple, 4-6" long, bright glossy green, obovate to oval, with 5-9 deep lobes, bristle-tipped ends, wide circular sinuses between, pale glabrous undersides, and bright scarlet red fall color deepest in colder areas. Insignificant yellowish tan, hanging flower tassels occur in spring, with females from axils of emerging yellowish new leaves. Acorns are reddish brown, oval, sessile or very short-stalked, 3/4" long, with rounded ends, and a short, blunt tip, maturing in fall of the first season, cups with tightly spaced, lustrous scales covering 1/3-1/2 of the nut. Twigs are slender, reddish brown, and glaucous. Young bark is smooth with a whitish cast, becoming grayish brown and furrowed, with narrow vertical ridges in age.

A desirable, deep-rooted street, park, parkway, or lawn shade tree, and one of the more commonly used ornamental oaks, with exceptionally fine fall color in cooler climates. Deep watering promotes establishment and more vigorous growth. Does best with moderate moisture. Longevity estimated to be 200-450 years.

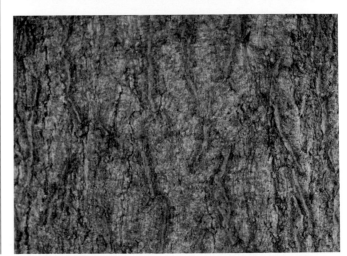

w N.A. (Ca only) California Native: VG, OW, FW (*NCoRI, CaRF, SNF, Teh, n SnJV, SnFrB, SCoRI, WTR n slope*); SHD, NAT; (IRR: VL/VL/VL/L/-/-)

Quercus douglasii

BLUE OAK

Fagaceae. Sunset zones 3-11, 14-24, Jepson *DRN,SUN:1-3, 7,8,**9**,11,**14-16,18-21**,22-24&DRY:4-6,10,17. Deciduous. Native to California's Central Valley and foothills. Growth rate slow to 30-50' tall or more with a 40-70' spread, developing a broad, low-branching, rounded canopy, either leaning or slightly bent, and a heavy clear trunk. Leaves are alternate, simple, 1 to 1-1/2" long by 1/2-3/4" wide, dull bluish green, variable from broad oval to almost squarish, with scalloped edges, blunt, often bristle-tipped ends, sparsely covered with minute star-shaped hairs, often with harmless small, red, gall-like warts on the uppersides, lighter undersides, with very fine tiny soft hairs at midveins and branches, and pastel pinkish orange or dull yellow fall color. Leaves are smaller and bluer than *Quercus lobata*, which grows alongside. Yellowish green, tassel-like flowers occur in early spring. Acorns are deep brown, conical, 1/2 to 1-5/8" long by 3/8-3/4" wide, with a rounded, short-tipped end, a thin scaled cup covering 1/8-1/4 of the base of the nut, from a short-stalked base, maturing in fall of the first season, and occurring in profusion on healthy trees. Brittle young twigs are dull gray to reddish brown with a slight minute hairiness. Bark is rather thin, light ashy gray with narrow ridges, peeling off easily in thin flakes.

A highly desirable native oak that tolerates valley and foothill heat and seasonal drought. Sensitive to any disturbance within the dripline. Naturally occurring trees should not be irrigated, and existing leaf mulch should be left in place. Slow to mature but often planted to replace trees removed, either from containers or as liner seedlings or acorns. Longevity estimated to be 200-300 years. Series associations: Nodneedlegra herbaceous series; **Bluoak**, Coalivoak, Foopine, Intlivoak, McNcypress, Mixoak, Valoak, Piucypsta.

sw N.A. California Native: CS, CH (*SCo*); EVG, SHD; (IRR: VL/VL/VL/VL/L/-)

Quercus dumosa

NUTTALL'S SCRUB OAK

Fagaceae. No Sunset zones. Evergreen. Native to central California and Baja California, most common in the Coast Ranges, coastal islands, and the San Bernardino Mountains of southern California. Growth rate slow to 6-10' tall and wide, forming an irregular oval canopy from multiple or low-branching trunks. Dense, fine, stiff, twiggy growth at branch ends becomes nearly impenetrable, with branching low to the ground, unless pruned to expose irregularly gnarled small trunks. Leaves are alternate, simple, 5/8 to 1" long, shiny, dark green, often coated with whitish hairs, elliptical to ovate, with sparsely to densely spiny-toothed crinkled margins slightly lobed and curved under slightly. Insignificant yellowish green, tassel-like flowers occur in early spring. Acorns are brown, conical to oblong, 1/2 to 1-3/4" long by 1/2-5/8" wide, sharply tapering to a rounded short-tipped end, with a thin scaled cup covering 1/6-1/3 of the base of the nut, a sessile base, maturing in fall of the first season. Brittle young twigs are dark brown. Bark is light ashy gray and rather thin, with shallow cross-checked fissures, peeling in thin plates or broad strips.

Uncommon in its native habitat and seldom cultivated. Deep, strong roots penetrate and hold in rocky soils and crevices. Often sprouts from underground roots, providing cover for exposed dry locations. Adaptable to dry, well-drained soils in a sunny location with occasional deep watering to become established. Longevity estimated to be 90-100 years. Series association: Torpinsta.

sw N.A. EVG, SHD; IRR: No WUCOLS

Quercus emoryi

EMORY OAK

Fagaceae. Sunset zones 10-13. Evergreen. Native to dry foot-hill and mountain slopes of southeastern Arizona to south-western New Mexico, western Texas, and northwestern Mexico. Growth rate slow to 50' tall and 40' wide, developing an oval, round-topped form with stout branches drooping to the ground. Leaves are alternate, simple, 1-1/2 to 3" long, shiny, dark green, leathery, oblong-lanceolate, margins entire or sparsely toothed, edges slightly wavy, with an acute tip, pale, glabrous undersides only slightly hairy if at all, persisting 1 year until new leaves develop, emerging leaves reddish, appearing velvety. Insignificant yellowish green flower tassels occur in spring. Acorns are brown to black, oblong to ovoid, sessile or short-stalked, 1/2 to 3/4" long, with a cup covering 1/2 to 1/3 of the nut, with brown hairy scales, also hairy on the inner surface of the cup, maturing in fall of the first year. Twigs are slender, reddish, hairy at first, with small, pointed brown buds also somewhat hairy. Bark is black or whitish-cast, thickening and developing deep vertical furrows, scaly plated on older trees.

A handsome small evergreen oak with distinctive leaves, rarely cultivated but sometimes seen in arboretums, where it makes an interesting specimen. Tolerates heat and dry soils in its native habitat, often hybridizing with other species. Longevity estimated to be 150 years or more in habitat.

sw N.A. California Native: (s Calif.) CH, OW (*SCo, s ChI Santa Catalina Island, SnGb, PR*); EVG, SHD, NAT; (IRR: -/L/L/L/-/-)

Quercus engelmannii

ENGELMANN OAK

Fagaceae. Sunset zones 7-9, 14-24, Jepson *DRN,SUN: 3,5,7, **14-16**,17,**18-21**,22-24&IRR:8,9. Semi-deciduous to evergreen. Native to low, dry hills and coastal regions of southern and northern extremities of Baja California. Growth rate slow to 40-50' tall or more with a spread sometimes twice as wide, developing a round-topped, oval canopy with short, medium-sized, upright trunks usually clear at the base. Leaves are alternate, simple, 1-3/4 to 4" long, thick, deep bluish green with a dull finish, elliptical, variably edged with sparsely to densely dentate, slightly recurved margins, slightly rounded or pointed ends, the smooth upper surface sometimes having sparse, star-shaped hairs, lighter undersides minutely hairy. New twigs are reddish brown, covered with minute hairs that gradually disappear. Insignificant, sparse, yellowish green, tassel-like flowers occur in spring. Acorns are brown, rounded, 1 to 1-1/4" long by 1/2-3/4" wide, short-stalked to sessile, with tight scales with sharp hairy points, on cups covering 1/4 of the shiny nut, which may be minutely hairy. Bark is pale grayish brown, deeply furrowed with wide, flat ridges, peeling in narrow strips.

Prized for its bluish cast in its native southern California habitat, where it often grows with coast live oak (*Quercus agrifolia*) and often hybridizes with Nuttall's scrub oak (*Q. dumosa*). Tolerates heat. Adaptable to drier locations in parks and higher, well-drained areas in riparian or semi-arid settings. Longevity estimated to be 50-60 years or more. Series associations: Coalivoak, **Engoak**.

sw N.A. SHD, NAT, SPC; IRR: No WUCOLS

Quercus gambelii

ROCKY MOUNTAIN WHITE OAK

Fagaceae. Sunset zones 1-3, 10. Deciduous. Native to dry foot-hills and river canyons in the southwestern U.S. from Nevada to Colorado south to the Mexican border and southwestern Texas. Growth rate slow to 15-30' tall (rarely 50') and half as wide, developing a shrublike or low, broadly rounded dome shape, often forming thickets in its native habitat, where it is a tenacious grower, often suckering from spreading surface roots. Leaves are alternate, simple, 2-1/2 to 7" long, thick, dark green, oblong to obovate, with 7-9 deep rounded lobes, pale, hairy undersides, and yellow to orange or red fall color. Insignifi-cant yellowish green, drooping, tassel-like flowers occur in spring. Acorns are brown, ovoid, 2/3-3/4" long, sessile or very short-stalked, with a thick-scaled hairy cup covering 1/4-1/2 of the base of the nut, maturing in fall of the first season. Twigs are stout, reddish brown, and hairy. Bark is light ashy gray, peeling in irregular thin, vertical plates.

An aggressive, small, drought-tolerant western native oak, the only abundant oak of the southern Rocky Mountains and Colorado. Rarely cultivated outside its habitat, though often found in arboretums, which indicates adaptability. Longevity estimated to be 100 years before losing vigor.

w N.A. California Native: (n Calif.) OW, ME, CF (*NW, CaRF, SnFrB*); (IRR: No WUCOLS)

Quercus garryana

= *Quercus garryana* var. *garryana*

OREGON OAK

Fagaceae. Sunset zones 2-11, 14-23, Jepson *DRN,SUN:**4-6,15-17**&IRR or part SHD:1-3,**7**,8,9,**14,18**,19-23. Deciduous. Native to valleys, prairies, and dry, gravelly slopes from British Columbia to northern central California at near sea level to 4,000' elevation. Growth rate slow to 40-90' tall with a 30-60' spread, forming a broad, round-topped crown, with a short, heavy trunk, large, upright upper limbs, and horizontal, thinner lower limbs drooping somewhat at the ends. More shrublike in exposed coastal situations. Leaves are alternate, simple, 3-6" long, deep green with a bluish cast, oblong to ovate, 5-9 lobed with deep, nearly closed indentations between wider lobes, often indented at the ends, smooth uppersides, pale undersides, petioles minutely hairy, with a dull orange to yellow-brown fall color. Insignificant yellowish green, tassel-like flowers occur in early spring. Acorns are 1 to 1-3/4", brown, conical, bulging slightly at the sides, broadest at the base, with a rounded, short-tipped end, a thickened, hairy-scaled cup covering 1/4 of the base of the nut, a sessile or short-stalked base, maturing in fall of the first season. Young twigs are hairy, and stout buds are covered with pale rust-colored hairs. Bark is scaly, light gray-brown with wide ridges and narrow, shallow furrows, similar to *Quercus alba*.

Somewhat adaptable outside its range, which begins where *Quercus lobata* terminates in the northern part of the state, becoming common in Oregon. Tolerates inland conditions of the Central Valley, but does not reach great size there. Benefits from deep watering until established, and seems adaptable to slightly moist sites. Longevity estimated to be 250-350 years in habitat. Series associations: Caloatgrass, Breoak herbaceous & shrub series; Blaoak, Canlivoak, Doufir, Doufirponpin, Mixoak, **Orewhioak**, Wesjuniper.

Mediterranean. EVG, SHD; IRR: L/L/L/L/M/M

Quercus ilex

HOLLY OAK

Fagaceae. Sunset zones 4-24. Evergreen. Native to the Mediterranean. Growth rate moderate to 30-60' tall and wide, developing a dense, oval, round-topped form from a heavy trunk with thick ascending limbs and twiggy ends. Leaves are alternate, simple, 1-1/2 to 3" long by 1/2-1" wide, dark glossy green, elliptical to oblong-lanceolate, varying in size and shape with either smooth or sparsely toothed, often wavy edges, recurving slightly, tapering to slender pointed ends, new leaves often yellow or silvery tomentose on the underside, persisting 2 years. Insignificant yellowish, tassel-like flowers occur in late spring from axils of new leaves. Acorns are dark brown, 1/2-3/4" long, often profuse, on 1/4-1/2" long stems, with a tightly scaled cup slightly curved away from the nut at the top edge and covering 1/2-2/3 of the nut, maturing in late summer. Twigs are fuzzy and grayish. Bark is dark brown to nearly black, somewhat thin, with many cracks and fissures, breaking into a maze of small squarish plates and ridges.

An attractive, deeply rooted evergreen lawn, park, or street tree. Tolerates wind and coastal conditions as well as heat inland, where it does best with moderate moisture, often in lawns. Growth is stunted in drought. Tolerates clipping into hedges or other forms. Longevity estimated to be 100-200 years.

w N.A. California Native: CH, OW, PF (*CA-FP exc GV, SCo, ChI*); SHD, NAT; (IRR: L/M-/M/-/-)

Quercus kelloggii

CALIFORNIA BLACK OAK

Fagaceae. Sunset zones 5 (inland areas), 6, 7, 9, 14-21, Jepson *DRN:1,4,5,**6,15,16**,17&IRR or part SHD:2,3,**7**,8,9,**14-18**,19-21. Deciduous. Native to California and southern Oregon mountains at roughly 2,000 to 3,500' elevation. Growth rate initially slow, with faster growth as its deep root system develops, to 30-80' tall and wide with a large main trunk and vertically ascending branches, bowed at the ends, forming a rounded, oval canopy. Leaves are alternate, simple, 3-6" long by 2-3" wide, bright shiny green, obovate, deeply 5-7 lobed, with multiple bristle-tipped ends and narrow sinuses, smooth uppersides sometimes covered with minute star-shaped hairs, duller, lighter undersides smooth or minutely hairy, new leaves pinkish red, and fall color deep yellow to dull orange. Pendulous yellowish green flower tassels occur in spring. Acorns are dark tan to brown, ovate, 1 to 1-1/2" long by 1/2-3/4" wide, maturing in fall of the second year, with tawny brown, short-stalked cups with shiny, elongated, thin, flat scales, thickened at the base, covering 1/2-1/3 of the nut. Twigs are reddish brown, smooth, minutely hairy, or with a whitish tinge, and prominent scaly buds, hairy at the ends. Bark is dull grayish brown, hardening and becoming dark brown, roughly furrowed, with shallow seams near the base of the trunk.

A widespread native oak in the ponderosa pine belt. Orange fall color creates a spectacular scene mixed with green conifers of the lower Sierra Nevada. Seldom grown in lower valley locations, where dry summer heat stunts growth. Longevity estimated to be 175-300 years in habitat. Series associations: Breoak shrub series; Bigdoufir, Bigdoufircan, **Blaoak**, Canlivoak, Coalivoak, Coupine, Coupincanliv, Doufir, Doufirponpin, Doufirtan, Engoak, Foopine, Giasequoia, Intlivoak, Jefpine, Jefpinponpin, Mixconifer, Mixoak, Moujuniper, Orewhioak, Ponpine, Tanoak, Valoak, Wesjuniper, Whifir.

se N.A. SHD, FAL CLR; IRR: No WUCOLS

Quercus laurifolia

LAUREL OAK

Fagaceae. No Sunset zones. Semi-deciduous. Native to the southeastern U.S. from Louisiana to Florida and the Tennessee seaboard. Growth rate moderate to 30-40' tall with equal or greater spread when grown in California, forming an upright, broad, round-topped canopy, a short clear trunk, and upward-pointing branches with twiggy ends. Leaves are alternate, simple, 2-4" long, glossy bright green, elliptical to oblong-lanceolate, entire or slightly wavy edges, occasionally shallowly 3-lobed near the bristle-tipped or rounded apex, glabrous lighter undersides, and dull yellow color through winter, leaves dropping in spring as new leaves appear. Insignificant yellowish green, tassel-like flowers occur in late spring. Acorns are small, dark brown to nearly black, 1/2" long, maturing in fall of the second season, with a shallow cup with thin, reddish brown, hairy scales. Bark is dark grayish brown, becoming deeply furrowed, with broad flat ridges and reddish tones in the furrows.

A good park, shade, or street tree, occasionally seen but not commonly used. Tall and leggy if planted closely or in shaded areas. Prefers moderate moisture and may develop some surface rooting if not deeply watered. Longevity estimated to be 150-200 years.

w N.A. (Ca only) California Native: VG, CH, OW, FW (*NCoR, CaRF, SNF, Teh, GV, SnFrB, SCoR, nw SCo, ChI [Santa Cruz, Santa Catalina Is] WTR, w SnGb*); SHD, NAT; (IRR: L/L/-/M/-/-)

Quercus lobata

VALLEY OAK

Fagaceae. Sunset zones 3-9, 11-24, Jepson *SUN:**4-6,14-16**,17 &IRR,DRN:1-3,**7-9,18-21**,22-24. Deciduous. Native to open valleys, Sierra Nevada foothills, and inland Coast Ranges of California. Growth rate slow to 70' tall or more with an equal or greater spread, developing a broad, oval, round-topped form with a massive trunk, large, arching limbs, and somewhat twisted, irregular outer branches, drooping and twiggy at the ends. Leaves are alternate, simple, 3-4" long, deep green to grayish green, oblong to ovate, varying in form, with 7-11 deep rounded lobes, minute star-shaped hairs on the uppersides, minutely hairy on the pale undersides and petioles. Insignificant, pendulous, yellowish green flower tassels occur in spring. Acorns are bright chestnut brown, 1-1/4 to 2-1/4" long, elongated and conical, often profuse, maturing in fall of the first year, the cup covering 1/3 of the nut, and the thin overlapping scales slightly hairy, thickened at the base and free at the tips. Twigs are reddish brown with small, tightly scaled round buds. Bark is light grayish brown, becoming deeply furrowed or broken into small square plates.

A handsome large native oak that tolerates valley and foothill heat, alkaline soils, and seasonal drought in its native habitat. Sensitive to any disturbance within the dripline. Naturally occurring trees should not be irrigated, and existing leaf mulch should be left in place. Slow to mature but often planted to replace trees removed, either from containers or as liner seedlings or acorns. Often has harmless, puffy, round oak gall balls. Longevity estimated to be 300-400 years or more. Series associations: Mexelderberr shrub series; Blaoak, Bluoak, Calsycamore, Mixoak, **Valoak**.

e N.A. SHD; IRR: No WUCOLS

Quercus macrocarpa

MOSSYCUP OAK

Fagaceae. Sunset zones 1-11, 14-23. Deciduous. Native to the northeastern and central U.S. and southeastern Canada. Growth rate moderate to 60-75' tall (150' in habitat) and half as wide in youth, equally wide at maturity, developing an upright, rounded canopy from a heavy trunk and thick, upward-arching limbs. Leaves are alternate, simple, 8 to 10" long, glossy dark green, oblong to ovate, broadest at the tip, with 5-9 rounded lobes, the 2 center lobes very deep, an indentation at the end of the central midrib, pale, hairy undersides, and dull orange-yellow fall color. Insignificant yellowish green flower tassels occur in spring. Acorns are green, 1-2" long, sessile or short-stalked, ellipsoidal, with a distinctive fringed scale cup covering 1/3 of the nut, maturing to brown in fall of the first season. Bark is thick, dark grayish brown, becoming deeply furrowed and ridged.

A tall, vigorous grower, occasionally seen as a large park or shade tree. Needs ample room. Deep-rooted, with wide-branching subsurface side roots. Well suited for large lawn areas with ample moisture. Tolerates moderate short-term drought when established. Longevity estimated to be 500 years.

w N.A. (Ca only) California Native: CH, OW, FW (*No region listing in* The Jepson Manual); SHD, NAT; (IRR: No WUCOLS)

Quercus morehus

ORACLE OAK

Fagaceae. No Sunset zones. Semi-evergreen. Native to upper foothill canyons, valley-facing mountain slopes, and the inland side of the Coast Ranges in scattered locations from 1,000-5,000' elevation from Lake and Marin counties east to mid-elevation western Sierra Nevada and south to Fresno County and the San Bernardino Mountains. Growth rate slow to 20-40' tall and wide, with fairly dense, upright growth, becoming broad-spreading and heavy-branched, usually multi-trunked, often wider than tall. Leaves are alternate, simple, 2-4" long by 1-2" wide, stiff, glossy, smooth, dark green, oblong to elliptical, with shallow spine-tipped lobes and rounded sinuses, finely hairy undersides, persisting 1 year. Insignificant cream-colored flower tassels occur in spring. Acorns are dark brown, ellipsoidal to ovate, 3/4 to 1-1/4" long, with a thin, tightly scaled, sessile cup covering 1/3-1/2 of the nut, maturing in fall of the second season. Young twigs and buds are dark reddish brown with fine woolly hairs. Bark is smooth, grayish brown, becoming dark brownish black, rough, fissured, and scaly.

A relatively uncommon, small, somewhat variable native oak, occurring in scattered locations, rarely in great quantity. Thought to be a naturally occurring hybrid between *Quercus kelloggii* and *Q. wislizeni*. Small, dense, spiny leaves make shrubby young trees nearly impenetrable. Trunks more visible on older trees. Longevity estimated to be 150-175 years in habitat.

n&c N.A. EVG, SHD; IRR: L/?/?/L/L/M

Quercus muehlenbergii

CHINQUAPIN OAK

Fagaceae. Sunset zones 2-12, 14-17. Deciduous. Native to dry soils of the northeastern and central U.S. from Maine to Alabama and Texas. Growth rate moderate to 40-50' tall or more by 50-60' wide, developing an upright, oval, irregularly round-topped canopy, with small, upward-arching or ascending branches. Leaves are alternate, simple, 4-7" long, shiny, yellowish green, obovate to oblong-lanceolate, with scalloped, or coarsely serrate edges with gland-tipped teeth, pale hairy undersides, and yellow to orange fall color. Pendulous, tassel-like, yellowish green flowers occur in spring. Sweet-tasting acorns are 1/2-3/4" long, sessile or very short-stalked, chestnut brown to black, ovoid, maturing in fall of the first season, the cup having small tight scales and covering 1/3-1/2 of the nut. Twigs are glabrous and slender, orange to gray-brown. Bark is ashy gray, with thin, loose, flaky scales.

A handsome, deep-rooted, small to medium-sized park or lawn tree with a high canopy. Rather uncommon and somewhat difficult to transplant or establish. Does best in well-drained soils with deep watering. Tolerates moderate alkalinity. Longevity estimated to be 100 years or more.

e Asia. EVG, SHD; IRR: No WUCOLS

Quercus myrsinifolia
JAPANESE LIVE OAK

Fagaceae. Sunset zones 4-7, 14-24. Evergreen. Native to southern China, Japan, Korea, and Laos. Growth rate slow to moderate to 30' (up to 50') tall and wide, with a dense, broad, oval canopy, resembling a hackberry from a distance, not readily recognizable as an oak until acorns are seen. Leaves are alternate, simple, 2-4" long by 1 to 1-1/2" wide, glossy light green, narrowly elliptical to lanceolate with a pointed tip, hairless, with a waxy feel, cuneate at the base, with finely serrate edges, and a glaucous, lighter underside, young leaves purplish red, slightly drooping at maturity, persisting into the second season after new leaves emerge. Yellowish tan, tassel-like flowers occur in spring from the axils of new leaves. Acorns are 1/4 to 3/4" long, ovoid to oblong, maturing in fall of the second year, with a shallow, tightly scaled, glabrous cup covering 1/3-1/2 of the nut. Bark is smooth, grayish tan, becoming slightly scaled and fissured on mature trees.

An uncommon, interesting small evergreen oak, attractive as a street tree or park specimen, shade tree, or in groves. Symmetrical, clean appearance. Requires little care. Does well with moderate moisture in valley heat, as in lawns. Longevity estimated to be 100-150 years in cultivation.

sw N.A. SHD; IRR: No WUCOLS

Quercus oblongifolia

MEXICAN BLUE OAK

Fagaceae. No Sunset zones. Evergreen. Native to dry foothill and mountain slopes in southeastern Arizona to southwestern New Mexico and northwestern Mexico. Growth rate slow to moderate to 30' tall and half as wide, with an upright, round-topped form. Leaves are alternate, simple, 1-2" long, blue-green, ovate to elliptical, with entire margins, edges rolled under, often with noticeable symmetrical indentations on each side toward the end, a blunt rounded tip, pale, glabrous undersides, persisting one year, until new leaves develop. Insignificant yellowish green flower tassels occur in spring. Acorns are bright chestnut brown, 1/2 to 3/4" long, ovoid to obovoid, sessile or short-stalked, maturing in fall of the first year, the cup covering 1/3 of the nut, with thin, overlapping, red-tipped, woolly scales. Twigs are slender, reddish gray, with small, blunt, brown buds, both devoid of hairs. Bark is thick, ashy gray with deep vertical furrows, exposing an orange underbark, breaking into small, square plates on older trees, with short horizontal cracking.

A handsome small evergreen oak with unusual leaves, rarely cultivated, usually seen only in arboretums or parks. Tolerates heat and dry soils. Effective in small groves. Tends to become tall and sparse at the base. Longevity estimated to be 100-150 years or more in habitat.

sw N.A. California Native: CH, OW, JP (*e NCoRI, nw SnJV, SCoR, SnGb, ePR, DMtns*); (IRR: No WUCOLS)

Quercus palmeri

PALMER'S OAK

Fagaceae. No Sunset zones. Jepson *DRN,SUN:**7**,8,**9**,10,**14-16**,17,**18-23**,24&IRR:11;STBL. Evergreen. Native to dry foothill canyons and mountain slopes of southern California into northern Arizona and southwestern New Mexico south to northern Baja California. Growth rate slow to 10-25' tall and wide, forming an irregular oval canopy, usually multi-trunked, with dense, stiff, twiggy, nearly impenetrable low branching, often to the ground unless pruned to expose the gnarled, irregular, slender trunks. Leaves are alternate, simple, 1/2 to 1-1/4" long, shiny dark green to gray-green, broadly elliptical, crinkled, with spiny-toothed margins, slightly recurved, and yellow-tinged undersides coated with whitish hairs. Sparse yellowish green, tassel-like flowers occur in early spring. Acorns are brown, conical to oblong, 3/4 to 1-1/4" long, short-stalked at the base, tapering to a rather pointed tip, with a shallow, flat, thin, golden-scaled cup covering 1/4 of the base of the nut, maturing in fall of the second season. Young twigs are stiff, hairy, and dark brown. Bark is dark gray-brown to black, scaly, rather thin, with shallow cross-checked fissures, becoming cracked and peeling in thin plates.

Closely related to *Quercus chrysolepis*, but much smaller. Rarely cultivated, but important as a habitat tree in southern chaparral and pinyon-pine woodland. Becomes leggy and shapeless when shaded by larger trees. Previously referred to as *Q. dunnii*. Longevity estimated to be 100 years in habitat. Series associations: Cupceafreoak, Mixscroak, Scroak, Scroakbirmou, Scroakcha, Scroakchawhi shrub series.

ne N.A. SHD, FAL CLR; IRR: M/M/-/M/-/-

Quercus palustris

PIN OAK

Fagaceae. Sunset zones 2-10, 14-24. Deciduous. Native to the northeastern U.S. west to Missouri. Growth rate moderate to fast to 50-80' tall and 30-40' wide, developing a pyramidal form, narrow at first, maturing to a round-headed, open canopy with lower branches drooping to the ground. Leaves are alternate, simple, 4-10" long by 2-6" wide, dark glossy green, obovate, 5-7 lobed with wide, rounded sinuses, pointed, bristle-tipped ends, pale, glabrous undersides except for minute brownish tufted hairs at vein axils, and bright red fall color, turning brown and persisting through winter, often until early spring. Insignificant, yellowish green, tassel-like flowers occur in spring. Acorns are reddish brown, 1/2" long, oval, maturing in fall of the second season, with a thin shallow cup with tightly appressed, free-tipped scales covering 1/4-1/3 of the nut. Twigs are slender, reddish brown, glabrous. Bark is smooth, grayish brown, becoming shallowly furrowed, with narrow flat vertical ridges.

Commonly planted in parkways, parks, and roadsides. Does well with moderate moisture and good drainage. Fills in quickly. Fall color is attractive, but some dislike the persistence of brown leaves into winter. Lower branches that droop to the ground on younger trees usually are left until trees reach maturity. Does not respond well to pruning, which destroys the natural shape. May become chlorotic in alkaline soils. Relatively pest free. Longevity estimated to be 200 years or more.

e N.A. SHD, FAL CLR; IRR: No WUCOLS

Quercus phellos

WILLOW OAK

Fagaceae. Sunset zones 2-4, 6-16, 18-21. Deciduous to semi-deciduous. Native to the eastern U.S. in low wetlands near streams and swamps. Growth rate moderate to 50-90' tall and 30-50' wide, developing a pyramidal form that broadens to a spreading, rounded canopy, with loose, slender, horizontal branching irregular and often drooping to the ground. Leaves are alternate, simple, 2-5" long, narrow, glossy green, elliptical, blunt-ended, quite willowlike, yellow as they emerge, with yellow to reddish fall color, some leaves remaining through winter in warm climates. Insignificant, small, sparse, yellowish green, tassel-like flowers occur in spring from axils of new growth, with females at branch ends. Acorns are brown, oval, 1/2" long, sessile or short-stalked, with tightly appressed scales on the cup, which covers nearly 1/2 of the nut. Twigs are greenish gray. Bark is grayish or reddish brown, rather smooth at first, becoming shallowly fissured, later furrowed, with broad shallow ridges.

Appealing as a specimen tree, delicate in texture with showy fall color. Sometimes grown as a large lawn or parkway tree, with good contrast to other broadleaf species. Should be planted where low branches will not be a problem. Relatively pest free. Longevity estimated to be 150 years or more in habitat.

e Asia. SHD, FAL CLR; IRR: No WUCOLS

Quercus phillyreoides

UBAME OAK

Fagaceae. No Sunset zones. Evergreen. Native to China, Korea, and Japan. Growth rate moderate to 10-15' tall (possibly 30') with an equal spread, pyramidal in youth, developing a spreading, rounded canopy in age, with loose, slender, horizontal branching rather irregular and drooping to the ground. Leaves are alternate, simple, 1-3" long, nearly as wide, bronzy to maroon as they emerge, becoming leathery, dark glossy green, elliptical to obovate, and inconspicuously toothed near the ends, which may be pointed or blunt. Insignificant, small, sparse, yellowish green, tassel-like flowers occur in spring from axils of new growth, with females near branch ends. Acorns are oval, brown, 1/2-3/4" long, sessile or short-stalked, with a woolly white cup covering 1/3-1/2 of the nut. Twigs are olive green, covered with small, scurfy, beige scales. Bark is rather smooth at first, becoming fissured, with broad, irregular flat ridges.

A small tree with unusual texture, resembling *Pittosporum*. Attractive as a specimen or background in contrast to other broadleaf species. Should be planted where low branches will not be a problem. Relatively pest free. Longevity estimated to be 100-150 years or more.

Europe, n Africa, w Asia. SHD; IRR: No WUCOLS

Quercus robur

ENGLISH OAK

Fagaceae. Sunset zones 1-12, 14-21. Deciduous. Native to Europe, northern Africa, and western Asia. Growth rate moderate to 50-60' tall with a 30' spread, developing an irregular, broad canopy, heavy, upward-arching limbs, and small twiggy branchlets. Leaves are alternate, simple, 3-4" long by 1-2" wide, oval, dark green, with little or no leaf stem, distinctly scalloped edges, little fall color, briefly turning a dull yellow-orange, dried brown leaves often remaining well into winter. Acorns are dark brown, oval, 3/4 to 1-1/4"long, with flattened rounded ends, a short blunt nib, drooping from branchlets on slender, glabrous stems, cup scales having a fat, warty surface and covering nearly 1/2 of the nut, maturing in fall. Bark is dark brown, furrowed, with shallow, flat, peeling ridgeplates.

Commonly cultivated around the world. Older trees are quite impressive. Less commonly grown on the west coast, but useful as a park, lawn, or shade tree. Does best with moderate deep watering. Relatively easy to grow. Noted for longevity, which is estimated to be 500-600 years or more. 'Ficifolia' has filigree leaf margins, 'Pendula' has weeping branches, 'Purpurascens' has purplish new growth, 'Atropurpurea' has dark purple leaves, and 'Variegata' has white-edged or variegated leaves.

Cultivar. SHD, VERT ACC; IRR: No WUCOLS

Quercus robur 'Fastigiata'

UPRIGHT ENGLISH OAK

Fagaceae. Sunset zones 1-12, 14-21. Deciduous. Variety of the species. Growth rate moderate to fast to 50' tall with a 15' spread, developing an upright columnar form, like a Lombardy poplar (*Populus nigra* 'Italica'), with many small, twiggy side branches from the main trunk, pointing upward, often slightly contorted, lower branches near the ground, becoming broader-based and pyramidal with age. Leaves are alternate, simple, 3-4" long by 1-2" wide, oval, dark green, with little or no stem, distinctly scalloped edges, persisting well into winter, with dull orange-yellow to tan fall color. Acorns are dark brown, oval, 3/4 to 1-1/4" long, with flattened, rounded ends and a short blunt nib, drooping from branchlets by long, slender, glabrous stems, cup scales having a fat warty surface and covering nearly 1/2 of the nut, maturing in fall. Bark is dark brown, furrowed, with shallow, flat, peeling ridgeplates.

An uncommon upright oak, effective in parkways, as a background screen, or in an allée along an entry drive. Slower growing than Lombardy poplar, but longer lived. Requires little care once established. Moderate moisture in good soils promotes sturdy growth and good form. Longevity estimated to be at least 300 years.

c & ne N.A. SHD, FAL CLR; IRR: M/M/-/M/-/-

Quercus rubra

RED OAK

Fagaceae. Sunset zones 1-10, 14-21. Deciduous. Native to the central and northeastern U.S. into southern Canada as far as Nova Scotia. Growth rate moderate to fast to 60-75' tall and 50' wide, forming a broad oval canopy and a symmetrically rounded shape with branches ascending from a large central trunk. Leaves are alternate, simple, 5-8" long by 3-5" wide, glossy dark green, ovate-oblong, deeply 5-7 lobed with bristle-tipped, pointed ends and rounded sinuses, pale, glabrous undersides, new growth reddish or yellow, and deep orange to red or ruddy brown fall color. Insignificant, yellow, tassel-like flowers occur in spring as new leaves emerge, from leaf axils at branch ends, with reddish females at terminal branch ends. Acorns are pale brown, 1/2-1" long, ovoid, sessile or short-stalked, with a tightly scaled cup covering 1/3 of the nut, maturing in fall of the second season. Twigs are stout, shiny, reddish brown. Bark is thick, dark brown to nearly black, developing shallow furrows and wide, flat ridges, with reddish inner bark showing within the splitting seams.

A desirable large-scale oak with consistent upright branching commonly used for streets, parks, or as a lawn shade tree. Deep-rooted with heavy subsurface lateral roots, which require room to spread. Requires deep watering and well-drained soil to become established. Initial growth very slow, within 10 years beginning to develop a sizable trunk and canopy suggesting eventual size. Longevity estimated to be 500 years or more.

sw N.A. SHD, SPC; IRR: No WUCOLS

Quercus rugosa

NETLEAF OAK

Fagaceae. No Sunset zones. Evergreen. Native to dry foothill and mountain slopes in southern Mexico north to central Arizona, southwestern New Mexico, and western Texas. Growth rate slow to 30-40' tall and half as wide, with an oval, upright, round-topped form and stout branches. Leaves are alternate, simple, 1-1/2 to 3" long, dark green and leathery, broadly ovate, margins entire or sparsely toothed, slightly wavy along the edges, a broad rounded tip, veins recessed on the upperside and raised on the underside, which is pale with yellowish hairs. Leaves persist one year, until rusty reddish orange new leaves develop in spring. Insignificant, yellowish green flower tassels occur in spring. Acorns are brown, oblong to ovoid, 3/4-1" long, in long-stalked clusters of up to 10, sparsely spaced, though many do not mature, and a cup with loose-tipped, fuzzy, reddish brown scales covering 1/3 to 1/2 of the nut. Twigs are slender, reddish, hairy at first, with small, pointed, brown, somewhat hairy buds. Bark is brown or with a whitish cast, thickening and becoming finely scaly-plated on older trees.

An unusual and uncommon evergreen oak, rarely cultivated, though sometimes seen in arboretums, where it makes an interesting specimen. Tolerates heat and dry soils. Adaptable to dry native garden settings. Longevity estimated to be 150 years or more in habitat.

sw Europe, n Africa. EVG, SHD; IRR: L/L/L/L/L/L

Quercus suber

CORK OAK

Fagaceae. Sunset zones 5-7 (with occasional winter damage), 8-16, 18-24. Evergreen. Native to Spain, Portugal, Algeria, and northern Africa. Growth rate moderate to 30-60' tall or more with an equal spread, forming a large dome canopy. Leaves are alternate, simple, 1-3" long by 1-1/2" wide, shiny dark green, elliptical to oblong-ovate, with variable wavy, entire, lobed, or sparsely toothed edges, gray-green, hairy, felted undersides, persisting 2-3 years. Insignificant, tiny, yellowish green, tassel-like flowers occur in late spring, with slender male catkins and short-stalked female clusters, from axils of current leaves. Acorns are 1 to 1-1/4" long, rounded, the fringed cups with heavy recurving scales covering 1/3-1/2 of the nut. Twigs are gray-green, tomentose. Bark is decorative, fissured, light tan, corky, porous, and the source of commercial cork for wine bottles. On trees grown for this purpose, bark is stripped every 8-10 years and usually grows back within that timespan.

Commonly grown as a handsome garden, park, or street tree for its interesting bark and foliage, though leaves drop constantly throughout summer. Tolerates many soils, but may become chlorotic in alkaline soils, and needs good drainage. Tolerates heat and drought when established. Longevity estimated to be 300-500 years.

w N.A. California Native: (s. California Islands) OW (*ChI*); (IRR: L/
?/L/-/-/-)

Quercus tomentella

ISLAND OAK

Fagaceae. Sunset zones 7-9, 14-17, 19-24, Jepson *DRN:5,**15-17**&IRR:7,**14,22-24**&SHD:8,9,**19-21**. Evergreen. Native to dry, rocky to gravelly canyons and slopes of the Channel Islands of California and the coastline of Mexico. Growth rate moderate to 25-40' tall and wide in habitat (in cultivation occasionally to 60' tall with a 20-30' spread), with upright symmetrical growth and a tall columnar form. Leaves are alternate, simple, 3" long, thick, stiff, leathery, elliptical, dark glossy green on the uppersides, depressed at the veining, with lighter undersides densely pubescent with minute, star-shaped, jointed hairs, which also cover young twigs. Leaves have a short petiole, sparsely toothed edges that recurve slightly, and persist for about 2 years. Acorns are broadly ovate, 1 to 1-1/4" long by 1/2-3/4" wide, sessile to short-stalked, tapering to a rounded, short-tipped end, with tawny to whitish-tomentose, chestnut-colored cups covering 1/4 of the nut, maturing in fall of the second season. Bark is smooth, fissured, gray-brown, and forms broad, thin, scaly plates.

Limited in range and rarely cultivated but strikingly handsome with symmetrical upright growth and distinctive dark, shiny foliage. Adaptable to dry garden settings in temperate zones. Young trees especially attractive, with foliage remaining close to the ground. Casts dense shade. Does not require frequent watering once established. Longevity estimated to be 60-80 years or more. Series associations: **Isloak**.

sw N.A. California Native: (s Calif.) JP (*e DMtns New York Mtns*); (IRR: No WUCOLS)

Quercus turbinella

SHRUB LIVE OAK

Fagaceae. Sunset zones 2, 3, 7-24, Jepson *DRN:7,8,9,10,11,**14**, **18-21**&SUN:5,15-17,22-24:STBL. Evergreen. Native to dry foothill and mountain slopes in southern California and Colorado to southwest New Mexico and Baja California. Growth rate slow to 10' tall and wide, usually shrublike, forming thickets, with nearly impenetrable foliage. Leaves are alternate, simple, 1/2 to 1-1/4" long, thick, stiff, dull gray to bluish green with a whitish bloom, elliptical to oblong, margins distinctly and sparsely spiny-toothed and slightly wavy, rounded at the base, with pale and finely hairy undersides, persisting 1 year until new leaves develop. Insignificant, yellowish green flower tassels occur in spring. Acorns are brown, narrowly oblong, 1/2 to 3/4" long, sometimes with 2-3 on the 1/4 to 1-3/4" long stalks, and a shallow, hairy-scaled cup covering 1/4 to 1/3 of the nut, maturing in fall of the first year. Twigs are reddish brown and hairy. Gray bark thickens, becoming fissured and scaly on older trees.

A small, dense, evergreen oak, rarely cultivated but can tolerate dry garden conditions as a small specimen, with colorful bluish foliage trimmed up to expose trunks. Tolerates heat and dry soils in its limited native habitat, where it is important in providing soil cover. Longevity estimated to be 100 years or more in habitat. Series associations: Cupceafreoak, Jostree, Mixscroak, Scroak, Scroakbirmou, Scroakcha, Scroakchawhi shrub series; Caljuniper, Twopinsta.

e N.A. EVG, SHD; IRR: M/M/M/M/M/M

Quercus virginiana
SOUTHERN LIVE OAK

Fagaceae. Sunset zones 4-24. Evergreen to semi-deciduous, fully deciduous only in coldest climates. Native to the eastern U.S. from southern Texas and along the eastern seaboard to Pennsylvania. Growth rate moderate to fast to 40-80' tall with a broad-spreading, heavy-limbed crown up to twice as wide. Leaves are alternate, simple, 1-1/2 to 5" long, shiny dark green, elliptical to oblong-ovate, with entire wavy margins or with sparse teeth, blunt rounded ends, and pale, sometimes hairy, whitish undersides. Insignificant, yellowish, tassel-like flowers occur in spring. Acorns are dark brown, ellipsoidal, 3/4-1" long, hanging down singly or combined at the end of a slender, 1-5" long stalk, with a reddish brown, woolly-scaled cup covering 1/3-1/2 of the nut, maturing in fall of the first season. Twigs are slender, gray-brown, hairy. Bark is dark reddish brown, becoming rather thick, with shallow furrows and flat, scaly ridges.

A desirable evergreen oak often seen in parks as a shade tree or background screen or as a residential lawn specimen. Too broad to be used as a street tree in most urban settings, though effective in parkway groupings. Tolerates heat and some dryness, but looks best with moderate moisture. Casts dense shade, but can be pruned to develop a higher branching canopy. Deep-rooted, but roots resist confinement. Longevity estimated to be 300-500 years.

sw N.A. California Native: VG, CH, OW, FW, PF (*NCoR, CaRF, SNF, Teh, SCoR, SW exc ChI*); EVG, SHD, NAT; (IRR: VL/VL/VL/VL/M/-)

Quercus wislizeni

= *Quercus wislizeni* var. *wislizeni*

INTERIOR LIVE OAK

Fagaceae. Sunset zones 7-9, 14-16, 18-21, Jepson *DRN,SUN:4-6,7,**14-17,22-24**&IRR:8,**9,18-21**. Evergreen. Native to central California foothills from 1,000-5,000' elevation, less common in southern California, and extending sparsely into northern Mexico. Growth rate slow to moderate to 30-75' tall, much wider than tall, often multi-trunked, with a dense, rounded crown. Leaves are alternate, simple, 1-3" long, glossy dark green, lanceolate to broadly elliptical, smooth with no surface hairs except occasionally on the petiole, spiny or sparsely toothed edges slightly wavy but not curving under, persisting for 2 years and dropping in summer or fall before new spring leaves, with the present year's growth remaining. Sparse, yellow-green, tassel-like flowers occur in spring. Acorns are slender, brown, 1 to 1-1/2" long by 1/2" wide, sessile or short-stalked, oblong with a short-tipped end, and closely overlapping ciliate scales on a thin cup that covers 1/2 of the nut or less, maturing in fall of the second year. Bark is dark brown, thinly fissured, darkening to nearly black, with wide, thin, flat, interconnected scaly ridges.

An evergreen oak, common in its foothill woodland habitat and along streams in the Central Valley. Young trees may adapt to lawn irrigation in well-drained soils, but older trees generally show some sensitivity. Tolerates moderately shaded conditions, but may become rangy or sparsely foliaged in heavy shade. Longevity estimated to be 50-80 years or more. Series associations: Bigmanzanita, Canlivoakshr, Chamise, Chabigman. Chawhitethor, Cupceafreoak, Deerbrush, Eastmanzanita, Intlivoakshr, Intlivoakcan, Intlivoakcha, Intlivoakcha, Ionmanzanita, Mixscroak, Scroak, Scroakcha, Scroakchawhi shrub series; Bluoak, Calbay, Calbuckeye, Coupine, Foopine, **Intlivoak**, Jefpine, Knopine, McNcypress, Mixoak, Sarcypress.

South America. EVG, SHD, FLW ACC; IRR: No WUCOLS

Quillaja saponaria
SOAPBARK TREE

Rosaceae. Sunset zones 8, 9, 14-24. Evergreen. Native to Chile. Growth rate moderate, then slowing, to 30-45' (occasionally 60') tall and nearly half as wide, with a dense, upright, columnar to semi-arching canopy and a shrubby appearance unless foliage is trimmed up. Leaves are alternate, simple, 1-1/2 to 2" long, shiny dark green, somewhat leathery and waxy, oval-oblanceolate to elliptical, with lightly wavy, entire edges often sparsely toothed. Insignificant flower panicles occur in dense 1" clusters of pale, greenish yellow flowers, with a semi-woody calyx, recurving at the ends, and fleshy yellow at the base. Small yellow petals quickly shrivel, and have conspicuous 1/4" long stamens with fattened anthers at the ends. Light tan, woody, pinwheel-like seed pods, with 5 elongated, radially attached capsules are covered with tiny stiff hairs. Each capsule splits along one side to expose many tightly packed, flattened, flaky, 1/8" long, reddish brown seeds, each with a broad, rounded, 3/16" long wing. Most are infertile. Bark is smooth, grayish brown, with swirls and faint crack lines, thickening with age, becoming dark brown and scaly, with many small squarish plates.

An uncommon but interesting garden specimen or background tree with a light-textured, semi-weeping appearance. Well suited to drier native plantings, where it looks best with infrequent deep watering. Seed husks can litter paving but make good mulch in planted areas. Longevity estimated to be 80-100 years.

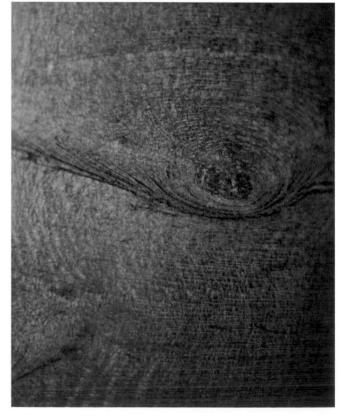

e N.A. SHD, FAL CLR, SPC; IRR: No WUCOLS Considered an invasive reseeder.

Rhus glabra

SMOOTH SUMAC

Anacardiaceae. Sunset zones 1-10, 14-17. Deciduous. Native to the eastern the U.S. and has become established sporadically throughout the west. Growth rate initially rapid, then slowing, to 10'-20' tall and about half as wide, with broad, horizontal branching from a single trunk or more broadly oval as a multi-trunk tree, commonly forming thickets and spreading by underground roots. Leaves are alternate, pinnately compound, 12-18" long, dull dark green, glabrous, with 11-31 pointed, 2-5" long, oblong to lanceolate leaflets with lightly serrate edges along a stiff central stem, drooping slightly in tufts at branch ends, with brilliant orange-scarlet fall color. Insignificant, greenish white flowers are dioecious or polygamous, regular, occurring in dense, hairy, 6-12" long compound terminal panicles, in late spring after new leaves appear. Fruits are compact, upright, conical clusters of 1/8-1/4" round, fleshy, red, hairy drupes, sour-tasting but edible, noticeable in summer, containing tiny, dark, hard-shelled seeds. Twigs are stout, glabrous, with a milky sap when cut. Bark is thin, smooth, gray, darkening and becoming scaly with age.

A relatively short-lived shrub or small accent tree useful for drier native plantings. Allelopathic chemicals may discourage growth of other plants. Requires little watering. Occurs in various locations throughout all 48 states, possibly through naturalization. Longevity estimated to be 25-40 years.

s Africa. EVG, SHD; IRR: L/L/L/L/M/M

Rhus lancea

AFRICAN SUMAC

Anacardiaceae. Sunset zones 8, 9, 12-24. Evergreen. Native to South Africa. Growth rate moderate to fast to 20-30' tall by 20-35' wide, with a dense, oval, spreading habit, often multi-trunked, with graceful, weeping branchlets and drooping foliage. Leaves are alternate, palmately compound, trifoliate, somewhat glossy or waxy, smooth, medium to dark green, and divided into 3 divergent, slender, willowlike leaflets, 3-4" long, with pale undersides. Insignificant clusters of yellowish flowers are followed by pea-sized, yellow-tan to reddish fruits in loose, long-stemmed, terminal clusters on female trees. Dustlike film covering reddish brown twigs eventually rubs off. Bark is rough, grayish brown, developing many fissures and becoming almost scaly or finely plated, exposing reddish orange underbark at the fissures.

Widely planted, especially in the southwest, as a medium-sized street, courtyard, or patio tree or screen with an airy, fine texture. Tolerates desert heat, and quite drought tolerant once established. Can be pruned up as a multi-trunk specimen, exposing attractive bark. Longevity estimated to be 60-75 years or more.

ne N.A. SHD, FAL CLR, SPC; IRR: L/L/L/?/L/-

Rhus typhina

STAGHORN SUMAC

Anacardiaceae. Sunset zones 1-10, 14-17. Deciduous. Native to the northeastern U.S. Growth rate slow to 15' (sometimes 30') tall, with broad, horizontal branching, usually a multi-trunk tree or large shrub with an irregular broad shape, often forming thickets as it spreads by suckering underground roots. Young branches have short, brown, velvety hairs, distinguishing it from *Rhus glabra*, which has smooth branches. Leaves are alternate, pinnately compound, 12-18" long, light green, with 2-5" long, oblong-lanceolate, pointed leaflets with serrate edges, along a stiff central stem, drooping slightly in tufts at branch ends, finely hairy on all surfaces, with brilliant orange-scarlet fall color. Insignificant greenish white flowers, dioecious or polygamous, regular, in dense, hairy, 6-12" long, compound terminal panicles, appear in late spring after new leaves. Fruits on female trees are dense, downy, upright, conical clusters of 1/8-1/4" round, fleshy, red, hairy drupes containing tiny, dark, hard-shelled seeds, noticeable in summer and showy in fall. Twigs are stout, hairy, with a milky sap when cut. Bark is dark reddish brown, becoming furrowed and scaly near the base.

A small, short-lived accent tree, useful in groupings and in drier native plantings, prized for its brilliant fall color, showy flower clusters, and twisted, contorted, fuzzy brown branching. Looks best with infrequent deep watering, but survives moderate drought. Most commonly cultivated in foothill and mountain settings, where it is hardy, with spectacular fall color in colder climates. Longevity estimated to be 25-40 years. 'Dissecta' (often sold as 'Laciniata') is smaller with finely cut, fernlike foliage.

c&e N.A. SHD, FLW ACC; IRR: L/L/L/L/L/L It is invasive, and its use is cautioned.

Robinia pseudoacacia

BLACK LOCUST

Leguminosae. Sunset zones 1-24. Deciduous. Native to central and eastern U.S. and naturalized throughout North America and elsewhere. Growth rate fast to 40-75' tall and 30-60' wide with an irregular, oval, open canopy. Leaves are alternate, pinnately compound, 10-16" long, dark blue-green, with 7-21 paired, ovate-oblong, glabrous, 1-2" leaflets along the main rib, lighter undersides, entire margins, a notched, spined tip, and yellow fall color. Showy clusters of sweetly fragrant, white, 1/2-3/4" long, papilionaceous, or pealike, perfect flowers in 4-8" long, hanging racemes occur in spring after new leaves. Seed pods are flattened, dark brown, 2-4" long, linear-oblong legumes, each containing 4-8 flat, brown, shiny, kidney-shaped seeds and hanging in clusters into winter. Twigs are reddish brown with 1/2" long, sharp spines, which persist until grown over by bark. Wood is dense and hard with reddish brown to grayish black bark, which develops deep fissures, with rounded scaly ridges.

Considered a weed tree, since it readily reseeds and spreads by root suckers, often becoming a pest if not controlled. Attractive flowers. Drought tolerant, but invades riparian areas and adjacent flatlands. Longevity estimated to be 50-80 years.

Cultivar. SHD, FLW ACC, FOL CLR; IRR: L/L/L/L/L/L

Robinia pseudoacacia 'Frisia'

GOLDEN LOCUST

Leguminosae. Sunset zones 1-24. Deciduous. Selected variety of the species. Growth rate fast and vigorous to 50' tall and 25' wide, forming an upright, open, oval canopy with upright to semi-arching branches, becoming twiggy at the ends. Leaves are alternate, pinnately compound, 12" long or more, yellow-green, with 7-10 paired, ovate-oblong, glabrous, 1-2" long leaflets along the main rib. Flowers are lightly fragrant, white, 1/2-3/4" long, pealike, in 8" long, hanging clusters, setting fewer seed pods than the species. Often develops red thorns.

Grown mostly for attractive, lush, yellow foliage, which holds its color throughout summer in full sun but may fade to green in shade.

Cultivar. SHD, FOL ACC, FLW ACC, SPC; IRR: L/L/L/L/L/L

Robinia pseudoacacia 'Tortuosa'

TWISTED LOCUST

Leguminosae. Sunset zones 1-24. Deciduous. Selected variety of uncertain origin, with the first known use in France in the 1800s. Growth rate slow to 50' tall and 30' wide, usually remaining a miniature-sized specimen for years, eventually developing a strong trunk and forming a broad, oval canopy. Branching is twisted, zigzag, semi-pendulous at the ends. Leaves are alternate, pinnately compound, 6-12" long, dark green, with 7-10 paired, ovate-oblong, glabrous, 1-2" long leaflets along the main rib, lighter undersides, and yellow fall color. Few, if any, noticeable white flowers or seed pods, usually on older trees only.

An unusual small specimen with multiple low branches as a small tree. Takes many years to reach significant size in a garden setting.

Cultivar. SHD, FLW ACC; IRR: L/L/L/L/M/M

Robinia x *ambigua* 'Idahoensis'

IDAHO PINK LOCUST

Leguminosae. Sunset zones 2-24. Deciduous. Selected variety developed in Utah and introduced about 1940. Growth rate moderate to rapid to 40' tall and 30' wide, with an upright, open, oval canopy and upright to semi-arching branches, becoming very twiggy at the ends. Leaves are alternate, pinnately compound, 12-18" long, dark green, with 9-15 paired, ovate-oblong, glabrous, 1-2" long leaflets along a main rib, with entire margins, lighter undersides, and yellow fall color. Flowers are lightly fragrant, bright pink to magenta rose, pealike, 1/2-3/4" long, in 8" long, hanging clusters, showy in spring after new leaves emerge. Flattened, dark brown legume seed pods are 2-4" long, linear-oblong, with 4-8 flat, brown, shiny, kidney-shaped seeds, hanging in clusters into winter. Twigs are hairy, reddish brown, with few if any spines, which are only on new growth and soon disappear. Wood is dense and hard. Bark is reddish brown to grayish black with deep fissures and rounded, scaly ridges.

A somewhat more desirable form of flowering locust, more adapted for use in street parkways and as a flowering accent tree. Tolerates heat and drought, and becomes well established with deep watering, but does not like to be overwatered. Longevity estimated to be 50-80 years.

Cultivar. SHD, FLW ACC; IRR: L/L/L/L/M/M Its use is cautioned, as it may become weedy.

Robinia x *ambigua* 'Purple Robe'

PP #2454 (1964)

PURPLE FLOWERING LOCUST

Leguminosae. Sunset zones 2-24. Deciduous. Patented hybrid between *Robinia ambigua* 'Decaisneana' and *R. hispida* 'Monument'. Growth rate fast to 40' tall and 30' wide, with an upright, open, oval canopy and upright to semi-arching branches, twiggy at the ends. Leaves are alternate, pinnately compound, 12-18" long, dark green, with 17-21 paired, ovate-oblong, glabrous, 1-2" long leaflets along a main rib, reddish bronze new growth, and yellow fall color.. Flowers are lightly fragrant, pea-like, dark purple to pink, 1/2-3/4" long, in 8" long, hanging clusters, showy in spring and occurring sporadically and sparsely throughout summer. Legume seed pods are hidden among leaves, 1-1/2 to 2-1/4" long by 1/2 to 3/4" wide, darkening to deep brown, with a sandpapery texture from dense, short, spiny hairs, linear-oblong, bulging around the 1-3 flat oval seeds, and tapering to a narrow tail at the end. Twigs are reddish brown with few if any spines, which occur only on juvenile sucker growth. Wood is dense and hard. Bark is reddish brown to gray, developing fissures and cracks with rounded, scaly ridges.

An attractive flowering accent tree. Tolerates heat and moderate drought. Overwatering produces rank, rangy growth. Pruning encourages lanky, vigorous new growth, often destroying natural branching structure. Wood is brittle, and limbs tend to break on older trees. May produce unwanted root-sucker growth. Reseeds easily and may become a nuisance. Longevity estimated to be 50-80 years.

e Asia. SHD, WPG ACC; IRR: H/H/H/H/H/H

Salix babylonica

WEEPING WILLOW

Salicaceae. Sunset zones 3-24. Deciduous. Native to China. Growth rate fast to 30-50' tall and wide or wider, forming a broad, weeping, round-headed crown, with a graceful outline and long, slender, wandlike branchlets that droop from heavy limbs. Leaves are alternate, simple, 3-6" long, shiny green, narrowly lanceolate to pointed ends, glabrous, with finely serrate edges, and deep yellow fall color. Flowers are dioecious, regular, yellowish green, catkinlike, in drooping aments in spring as new leaves appear. Seed capsules are one-celled, pointed, 2-valved, and contain tiny, dark brown seeds covered with long, silky hairs, easily carried in the slightest breeze, in early summer. Twigs are green or yellowish, turning reddish brown. Slender, drooping, yellowish to orange branches create an interesting weeping winter effect. Bark is thin, smooth, grayish green, darkening and thickening on mature trees, developing wide, shallow, dark brown, irregular furrows with grayish, flat, scaly ridges.

A classic waterside tree for parks and large lawns. Requires ample moisture. Roots are extremely invasive. Wood is quite brittle. Pruning and staking may be required on young trees to develop a strong trunk and branching structure, but older trees are best left unpruned. Commonly cultivated in the southeast, less commonly in the west, where it is used as a special feature tree. Longevity estimated to be 50-75 years.

sw&s N.A. California Native: CP, VG, OW (Riparian) (*NCoRI, CaRF, SNF, GV, SCo, PR, D – esp GV, D*); (IRR: H/H/H/H/H/H)

Salix gooddingii

GOODDING'S BLACK WILLOW

Salicaceae. No Sunset zones. Jepson *WET:5,6,**7-9**,10,11,**12-14**,15-17,**18-24**:STBL. Deciduous. Native to Kansas and southern Texas, west to California in the Central Valley along streams or wetlands from Mt. Shasta south to northern Mexico. Growth rate initially fast and vigorous, then slowing, to 20-40' tall, with a variable, irregular, upright to oval form and single or multiple low-forked trunks often leaning to one side. Leaves are alternate, simple, 2-6" long by 1/4-3/4" wide, narrowly lanceolate, shiny green, hairless, with finely serrate edges, pale undersides, conspicuous glandular stipules, and dull yellow fall color. Flowers are dioecious catkins, 1-3" long, with yellow, hairy or cottony scales, in late spring. Seed capsules are 3/16" long, reddish brown, maturing in early summer. Twigs are yellowish green, turning dark brown in fall. Bark is rough, grayish, with many forked furrows and flat, platy ridges.

The predominant large willow in otherwise treeless areas of the Central Valley, occurring in wet soils and riparian areas, where it is a valuable habitat tree. Sometimes used for revegetation and erosion control. Similar to *Salix nigra*, which occurs mostly in the south, occasionally in the southwest, but with paler yellowish gray twigs and female flowers sometimes slightly hairy. Longevity estimated to be 40-50 years in habitat. Series associations: Narwillow, Sanwillow shrub series; **Blawillow**, Blupalveriro, Calsycamore, Fanpalm, Hoowillow, Mixwillow, Pacwillow, Sitwillow.

w N.A. California Native: VG, OW, FW, PF (Riparian) (*CA*); (IRR: H/H/H/H/H/H)

Salix hindsiana

SANDBAR WILLOW

Salicaceae. No Sunset zones. Deciduous. Native to southwestern Oregon throughout California to Baja, usually along moist streambanks, sandbars, and ditches. Growth rate fast to 10-30' tall, often shrubby, with many slender trunks and forming thickets, only occasionally single-trunked. Leaves are alternate, simple, 1-1/2 to 3-1/2" long by 1/8-1/4" wide, dull gray-green, narrowly linear-lanceolate, nearly stalkless, covered with silvery gray silky hairs, usually with smooth edges, and yellow fall color. Insignificant, dioecious, male and female catkin flowers are 3/4 to 1-1/2" long, with yellow hairy or cottony scales, occurring in late spring after new leaves appear. Seed capsules are 1/4" long, light brown, densely hairy, maturing in early summer. Twigs are grayish to silvery, covered with white hairs when young, later becoming smooth and reddish brown. Bark is smooth, thin, gray-green, thickening and becoming brownish gray, scaly, and furrowed.

A useful riparian habitat tree or large shrub. Good soil binder along streams, but may become invasive and clog waterways. Soft gray-green foliage is colorful in the native landscape. Sometimes becomes large and treelike, the trunk usually hidden by lower foliage. Longevity estimated to be 40-50 years in habitat.

w N.A. California Native: CO, CM, CP (*Nco, NcoRO*); (IRR: H/H/H/H/H/H)

Salix hookeriana

HOOKER WILLOW

Salicaceae. No Sunset zones. Jespon *IRR or WET:**4,5**,6,7,**14-17**,19-21,**22-24**:STBL. Deciduous. Native to coastal lowlands from Alaska to northwestern California. Growth rate fast to 30' tall or more, often with many trunks and a broad, rounded crown. Leaves are alternate, simple, 1-1/2 to 4-1/2" long by 3/4-2" wide, dark green to yellowish green, somewhat leathery, broadly lanceolate, with whitish and hairy undersides, cordate at the base, with a stout, hairy, 1/4-3/8" long petiole, edges usually smooth or sparsely wavy toothed, and yellow fall color. Insignificant, dioecious, male and female catkin flowers are 3-4" long, with blackish scales, covered with long white hairs, occurring in midspring. Seed capsules are light brown, hairless, 1/4" long, in early summer. Twigs are stout, dark brown, brittle, densely covered with gray woolly hairs when young. Bark is thin, smooth, gray, thickening and becoming rough and scaly.

A common native in northern lowland coastal regions, either forming thickets among other willows or as large individual trees in grasslands bordering marsh areas. Does not often extend into nearby coniferous forests. Large, smooth leaves are quite distinctive. Longevity estimated to be 75-90 years in habitat. Series associations: Pacreedgrass, Sanverbeabur, Narwillow, Sanwillow shrub series; Blacottonwoo, Blawillow, **Hoowillow**, Mixwillow, Pacwillow, Redalder, Sitwillow.

w&sw N.A. California Native: VG, OW, FW, PF (Riparian) (*CA*); (IRR: H/H/H/H/H/H)

Salix laevigata

RED WILLOW

Salicaceae. No Sunset zones. Deciduous. Native to California along streams below 5,000' elevation west of the Sierra Nevada, from the Oregon border to southern California, and east to Arizona, Utah, and Nevada. Growth rate fast to 15-45' tall, forming a broad, billowy canopy, with foliage drooping to the ground, usually shrublike, with one or more large, thick trunks usually hidden by lower branches. Leaves are alternate, simple, 3-4" long, dark green, oblong-lanceolate, finely toothed, with pale undersides, long-pointed, tapered tips, rounded at the base, with conspicuous red-colored stipules near the end of the petioles, and yellow fall color. Flowers are male and female catkins on separate trees, 1-4" long, with glabrous scales, from March to May, producing many tiny brown seeds in late summer. Twigs are shiny, reddish to yellow-brown, slightly hairy at first, later becoming glabrous. Bark is rough, grayish, with dark brown furrows.

An interesting multi-trunk tree in riparian settings throughout the Central Valley and foothills. Longevity estimated to be 50-60 years. Series associations: Narwillow, Sanwillow shrub series; Blawillow, Calsycamore, **Redwillow,** Frecottonwoo, Hoowillow, Mixwillow, Pacwillow, Sitwillow.

w&sw N.A. California Native: VG, OW, FW, ME, CF, PF (Riparian) (*CA*); (IRR: H/H/H/H/H/H)

Salix lasiandra

= *Salix lucida* ssp. *lasiandra*

YELLOW WILLOW

Salicaceae. No Sunset zones. Deciduous. Native from British Columbia to New Mexico and throughout much of California along valley rivers and moist foothill and mountain streambanks. Growth rate fast to 15-45' tall, usually shrubby with many trunks, forming thickets, or with single, heavy, many-forked trunks and an open, rounded crown. Leaves are alternate, simple, 2-5" long by 1/2-1" wide, shiny green, narrow to oblong-lanceolate, long pointed, rounded at the base, with finely serrate edges, becoming nearly hairless, with whitish undersides, conspicuous glands at the upper end of leafstalks, and yellow fall color. Flowers are dioecious male and female catkins, 1-1/4 to 3" long, with shaggy, hairy, yellow or brown scales, occurring in late spring, followed by light brown, 1/4" long, hairless seed capsules in early summer. Twigs are shiny, reddish or yellow, hairless. Bark is dark grayish brown, becoming rough and deeply furrowed, with flat scaly ridges

A common riparian tree or large shrub, effective as a soil binder along streams. Rarely cultivated, but available for use in restoration plantings, often in the form of "wattles" or cuttings. Longevity estimated to be 40-50 years in habitat. Series associations: Buttonbush, Narwillow, Sanwillow shrub series; Blawillow, Hoowillow, Mixwillow, Pacwillow, Sitwillow, Watbirch; Monwetshrhab habitat.

w&sw N.A. California Native: CP, VG, OW, FW, ME, PF (Riparian) (*CA*); (IRR: H/H/H/H/H/H)

Salix lasiolepis

ARROYO WILLOW

Salicaceae. No Sunset zones. Jepson *WET,SUN:1-5,**6-9**,10-13,**14-24**;STBL; INV. Deciduous. Native throughout the California Coast Ranges, Central Valley, and Sierra Nevada foothills, including southern California, along rivers and streambanks, and from Washington and Idaho to western Nevada, New Mexico, and Texas into Mexico. Growth rate fast to 10-30' tall, shrubby, with many trunks, often forming thickets, or with a single, heavy, many-forked trunk, and an open, rounded crown. Leaves are alternate, simple, 2-5" long by 1/2-1" wide, shiny green, rather thick and leathery, oblanceolate, with glabrous uppersides, whitish pubescent and slightly hairy undersides, entire or with finely, barely serrate edges, without conspicuous glands at the upper end of leaf stalks, and yellow fall color. Flowers are dioecious male and female catkins, 1-2" long, with black to brown densely hairy scales, occurring in late spring, followed by light brown, 1/4" long, hairless seed capsules in early summer. Twigs are shiny, reddish or yellow, hairless. Bark is thin, pale, blotchy, grayish brown, darkening and becoming rough and deeply furrowed, with broad ridges.

A common and useful riparian habitat tree or large shrub, effective as soil binder along streams. Smaller than yellow willow (*Salix lasiandra*). Attractive trimmed up to expose smooth gray bark, which often has blotchy white colorations in areas that receive fog. Longevity estimated to be 40-50 years in habitat. Series associations: Mulefat shrub series; **Arrwillow**, Blawillow, Calsycamore, Fanpalm, Frecottonwoo, Mixwillow, Redalder, Watbirch; Monwetshrhab habitat.

w N.A. California Native: CM, CP, ME, CF, PF (Riparian) (*NW, CaRH, SN, n CCo, SnGb, SnBr, SnJt, MP*); (IRR: H/H/H/H/H/H)

Salix scouleriana

SCOULER'S WILLOW

Salicaceae. No Sunset zones. Jepson *IRR or WET,SUN:**1-3**,4-**6**,**7**,8-10,14,**15-18**,19-24;STBL. Deciduous. Native throughout much of western North America, in coastal and transitional mountain zones, though mostly in Canada, extending through California to New Mexico and east to Wyoming. Growth rate initially fast to 10-30' tall, compact, with a small rounded crown, or shrubby, often forming thickets, with many trunks, a single heavy trunk, or a many-forked trunk, branches twiggy at the ends. Leaves are alternate, simple, 1-1/4 to 2-1/2" long by 1/2 to 1-1/2" wide, yellow-green, narrow to oblanceolate to obovate, variable in shape, broadest near the end, blunt tipped, tapering to the base, with a short petiole, stipules usually lacking, entire margins or barely toothed, lighter, slightly hairy undersides, and yellow fall color. Flowers are dioecious male and female catkins, 1-2" long, with hairy black scales, in late spring, followed by light brown, 3/8" long, woolly seed capsules in early summer. Twigs are stout, shiny, reddish or yellow, hairy when young. Bark is thin, smooth, grayish brown, darkening and becoming dark brown, rough and deeply furrowed, with broad, flat ridges

A useful and attractive riparian habitat or lawn tree or large shrub. Good soil binder along streams. Longevity estimated to be 40-50 years in habitat. Series associations: Moualder, Sitalder shrub series; Blacottonwoo; Monwetshr habitat.

Cultivar. SHD, FOL ACC; IRR: H/H/H/H/H/H

Salix matsudana 'Tortuosa'

CORKSCREW WILLOW

Salicaceae. Sunset zones 3-11, 14-24. Deciduous. Selection of the species native to China. Growth rate moderate to fast to 30' tall with a 20' spread, forming an upright, oval canopy and twisted or contorted, upright, spiral branches with weeping ends. Leaves are alternate, simple, 2-4" long by 1/2" wide, shiny green, twisted, narrowly lanceolate and narrow pointed, glabrous, with finely serrate edges and slight yellow fall color. Insignificant, yellowish green, catkinlike flowers occur in drooping aments in spring as leaves appear, followed by one-celled, pointed, 2-valved capsules containing tiny dark brown seeds with long, silky hairs. Twigs are greenish, slender, drooping, with a somewhat irregular weeping effect in winter. Bark is thin, smooth, greenish, thickening and darkening to brown on mature trees, with wide, shallow, dark brown, irregular furrows and grayish, flat, scaly ridges.

A rather compact weeping willow, often grown as a lawn accent in parks or as a courtyard specimen. Effective in small groves, especially near water features. Tolerates valley heat with adequate moisture. Roots can be quite invasive. Interesting but odd appearance can be tiresome if overused. Longevity estimated to be 50-75 years.

w N.A. California Native: VG, OW, ME, PF, FF (Riparian) (*CA-FP, GB*); SHD, NAT; (IRR: L/L/L/L/M/M)

Sambucus mexicana

BLUE ELDERBERRY

Caprifoliaceae. Sunset zones 2-24, Jepson *4,5,**6,7,14-17,24**, IRR:1-3,**8,9**,10,**18-23**. Deciduous. Native to the western U.S. from the Rocky Mountains to British Columbia, usually in moist soils near streams in mountain and foothill canyons and open valleys. Growth rate initially fast to 10-30' tall by 8-20' wide, forming an irregular, broad-spreading canopy, with many trunks or with a large main trunk and many vigorous, thin branches covered by dense foliage. Leaves are opposite, pinnately compound, 5-8" long, leathery, dull green, divided into 5-9 leaflets, 1-6" long, ovate to narrowly oblong, with serrate margins, smooth uppersides, sparsely to densely hairy, pale undersides, and dull reddish to yellow fall color. Flowers are regular, perfect, small, creamy white, in large, flat, 10" round corymb clusters intermittently through late spring and summer, often simultaneously with dense clusters of 1/8-1/4" dark blue or purple, sweet-tasting, juicy, drupelike berries, with a white powdery film on the skins and 3-5 hardshelled seeds. Twigs are reddish brown, densely covered with fine whitish hairs in the first year only, with a hollow pithy core. Bark is thin, smooth, reddish to yellowish brown, becoming fissured and grooved, thickening and darkening to deep reddish brown.

An important habitat plant, providing food and cover for birds and other wildlife. Seldom cultivated, but with careful pruning can develop strong trunks as a flowering and fruiting garden tree or large shrub. Naturally quite sprawling and rarely attains great size. Hardy and tolerates heat and poor, wet soils. Longevity estimated to be about 35 years, then loses vigor. Series associations: Calencelia, Calsagebrush, Coapripea, **Mexelderberr**, Mixsage, Pursage, Scalebroom, Sumac shrub series; Arrwillow, Blawillow, Calwalnut, Pacwillow, Redwillow, Sitwillow.

e Asia. SHD, FAL CLR; IRR: M/M/M/M/-/- May be invasive and weedy in moist areas.

Sapium sebiferum

CHINESE TALLOW TREE

Euphorbiaceae. Sunset zones 8, 9, 12-16, 18-21. Deciduous. Native to China and naturalized in the southeastern U.S. Growth rate moderate to 30-40' tall and 25-30' wide, forming a spreading, round-topped, loose canopy. Leaves are alternate, simple, 2-3" long, light to medium green, widest at the base, deltoid, poplarlike, with a long, pointed, tailed end, 2 glands at the base, entire margins slightly wavy, fluttering in the breeze like an aspen, with glowing red, orange, yellow, or plum-purplish fall color, often in varied combinations. Tiny yellow flowers are in attractive clusters of slender, 4" long, drooping, tassel-like spikes, males at the tips and females near the base, at the ends of new growth in midsummer. Fruits are rounded, oblong capsules, 3-lobed, 1/2-3/4" long, in loose panicle clusters, maturing and hardening in late fall, the outer skin falling away, becoming whitish and splitting to expose small waxy seeds. Distinctive bark is tan and fissured, darkening with age.

A popular shade tree or residential lawn specimen that thrives in warm, moist climates. Tolerates only brief periods of drought, which may limit invasive tendencies in the western U.S. Resists oak root fungus. Sap is poisonous. Single-trunk trees are more upright, multi-trunk trees broader with an interesting, irregular shape. Half-hardy, but branch ends damaged by frost usually regrow. Longevity estimated to be 80-125 years.

South America. EVG, FOL ACC, SHD; IRR: VL/L/VL/L/M/M

Schinus molle

PEPPER TREE

Anacardiaceae. Sunset zones 8, 9, 12-24. Evergreen. Native to Brazil, Peru, Chile, and Argentina and naturalized in California (often called California pepper tree). Growth rate fast to 25-40' tall and wide, forming an oval canopy with heavy irregular main branching and drooping branchlets. Leaves are alternate, pinnately compound, 6-10" long, delicate and fine-textured, with 11-31 narrow lanceolate, 1-2" long leaflets, shiny light green, with minutely serrate edges, persisting until the following year. Flowers are tiny, greenish yellow, in numerous, loose, hanging, many-branched clusters at branch tips in late spring. Fruits are round, reddish, 1/4" drupes, with a tiny, hard-shelled seed, on female trees, from late summer into winter. Bark is rough, reddish brown or grayish, deeply fissured, with scaly, peeling ridges.

A strikingly handsome parkway or lawn tree, commonly grown in southern California, with a fine-textured weeping silhouette that casts filtered shade. Tolerates heat, drought, and poor soils once established. Produces litter, may develop invasive surface roots, and may suffer from scale infestations. All parts of the tree have a distinctive odor. Longevity estimated to be 100-175 years.

South America. EVG, SHD; IRR: No WUCOLS

Schinus peruviana
PERUVIAN PEPPER

Anacardiaceae. No Sunset zones. Evergreen. Native to Peru. Growth rate moderate to 35-50' tall and wide or wider, forming a broad, semi-weeping, oval canopy with heavy, irregular, main branching and drooping, twiggy branchlets. Leaves are alternate, pinnately compound, 4-6" long, shiny dark green, with 3-5 lanceolate-elliptical, pointed, somewhat sparsely spaced, sessile leaflets, 1" long by 1/8-1/4" wide, with smooth edges, lighter undersides, long petioles, persisting 2-3 years. Insignificant, small, loose clusters of tiny, creamy yellow flowers occur at branch ends in June. Small panicles of round, red, 1/4" drupes, with a tiny, hard-shelled seed, occur on female trees, ripening in September and usually quickly eaten by birds. Twigs are slender, smooth, reddish brown. Bark is grayish brown, becoming shallowly fissured with thin, scaly ridges and mottled reddish orange to brown and black as flakes fall off.

A useful parkway, shade, or background tree that doesn't seem out of place in a native garden setting, Tolerates heat, drought, and alkaline soils, but more vigorous with occasional deep watering or in lawns. Flower odor may attract flies. Surface roots seek water. Longevity estimated to be 50-75 years, losing attractiveness with age.

South America. EVG, SHD; IRR: M/M/M/M/-/M It may become weedy or invasive, by reseeding. Its use is cautioned in favor of other suitable species.

Schinus terebinthifolius

BRAZILIAN PEPPER

Anacardiaceae. Sunset zones 13, 14 (with protection), 15-17, 19-24. Evergreen. Native to Brazil and naturalized in California. Growth rate moderate to rapid to 30' tall and wide, forming a dense, dome-shaped canopy with heavy, irregular main branching, often multi-trunked or low-forked, and leaning from the canopy weight. Leaves are alternate, pinnately compound, 6-8" long, dark glossy green, the rachis usually winged along each side, with 5-7 pairs of closely spaced, oblong-lanceolate leaflets, 1" long by 1/4-1/2" wide, with smooth or sparsely toothed margins, lighter undersides, new growth bronzy red. Short clusters of tiny white flowers tucked among leaves occur in July. Female trees set clusters of 1/8" red berries in fall and early winter. Twigs are smooth, fleshy, reddish green. Bark is reddish brown, furrowed, with irregular, interconnected, thin, flat, scaly ridges.

Commonly used as an evergreen street or shade tree with dense foliage casting heavy shade. Tolerates drought and moderate inland or wind-protected coastal conditions. Irregular, angular branching habit defies attempts to prune misshapen trees. Subject to verticillium wilt and occasional aphids. Flower odor may attract flies. Slightly invasive in California, appearing along roadsides and streams, where it is shrublike, usually less than 6-8' tall, with dull, leathery leaves, smaller than those in cultivation. Longevity estimated to be 75-100 years.

e Asia. EVG, FOL ACC; IRR: No WUCOLS

Sciadopitys verticillata

JAPANESE UMBRELLA PINE

Taxodiaceae. Sunset zones 4-9, 14-24. Evergreen. Native to Japan. Growth rate very slow to 25-40' tall in cultivation (to 100-120' in habitat) and a 25-30' spread, with a dense pyramidal shape in youth, developing a more rangy, irregular, open form with age, often leaning and with branches drooping outward. Needles are shiny dark green, flattened, fleshy, 3-6" long, furrowed along the sides, grooved on the uppersides, white stoma on the undersides, a rounded notched tip, whorled around branches in tufts of 10-25, radially, in umbrella- or spokelike fashion. Older trees bear compact, clustered, yellow-green male flowers and solitary conelike female flowers with lanceolate bracts at branch ends, in late spring. Brown cones, which may develop on older trees, are ovate-oblong, 3-5" long, resembling those of a sequoia, with peltate woody scales, maturing in fall of the second season, breaking easily when handled. Seeds are compressed, 1/2" long, ovoid, narrowly winged. Bark is dark reddish brown, deeply fissured, peeling in thin, narrow strips.

A distinctive conifer, not often cultivated, in form resembling a podocarpus. Attractive as a young specimen tree for garden and patio use, especially in Japanese-style gardens. Prefers moist, well-drained, fertile, neutral or slightly acid soil and ample water. Does well in coastal areas. Requires some shade in hot interior valleys, where mites may be a problem. Tolerates clipping and thrives in containers. Long-lasting in floral arrangements. Longevity estimated to be 150-200 years.

w N.A. California Native: RW (*NCo, w KR, NCoRI, CCo, SnFrB, n ScoRO*); EVG, CNF, SHD; (IRR: H/H/H/H/-/-)

Sequoia sempervirens

COAST REDWOOD

Taxodiaceae. Sunset zones 4-9, 14-24. Jepson *SUN:4,**5**,6, **16,17**&IRR:**7-9,14,15,18-24**;CVS. Evergreen. Native to the Pacific coast of California and Oregon. Growth rate fast to 70-90' in cultivation (150-300' in habitat) with a 15-30' spread, developing an open pyramidal form and horizontal branching from a large main trunk with a broad base. Foliage sprays of feathery, flat, glossy green, needlelike leaves, 1/2-1" long, tightly spaced in an alternate, opposite, flat plane along green stems, have slightly prickly ends, lighter undersides, and persist for 3-4 years, clinging to branches 1-2 years after drying to dull brown. Inconspicuous male and female flowers occur on the same tree, appearing as yellowish, thickened, scaly bodies at ends of branchlets. Cones are brown, oval, 3/4-1" long, with densely spaced woody scales with crape-like, peltate thickened ends, maturing and opening in fall of the first year, but may persist for several months afterward. Seeds are linear, pale brown to tan, 1/16" long, encircled by a papery thin, 1/8" wing. Bark is fibrous, dark brown to cinnamon red, becoming very thick and spongy with age, with deep wide furrows and vertical ridges.

Commonly cultivated as a long-lived, fast, evergreen lawn, shade, or screen tree. Requires deep watering, and sulks in drought or heavy alkaline soils, but tolerates inland heat with consistent moderate moisture. Seedling stock is variable, and named varieties grown from cutting stock are more consistent and true to form. 'Los Altos' has lush green foliage, 'Soquel' is somewhat bluish green, 'Aptos Blue' is densely blue-green, and 'Filoli' and 'Woodside' are distinctly bluish. Longevity estimated to be 2,000 years or more, with the largest known specimen over 375' tall with a 50' diameter trunk. Series associations: Caloatgrass, pamgrass herbaceous series; Bispine, Calbay, Doufir, Grafir, Monpine, Pygcypress, Redalder, **Redwood**, Sitspruce, Weshemlock.

w N.A. (Ca only) California Native: CC, PF (*c&s SNH*); EVG, CNF, SHD, SPC; (IRR: M/M/-/M/-/-)

Sequoiadendron giganteum

GIANT SEQUOIA

Taxodiaceae. Sunset zones 1-9, 14-23, Jepson *DRN,SUN:1,4-6,17&IRR:2,3,7,14-16,18-23;CVS. Evergreen. Native to western slopes of the Sierra Nevada from Placer to Tulare counties at 4,500-8,000' elevation. Growth rate very slow (less than a foot per year) to 60-100' tall in cultivation (150-325' in habitat) and 30-50' wide, developing a dense, pyramidal form, a large central trunk with a broad, buttressed base, and symmetrical horizontal branching. Narrow, gray-green, flat, lancelike leaves have prickly ends and are arranged alternately and oppositely, spreading in two lines from opposite sides of branchlets, forming round, ropelike foliage sprays of short, overlapping, scalelike needles, remaining on branches for 3-4 years. On older trees inconspicuous male and female flowers occur on the same tree at ends of branchlets as thickened, yellowish, scaly bodies, in fall, with fertilization occurring the following summer. Cones are dark reddish brown, 2 to 3" long by 1-1/2 to 2" wide, oval, with densely spaced woody scales with crape-like, peltate, thickened ends, maturing and opening in fall of the second year but may persist on the tree through the following year. Seeds are stiff, brown, 1/8" long, papery thin, with short winged sides. Bark is thick, cinnamon red to dark reddish brown, spongy, fibrous, with large ridges and deep, vertical, interconnected furrows.

Limited to scattered groves in rocky or granitic soils of west-facing Sierra Nevada foothills, in drier habitat than coast redwood, which grows much faster. In fireprone regions only older trees with thick trunks usually survive. In cultivation a tall stately tree with beautiful form and foliage, well-suited as a lawn specimen.. Tolerates shade. Prefers moderately moist, well-drained soils. Isolated groves in the Sierra include the world's tallest tree (325' tall), the largest trunk (30' diameter), and one of the oldest living trees (over 2,000 years). Series associations: **Giasequoia**.

e Asia. FLW ACC, SHD; IRR: L/L/M/M/M/M

Sophora japonica

JAPANESE PAGODA TREE

Fabaceae. Sunset zones 2-24. Deciduous. Native to China. Growth rate moderately fast to 50-70' tall and wide with a broad oval shape and a rounded or tapered crown. Leaves are alternate, bipinnately compound, 6-10" long, dark yellowish green, glabrous, with 7-17 rather stiff, oblong-elliptical, 1-2" long, smooth-edged leaflets, with pale undersides and yellow fall color. Large, open racemes of lightly fragrant, 1/4" pealike flowers are showy in late summer. Seed pods are hanging, green "string-of-pearls" legumes, 6-12" long, constricted between 1/4-1/2" long, poisonous, pealike seeds. Pods turn yellow briefly in fall after leaves drop, then turn grayish brown and hang into winter. Twigs are slender, greenish, hairy at first, later turning light brown. Bark is thin, reddish brown, developing a scaly, rough surface between thin, reddish furrows.

An excellent lawn, shade, or patio tree casting light airy shade and effective in random, small groves in parklike lawn settings. Flower, seed, and leaf drop may litter paved areas. No special watering or soil requirements. Blooms best in hot summer climates with moderate moisture. Relatively pest free, and resists oak root fungus. 'Regent' has fast uniform growth. Longevity estimated to be 50-75 years or more.

Europe, n Africa, w Asia. EVG, SHD, FLW ACC, FRU; IRR: -/M/-/-/M/M

Sorbus aucuparia

EUROPEAN MOUNTAIN ASH

Rosaceae. Sunset zones 1-10, 14-17. Deciduous. Native to Europe, Great Britain, North Africa, and Asia Minor. Growth rate slow to moderate to 20-40' tall or more with a 15-25' spread, developing a tall oval form, a vertical trunk, upright branching, and a rounded crown, with sparse growth if heavily shaded. Leaves are alternate, pinnately compound, 6-8" long, dark dull green, divided into 9-15 oblong-oval to lanceolate, glabrous, 1-2" long leaflets on short, stout pedicels, with lightly serrate edges, gray-green undersides, and orange, yellow, or red fall color, especially showy in colder climates. Showy clusters of tiny, white, perfect flowers with noticeably long stamens, in 3-5" wide cymes, appear in spring after new leaves. Dense clusters of 1/4-1/2" round, orange-red, berrylike pome fruits mature in late summer, with 1-2 narrow tiny dark seeds, and may hang on branches into winter until eaten by birds. Twigs are reddish brown, hairy, and buds are sticky in winter. Bark is thin, smooth, gray, becoming scaly with age.

An attractive but uncommon small lawn, patio, or garden tree with fernlike foliage. Prefers cooler climates. Thrives in foothill locations, but tolerates moderate inland heat in semishade with moderate moisture in fertile, well-drained soils. Subject to canker, borers, and fireblight. Longevity estimated to be 75-100 years.

Australia. EVG, SHD; IRR: M/M/M/M/-/-

Syzygium paniculatum

BRUSH CHERRY

Myrtaceae. Sunset zones 16, 17, 20-24. Evergreen. Native to Australia. Growth rate fast to 30-60' tall and 10-20' wide, developing a dense, upright canopy, more widely spreading if multi-trunked. Leaves are opposite, simple, 1-1/2 to 3" long, thick, smooth, shiny green, oblong-elliptical, with slightly recurving margins, pale undersides, and reddish bronze new growth. Tufts of creamy white, many-stamened, 1/2" wide flowers occur among leaves in spring. Fruit is rosy purple, 3/4" long, oblong, fleshy and juicy, edible but not tasty, containing many dark, tiny seeds, and hanging closely to shiny, slender, reddish brown twigs. Bark is light brown, becoming cracked and scaly.

A fast evergreen screen or background tree, popular for its attractive shiny foliage. Vigorous without much care in coastal regions, but not hardy below 26 degrees. Aphids can turn an entire tree into an unsightly mess of gummy, shriveled leaves. Fruit drop stains pavement. Longevity estimated to be 50-75 years. 'Compacta', 'Brea', and 'Globulus' are compact forms. 'Red Flame' has showy, bright red new growth.

w Asia. EVG, SCR, SHD; IRR: VL/VL/-/L/L/L It is very invasive, and its use is discouraged.

Tamarix aphylla

ATHEL TREE

Tamarixaceae. Sunset zones 7-24. Evergreen. Native to western Asia. Growth rate very fast to 30-50' tall and 25-50' wide in 15 years, upright at first, then bending with the weight of the limbs and becoming irregular with a heavy trunk, usually with a densely bushy, broad foliage crown and many slender, arching branches. Branchlets are scalelike, jointed much like casuarina, and greenish or bluish green in saline soils. Tiny leaves are barely visible as a swollen sheath at each joint, with protruding, tiny, hooklike ends. Panicled spikes of tiny, whitish pink, five-petaled flowers occur in summer, among the present year's growth, in sparse clusters, not nearly as showy as *Tamarix parvifolia*. Fruits are 3/16" long, brown, pointed capsules with 3 narrow valves, maturing in late summer and containing many minute seeds with tufts of tiny white hairs at the ends. Bark is smooth, brown, thickening and aging to grayish or reddish brown, becoming deeply fissured, with broad, thick, scaly ridges.

Often used like casuarina, which it resembles, as a windbreak or for shade in hot, dry climates. Tolerates heat, drought, poor and alkaline soils. Deep tap root, but surface roots are extremely invasive. Easily escapes cultivation. Small indigenous groves occur along freeways near the west end of the Mojave Desert. Sometimes seen in the Central Valley as a relic of previous agricultural use as a shade or windbreak tree, or in riparian washes. Wood is brittle, and branches may break off in strong winds. Longevity estimated to be 30-50 years, losing vigor and shape with age.

se Europe, Asia. FLW ACC, SHD; IRR: VL/VL/-/L/L/L It is very invasive, and its use is discouraged. Programs are ongoing for its removal.

Tamarix parviflora

TAMARISK

Tamarixaceae. Sunset zones 2-24. Deciduous. Native to southeastern Europe and Asia and more or less naturalized in the southern U.S. and parts of the central Southwest. Growth rate very fast to 6-15' tall and wide, becoming sprawly and arching in form, usually multi-trunked or shrublike. Leaves are tiny, alternate, feathery, scalelike, green or grayish, sparsely spaced, barely visible along branchlets. Tiny, pink, four-petaled flowers in elongated clusters resembling pussy willow are showy in spring, before foliage emerges, from the previous year's growth. Fruits are small brown capsules with 3-5 valves, containing many minute seeds. Bark is rather thin, dark brown, smooth, becoming nearly black, with numerous small fissures creating tiny, squarish, woody, scalelike plates.

Once cultivated as a lawn or garden specimen, but now a serious pest in some desert areas, reseeding profusely along washes. Deep tap root but extremely invasive surface roots. Requires adequate room for its graceful arching branches in bloom. Needs pruning or staking to develop strong trunks required for tree canopy branching. Relatively pest-free. Tolerates heat, poor soil, and drought. Longevity estimated to be 25-30 years, when it loses vigor and shape. Series associations as an indigenous but often an undesirable shrub in: Arrweed, Tamarisk shrub series.

e N.A. SHD, FOL ACC, SPC; IRR: M/-/M/M/-/-

Taxodium distichum

BALD CYPRESS

Taxodiaceae. Sunset zones 2-10, 12-24. Deciduous. Native to swamps of eastern U.S. from Delaware to Florida. Growth rate fast in wet soils to 100-120' in habitat, usually 50-70' tall and 20-30' wide in cultivation in the west. Develops a tall, open, narrow, pyramidal form, broadening with age to a rounded, irregular canopy. Large trunks develop "knees" that bring air to flooded roots. Foliage is soft, yellow-green, needlelike, simple, linear leaflets, 1/2-3/4" long, whorled but lying flatly to each side of a central rib on lateral branchlets, appearing 2-ranked, turning dull orange to brown and dropping along with branchlets in fall. May also appear scalelike, and appressed, to 1/2" long, on upright, fertile terminal branchlets. Flowers are inconspicuous, monoecious, the greenish yellow male catkins with 6-8 stamens occurring in drooping panicles at ends of the previous year's growth and females forming 3/4" round, green, pendent, short-stalked conelets just below the base of the males. Cones are woody, brown, with several wrinkled pelted scales, maturing in early fall and disintegrating quickly. Seeds are 1/4", brown, 2 per cone scale, with 3 angled-winged sides. Bark is dark reddish brown, fibrous, peeling in thin strips or grayish brown, thin, and scaly. Trunks develop wide, tapered, fluted bases.

A fine-textured tree for bogs, lawns, or pond settings in parks. Similar to *Metasequoia* but not nearly as common. Relatively pest free, surprisingly hardy, and heat tolerant with adequate moisture. Shallow, wide-spreading roots make it stable in moderately windy areas, even in wet soils. Longevity estimated to be 1,000-3,000 years.

sw N.A. EVG, SHD, FOL ACC, SPC; IRR: M/?/M/M/-/-

Taxodium mucronatum

MONTEZUMA CYPRESS

Taxodiaceae. Sunset zones 5-9, 12-24. Evergreen to semi-deciduous. Native to Mexico. Growth rate initially fast to 40' tall (75' tall with regular moisture) and 50' wide, developing an open, narrow, pyramidal form, broadening with age to a rounded, irregular canopy. Foliage is soft, yellow-green, needle-like, simple, linear-lanceolate leaflets, 1/2-3/4" long, whorled, but lying flatly to each side of the central rib on lateral branchlets, appearing 2-ranked, turning dull brown before dropping in early spring before new leaves. May also appear scalelike, and appressed, to 1/2" long, on fertile terminal branchlets, persisting 2 years. Flowers are inconspicuous, monoecious, the loose, greenish yellow male catkins with 6-8 stamens occurring in drooping panicles at the ends of year-old growth and females forming 3/4" round, green, pendent, short-stalked conelets just below the base of the males. Cones are woody, 1", brown, with several wrinkled pelted scales, maturing in early fall and disintegrating quickly. Seeds are 1/4", brown, 2 per cone scale, with 3 angled-winged sides. Bark is reddish brown, fibrous, peeling in thin strips or gray-brown, thin, and scaly. Trunks develop wide, tapered, fluted bases.

Not widely cultivated, except in moist lawns of older parks, where it is an unusual tall specimen with a noticeably weeping habit and slightly less delicate appearance than bald cypress. Relatively pest free and heat tolerant with adequate moisture. Longevity is estimated to be 1,000-3,000 years. The oldest known specimen is 4,000 years.

w N.A. California Native: RW, CF, ME, PF (*NW, CaR, n&c SN, SnFrB*); (IRR: No WUCOLS)

Taxus brevifolia

WESTERN YEW

Taxaceae. Sunset zones 3-7, 14-17, Jepson *IRR,DRN,SHD: 1,2,3,7,14,15-17,18-23,24;SUN:4-6. Evergreen. Native to forested Pacific Coast mountain regions from the southern tip of Alaska to central California, in the Coast Ranges and the Sierra Nevada, at 2,000-8,000' elevation. Growth rate moderate to 50-60' tall and 30-40' wide, developing an open, conical crown, denser in sunny locations, with long, slender, drooping horizontal branches, often with multiple leaders from the base. Attractive needlelike leaves are 1/2-3/4" long, deep yellow-green, dull-pointed to blunt-ended, simple, linear-lanceolate, spirally arranged but lying nearly flat along each side of the twigs, appearing 2-ranked, with pale undersides, persisting 5-9 years. Inconspicuous, solitary, dioecious flowers occur at leaf axils, the yellowish males in rounded heads with 6-14 stamens and single females a greenish, pointed ovule with a fattened basal disk. Fruits are single, ovoid-oblong, maturing in September and falling shortly thereafter. Bony-shelled seeds are surrounded at the base but separate from the scarlet to coral-red, fleshy, basal flower disk, which attracts birds to the sweetish fleshy sheath. Seeds are not affected by digestion and are distributed freely. Bark is thin, dark reddish purple, rather papery, peeling in loose scales, exposing redder layers underneath, becoming bright red when wet.

An attractive, light-textured, evergreen understory tree of coniferous forests and canyons along cool shaded streams. Thrives in moist, loose, slightly acidic soils, and tolerates dense shade. Rarely grows in pure open stands and usually is concealed by larger conifers and broadleaf trees. Can regrow from the base of trunks left behind after logging. According to Sunset, plants sold under this name are often *Taxus cuspidata* 'Nana'. Rarely available for cultivation otherwise, and difficult to grow, but may be adaptable to northwest gardens. Longevity estimated to be 200-300 years or more in habitat. Series associations: Doufirtan, Engspruce, Subfir, Alayelcedsta, KlaMouEnrsta.

e N.A. EVG, SHD, FOL ACC; IRR: M/M/-/M/M/M

Thuja occidentalis

AMERICAN ARBORVITAE

Cupressaceae. Sunset zones 1-9, 15-17, 21-24. Evergreen. Native to the eastern U.S. from North Carolina into Canada and west into Minnesota. Growth rate slow to moderate to 30-60' tall by 10-15' wide in cultivation, developing an upright, ovate to elliptical form, usually with a pointed tip and thin horizontal branches turning upward at the ends. Foliage is dull green, scalelike, decussate, flattened, glandular-pitted, in dense flat sprays, twisting irregularly with branches, with yellow coloration at the ends of new growth, persisting 2-5 years. Flowers are yellowish brown, monoecious, occurring in spring at branch tips, males with 3-6 stamens and females with 8-12 scales. Cones are 1/4-1/2" round, green, with only 4 fertile scales, maturing in one season, turning brown, and persisting 1-2 years, tucked among foliage. Bark is thin, fibrous, cinnamon red, with interlaced ridges.

A small evergreen, shrublike conifer on the west coast that grows to 100' tall in humid eastern states. Older specimens may become treelike. Many named varieties of different sizes and shapes. May suffer viral dieback on some tips. Thrives in lawns or mulched shrub planters with regular moisture and in northern coastal areas. Not as heat and drought tolerant inland as juniper. Longevity estimated to be 75-100 years.

w N.A. California Native: CF (*NCo, KR, NCoRO; to AK, MT*); EVG, SHD, SPC; (IRR: No WUCOLS)

Thuja plicata

WESTERN RED CEDAR

Cupressaceae. Sunset zones 1-9, 14-24, Jepson *4-6&IRR:1, 2,**7**,**14-24**&SHD:**3**,8,9. Evergreen. Native to the northwestern U.S. from coastal California to Alaska east to Montana. Growth rate moderate to 50-100' tall (to 150-200' in habitat) and 20-60' wide in cultivation, with a pyramidal form, a heavy buttressed main trunk, somewhat drooping branches arching gracefully upward, foliage drooping at the ends, and lower limbs usually retaining foliage to the ground. Flat, hanging, lacelike, green foliage sprays of small, decussate, scalelike leaves occur in alternate, linear fashion along a central rib, persisting 2-5 years, with smaller side branches browning after about 2 years. Inconspicuous, monoecious, yellowish brown flowers occur at branch tips in spring, males with 3-6 stamens and females with 8-12 scales. Cones are 1/2" round, reddish brown, with 6 fertile scales, maturing and releasing 2-3 seeds each in late summer of the first year, but may remain into the following spring or summer. Seeds are narrow, 1/8" long, with small wings on each side. Bark is thin, reddish brown, fibrous, with shallow furrows and long vertical ridges interconnected by thinner diagonal seams and ridges, peeling in long strips and taking on a grayish cast with age. Trunk base is often fluted.

A common evergreen tree in northwestern coniferous forests, surprisingly adaptable as a magnificent lawn specimen, especially near water, and in parks in the central and northern inland parts of the state. Usually shades out lawns beneath, and branches usually are best left sweeping to the ground. Prefers moist but not soggy soil. Tolerates heat once established, but not prolonged drought. Longevity estimated to be 200-500 years or up to 800 years in habitat.

Cultivar. EVG, SCR, VERT ACC; IRR: No WUCOLS

Thuja plicata 'Emerald Cone'

COLUMNAR RED CEDAR

Cupressaceae. Sunset zones 1-9, 14-24. Evergreen. Recently introduced clonal selection of the species. Growth rate moderate to an estimated height of 30-50' with a 10-20' spread, developing a graceful, dense, upright, columnar form, with a narrow pointed peak, lower limbs retaining foliage to the ground. Bright green foliage sprays of small, decussate, scale-like leaves are larger than those of the species.

An attractive new cultivar, dependable and relatively pest-free, highly touted as a more reliable replacement for Leyland cypress (x *Cupressocyparis leylandii*), which is rapidly dying from disease to which this introduction seems resistant.

Cultivar. EVG, FOL ACC, SPC; IRR: No WUCOLS

Thuja plicata 'Gracilis'

GIANT ARBORVITAE

Cupressaceae. Sunset zones 1-9, 14-24. Evergreen. Selection of the species. Growth rate moderate to 30-50' tall and wide, forming a dense, rounded, oval canopy, with upward-arching branches from multiple main trunks, foliage slightly drooping at the ends and lower limbs retaining foliage to the ground. Flat, hanging, green foliage sprays of small, decussate, scale-like leaves are rather lacelike, arranged in alternate, linear fashion along a central rib. Smaller side branches and foliage die after about 2 years as main branches grow and expand. Inconspicuous, yellowish brown, monoecious flowers occur in spring at branch tips. Cones are 1/4" round, reddish brown, 6-scaled, maturing in late summer of the first year and releasing 2-3 seeds each, remaining into the following spring or summer. Bark is thin, reddish brown, fibrous, with shallow furrows and long vertical ridges, peeling in strips and taking on a grayish cast with age.

Rarely cultivated as an evergreen park or residential lawn specimen, most planted long ago. Young trees have a fairly symmetrical, rounded canopy. Older trees usually become devoid of foliage within the dense outer foliage mass, developing large, wide-spreading limbs within the airy space underneath. Longevity estimated to be 200 years. 'Gracilis Aurea' has faint yellow foliage variegation.

e Asia. EVG, SPC; IRR: No WUCOLS

Thujopsis dolobrata

DEERHORN CEDAR

Cupressaceae. Sunset zones 3-7, 14-17. Evergreen. Native to Japan. Growth rate slow to 30-50' tall and 10-20' wide, forming a dense, rounded canopy, often shrublike, or a multi-trunked small tree, with attractive frondlike branchlets, foliage slightly drooping at the ends and lower limbs retaining foliage to the ground. Flat, lacelike, bright shiny green foliage sprays of small, scalelike leaves, broadly decussate, occur in alternate, linear fashion along a central rib, with 2 whitish bands on the undersides, persisting 2-5 years. Smaller side branches die and fall off after about 2 years as main branches grow and expand. Inconspicuous, monoecious flowers occur at branch tips in late spring, males dark green and females bluish green. Cones are 1/4-1/2" by 3/4", dark bluish brown, ovoid, 6-scaled, with prickles, opening with wide gaps between and releasing tiny seeds. Bark is thin, dark reddish brown, fibrous, with shallow furrows and long vertical ridges interconnected by thinner diagonal seams and ridges, peeling in long strips and taking on a grayish cast with age.

Rarely cultivated except in arboretums and Japanese-style gardens, often as a featured specimen. Interesting foliage resembles thuja, but is noticeably heavier and thicker. Prefers cool, semi-moist conditions and requires shade in hot climates. Longevity estimated to be 200-500 years. 'Hondai' is taller with smaller leaves, 'Nana' is dwarf with lighter-colored leaves, and 'Variegata' has white-tipped branchlets that revert to green.

Europe. SHD; IRR: M/M/-/-/-/-

Tilia cordata

LITTLE-LEAF LINDEN

Tiliaceae. Sunset zones 1-17. Deciduous. Native to Europe.
Growth rate slow to 30-50' tall with a 15-30' spread, developing a symmetrical pyramidal form, broadening at the base,
with upright, regular branching. Leaves are alternate, simple,
1-1/2 to 3" long and nearly as wide or wider, dark green, broadly
ovate, heart-shaped at the base, glabrous except for pale downy
tufts in vein axils on silvery, light green undersides, with serrate edges and yellow fall color. Noticeable clusters of fragrant,
regular, perfect, creamy white and yellow flowers in loose
drooping cymes of 4-10 are attached to a long slender stalk,
extending from the midpoint of a leafy, yellowish green, tail-
like bract, and appear in late spring among leaves. Fruits are
woolly, gray, nutlike drupes, 1/3-1/2" in diameter, maturing
in fall with accompanying bracts. Twigs are reddish brown.
Bark is dark gray-brown, developing narrow, shallow furrows,
with narrow, flat, scaly ridges.

A medium-sized lawn, park, or street tree. Flowers attract
bees. Aphids produce honeydew drip, limiting use in pedestrian and parking areas. Otherwise relatively pest-free. Prefers
moderately moist, well-drained soils. Longevity estimated to
be 150-200 years or more. 'Greenspire' and 'Rancho' are more
columnar, upright forms.

nw Europe. SHD; IRR: No WUCOLS

Tilia europaea

EUROPEAN LINDEN

Tiliaceae. No Sunset zones. Deciduous. Native to northwestern Europe, reportedly a natural hybrid between *Tilia cordata* and *T. platyphyllos*. Growth rate slow to 30-50' tall with a 15-30' spread, developing a dense, broad pyramidal form from a strong central trunk, with short angular or irregular branching and slender twiggy ends, becoming more irregular with age. Leaves are alternate, simple, 1-1/2 to 3" long and wide, dark green, broadly ovate, heart-shaped at the base, glabrous except for pale downy tufts in vein axils on silvery, light green undersides, with finely serrate edges and yellow fall color. Showy clusters of fragrant, regular, perfect, creamy white and yellow flowers occur in late spring among leaves, in loose drooping cymes of 4-10, attached to a long slender stalk extending from the midpoint of a leafy, green, tail-like bract. Fruits are woolly, gray, nutlike drupes, 1/3-1/2" round, maturing in fall with accompanying bracts. Twigs are reddish brown. Bark is smooth, gray-green, becoming dark brownish gray with multiple shallow furrows and scaly ridges.

Sometimes used as a medium-sized lawn tree and attractive in a parklike setting. Relatively pest-free, except for aphids, which exude a sticky honeydew, often covering the leaves. Flowers attract bees. Young trees often sucker near the base. Longevity estimated to be 100-200 years.

c N.A. SHD; IRR: No WUCOLS

Tilia tomentosa

SILVER LINDEN

Tiliaceae. Sunset zones 2-21. Deciduous. Native to the midwestern U.S. from Missouri to eastern Texas. Growth rate slow to 40-50' tall with a 20-30' spread, developing a tall oval shape with a somewhat pointed crown and upright branching. Leaves are alternate, simple, 4-6" long by 3-4" wide, dull dark green, broadly ovate, heart-shaped at the base, glabrous except for pale downy tufts in vein axils on the silvery, light green undersides, with finely serrate edges and yellow fall color. Showy clusters of fragrant, regular, perfect, creamy white and yellow flowers occur in loose, drooping cymes of 4-10, attached to a long, slender stalk extending from the midpoint of a leafy, green, tail-like bract, appearing in late spring among the leaves. Fruits are woolly, gray, nutlike drupes, 1/3-1/2" round, maturing in fall with accompanying bracts. Twigs are greenish brown. Bark is grayish, becoming furrowed with reddish orange inner bark and vertical, interconnected, flat, narrow ridges.

A useful large shade tree for streets, parks, and lawns where it has ample room to develop. Does best in moderately moist, well-drained soil. Aphids cause honeydew drip and accompanying sooty mildew. Otherwise relatively pest-free. Longevity estimated to be 200 years or more. 'Fastigiata' and 'Pyramidalis' have narrower, more upright branching.

South America. SHD, FLW ACC; IRR: M/-/M/M/-/-

Tipuana tipu

TIPU TREE

Fabaceae. Sunset zones 12 (warmest areas), 13-16, 18-24. Deciduous to semi-deciduous. Native to South America. Growth rate fast to 25-40' tall and 30-60' wide or larger, forming a broad, flattened canopy and upright to semi-arching branches. Leaves are alternate, pinnately compound, light green, 8-12" long, with 11-21 paired, ovate-oblong, glabrous, 1 to 1-1/2" long by 1/2" wide leaflets along a main rib, with lighter undersides and yellow fall color. Leaves may remain through winter in mild areas of southern California and may not leaf out until late spring in colder climates. Sparse axillary clusters of showy, orange-yellow, 1/2-3/4", pealike flowers occur in 8" long, hanging panicles in June and July when most other trees have stopped blooming. Seed pods are single-winged, 2-1/2" long, linear-oblong legumes that ripen to dull tan in fall. Twigs are green, darkening to reddish brown. Bark is reddish brown to grayish black with deep, wide fissures and flat, cracked or scaly ridges.

Well adapted for use as a street, parkway, or flowering accent tree in patio or garden settings. An alternative to flowering locust, though not as cold hardy and grown more often in southern California. Young trees become well established with deep watering, later tolerating heat and drought. Overwatering older trees produces rank, rangy growth. Pruning encourages lanky, vigorous new growth, destroying natural branching structure. Blooms heaviest in warm climates, not well along the coast. Hardy to about 25 degrees. Longevity estimated to be 50-80 years.

w N.A. (Ca only) California Native: CH, CF, PF (*NCo, NCoR, CaRF, SN, SCoRO*); (IRR: No WUCOLS)

Torreya californica

CALIFORNIA NUTMEG

Taxaceae. Sunset zones 4-9, 14-24, Jepson *4-6,17;IRR,SHD:2, 3,7,8,9,14-16,18,19,20-21,22,24. Evergreen. Native to California in cool canyons and on shaded slopes at 2,000-7,000' elevation. Growth rate slow to 15-20' tall (more in habitat) with a 12-15' spread, developing a 1-3' diameter trunk, often multi-trunked, with an open, broad, pyramidal crown, rounded in age, rangy in dense forest locations, with slender horizontal branches standing out straight from the trunk in symmetrical circles, drooping at the ends. Needlelike leaves are rigid, flat, linear-lanceolate, 1-1/4 to 2" long, shiny deep green, spirally arranged, appearing flatly 2-ranked, with sharp pointed ends, two whitish bands on the undersides, and a disagreeable resinous odor when crushed. Inconspicuous flowers are dioecious, 1/8" round, with yellow males grouped in rows along leaf axils, laden with pollen in late spring, and slightly larger, paired, greenish females on separate trees at the base of new branchlets. Fruits are 1 to 1-1/4" long, ovoid, plumlike, pale green, with purple markings, a blunt pointed end, and wrinkled skin when mature in fall of the second season. Seeds have a thin, hard, brittle shell and germinate after 2 years. Slender green branchlets turn reddish brown. Trunks are uneven in cross-section, rarely fully round. Bark is ashy yellowish brown, finely checked with narrow seams and short, narrow, loose, scaly, interconnected ridges.

An attractive understory tree, occurring sparsely in evergreen coniferous forests and canyons in moist shade, usually concealed within the canopy of larger conifers. Rarely occurs in pure open stands. Sometimes cultivated as a small specimen or screen tree. Requires summer moisture in semi-shade. Needles are razor-sharp. Longevity estimated to be 100-250 years in habitat.

e Asia. EVG, SHD; IRR: No WUCOLS

Torreya nucifera

JAPANESE NUTMEG

Taxaceae. No Sunset zones. Evergreen. Native to Japan. Growth rate slow to 25-50' with a 20-25' spread, developing a broad, pyramidal crown, rounded with age, from low-forked trunks with multiple upright leaders and slender horizontal branches tending to droop at the ends. Rigid, flat, needlelike leaves are linear, dark shiny green, 3/4 to 1" long, aromatic when crushed, spirally arranged, appearing flatly 2-ranked, with dull pointed ends, usually bowed and curving under at the ends. Inconspicuous, dioecious, soft-cone flowers occur in late spring on separate trees, the 1/8" round males, with pollen sacs, clustered at terminal leaf axils and slightly larger, paired, greenish females with a fleshy urn-shaped sac at the base of new shoots. Fruits are 1" long, plumlike, oblong-ellipsoidal, pale green with a slight purple tinge, a blunt pointed end, and wrinkled skin when they mature in fall of the second season. Edible seeds have a thin, hard, brittle shell and require an additional year before germinating. Bark is grayish brown, becoming shallowly fissured and flaky on older trees.

An uncommon specimen, park, screen, or lawn tree that casts dense shade. Prefers moisture. Benefits from partial shade of nearby trees, but loses upright form in heavy shade. Rarely cultivated, but older trees are sometimes seen in parks. Needles are not sharp, as are those of the native *Torreya californica*. Longevity estimated to be 100-250 years.

e Australia. SHD, FLW ACC; IRR: M/-/M/M/-/-

Tristaniopsis laurina
(*Tristania laurina*)

LAUREL LEAF BOX

Myrtaceae. Sunset zones 15-24. Evergreen. Native to eastern Australia. Growth rate slow to 10-45' tall (70' in habitat) and 5-30' wide, with a dense, rounded, upright form. Leaves are alternate, simple, 2-4" long by 1/2-1" wide, dark glossy green, lanceolate, nearly sessile, with smooth margins, barely visible veins, a glabrous, waxy appearance, light yellow-green undersides, densely spaced along branches, persisting 2 years, and aromatic when crushed. Clusters of 1/8-1/4", 5-petaled, bright shiny yellow flowers with many short stamens, in bundles of 5, are showy in early summer. Small, green seed capsules dry to dark brown in late summer, when they split and open, cup-like, along 3 or more seams, releasing tightly packed, elongated, flat, 3/16" long by 1/16" wide, light brown to yellowish tan, winged seeds. Twigs are light green, tomentose, becoming shiny brown. Attractive bark is mottled brown, peeling in irregular, thin flakes to reveal light tan or whitish underbark.

An attractive small evergreen accent tree, useful as a standard with staking and pruning. Tolerates inland heat and moderate dryness in semi-shade, but becomes leggy in full shade. Best in full sun near the coast. Longevity estimated to be 80-100 years. 'Elegant' has broader leaves and reddish new growth.

Tropicals

N.Z. EVG, FOL ACC; IRR: L/M/L/M/M/M

Cordyline australis

GREEN DRACAENA

Liliaceae. Sunset zones 5, 8-11, 14-24. Evergreen. Native to New Zealand. Growth rate slow to 20-30' tall and 6-12' wide, with a broad-based trunk and secondary, shorter, ascending branches. Leaves are thick, upright-arching, lanceolate, long-pointed, dark grayish green, 18-36" long by 1-2" wide, in erect tufts at branch ends. Older foliage tends to droop at the base of tufts and dries to a brown thatch, persisting for a few years beneath new foliage. Showy, erect or drooping, 2-3' long racemes of many 1/4" creamy white flowers occur sporadically in late spring and summer. Fruits are white to bluish white, 1/4" berries with small round seeds. Bark is grayish brown with small, corky ridge plates, becoming exposed after the tree is at least 5' in height and dried leaf thatch skirts are trimmed off.

Commonly cultivated for a Spanish, desert, or tropical effect in terrace, yard, or foundation plantings in groupings or as a single specimen. Effective in accentuating architectural features. Tolerates drought and heat when established. Benefits from occasional deep watering, but tolerates only moderately moist soils. Unsightly long stalks usually are trimmed off before fruits ripen. Generally hardy to 15 degrees. Longevity estimated to be 60-80 years or more.

Tropicals

Australia. FOL ACC; IRR: H/H/H/H/-/-

Cyathea cooperi

AUSTRALIAN TREE FERN

Dicksoniaceae. Sunset zones 15-24. Evergreen. Native to sub-tropical Australia. Growth rate slow to 6-8' (up to 20') tall by 12' wide, developing an arching form from a stout, hairy trunk, with knobs remaining from cut frond bases. Fronds are finely cut, soft, light yellow-green, with brown spores on the under-sides, 3-6' long or more, spreading outward, lasting roughly one season, turning brown after the new year's growth.

Fastest growing and most commonly cultivated tree fern, often used as a tropical accent or garden or courtyard speci-men. Prefers adequate light, but not direct sun, in moderately moist, fertile soils. Hardy to about 32 degrees for short peri-ods, but at 20 degrees probably will not recover. Longevity estimated to be over 60 years in frost-free, temperate climates.

e Asia. FOL ACC; IRR: M/M/M/M/M/M

Cycas revoluta

SAGO PALM

Cycadaceae. Sunset zones 8-24. Native to the West Indies, Ja-pan, and China. Growth rate extremely slow to 20' tall by 16' wide or wider, forming a short cylindrical trunk with whorled stumps left behind after dead fronds are removed. Develops offshoots from the base, forming upward-curving multiple trunks with age, terminating in large, conspicuous, fuzzy buds at the center of foliage rosettes. Feathery fronds are rigid, dark glossy green, glabrous, 2-3' long, with many stiff, narrow, lin-ear-paired pinnae with sharply pointed ends, arising from a short-sheathed stalk, and persisting until cut off. Dioecious flowers occur on separate plants, with yellowish conelike males producing pollen and larger female rosettes covered with thick down, maturing into a cylindrical dark cone, with 1/2-3/4" edible red seeds. Bark is scaly, cinnamon-red to dark brown.

A prized tropical accent specimen or large container plant. Durable, tough, and long-lived. Prefers part shade in hot inte-rior climates, with moderate moisture in fertile, well-drained soils. One of the hardiest cycads, generally tolerating light frosts to about 15 degrees. Longevity estimated to be over 500 years.

Tropicals

Australia. FOL ACC; IRR: H/H/H/H/-/-

Dicksonia antarctica

TASMANIAN TREE FERN

Dicksoniaceae. Sunset zones 8, 9, 14-17, 19-24. Evergreen. Native to southeastern Australia and Tasmania. Growth rate slow to 15' tall, developing an arching form from a stout, hairy trunk with knobby cut frond bases. Fronds are finely cut, soft, light green with dark spores on the undersides, 3-6' long, spreading upward and outward, lasting roughly one season, turning brown after the new year's growth.

Hardiest of the tree ferns, established specimens becoming slightly more tolerant with age, to about 22 degrees for short periods. Tolerates full sun along the coast, protected from strong winds. Often grown as an atrium specimen in inland climates, where it prefers moderate moisture and semi-shade. Longevity estimated to be over 100 years in frost-free, temperate climates.

e Asia. FOL ACC; IRR: L/L/M/M/M/M Roots are highly invasive.

Phyllostachys bambusoides

TIMBER BAMBOO

Gramineae. No Sunset zones. Evergreen. Native to China and Japan. Growth rate fast to 15-35' tall with an 8-10' spread if controlled, with multiple crowded, slender trunks at the base, developing an arching, vaselike canopy. Leaves are pale green, 3-6" long, with a long tapered tip, minutely serrate with surprisingly sharp edges, bright uppersides, dull, pale bluish white undersides, sparsely spaced at 90 degrees to branchlets, with a hairy, clasping, basal leaf sheath, the base of the leaves sharply constricting nearly at right angles to the short petiole stem. Blooms once every hundred years or so, after which the plant dies. Attractive trunks are hollow, fibrous, woody, smooth, yellowish green, the hollow pithy core closed off inside at narrowly constricted joints where leaf sheaths were previously attached.

A tropical accent specimen, forming clumps or small groves, often used for screening or in large containers. Drought tolerant but looks best with moderate moisture. Roots are highly invasive. Needs containment to constrict underground runners. Generally hardy to 0 degrees. Longevity is indefinite, as plants continually send up new growth. However, many plants were lost after flowering in the 1960s and 1970s.

Tropicals
Asia. FOL ACC; IRR: H/H/H/H/-/M

Musa x *paradisiaca*
(*Musa sapientum*)

BANANA

Musaceae. Sunset zones 9, 12-16, 19-24. Evergreen. Native to India. Growth rate fast to 20' tall or more and half as wide, from multiple fat, fleshy, green trunks, forming an arching, vaselike canopy. Leaves are shiny green, to 10' long by 12-18" wide, with oblong blades on a fleshy, short-sheathed stalk, either erect or ascending, drooping or arching at the ends, becoming lighter textured and easily torn by winds, due to the parallel veining. Curious flower spikes occur on drooping stems, with long, reddish purple, lanceolate bracts opening around a large, conical central bud, surrounded by small yellow flowers in summer. Fruit develops without fertilization, forming fists of 7-15 small, green bananas, seedy and bland-tasting but edible. Trunks are not woody, the skin consisting of tightly sheathed bases of succulent leaf stalks, green with light yellow and purplish striations, and older, brown, dried sheaths remaining near the base.

Commonly cultivated as a tropical accent, mostly in southern California. Not reliably hardy below 30 degrees. Foliage is unsightly when damaged by frost, but regrows from trunks or if removed during winter. Prefers moist, rich soil, protected from winds, in a sunny location. Longevity is indefinite, over 100 years, since plants continually send up new growth.

Tropicals

Australia. EVG, SHD, FOL, FLW ACC; IRR: -/-/M/M/-/-

Stenocarpus sinuatus

FIREWHEEL TREE

Proteaceae. Sunset zones 16, 17, 20-24. Evergreen. Native to Australia. Growth rate slow to 30' tall with a 15' spread, forming a narrow, rounded canopy, with heavy, dense foliage. Leaves are alternate, simple, glossy dark green, leathery, and variable, juvenile leaves 12-18" long by 2-3" wide with 1-4 oblong lobes and adult leaves shorter, without lobes, linear-elliptical, to 1 to 1-1/2" wide, with recurving, entire, or slightly wavy edges. Showy clusters of brilliant scarlet flowers with 1" long, yellow-tipped spokes, like a pinwheel, occur in dense umbels, sporadically, mostly in summer and early fall on older, well established trees. Fruit is a long, thin follicle. Bark is smooth, light brown, becoming furrowed with age.

A small flowering accent tree with attractive foliage, often grown as a standard in southern California, less often in north coastal areas. Hardy to about 32 degrees for short periods. Prefers fertile, well-drained soils with moderate moisture. Longevity estimated to be over 60 years in frost-free, temperate climates.

se Asia. FOL ACC; IRR: L/M/M/M/-/M

Tetrapanax papyriferus

(*Aralia papyrifera*)

RICE PAPER PLANT

Araliaceae. Sunset zones 15-24. Evergreen. Native to Formosa. Growth rate fast to 10-15' tall, usually with long, slender, multiple exposed trunks, unless concealed by sucker growth from the base, often forming thickets. Leaves are dark green, 1-2' in diameter, with many deep lobes, white, felty undersides, from long stems, sheathed to the central trunk, older leaves dropping off after 1 year. Creamy white flower clusters on fuzzy brown stems extend loosely above the foliage canopy in late winter, flower parts in 4s rather than 5s like other plants of this family. Trunks are smooth, brownish gray, with leaf scars, often forked, and leaning from the weight of the foliage.

A half-hardy, small tropical accent or silhouette tree. Needs protection in windy locations. Often damaged by temperatures below 22 degrees, but may resprout from the base. Cultivation around the base promotes suckering. Fuzzy hairs are irritating to the eyes, throat, or skin if brushed loose. Longevity estimated to be 20-30 years, extended indefinitely with new sucker growth.

See also: PALMS

w N.A. California Native: CF, ME, RW (*NCo, w KR, NcoRO*); (IRR: No WUCOLS)

Tsuga heterophylla

WESTERN HEMLOCK

Pinaceae. Sunset zones 2-7, 14-17, Jepson *IRR,DRN:**4-6,15-17**&SHD:**3,7**,14;CVS. Evergreen. Native to the Pacific Coast from Alaska to northern California at 2,000-5,000' elevation. Growth rate slow to moderate to 70-130' tall (100-150' or more in habitat) by 20-30' wide, developing a tall pyramidal form, with drooping branchlets on slender, horizontal branches, and a distinctly nodding top. Needles are glossy, yellow-green, 1/4-3/4" long, appearing to grow from opposite sides of branchlets, appearing 2-ranked, or in a flattened profile, from a short stem, with grooved uppersides, persisting 3-4 years. Cones are light brown, elliptical, 3/4 to 1-1/4" long, with thin papery scales, hanging along ends of branchlets, maturing in summer of the first season. Seeds are 1/16" with a 3/8" wing. Bark is thin, brown, finely scaled, becoming thicker and harder, dark brown and red-tinged, deeply furrowed, with flat, wide ridges irregularly interconnected with narrow cross-ridges.

Often rather spindly and shapeless in its densely shaded forest habitat, mixed with Douglas-fir, redwood, and Sitka spruce. In recent years found to be a valuable forest tree in the northwest. In sunnier exposed locations, trees become more shapely. Somewhat adaptable to landscape use, as stunning specimens for lawn or garden, in moderately moist, acidic soils, in north coastal areas. Longevity estimated to be 200 years or more in habitat. Series associations: Beapine, Doufir, Grafir, Redalder, Redwood, Sitspruce, **Weshemlock**.

nw N.A. California Native: FF, AL (*KR, NcoRH, CaRH, SNH, MP, n SNE*); (IRR: No WUCOLS)

Tsuga mertensiana

MOUNTAIN HEMLOCK

Pinaceae. Sunset zones 1-7, 14-17, Jepson *DRN,IRR:1,**2**,4,5 &SHD:**6**,7,14-17. Evergreen. Native to the Pacific Coast from Alaska to northern California and east to Idaho and Montana at 4,000-11,000' elevation. Growth rate slow to moderate to 50-90' tall, usually 20-30' tall and half as wide in cultivation, developing a tall, sharply pyramidal form, with a distinct nodding top and short drooping branchlets from slender horizontal branches often drooping to the ground. Foliage is dense, dark green to blue-gray, bluntly pointed, rounded and plump looking, with small distinct stems, growing from all sides of branchlets but appearing thicker on the uppersides, persisting for 3-4 years. Cones are yellowish green to bluish purple, elliptical, 1-3" long, with thin papery scales, hanging at the ends of branchlets, maturing in summer of the first season, opening in fall. Seeds are 1/8" long with a 3/8" wing. Bark is relatively thin, finely scaled, becoming thicker and harder, dark reddish brown to grayish brown, deeply and narrowly furrowed, with narrow rounded ridges.

A beautiful and valuable timberline forest tree, not adaptable outside its native habitat, where it does well on high, cold, mountain slopes and valleys with moist, well-drained, loose soils. Longevity varies, estimated to be 100-200 years in habitat. Series associations: Engspruce, Foxpine, Lodpine, Mixsubfor, **Mouhemlock**, Redfir, Subfir, Whipine, Alayelcedsta, KlaMouEnrsta, Pacsilfirsta.

e N.A. SHD; IRR: M/M/-/-/-/-

Ulmus americana

AMERICAN ELM

Ulmaceae. Sunset zones 1-11, 14-21. Deciduous. Native to the U.S. east of the Rocky Mountains. Growth rate rapid to 100' tall or more with an equal or greater spread, developing an upright, vaselike form, often with multiple trunklike main branches gradually arching upward from a large basal trunk. Leaves are alternate, simple, 3-6" long by 2-3" wide, dark green, oblong-ovate to oval, rough-surfaced and crinkled, oblique at the base, with doubly serrate edges, pale undersides, tufted hairs at vein axils, and yellow fall color. Clusters of pale yellowish green, long-stalked, fascicled flowers occur in early spring. Fruits are light brown, thin, papery samaras, 1/2" long, oval to ovate, with a ciliate-edged wing encircling the seed, maturing in summer. Twigs are slender, reddish brown, slightly hairy at first, becoming smooth. Thick gray bark becomes furrowed, with narrow, flat ridges.

A classic street or shade tree for large areas. Dutch elm disease has restricted its use and placed all existing trees in jeopardy. Grows best in deep soil with ample room for heavy surface roots. Subject to aphids, which produce honeydew drip, sooty mildew, and leaf and bark beetles. Longevity estimated to be over 200 years in ideal conditions.

e Asia. SHD; IRR: M/M/M/M/M/M

Ulmus parvifolia

CHINESE ELM

Ulmaceae. Sunset zones 3-24. Deciduous. Native to China and Japan. Growth rate fast or very fast to 40-60' tall with a 50-70' spread, forming a broad, arching or oval canopy, with branch ends slightly weeping. Leaves are alternate, simple, 1 to 2-1/2" long by 1/2-1" wide, dark green, elliptical, pointed, with lighter undersides, finely toothed edges, and yellow to reddish orange fall color in colder areas. Often only briefly deciduous in colder areas, retaining most foliage through mildest winters. Tiny yellow-green flowers occur in fall, usually unseen, but the tiny brown petals fall to the ground, forming dustlike litter, followed by small, flat, papery seeds in winter. All other elms flower in spring. Bark is rather smooth, light brown, with thin, flaky scales.

A widely used shade tree for streets, parks, parking lots, and residential patios or lawns. Large older specimens are quite attractive with their mottled trunks and broad, irregular branching. A carrier of Dutch elm disease, but does not suffer visible effects. Needs adequate room for vigorous roots. Leaf and seed litter may be objectionable in some situations. Longevity estimated to be 100 years or more. 'Brea' has large leaves and upright form, 'Drake' has small leaves and is more weeping, 'True Green' has small dark green leaves, a round-headed form, and is the most evergreen.

e&s N.A. SHD; IRR: No WUCOLS

Ulmus rubra

SLIPPERY ELM

Ulmaceae. No Sunset zones. Deciduous. Native to the eastern U.S. west to the eastern Great Plains and central Texas. Growth rate rapid to 40-70' tall or more with an equal or wider spread, developing an upright, round-headed, vaselike form and multiple branches with long drooping ends gradually arching outward from a large basal trunk. Leaves are alternate, simple, 4-7" long by 2-3" wide, dark green, oblong-ovate to oval, short-stalked, oblique at the base, with coarse, doubly serrate edges, rough textured on both the upperside and the pale underside, and yellow fall color. Clusters of pale yellowish green, short-stalked, fascicled flowers occur in March to May. Fruits are light brown, 1/2" long, oval, thin, papery samaras, tightly bunched, brown and woolly at the distinct seed cavity, otherwise smooth, with a wing encircling the seed at the center, notched at the end, maturing in spring. Twigs are stout, rough, hairy, reddish brown, with red to dark brown hairy buds. Bark is thick, dark reddish brown, becoming dark gray, with nearly parallel furrows and narrow, flat ridges.

An attractive elm, once commonly used as a park, street, or shade tree. Smaller, broader, with larger leaves and more drooping, arching, vaselike appearance than American elm (*Ulmus americana*). Longevity estimated to be over 200 years.

w N.A. California Native: CP, VG, OW, FW, RW, CF, ME, PF (Riparian) (*NW, CaRF, SNF, SnFrB, SCoRO, scattered in TR, PR*); EVG, SCR; (IRR: M/M/M/M/-/-)

Umbellularia californica

CALIFORNIA BAY

Lauraceae. Sunset zones 4-9, 14-24, Jepson *4,**5,6,14-17**,IRR:1-3,**7-9,18-24**. Evergreen. Native to southern Oregon, the California Coast Ranges, and lower elevations of the Sierra Nevada. Growth rate slow to 75' tall by 100' wide in forest habitats, usually 20-25' tall and wide in cultivation, developing a dense, oval, upright billowy form, often multi-trunked, with foliage to the base. Denser in sunny locations, often exhibiting shrubby, stunted growth in windy coastal areas. Leaves are alternate, simple, 3-5" long by 3/4-1" wide, thick, leathery, smooth, shiny, dark to light yellowish green, lanceolate to elliptical, glabrous, with a blunt, pointed tip, smooth edges, dull, pale undersides, strongly aromatic, and continuously produced as branches grow throughout the year, persisting 2-6 years. Small, dense umbels of tiny, yellow, perfect flowers are noticeable in spring before new growth begins. Fruits are inedible, acrid, 1/2-3/4" round, olivelike, green drupes with fleshy skin, enclosing a hard-shelled seed, turning purple in winter and dropping shortly thereafter. Bark is thin, smooth, grayish, thickening and darkening to brown, with a pockmarked, scaly surface.

A common riparian tree in foothill and mountain canyons near streams or in open bottomlands with nearby rivers and high water table. Reseeds readily in habitat. Dense, clean appearance. Does best in fertile soil with ample water, but tolerates heat, drought, and coastal conditions. Leaves are a more potent substitute for sweet bay (*Laurus nobilis*) leaves in cooking. Longevity estimated to be 150-200 years. Series associations: Canlivoakshr shrub series; BigDoufircan, Blaoak, **Calbay**, Calbuckeye, Calsycamore, Calwalnut, Canlivoak, Coalivoak, Doufir, Doufirtan, Mixoak, PorOrfced, Redwood, Sarcypress, Tanoak, Weshemlock.

sw N.A. California Native: MD (*Dmoj*); EVG, DES, FOL ACC, SPC; (IRR: L/L/L/L/L/L)

Yucca brevifolia

JOSHUA TREE

Liliaceae. Sunset zones 7, 9-16, 18-23, Jepson *DRN,SUN,DRY: 2,**3**,7,9,**10**,11,12,14-16,**18-21**,22,23. Evergreen. Native to dry soils of the Mojave Desert of California, Nevada, Utah, and Arizona at 2,000-6,000' elevation. Growth rate slow to 15-30' tall and 30' wide, developing a columnar trunk and many irregular, ascending branches with tufted foliage at the ends. Leaves are thick, leathery, lance-shaped, dark green, 6-10" long by 1/4-1/2" wide, somewhat flexible but not drooping, dark green to bluish green, clasping the trunk in a densely whorled fashion, with finely toothed margins without threadlike fibers, short, sharp-pointed ends, and dried, brown foliage hanging skirtlike below current green growth. Flowers are showy, nodding, greenish white, bell-shaped, 2-3" long, with 6 leathery, broad-pointed sepals, arising from a short, hairy stalk, densely spaced on a 1' long vertical spike, in early spring, with a mushroom-like odor. Fruits are cylindrical, 2-4" long by 2" wide, green, blunt-ended berries with fleshy pulp, containing many small, flat, rough, dull black seeds, drying when mature but not splitting open, falling the following spring. Dried foliage may remain attached to trunks or fall off to expose scaly brown or grayish bark.

The largest of the native yuccas and a characteristic plant of Mojave Desert regions with a picturesque silhouette of irregularly upright limbs. Not commonly cultivated, but often the predominant tree where it grows naturally, often occurring in sparse groves. Longevity estimated to be 100-150 years. Series associations: Beasedge, Biggalleta, Bigsagebrush, Bitterbrush, Blabush, Crebush, Crebuswhibur, Hopsage shrub series; **Jostree**, Caljuniper.

se N.A. EVG, FOL ACC, FLW ACC; IRR: L/L/L/L/L/L

Yucca gloriosa

SPANISH DAGGER

Liliaceae. Sunset zones 7-9, 12-24. Evergreen. Native to the southeastern U.S. from North Carolina to Florida. Growth rate slow to 10' tall and 8' wide, developing a broad-based trunk and upward-ascending branches with tufted foliage at the ends. Leaves are thick, leathery, lance-shaped, dark green, 1-2' long, somewhat flexible but not drooping, bluish green, clasping the trunk in densely whorled fashion, with smooth reddish margins, without teeth or threadlike fibers, pointed ends, a short, stout spine not sharp enough to pierce skin, and dried brown foliage hanging skirtlike below current green growth. Showy, nodding, pure white, 2-3" long, bell-shaped flowers, with 6 broadly pointed sepals, from a short, hairy stalk, are densely spaced on a 1 to 2-1/2' vertical spike, in summer. Fruits are cylindrical, 3-5" long by 1 to 1-1/2" wide, green, blunt-ended berries with a sweet-tasting pulp, containing many small flat, rough, dull black seeds, maturing to black in fall and persisting briefly into early winter. Dead foliage may be removed to expose trunks, which have scaly brown or grayish bark.

A desert accent or lush tropical specimen tree also effective in small groupings. Tolerates heat, drought, and poor soil, but not excessive frost or overwatering. Longevity estimated to be 80-100 years. A variegated form exists, but is rarely seen. According to Sunset, plants sold under this name are often a form of *Yucca elephantipes* or a hybrid between that species and *Y. gloriosa*.

e Asia. SHD, STR, FAL CLR; IRR: M/M/L/M/M/M

Zelkova serrata

SAWTOOTH ZELKOVA

Ulmaceae. Sunset zones 3-21. Deciduous. Native to Japan. Growth rate moderate to fast to 60' tall or more with an equal spread, forming a broad, arching canopy with a strong framework of multiple branches from a large, short trunk. Leaves are alternate, simple, 2 to 3-1/2" long by 1 to 1-1/2" wide, dark green, short-stalked, ovate-elliptical, with 8-14 straight parallel veins on each side of the midvein, sharply sawtoothed edges, a sandpapery texture from the short stiff hairs, flatly pointed toward the leaf tip on the upperside, pale undersides, usually glabrous, and yellow to orange, red, or reddish brown fall color. Insignificant clusters of 1/8" long, greenish yellow flowers occur in spring, males clustered at axils of new lower leaves and the few solitary females at upper leaves. Fruits are small, smooth, 3/16" egg-shaped, oblique drupes, nearly stalkless, maturing in fall. Twigs are slender, smooth, reddish brown. Smooth gray bark often has curious horizontal brownish warty lines, which disappear as bark ages to a dark grayish brown, becoming vertically fissured, with scaly, thin, peeling plates.

An excellent large street or park shade tree, generally preferring climates with seasonal change and colder winters. Deep-rooted, but requires adequate room for its canopy and large roots. Tolerates wind and heat. Fairly pest resistant, except for occasional red spider mites and black scale. Closely related to elms, and though resistant to Dutch elm disease, can be considered a carrier. Longevity estimated to be 200-300 years.

se Europe, e Asia, Africa. SHD, SPC, FAL ACC, FRU; IRR: L/L/L/M/M/M

Zizyphus jujuba
CHINESE JUJUBE

Rhamnaceae. Sunset zones 6-16, 18-24. Deciduous. Native to southeastern Europe, southern and eastern Asia, and Africa. Growth rate slow to moderate to 15-20' tall (possibly 30') with a 10-15' spread, developing a tall oval form with gnarled, spiny, erratic, very twiggy branches drooping at the ends. Leaves are alternate, simple, 1-2" long, bright shiny green, oblong-ovate to ovate-lanceolate, unevenly sided, with 3 prominent veins, finely crenate edges, prickly spines at the base, blunt-tipped, short petioles giving the appearance of a compound pinnate leaf, and deep yellow fall color. Clusters of small, short-stalked, yellowish flowers occur in May and June. Fruits are small, elongated, ovoid to oblong, edible, datelike drupes, maturing to shiny dark red to black in September and October. Bark is dark brown, furrowed, with narrow, flat, scaly ridges.

Mostly a curiosity tree, though sometimes seen as a lawn or parkway tree with an attractive semi-weeping character when well-maintained. Deep-rooted and relatively pest free. Extremely tolerant of heat, including desert conditions and alkaline or salty soils. Thrives in fertile soil with moderate watering. Minor pruning may be necessary to maintain desired shape. Sometimes suckers heavily from the base. Fruit drop can be messy. Longevity estimated to be 50-75 years.

TAXONOMY

BASIC PLANT CLASSIFICATION AND

IDENTIFICATION TOOLS

ELEMENTS OF TAXONOMY

When we see a tree, how do we know what we're looking at or describe to others what we see? In order to do this in any meaningful detail, we need an understanding of botanical terms and principles of classification. We can more effectively recognize and describe detailed aspects of a tree's leaves, flowers, and seeds by utilizing established botanical terminology. Classification leads us to a botanical name and often to at least one common name as well. Together, these elements provide the tools necessary to identify the plant in question, and familiarity with these tools facilitates the process.

The taxonomy of plants, or the science of systematic botany, involves the terminology, classification, nomenclature, and identification of plants. Terminology refers to words or phrases used to describe the characteristics of plants. Specific aspects of a leaf or flower can be described in a single widely understood and well defined word. The glossary provided in this book contains most of the terminology one will encounter at a basic level.

Classification of plants is a means of organizing plants according to recognizable features by categorized groups sharing similar characteristics and conforming to an accepted nomenclatural system or hierarchy. Classification systems may be based on genetic relationships (phylogeny) or on observed physical characteristics. Systems based on phylogeny classify plants according to descent and evolution, where present-day species are considered to derive through mutation or ecological adaptation over thousands or millions of years from evolutionary ties to plants that are now extinct. Much of this is simulated from fossils or relics of plants that existed millions of years ago and is often subject to conjecture. Classification by observed characteristics is the approach most commonly used today, where groupings of plants (taxa) are categorized in a hierarchy according to their physical characteristics. Many such systems exist, but the hierarchy typically begins at the highest level of Division followed by Class, Subclass, Group, Order, and Family. The Key to Tree Genera and the Key to Tree Species provided herein follow this hierarchy in outline format. Charts depict groupings by family and by genus within families.

Nomenclature refers to the accepted naming of a plant species, representing the final step in identification. Species are the basic unit of classification. Species names consist of two parts: the genus name (capitalized) and the specific epithet (lower case). Variations below the species level may be assigned subordinate definitions of subspecies (ssp.) and variety (var.) to designate significant variations from the species. In this book subspecies is used where it applies to native trees, which more often occur in evolutionary settings and exhibit a distributional range. Varietal names are more often used herein for ornamental trees that are artificially induced and may not bear true from seed (many are grafted stock from single genetic mutations).

The science of taxonomy can involve complex and highly detailed study. This section presents basic elements that can assist in identifying the trees presented in this book. Since leaves are usually the starting point for identification, a logical first step is to determine whether the leaf is from an evergreen or a

A piece of typing paper provides a white background for photographing a spray of *Abies procera* in an arboretum where specimens cannot be removed for study.

deciduous tree. Broadleaf evergreen leaves generally are more glossy and leathery than deciduous leaves and also differ in obvious ways from needlelike or scalelike leaves typical of conifers. This narrows the search considerably. If the leaf is not coniferous, determining the leaf shape is a logical next step. Whether a leaf is simple (a single leaf) or compound (a leaf with multiple leaflets) further narrows the search. Simple leaves may be palmate or lobed, and compound leaves are either pinnate (leaflets along a central stalk) or compound pinnate (leaflets along secondary stalks). From this point, the outline or shape of the leaf (round, linear, oval, oblong, etc.) and the type of margin (lobed, serrated, entire, etc.) can generally lead to the genus name. The full species name may require further investigation into other aspects about the tree, such as bark, cones, flowers, height, or habit for species that have similar leaf characteristics.

The keys provided in this book are based on discernible features of leaves, flowers, and seeds, most of which are also illustrated for reference. Using the main headings of leaf types, one can quickly skip to the applicable heading in the Key to Tree Genera. For quick reference where the genus is already known, species are listed alphabetically according to genus names in the Key to Tree Species.

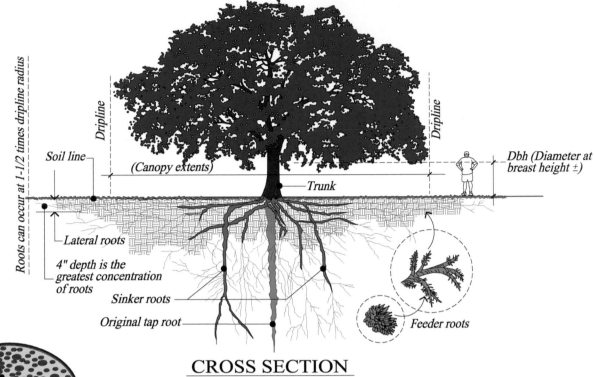

CROSS SECTION
Rootzone of native oak tree

Redrawn and adapted from an original drawing by Martin C. Hughes,
from *Native Oaks - Our Valley Heritage,* Heritage Oaks Committee, 1988

Roots can occur at 1-1/2 times dripline radius

Dripline

Dripline

Soil line

(Canopy extents)

Trunk

Dbh (Diameter at breast height ±)

Lateral roots

4" depth is the greatest concentration of roots

Sinker roots

Original tap root

Feeder roots

Scattered vascular bundles, not symmetrical (No growth rings)

CROSS SECTION
Monocotyledons

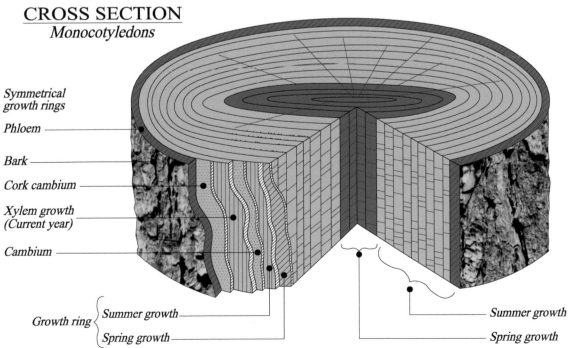

Symmetrical growth rings

Phloem

Bark

Cork cambium

Xylem growth (Current year)

Cambium

Growth ring { Summer growth

Spring growth

Summer growth

Spring growth

CROSS SECTION
Dicotyledons

GENERAL CHARACTERISTICS

BARK TYPES

Smooth
Usually thin bark

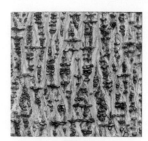

Warty
Smooth, with bumps

Scaly
Cracks into plates

Papery
Peeling, recurving

Fibrous
Matted fibers, threadlike

Fissured
Shallow, thin, vertical cracks

Furrowed
Deep depressions, ridged

Platy
Cracked, horiz. & vert.

HABIT

Trunk

Excurrent / Decurrent
Trunk continuous   Trunk splits
(Deliquescent)

Foliage

Conifer / Broadleaf
Needles   Flat leaves

Dormancy

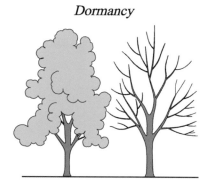

Evergreen / Deciduous
Leaves remain   Dormant period

LEAF VEINING

Arcuate
In parallel arcs

Palmate
Radial, from center

Parallel
Parallel to margin

Pinnate
Parallel, straight
(Penniveined)

LEAF PARTS

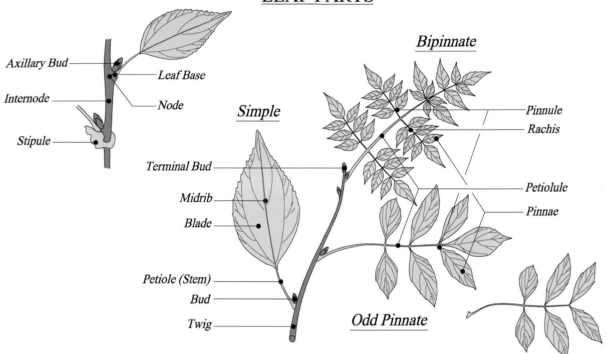

Axillary Bud
Leaf Base
Internode
Node
Stipule

Simple

Bipinnate

Terminal Bud
Midrib
Blade

Pinnule
Rachis

Petiolule
Pinnae

Petiole (Stem)
Bud
Twig

Odd Pinnate

Even Pinnate

LEAF ARRANGEMENT

Some examples redrawn and adapted from *Pacific Coast Trees*, McMinn & Maino, UC Press 1963

Alternate
Single leaf each node,
aligned alternately along stem

Opposite
2 leaves each node,
aligned oppositely along stem

Whorled
3 or more leaves per node,
arising in circular fashion around stem

MODIFIED LEAF SHAPES

Scalelike
Short jointed, either
flattened or ropelike

Awl-like
Short, pointed, needlelike
closely spaced, usually whorled

Needlelike
Sheathed, in bundles
at base

Needlelike
Unsheathed, sessile or
short-stalked

Vestigial
Modified remnant leaf,
as if not fully developed

LEAF TYPES

LEAF APICES

Acute
Sharply tapered,
to a pointed end

Acuminate
Gradually tapered to point,
slightly elongated

Apiculate
Acute tip with small, sharp
point, not of midrib

Aristate
Slender, bristlelike
elongated tip

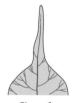
Caudate
Slender, elongated
tail-like tip

Cuspidate
Concave, constricted end
with sharp, stiff tip

Emarginate
Medium inset apex,
notched, not lobed

Mucronate
Short, pointed tip,
extension of midrib

Mucronulate
Very small tip projection,
extending from midrib

Obcordate
Deeply inset lobed
tip

Caudate
Slender, elongated
tail-like tip

Cuspidate
Concave, constricted end
with sharp, stiff tip

LEAF BASES

Attenuate
Constricting concavely,
with sides extended

Auriculate
Lobes each side of petiole,
narrow sinus between

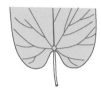
Cordate
Lobes each side of petiole,
heart-shaped

Cuneate
Wedge-shaped, angled
sides to base

Truncate
Flat-angled base, roughly
perpendicular to petiole

Oblique
Connecting unevenly, or
offset at base

Obtuse
Rounded, constricted
abruptly at base

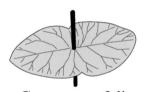
Connate-perfoliate
Two opposite stemless leaves
fused at base around stalk

COMPOUND LEAF SHAPES
(2 or more leaflets)

Pinnate
Multiple pinnae
along central stalk

Trifoliate
3 leaflets radiating
from central stalk

Palmate
5 or more leaflets radiating
from central stalk

Bipinnate
Twice compound multiple
leaflets along substalks

SIMPLE LEAF SHAPES
(Single Leaf)

Deltoid
Equilateral,
triangle shaped

Elliptical
Broadest at middle,
half as wide as long

Lanceolate
Broad base, tapered
end, lance shaped

Linear
Narrow, flattened,
parallel sides

Oblanceolate
Broadest near end,
tapered to base

Oblong
Parallel margins,
2-3 X longer
than wide

Obovate
Very broad near end,
tapered to base

Oval
Circular,
or nearly so

Ovate
Egg shaped, broadest
near base

Palmate
5 lobed margins, handlike,
arise from central point

Orbicular
Lobed, with tips in
circular outline

Reniform
Very broad at end, tapers
to base, fan shaped

Rhomboid
Square shaped,
on 45 degree axis

LEAF MARGINS

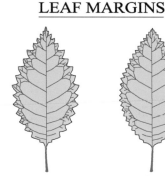

Cleft
Shallow, pointed
sinuses, angled
towards tip

Crenate
Broad, rounded teeth,
angled towards tip

Dentate
Coarse, sharp teeth,
pointing directly outward

Denticulate
Finely dentate, with
sharp teeth, pointing
outward

Double-Serrate
Serrated edges, with
serrations on teeth

Entire
Smooth edges, without
lobes or teeth

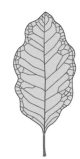

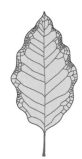

Lobed
Shallow sinuses,
usually rounded

Pinnatifid
Deeply lobed margins,
nearly pinnate

Serrate
Sharp sawlike teeth,
pointing forward

Serrulate
Finely serrated margins,
pointing forward

Sinuate
Edges wavy, in and out,
too shallow to be lobed

Undulate
Edges wavy, upward
and downward

SIMPLE-ENTIRE

Acacia
melanoxylon

Acacia longifolia

Acacia
retinoides

Agonis flexuosa

Agathis robusta

Brachychiton
acerifolius

Brachychiton
populneus

Arbutus
menziesii

Callistemon
citrinus

Catalpa speciosa

Cercis canadensis

Cercis canadensis
'Forest Pansy'

Callistemon
viminalis

Cercis occidentalis

Cercis reniformis

Chilopsis
linearis

Chrysolepis
chrysophylla

X Chitalpa
tashkentensis

Cornus
capitata

Cornus florida
'Cherokee Sunset'

Cornus kousa

Cinnamomun
camphora

Cornus florida

Cornus mas

Cornus nuttallii

SIMPLE-ENTIRE

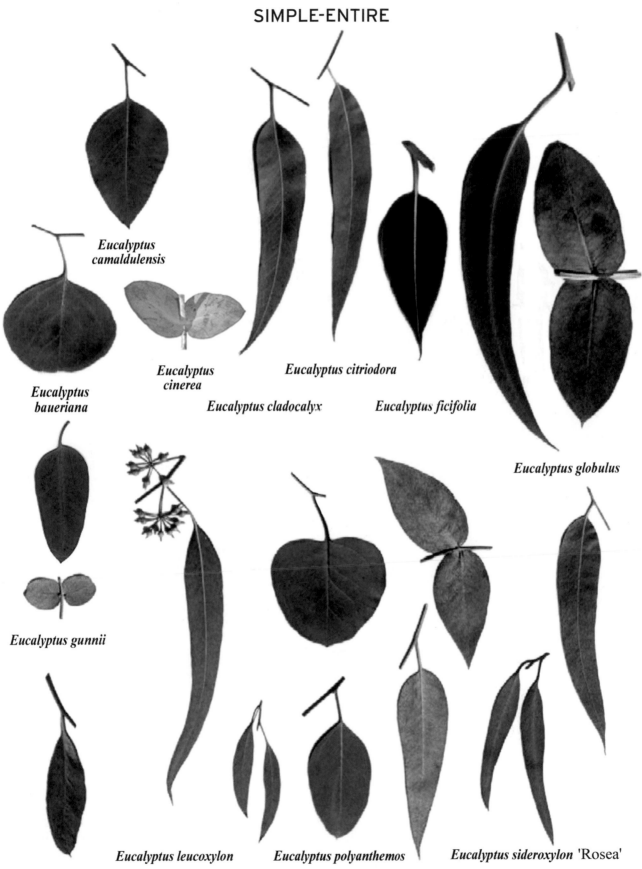

Eucalyptus camaldulensis

Eucalyptus baueriana

Eucalyptus cinerea

Eucalyptus cladocalyx

Eucalyptus citriodora

Eucalyptus ficifolia

Eucalyptus globulus

Eucalyptus gunnii

Eucalytpus conferruminata

Eucalyptus leucoxylon

Eucalyptus nicholii

Eucalyptus polyanthemos

Eucalyptus pulverulenta

Eucalyptus sideroxylon 'Rosea'

Eucalyptus viminalis

LEAVES

SIMPLE-ENTIRE

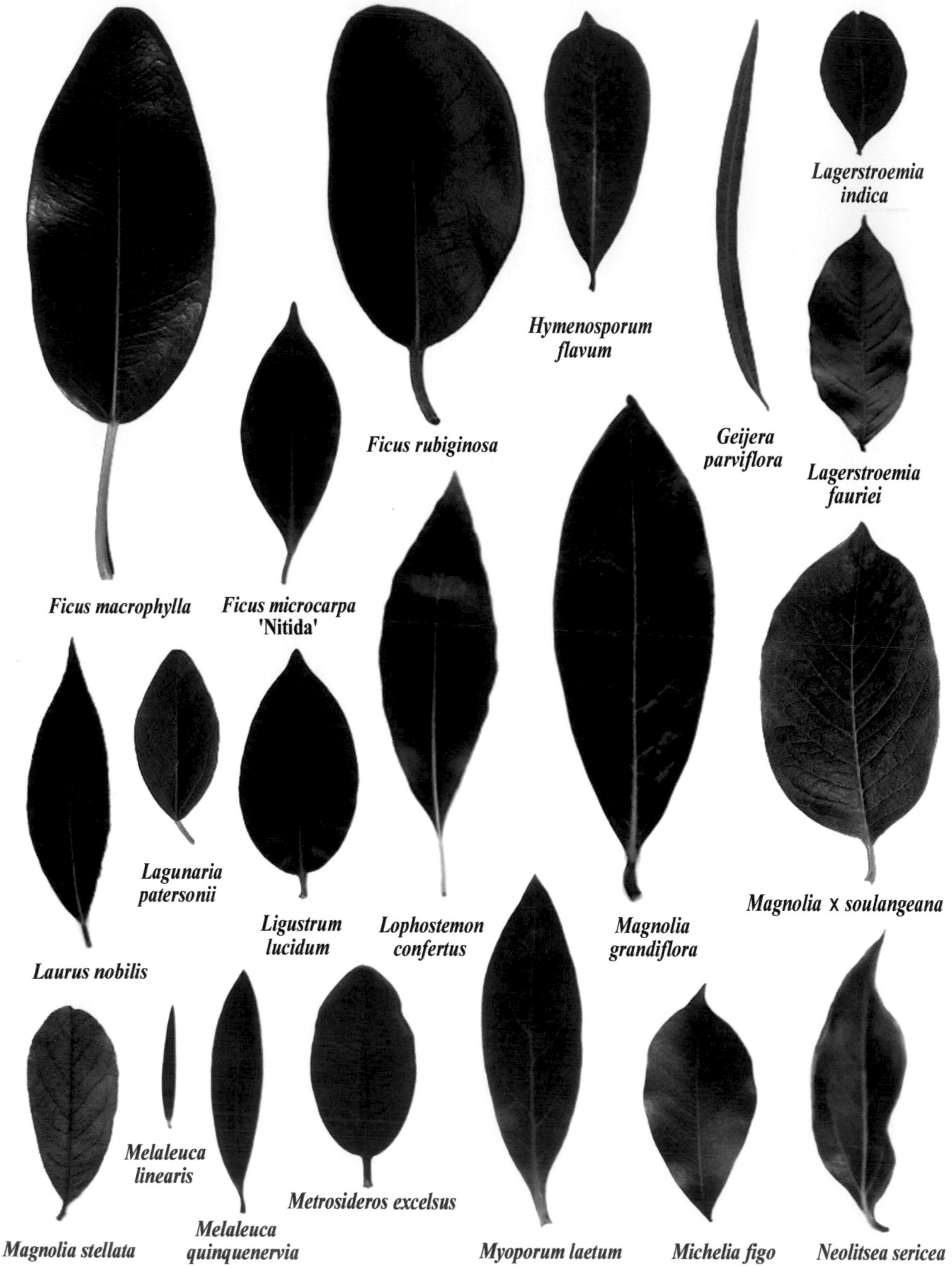

Lagerstroemia indica

Hymenosporum flavum

Ficus rubiginosa

Geijera parviflora

Lagerstroemia fauriei

Ficus macrophylla

Ficus microcarpa 'Nitida'

Lagunaria patersonii

Ligustrum lucidum

Lophostemon confertus

Magnolia grandiflora

Magnolia × soulangeana

Laurus nobilis

Melaleuca linearis

Metrosideros excelsus

Magnolia stellata

Melaleuca quinquenervia

Myoporum laetum

Michelia figo

Neolitsea sericea

SIMPLE-ENTIRE

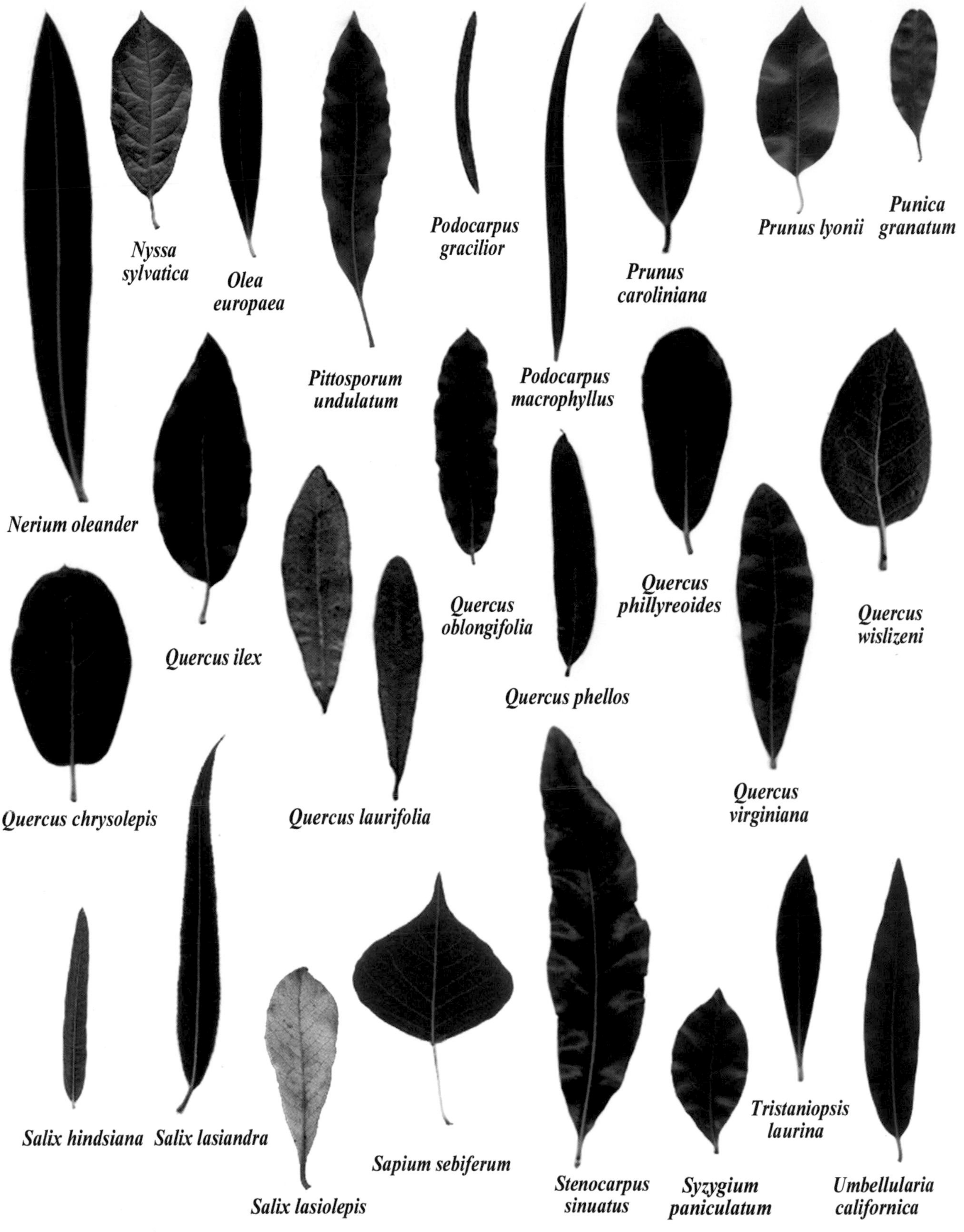

Nyssa sylvatica

Olea europaea

Podocarpus gracilior

Prunus caroliniana

Prunus lyonii

Punica granatum

Pittosporum undulatum

Podocarpus macrophyllus

Nerium oleander

Quercus ilex

Quercus oblongifolia

Quercus phillyreoides

Quercus wislizeni

Quercus phellos

Quercus chrysolepis

Quercus laurifolia

Quercus virginiana

Salix hindsiana **Salix lasiandra**

Sapium sebiferum

Salix lasiolepis

Tristaniopsis laurina

Stenocarpus sinuatus

Syzygium paniculatum

Umbellularia californica

LEAVES

SIMPLE-TOOTHED

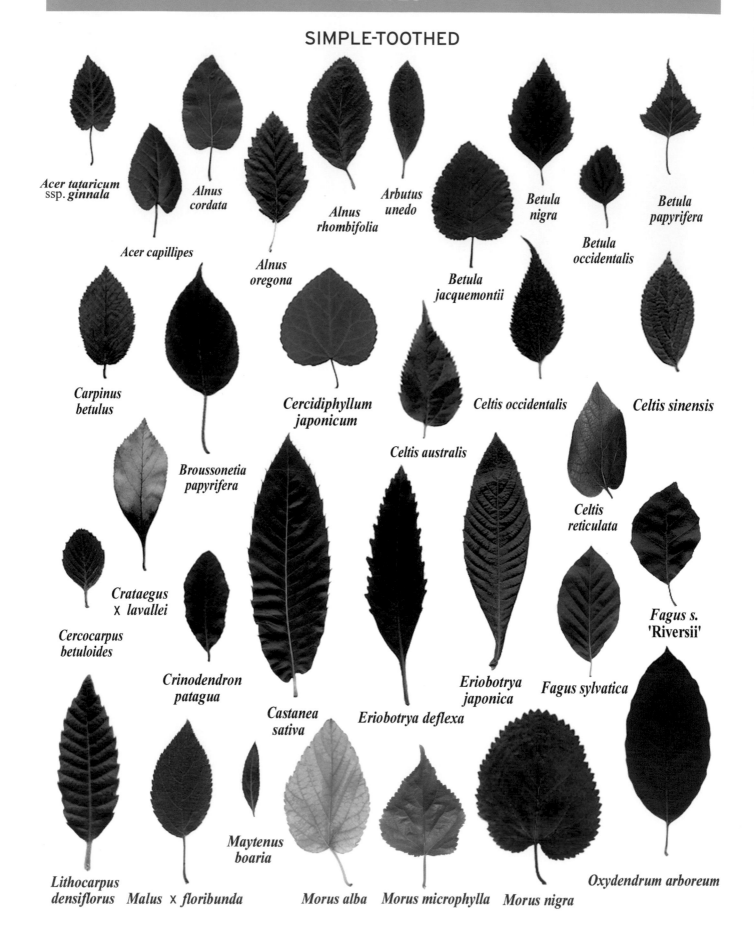

Acer tataricum ssp. ginnala

Acer capillipes

Alnus cordata

Alnus rhombifolia

Alnus oregona

Arbutus unedo

Betula jacquemontii

Betula nigra

Betula occidentalis

Betula papyrifera

Carpinus betulus

Broussonetia papyrifera

Cercidiphyllum japonicum

Celtis australis

Celtis occidentalis

Celtis sinensis

Celtis reticulata

Cercocarpus betuloides

Crataegus x lavallei

Crinodendron patagua

Castanea sativa

Eriobotrya deflexa

Eriobotrya japonica

Fagus sylvatica

Fagus s. 'Riversii'

Lithocarpus densiflorus

Malus x floribunda

Maytenus boaria

Morus alba

Morus microphylla

Morus nigra

Oxydendrum arboreum

LEAVES

SIMPLE-TOOTHED

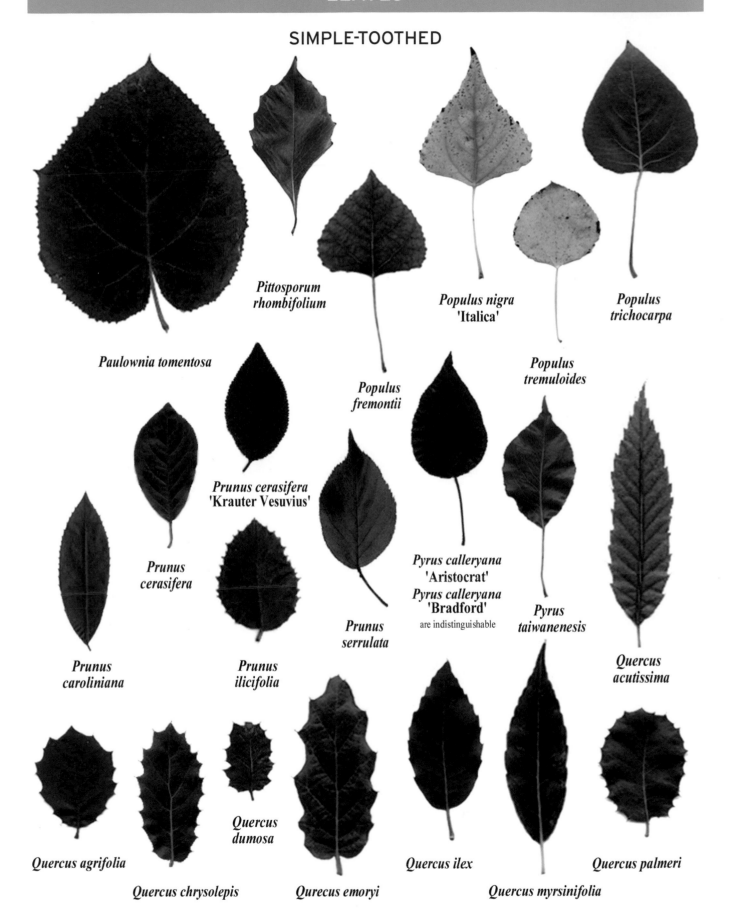

Paulownia tomentosa

Pittosporum
rhombifolium

Populus nigra
'Italica'

Populus
trichocarpa

Populus
tremuloides

Populus
fremontii

Prunus cerasifera
'Krauter Vesuvius'

Prunus
cerasifera

Prunus serrulata

Pyrus calleryana
'Aristocrat'
Pyrus calleryana
'Bradford'
are indistinguishable

Pyrus
taiwanenesis

Quercus
acutissima

Prunus
caroliniana

Prunus
ilicifolia

Quercus
dumosa

Quercus agrifolia

Quercus chrysolepis

Qurecus emoryi

Quercus ilex

Quercus myrsinifolia

Quercus palmeri

LEAVES

SIMPLE-TOOTHED

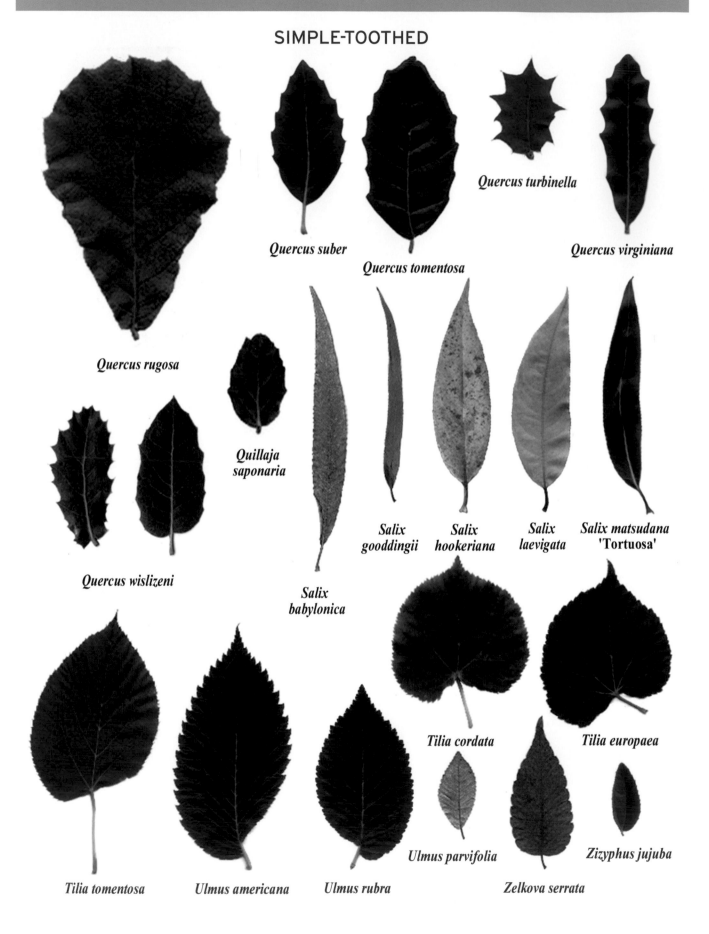

Quercus rugosa

Quercus suber

Quercus tomentosa

Quercus turbinella

Quercus virginiana

Quillaja saponaria

Quercus wislizeni

Salix babylonica

Salix gooddingii

Salix hookeriana

Salix laevigata

Salix matsudana 'Tortuosa'

Tilia cordata

Tilia europaea

Ulmus parvifolia

Zizyphus jujuba

Tilia tomentosa

Ulmus americana

Ulmus rubra

Zelkova serrata

SIMPLE-LOBED

Bauhinia variegata

Ginkgo biloba

Populus fremontii

Quercus arizonica

Quercus alba

Quercus coccinea

Quercus douglasii

Quercus engelmannii

Quercus gambelii

Quercus garryana

Quercus kelloggii

Quercus lobata

Quercus laurifolia

Quercus muehlenbergii

Quercus morehus

Quercus palustis

Quercus macrocarpa

Quercus robur

Quercus rubra

Stenocarpus sinuatus

LEAVES

SIMPLE-PALMATE

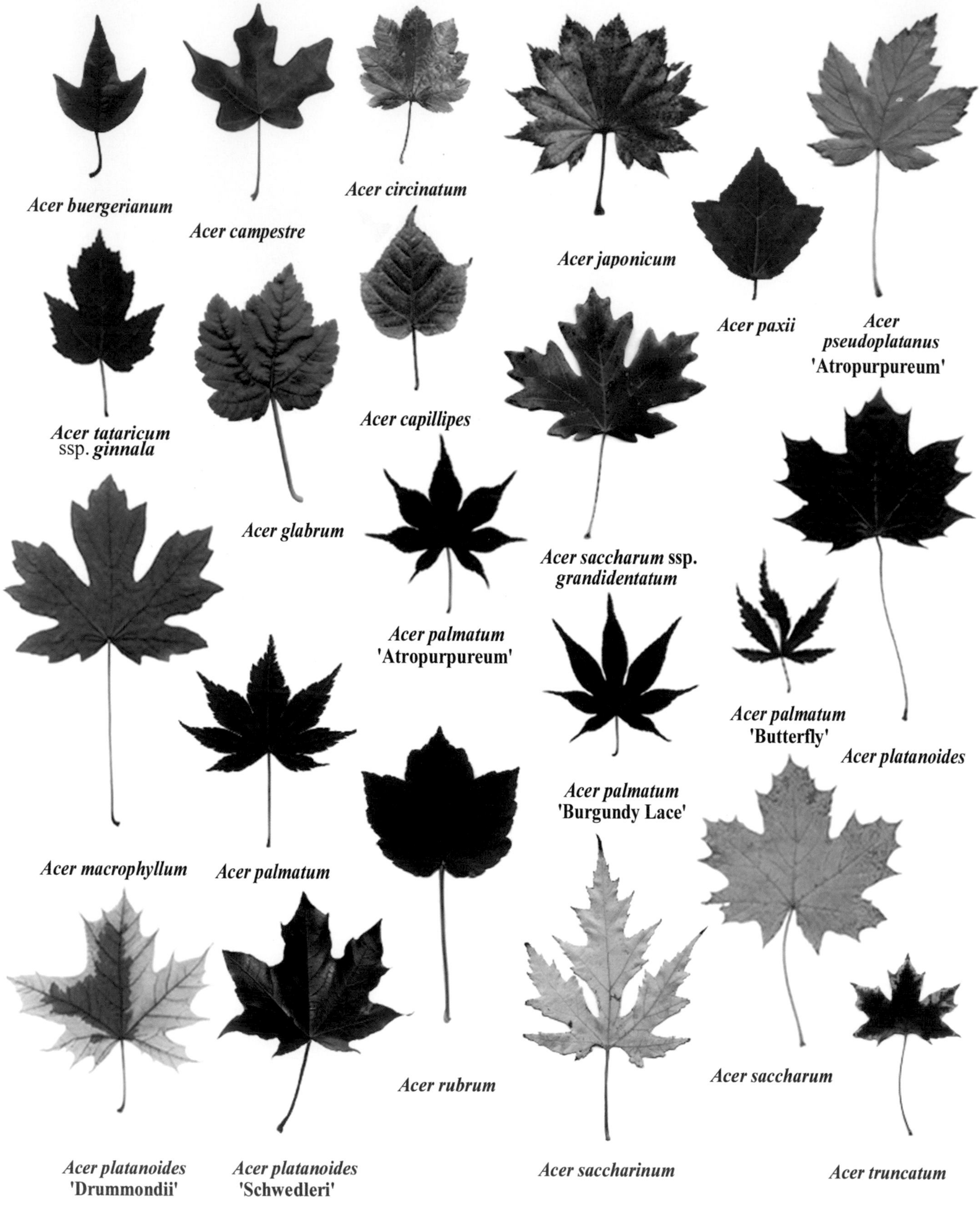

Acer buergerianum

Acer campestre

Acer circinatum

Acer japonicum

Acer paxii

*Acer
pseudoplatanus*
'Atropurpureum'

Acer tataricum
ssp. *ginnala*

Acer capillipes

Acer glabrum

Acer saccharum ssp.
grandidentatum

Acer palmatum
'Atropurpureum'

Acer palmatum
'Butterfly'

Acer platanoides

Acer macrophyllum

Acer palmatum

Acer palmatum
'Burgundy Lace'

Acer rubrum

Acer saccharum

Acer platanoides
'Drummondii'

Acer platanoides
'Schwedleri'

Acer saccharinum

Acer truncatum

SIMPLE-PALMATE

**Betula pendula
'Crispa'**

Brachychiton acerifolius

Brachychiton populneus

Broussonetia papyrifera

Firmiana simplex

**Crataegus
laevigata**

**Liquidambar
formosana**

**Liquidambar styraciflua
'Rotundiloba'**

**Liriodendron
tulipifera**

**Liriodendron
tulipifera
'Aureomarginata'**

**Crataegus
phaenopyrum**

Liquidambar styraciflua

Morus alba
(Adult)

Morus alba
(Juvenile)

Platanus occidentalis

Platanus x acerifolia

Platanus racemosa

Populus alba

LEAVES

COMPOUND-PALMATE

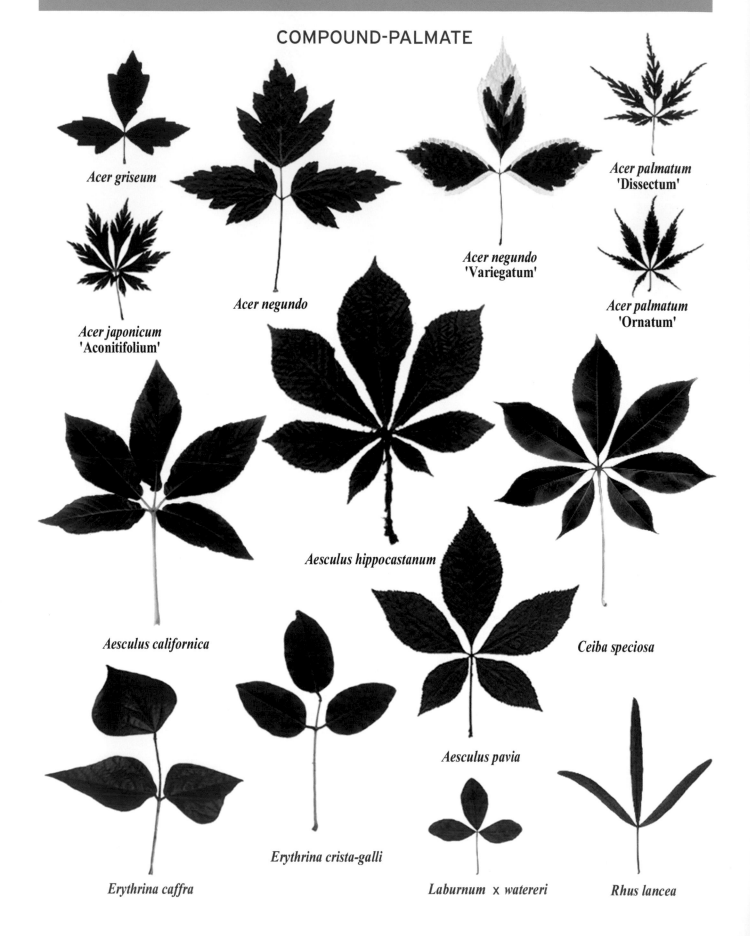

Acer griseum

Acer negundo

Acer negundo
'Variegatum'

Acer palmatum
'Dissectum'

Acer japonicum
'Aconitifolium'

Acer palmatum
'Ornatum'

Aesculus hippocastanum

Aesculus californica

Ceiba speciosa

Aesculus pavia

Erythrina crista-galli

Erythrina caffra

Laburnum x watereri

Rhus lancea

LEAVES

COMPOUND-PINNATE

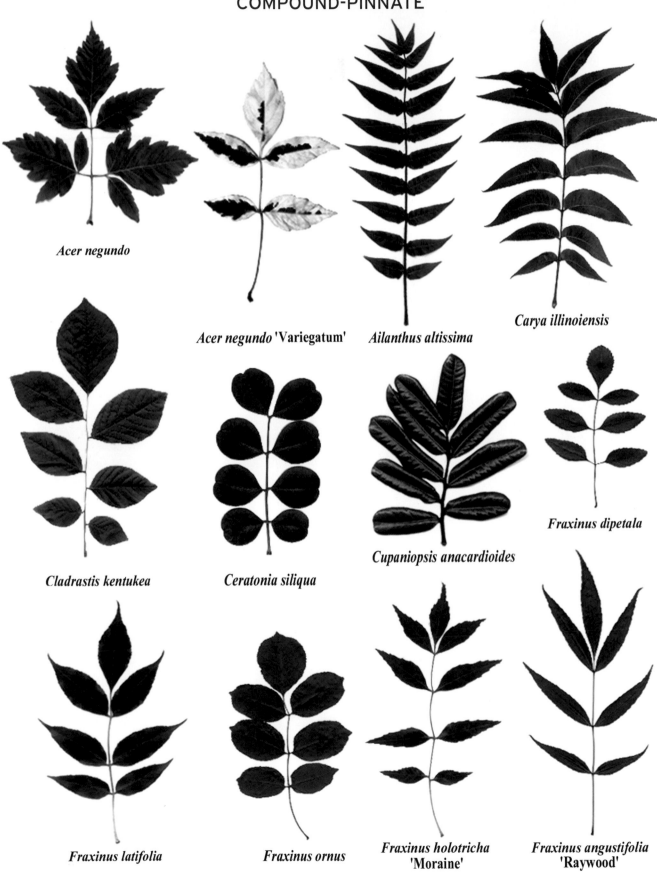

Acer negundo

Acer negundo 'Variegatum'

Ailanthus altissima

Carya illinoiensis

Cladrastis kentukea

Ceratonia siliqua

Cupaniopsis anacardioides

Fraxinus dipetala

Fraxinus latifolia

Fraxinus ornus

Fraxinus holotricha 'Moraine'

Fraxinus angustifolia 'Raywood'

LEAVES

COMPOUND-PINNATE

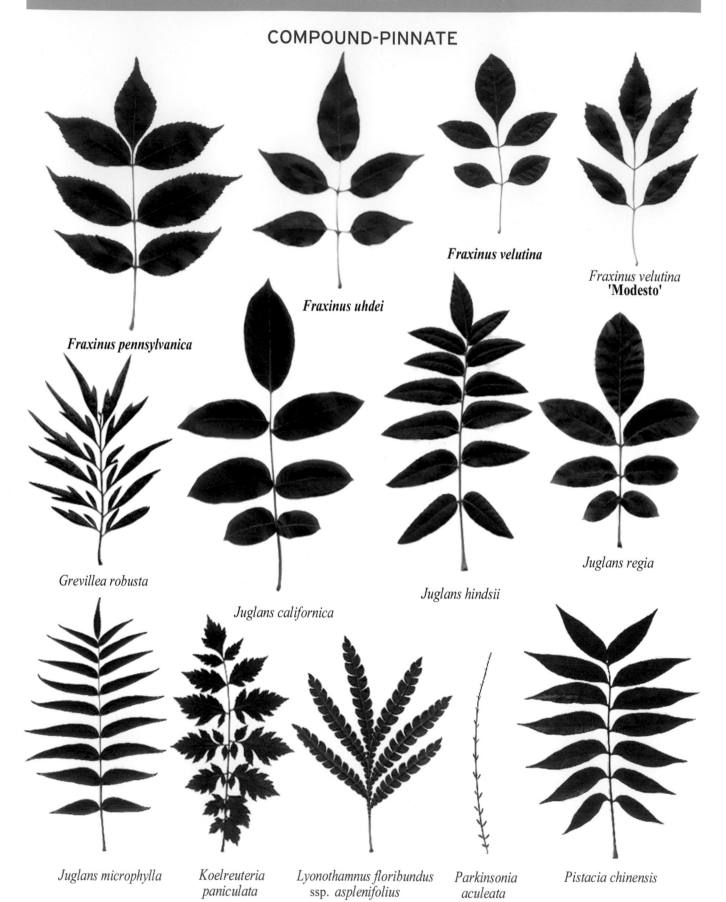

Fraxinus pennsylvanica

Fraxinus uhdei

Fraxinus velutina

Fraxinus velutina
'Modesto'

Grevillea robusta

Juglans californica

Juglans hindsii

Juglans regia

Juglans microphylla

Koelreuteria
paniculata

Lyonothamnus floribundus
ssp. *asplenifolius*

Parkinsonia
aculeata

Pistacia chinensis

COMPOUND-PINNATE

Pterocarya stenoptera

Rhus glabra

Rhus typhina

Rhus typhina
'Laciniata'

Robinia pseudoacacia

Sambucus mexicana

Schinus molle

Schinus peruviana

Schinus terebinthifolius

Sorbus aucuparia

Sophora japonica

Tipuana tipu

LEAVES

COMPOUND-BIPINNATE

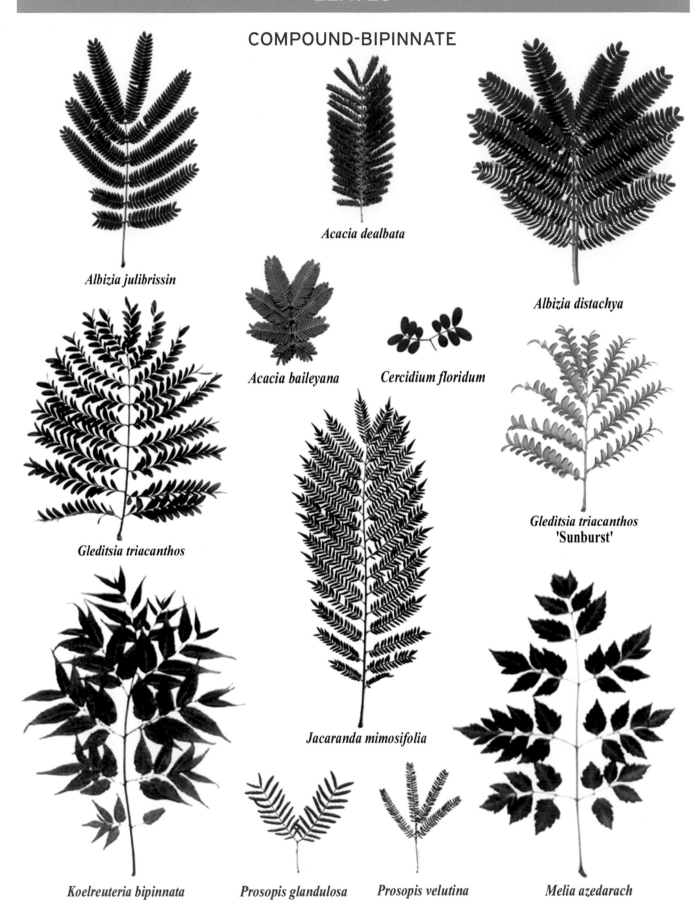

Albizia julibrissin

Acacia dealbata

Albizia distachya

Acacia baileyana

Cercidium floridum

Gleditsia triacanthos

Gleditsia triacanthos
'Sunburst'

Jacaranda mimosifolia

Koelreuteria bipinnata

Prosopis glandulosa

Prosopis velutina

Melia azedarach

NEEDLELIKE

Abies alba

Abies amabilis

Abies bracteata

Abies concolor

Abies grandis

Abies lasiocarpa

Abies magnifica

Abies magnifica 'Glauca'

Abies magnifica var. shastensis

Abies nordmanniana

Abies pinsapo

Abies procera

Araucaria araucana

Araucaria bidwillii

Araucaria heterophylla

COMPOUND-NEEDLELIKE

Casuarina equisetifolia

Cedrus atlantica
'Glauca'

Cedrus deodara

Cedrus deodara
'Descanso Dwarf'

Cryptomeria japonica

Cryptomeria japonica
'Elegans'

Cunninghamia lanceolata

Larix kaempferi

Melaleuca linariifolia

Metasequoia glyptostroboides

Picea abies

Picea breweriana

Picea engelmannii

Picea glauca **'Conica'**

Picea glauca **'Pendula'**

Picea jezoensis

NEEDLELIKE

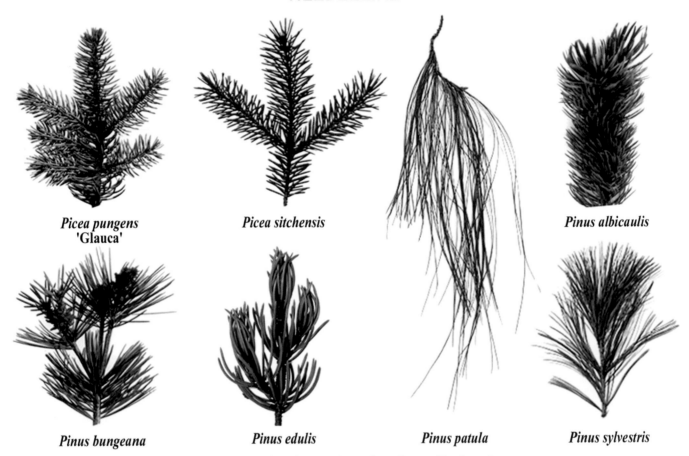

Picea pungens 'Glauca'

Picea sitchensis

Pinus albicaulis

Pinus bungeana

Pinus edulis

Pinus patula

Pinus sylvestris

Differentiating pines by number of needles and/or length

Not all pine species in the compendium are shown above, as many are difficult to differentiate visually in this context.
Identification of pines often begins with the needles or cones, as listed in the chart below.
Pages 315-343 of the compendium show the needles in more detail, and either list or show the shape of the cones.

Needles usually in 5s

Pinus albicaulis
Needles 1 to 2-1/2" long;
cones 1-3" long
Pinus flexilis
Needles 1 to 2-1/2" long;
cones 3-8" long
Pinus lambertiana
Needles 2 to 3-1/2" long;
cones 11-20" long
Pinus monticola
Needles 1 to 3-1/2" long;
cones 6-8" long
Pinus strobus
Needles 3-5" long;
cones 3-8" long
Pinus torreyana
Needles 7-11" long;
cones 4-6" long

Needles usually in 4s

Pinus quadrifolia
Needles 3/4 to 1-1/2" long,
occasionally in 1s to 5s;
cones 1-1/4 to 2" long

Needles usually in 3s

Pinus attenuata
Needles 3-5" long;
cones 3-6" long
Pinus bungeana
Needles 2 to 5-1/2" long,
usually in 3s, sometimes 5s;
cones 1-1/2 to 3" long
Pinus canariensis
Needles 9-12" long;
cones 4-8" long
Pinus coulteri
Needles 5-12" long;
cones 9-14" long
Pinus jeffreyi
Needles 5-10" long;
cones 5-8" long
Pinus patula
Needles 4-12" long, usually
in 3s. sometimes in 4s or 5s;
cones 2-1/2 to 4-3/4" long
Pinus ponderosa
Needles 5-10" long;
cones 3-5" long
Pinus radiata
Needles 3-1/2 to 6" long,
usually in 3s, rarely in 2s;
cones 2-5" long
Pinus sabiniana
Needles 7-13" long;
cones 6-10" long

Needles usually in 2s

Pinus brutia
Needles 4-7" long;
cones 2-3/8 to 4" long
Pinus brutia var. *eldarica*
Needles 2-1/2 to 6" long,
usually in 2s, rarely 3s;
cones 2 to 4-1/2" long
Pinus contorta ssp. *contorta*
Needles 1-1/4 to 2" long;
cones 1-1/4 to 2" long
Pinus contorta ssp. *murrayana*
Needles 1 to 2-1/2" long;
cones 1 to 1-3/4" long
Pinus densiflora
Needles 2-1/2 to 5" long;
cones 1-1/4 to 3" long
Pinus densiflora 'Umbraculifera'
Needles 1-1/2 to 3" long;
cones 1/2 to 3/4" long
Pinus edulis
Needles 1-2" long;
cones 1-2" long
Pinus halepensis
Needles 2-1/4 to 5-1/2" long,
usually in 2s, rarely in 3s;
cones 2-1/2 to 4" long
Pinus mugo
Needles 1-2" long;
cones 3/4 to 1-1/2" long

Needles usually in 2s

Pinus muricata
Needles 4-6" long;
cones 2 to 3-1/2" long
Pinus pinea
Needles 3-1/2 to 7" long,
usually in 2s, seldom in 3s;
cones 3-1/2 to 5" long
Pinus sylvestris
Needles 1-1/2 to 2-3/4" long;
cones 2-3" long
Pinus thunbergii
Needles 3 to 4-1/2" long;
cones 1-1/2 to 3" long

Needles usually single

Pinus monophylla
Needles 1-2" long;
cones 2-1/2 to 3-1/2" long

LEAVES

NEEDLELIKE

Podocarpus gracilior

Podocarpus macrophyllus

Pseudotsuga macrocarpa

Pseudotsuga menziesii

Sciadopitys verticillata

Sequoia sempervirens

Tamarix aphylla

Tamarix parviflora

Taxodium distichum

Taxus brevifolia

Torreya californica

Torreya nucifera

Taxodium mucronatum

Tsuga heterophylla

Tsuga mertensiana

SCALELIKE

Calocedrus decurrens

Chamaecyparis lawsoniana

Chamaecyparis lawsoniana
'Stewartii'

Chamaecyparis nootkatensis

× *Cupressocyparis leylandii*

Cupressus arizonica

Cupressus funebris

Cupressus guadalupensis

Cupressus macnabiana

Cupressus macrocarpa

Cupressus macrocarpa 'Lutea'

Cupressus sargentii

Cupressus sempervirens

Cupressus sempervirens
'Stricta'

Cupressus sempervirens
'Swanes Golden'

Juniperus californica

SCALELIKE

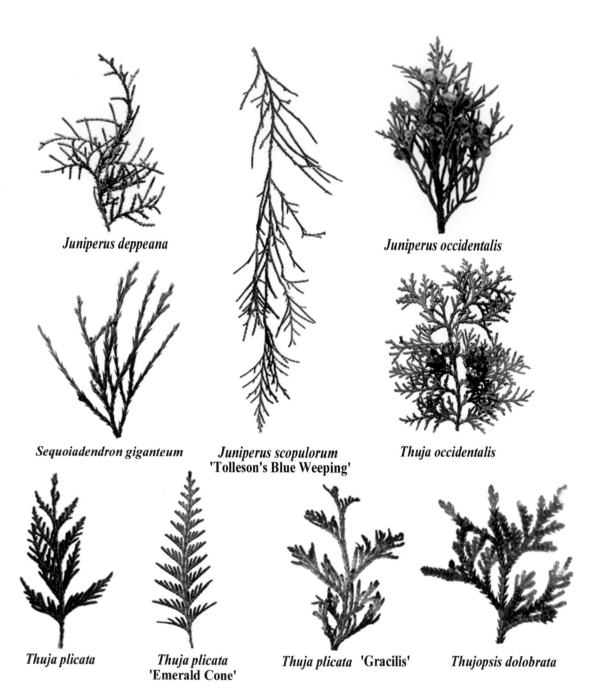

Juniperus deppeana

Juniperus occidentalis

Sequoiadendron giganteum

Juniperus scopulorum
'Tolleson's Blue Weeping'

Thuja occidentalis

Thuja plicata

Thuja plicata
'Emerald Cone'

Thuja plicata **'Gracilis'**

Thujopsis dolobrata

FLOWER PARTS

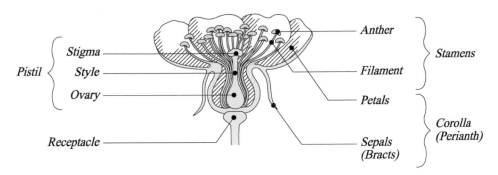

Pistil { Stigma — Anther

Style — Filament } Stamens

Ovary — Petals } Corolla (Perianth)

Receptacle — Sepals (Bracts)

PERFECT FLOWER

Perfect flower, without any missing parts,
though positions and parts may vary
Imperfect flowers have some parts missing
or modified

INFLORESCENCES

Arrangement of flowers on stems

   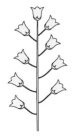 

Conelet
flowers enclosed
between woody scales

Ament (Catkin)
Sessile, tightly spaced
on usually drooping stalk
(unisexual flowers)

Spike
Sessile, spaced along
central stalk

Raceme
Short stems
from central stalk

Panicle
Branched stems from
central stalk

Umbel
Rounded, flat head,
unbranched stems

Cyme
Flat-topped head,
unbranched stems, flowers
open outward, from center

Fascicle
Long-stemmed,
from sheathed base,
no stalk

Corymb
Rounded, flat
head, branched stems,
from central stalk

VARIOUS COMMON TYPES

Petals —

Stamens —

Pistil —

Modified bracts (not true petals)

Flower cluster (vestigial petals)

Magnolia stellata

Cornus florida

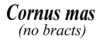

Cornus mas
(no bracts)

Multiple racemes of legume (pea-shaped) flowers —

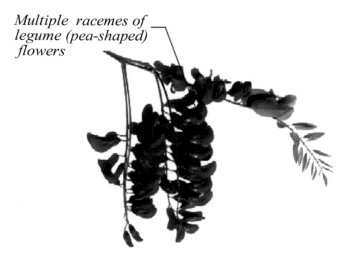

Flower cluster of urn-shaped flowers from thick central stalk

Terminal bud

Robinia 'Purple Robe'

Arbutus menziesii

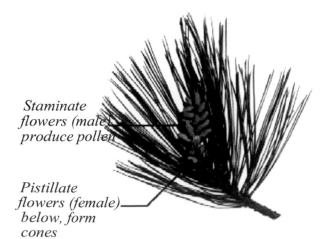

Staminate flowers (male) produce pollen

Pistillate flowers (female) below, form cones

Staminate flower (stamens expanded)

Stamens unopened

Pistillate flower (fruit) unopened

Leaf buds below

Pinus radiata

Salix lasiandra

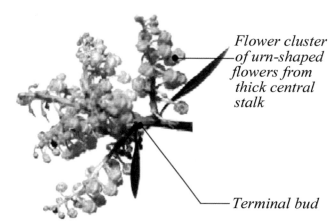

VARIOUS COMMON TYPES

Staminate
flowers (male)
produce pollen

Pistillate
flowers (female)
form cones

Calocedrus decurrens

Flowers in starlike
rosettes (corymb),
ripen into single-
celled drupe
berries

Sambucus mexicana

Male and
female flowers
on separate
trees, before
leaves emerge

Acer platanoides

Male and female
flowers together
in drooping
racemes

Acer macrophyllum

Staminate
flowers (male)
produce pollen,
then raceme
drops from
tree

Pistillate flowers
(female) form seedballs

Liquidambar styraciflua

Small greenish
flowers expand
into waferlike
seeds, appearing
flowerlike.

Seed

Ulmus americana

VARIOUS COMMON TYPES

SIMPLE

Achene
Dry 1-celled seed with no slits or valves

Acorn
Dry 1-celled nut with cuplike base

Aril
Dry 1-celled seed partially covered from base, inside pulpy sheath

Bean
Elongated fleshy pod splitting along 2 seams, containing multiple seeds

Berry
Fleshy fruit with 1 or more seeds

Capsule
Dry fruit containing multiple seeds, multiple seam splits

Cone
Woody scales, multiple separate seeds

Drupe
Fleshy fruit enclosing single hard-walled seed pit

Follicle
Dry 1-celled fruit, splits along 1 seam

Legume
Dry multiple seed fruit, usually flattened, 2 seams

Key
Samara type seed with persistent firmly attached wing

Nut
Dry shelled single seed fruit with outer fleshy skin clinging or separating

Pome
Fleshy skinned fruit with multiple seeds, apple-shaped

Samara
Dry 1-celled winged seed

Samara
Dry 1-celled winged seed

Samara
Dry 1-celled winged seed

Utricle
Inflated bladder pod enclosing seeds

AGGREGATE

Achenes
(Platanus)

Capsules
(Liquidambar)

Follicles
(Magnolia)

Samaras
(Liriodendron)

VARIOUS SEED TYPES

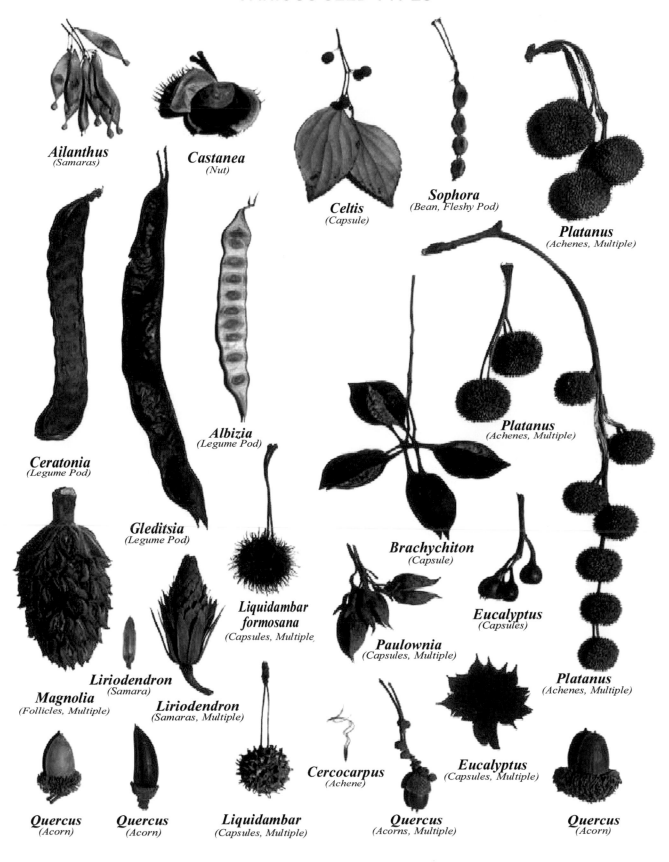

Ailanthus
(Samaras)

Castanea
(Nut)

Celtis
(Capsule)

Sophora
(Bean, Fleshy Pod)

Platanus
(Achenes, Multiple)

Ceratonia
(Legume Pod)

Gleditsia
(Legume Pod)

Albizia
(Legume Pod)

Platanus
(Achenes, Multiple)

Brachychiton
(Capsule)

**Liquidambar
formosana**
(Capsules, Multiple)

Eucalyptus
(Capsules)

Paulownia
(Capsules, Multiple)

Magnolia
(Follicles, Multiple)

Liriodendron
(Samara)

Liriodendron
(Samaras, Multiple)

Platanus
(Achenes, Multiple)

Quercus
(Acorn)

Quercus
(Acorn)

Liquidambar
(Capsules, Multiple)

Cercocarpus
(Achene)

Eucalyptus
(Capsules, Multiple)

Quercus
(Acorns, Multiple)

Quercus
(Acorn)

FRUIT TYPES

VARIOUS SEED TYPES

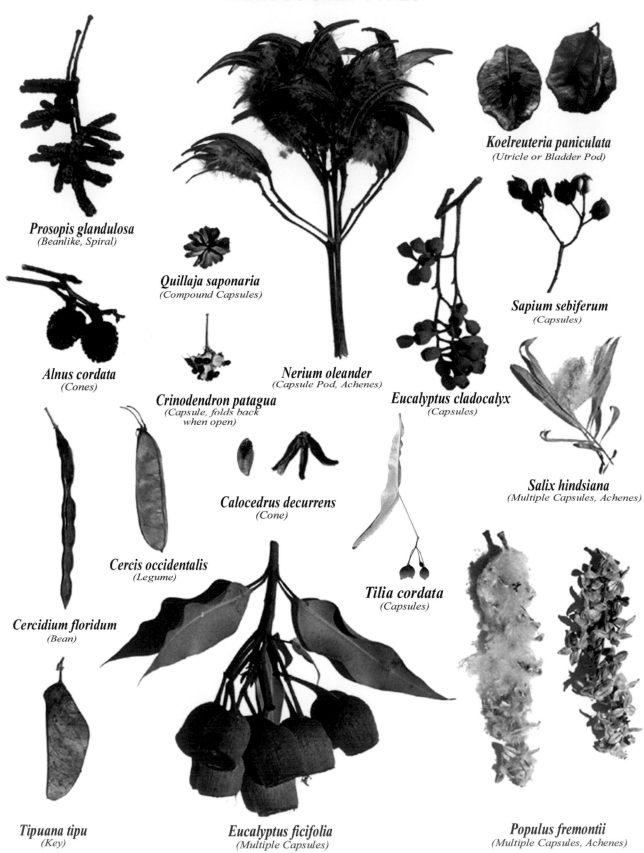

Prosopis glandulosa
(Beanlike, Spiral)

Quillaja saponaria
(Compound Capsules)

Koelreuteria paniculata
(Utricle or Bladder Pod)

Sapium sebiferum
(Capsules)

Alnus cordata
(Cones)

Crinodendron patagua
*(Capsule, folds back
when open)*

Nerium oleander
(Capsule Pod, Achenes)

Eucalyptus cladocalyx
(Capsules)

Salix hindsiana
(Multiple Capsules, Achenes)

Calocedrus decurrens
(Cone)

Cercis occidentalis
(Legume)

Tilia cordata
(Capsules)

Cercidium floridum
(Bean)

Tipuana tipu
(Key)

Eucalyptus ficifolia
(Multiple Capsules)

Populus fremontii
(Multiple Capsules, Achenes)

CONES & CONELIKE

Casuarina equisetifolia

Alnus cordata

Alnus rhombifolia

Cunninghamia lanceolata

Cupressus macrocarpa

Juniperus californica

Pinus pinea

Pinus bungeana

Cryptomeria japonica

Pinus contorta

Pinus radiata

Pinus attenuata

Cedrus deodara

Pinus lambertiana

Pinus jeffreyi

Pinus ponderosa

Pinus sabiniana

FRUIT TYPES

CONES & CONELIKE

Pseudotsuga macrocarpa

Pseudotsuga menziesii

Picea engelmannii

Picea sitchensis

Picea abies

Sequoia sempervirens

Pinus strobus

Pinus edulis

Sciadopitys verticillata

Metasequoia glyptostroboides

Sequoiadendron giganteum

Tsuga heterophylla

Tsuga mertensiana

Pinus densiflora 'Umbraculifera'

Pinus flexilis

Pinus sylvestris

Pinus coulteri

Larix kaempferi

acerifolius Maplelike leaves.

achene Dry 1-celled and 1-seeded fruit or carpel, with no slits or valves.

acicular Slenderly needle-shaped.

acorn Nut surrounded at the base by a fibrous or woody, usually scaled cup.

aculeate Covered with prickles.

acuminate Leaf tip gradually tapering to the apex, or long-pointed.

acute Leaf tip tapering to a point, sides straight or slightly convex.

adherent Unlike parts in close contact but not fused.

adpressus Pressed against.

aestival In spring.

aggregate Compound fruit developing from separate pistils of the same flower (e.g., *Morus*).

alate Winged.

albus White.

alpinus Alpine, or of the Alps.

alternate Leaf or twig arrangement, with one per node, instead of opposite or whorled.

altus Tall.

ament Scaly, bracted spike of a usually unisexual flower, frequently deciduous in one piece. Same as catkin.

androecium Male reproductive element of a flower, composed of one or more stamens.

androdioecious Having staminate flowers on one plant and bisexual flowers on another.

andromonoecious Having male and bisexual flowers on the same plant.

angiosperm Plant with seeds borne in an ovary.

angustifolia Having narrow leaves.

anther Pollen-bearing part of the flower stamen.

apetalous Lacking petals.

apex Uppermost point.

apiculate Leaf ending in a minute, short pointed tip, not an extension of midvein, but stiff or sharp to touch.

apophysis Exposed swollen part of a cone scale.

appressed Pressed close and flat against.

aquifolius Having spiny leaves.

arboreus Treelike.

arcuate Curved or bowed.

areolate Divided into irregular squarish or angular spaces.

argenteus Silvery.

asperous Rough to the touch, especially when rubbed in one direction.

attenuate Leaf base constricting concavely and gradually into a somewhat winglike petiole, with a widened base.

aureus Golden.

auriculate Leaf having a small earlike lobe or pair of separate flattened stipules on either side of the leaf base petiole (e.g., *Salix*).

australis Southern.

autumnal In fall.

awn Bristlelike tip or appendage.

berry Fleshy or pulpy fruit in which seeds are not encased in a stone.

bifid Divided into 2 parts, in the middle.

bipinnate Compound leaf having pinnate divisions.

bract Modified leaf supporting or belonging to a flower.

branchlet Ultimate division of a branch, twig, current year's growth.

burr Fruit with prickly skin.

buttressed Trunk swollen at base.

buxifolius Foliage resembling boxwood.

caducous Falling off early or easily.

caeruleus Dark blue.

callous-tipped Ends with hardened swellings.

campanulate Bell-shaped.

capitate Shaped like a head, or in dense headlike clusters.

capsule Dry fruit with more than one carpel, which splits at maturity to release seeds.

carpel Simple pistil, or part of a compound pistil.

catkin Scaly, bracted spike of a usually unisexual flower, frequently deciduous in one piece. Same as ament.

canadensis Of Canada.

canariensis Of the Canary Islands.

canescent Gray-pubescent, hoary, or becoming so.

capensis Of the Cape of Good Hope.

capitatus Headlike.

caudate Leaf with tail-like or long slender tip.

chartaceous Thin and papery but opaque.

chilensis Of Chile.

chimaera Graft-hybrid plant formed from the tissues of 2 different forms or species, usually originating at the graft point, showing parts of each parent.

chinensis Of China

ciliate Fringed with a row of fine hairs, resembling eyelashes.

cinereous Ash-colored.

circinate Circular.

citrinus Yellow.

clavate Club-shaped.

cleft Leaf margin sinuses extending more than halfway from margin to midvein, usually acute.

coaetaneous Flowering as the leaves expand.

coccineus Scarlet.

compactus Compact, dense.

compound Leaves with more than one individual segment.

concolor One color.

cone Fruit, usually woody, with overlapping scales.

confertus Crowded, pressed together.

coniferous Cone-bearing plant, or of the Order Coniferales.

connate United or joined.

connate-perfoliate 2 opposite leaves without a petiole, fused at the base, appearing as one, with a stem through the center, on a perpendicular axis.

contortus Contorted, twisted.

coppice Growth arising from the trunk; thicket.

cordate Heart-shaped.

coriaceous Leathery.

cornutus Horned.

corolla Inner part of the flower perianth, composed of petals.

corymb Flat-topped flower cluster, with flowers opening from outside in.

cotyledon Seed leaf, primary leaf, or leaves in the embryo.

crassus Thick, fleshy.

crenate Leaf margin with rounded teeth.

crown Upper part of a tree, including branches and foliage.

cultigen An introduced plant, under cultivation outside its native habitat.

cultivar Cultivated variety; originating and persisting under cultivation.

cuneate Wedge-shaped, or triangular with an acute angle downward.

cupule Cuplike structure at the base of some fruits, formed by dry enlarged floral envelopes (e.g., oaks and some palms).

cuspidate Tipped with a sharp rigid point.

cyme Flat-topped flower cluster, with flowers opening from the center, outward.

deciduous Falling at the end of one season of growth.

declinate Bent downward or forward, tips often recurved.

decumbent Branches reclining or lying on the ground, with ends ascending.

decurrent Leaf blades extending down to petioles.

decussate Paired leaves arranged at alternating right angles to the central stem.

dehiscent Opening of an anther or capsule, by slits or valves.

deliquescent Trunk dividing into several large branches, opposite of excurrent.

deltoid Delta-shaped, triangular, more-or-less equilateral.

dendron Tree.

dentate Leaf margin with sharp, rather coarse teeth pointing outward from the mid-vein.

denticulate Finely dentate leaf margin.

denticule Minute tooth.

depressus Pressed down.

dichotomous Forked, or in one or more pairs, as in a dichotomous key, which has pairs of opposing statement choices.

dicotyledonous Having 2 cotyledons.

dioecious Unisexual, with male and female flowers on separate plants.

dimorphic Occurring in two forms, or having juvenile and adult forms (e.g., *Acacia*, *Eucalyptus*).

discolor Two, or separate colors.

distichous Alternate arrangement of adjoining leaves on opposite sides of a central twig, spaced symmetrically at the midway spacing of the leaves opposite.

dorsal Lower surface of a leaf, or the back or outer surface.

double serrate Toothed-edged serrations.

drupe Fruit with a stonelike seed.

echinate Having stiff sharp spines (e.g., chestnut).

eglandular Without glands.

elegans Elegant, slender, willowy.

ellipsoidal Shaped like an elliptical solid object.

elliptical Shaped like an ellipse.

emarginate Apex notched, more so than obtuse, but not lobed.

endemic Restricted to a particular locale.

entire Uniform, smooth leaf margin without divisions, notches, lobes, or teeth. May be variously wavy in vertical plane, or have a row of hairs.

erose Irregularly toothed or eroded margin.

evergreen Foliage remains green and persists through the dormant season.

excurrent Trunk extending to top of tree; opposite of deliquescent. Also, midrib extending beyond the margin or leaf tip.

exfoliate Peeling off in shreds, thin layers, or plates, as in bark on a tree.

exotic Introduced.

falcate Shaped like a sickle.

fasciate Much widened and flattened abnormal stem growth.

fascicle Dense cluster or bundle, as in a flower cluster or pine needles.

fasciculate Leaves in clusters, enclosed at base in bractlike sheath (e.g., *Pinus*).

fastigiata Branches erect and close together; very upright.

ferruginous Rust-red.

filament Stalk of an anther.

flexuous Bent alternately in opposite directions, in a zigzag fashion.

floridus Free-flowering.

florum Flowers.

fluted Regularly spaced alternating ridges and grooved depressions, or trunks with extending swelled areas that radiate from center around base.

foliaceous Leaflike in texture or appearance.

follicle Dry 1-celled fruit from simple pistil, dehiscent by one suture.

fruit Seed-bearing part of a plant.

fugaceous Falling or withering away very early.

fulvous Tawny, dull yellow with gray.

funiculus Stalk of an ovule.

furcate Forked.

furrowed Having longitudinal grooves or channels.

fusiform Spindle-shaped, narrowed on each side from widened mid-region.

fuscous Grayish brown.

gibbous Swollen on one side.

glabrate Nearly or becoming glabrous.

glabrous Smooth, not hairy; opposite of pubescent.

gladiate Sword-shaped, either straight or curved.

glaucous Surface covered with fine waxy film, which can be rubbed off.

globose Spherical, or nearly so.

glutinous Sticky, or covered with a sticky exuded film.

grandiflorus Large-flowered.

gymnosperm Plant with naked seeds, not enclosed in an ovary.

habit General appearance or form, as seen from a distance.

habitat Place where a plant naturally grows.

hermaphroditic Containing both sexes, with functional ovary and stamens.

hiemal In winter.

hirsute Surface with short, erect, stiff, but not harsh hairs.

hirtellous Minutely hirsute.

hispid Surface with dense, erect, straight, harshly stiff but slightly flexible hairs.

hispidulous Diminutive of hispid.

hoary Covered with a whitish or grayish pubescence.

humilis Low, small, humble.

husk Thin dry covering on some fruits and seeds.

hybrid A genetic cross, usually between 2 related species.

hypogynous Ovary free from and inserted above calyx.

hysteranthous Leaves appear after flowers.

ilicifolius Having hollylike foliage.

imbricate Leaves overlapping, in shinglelike fashion (e.g., *Cedrus*).

imperfect Flower containing one sex, but not the other.

impressus Impressed upon.

incised Leaf margin cut jaggedly with very deep teeth.

incumbent Leaning or resting on.

indehiscent Remaining closed, not opening.

indicus Of India.

integument A covering or envelope, one or two, that cover, and are part of the ovule.

internode Portion of stem between 2 bud nodes.

indigenous Native to a country or region, or a plant from a foreign region that becomes thoroughly established and reproduces in its new surroundings.

indurate Hardened, as in spines or thorns that harden with maturity.

inermis Spineless, not prickly.

inferior ovary Appearing below attachment of perianth and androecium.

inflorescence Flowers appearing in clusters.

infrastipular Situated below the stipules.

involucre Circle of bracts surrounding a flower cluster.

irregular flower Bilaterally symmetrical, similar parts having different shapes or sizes.

japonicus Of Japan.

keeled With a central ridge, like a boat.

key Samara-type dry fruit with persistent attached wing (e.g., *Fraxinus*).

lacerate Leaf margin irregularly cut from one-half to two-thirds the distance to the midvein.

laciniate Leaf margin cut into narrow ribbonlike segments or pointed lobes.

lactiferous Latex-containing, or yielding a milky substance.

laevigate Smooth, as if polished.

lanate Woolly.

lanceolate Lance-shaped.

lanuginous Cottony or woolly.

lanulose Woolly.

lateral From the side, not an apex.

lateritous Brick red.

leaflet One of the individual small blades of a compound leaf.

legume Fruit of the pea family, podlike and splitting along 2 sutures.

lenticel Lens-shaped corky or spongy areas on young bark that admit air to the interior of a twig or branch.

ligneous Woody.

linear Long and narrow, with parallel edges.

lobed Leaf margin with rounded incisions not more than halfway from margin to the midvein.

lobulate Divided into small lobes.

lunate Crescent-shaped.

lustrous Glossy, shiny.

luteolus Yellowish, pale yellow.

maculate Blotched or spotted.

midrib Central vein of a leaf or leaflet.

miniate Dull vermilion.

monocarpic Fruiting once, then dying.

monoecious Stamens and pistils in separate flowers on the same plant.

montane Mountainous region, or growing in the mountains.

mucro A small and abrupt leaf tip.

mucronate Having a small abrupt leaf tip.

muricate Roughened surface with minute, firm epidermal proliferations, scarcely classifiable as hairs.

muticous Blunt, lacking a point.

nana Dwarf.

napiform Turnip-shaped.

native A naturally occurring, non-introduced species.

naturalized Introduced from a foreign area and reproducing naturally in the new environment.

nigrescent Becoming black, or nearly so.

nitid Shining.

niveous Snow white.

nuciform Nut-shaped.

nut Hard and indehiscent fruit, 1-seeded pericarp produced from a compound ovary.

ob- Latin prefix, inverted.

obconical Inverted cone-shaped.

obcordate Inverted heart-shaped, with notched apex.

oblanceolate Lanceolate, with broadest part near the apex.

oblique Slanting, or with unequal sides, leaf base with lowermost sides markedly unequal, usually one side offset, larger and lower where it meets petiole (e.g., *Ulmus*).

oblong About 3 times longer than broad, with nearly parallel sides.

obovate Ovate, with broader end toward apex.

obovoid Obovate.

obtuse Blunt or rounded at the apex.

ochroleucous Yellowish white to buff colored.

odd-pinnate leaf Pinnate with single terminal leaflet.

olivaceous Olive green.

opaque Dull surface, not shining, not transparent.

opposite Leaves paired directly opposite each other along a central twig.

orbicular Flat, circular in outline.

ovary Part of a pistil that contains the ovules.

ovate Shaped like a longitudinal cross-section of an egg, with the broader end at the base.

ovoid Egg-shaped in three dimensions, widest below the middle.

ovule The part of the flower that becomes the seed after fertilization.

paleaceous Chaffy in texture.

paleobotany Fossil botany, study of plants as they occurred and evolved in the geologic past.

palmate Having lobes or leaflets arising from one point, in hand-shaped configuration.

palmatifid Leaf margin palmately cleft or parted.

paludose Inhabiting marshes.

pandurate Fiddle-shaped.

panicle A loose, compound or branched flower cluster.

papilionaceous Resembling a butterfly, typical flower shape of leguminous plants.

papillate Bearing many nipplelike projections.

parted Leaf margin with sinuses nearly reaching the midvein.

parvifolius Having small leaves.

pectinate Divided like the teeth on a comb, closely spaced.

pedate Resembling a bird's foot.

pedicel Stalk of a single flower in an inflorescence.

peduncle Flower stalk supporting a cluster of flowers or a solitary flower.

peltate Shield-shaped, attached to central stalk from its lower surface, rather than from the end.

pendent Hanging.

pendulous Hanging.

penninerved Pinnately veined.

penniveined Secondary veins arranged parallel to each other, arising from a central main vein.

perfect flower Flower with both stamens and pistil.

perianth Calyx and corolla of a flower, considered as a whole.

persistent Remaining attached, not falling off.

petiolate Having a petiole.

petiole Leaf stalk, or footstalk of a leaf.

petiolule Footstalk of a leaflet.

phyllus Leaf.

pilose Hairy, with distinct soft hairs.

pinnate Compound leaf with leaflets arranged symmetrically along each side of a common petiole.

pinnatifid Leaf margin pinnately cleft or parted.

pistil Female organ of a flower, consisting of ovary, style, and stigma.

pistillate Female flowers, having one or more pistils and no functional stamens, unisexual.

pith Soft spongy central part of a woody stem.

plicate Folded, as in a fan shape, or nearly so.

plumose Having featherlike compound hairs.

pod Dehiscent dry fruit.

pollen Grains borne on the anther(s) of a flower.

polycarpic Bearing fruit many times.

polygamo-dioecious Flowers sometimes perfect, sometimes unisexual and dioecious.

polygamo-monoecious Flowers sometimes perfect and sometimes unisexual, the two forms borne on the same plant.

polygamous Flowers sometimes perfect and sometimes unisexual.

pome Inferior fruit of two or several carpels enclosed in thick flesh, as an apple.

populifolius Having poplarlike foliage.

protantherous Leaves appearing before flowers.

puberulous Surface with extremely short, fine, straight, dense, barely perceptible hairs.

pubescent Fuzzy or hairy leaf surface; opposite of glabrous.

pulverulent Surface covered by very fine powder, much like glaucous waxy film (e.g., *Eucalyptus pulverulenta*).

pulvinate Cushion-shaped.

pumilus Dwarf, small.

punctate Covered with minute impressions, with scarcely any depth, appearing as if made by a pinpoint.

pungent Ending in a sharp point. Also, acrid-tasting.

pyramidal Pyramid-shaped.

pyriform Pear-shaped.

raceme Simple inflorescence of stalked flowers on an elongated rachis or stem.

racemose In or resembling racemes.

rachis Axis bearing leaflets, as in a compound leaf, or bearing flowers.

ramiform Branching.

receptacle Expanded portion of an axis bearing the organs of the flower, or the collective flowers of the flower head.

reclinate Bent down or falling backward from the perpendicular.

recurved Curving downward or backward.

reflexed Turned abruptly downward.

regular flower Radially symmetrical. Similar parts are same shape and size.

reniform Kidney-shaped.

repand Having a slightly wavy margin.

resinous Containing or producing resin.

reticulate Netted. Surface with netlike pattern of weak grooves or color variation outlining veinlets beneath.

retrose Bent or turned backward or downward.

retuse With rounded apex very slightly notched at the terminus of the midvein.

revolute Rolled backward, as with margin rolled toward lower side.

rhombate Rhomboid.

rhombic Rhomboid.

rostellum Small beak.

rostrate Having a beak, or beaklike projection.

rosulate In rosettes.

rotund Orbicular, inclining to be oblong.

rubrum Red.

rufous Red-brown.

rugose Wrinkled. Surface with deeply grooved reticulation over the network of veinlets.

ruminous Mottled appearance.

saccate Bag-shaped, pouchy.

sagittate Leaf with basal lobes turned downward and inward.

salicifolius Having willowlike foliage.

samara Indehiscent winged fruit.

scabrous Surface with scattered harsh hairs, not erect, usually with bulbous bases, rough to touch.

scarious Thin dry and membranaceous, not green.

scurfy Skin covered with small branlike scales.

seed Ripened ovule.

sepal Division of the flower carpel, usually bractlike.

sericeous Surface silky, with straight, soft long hairs, often with a satiny sheen when dense.

serrate Toothed margin, with teeth pointing upward or forward.

serrulate Finely serrated leaf margin.

sessile Without a stalk, or grasslike.

setaceous Bearing bristles.

setose Surface less dense than hispid.

sheath Tubular enclosing structure around an organ or plant part.

sinuate With a strong wavy margin, inward and outward, but too shallow to be lobed.

sinus Cleft or space between two lobes.

soboliferous Spreading and forming clumps by underground roots (e.g., aspen, sumac).

spatulate Spatula-shaped.

spike Simple inflorescence of sessile flowers arranged on a common elongated axis.

spine Sharp woody outgrowth in the place of a leaf or stipule.

spinescent Having short rigid branches resembling spines.

spinose Having spines.

squamate Having small scalelike leaves or bracts, scaly.

stamen Pollen-bearing organ of the male flower.

staminate Male flowers, with stamens, but without pistils.

stellate Star-shaped.

sterigmata Short persistent leaf bases (e.g., spruce, hemlock).

stigma Part or surface of a pistil that receives pollen to fertilize the ovules.

stipe Stalklike support of a pistil or carpel.

stipule Appendage at the base of the petiole, usually one on each side.

stolon Horizontal stem at or below ground, or a shoot that bends to the ground that gives rise to a new plant at its tip.

stoloniferous Spreading or forming clumps by stolons.

striate Marked by longitudinal lines.

strigose Surface with harsh, straight, stiff, short hairs, either flattened or weakly ascending.

stoma Orifice in the epidermis of a leaf, used to connect internal cavities with air.

stomata Plural of stoma.

stomatiferous Having stomata.

strigose Having harsh, straight, stiff, short hairs, appressed or weakly ascending (e.g., *Ulmus*).

strobile Cone.

style Slenderly tapered portion of a pistil between ovary and stigma.

suberous Corky.

subinferior ovary Position more or less intermediate between perianth and androecium.

suborbicular Nearly round.

subtend To lie under or opposite to.

subulate Awl-shaped.

succulent Juicy, fleshy.

superior ovary Free from and inserted above calyx; hypogynous.

sulcate Surface furrowed, with longitudinal grooves.

suture Junction or line of dehiscence.

synantherous Flowers and leaves appear simultaneously.

syncarp Multiple fleshy fruit.

synonym Equivalent superseded name, a second name for a given taxon.

taproot Primary descending root.

taxon A general term applied to a taxonomic element, population, or group, irrespective of its classification level.

terete Circular in transverse section.

terminal At the end.

ternate In threes.

testa Outer coat of a seed, which develops from an integument.

thorn Sharp woody outgrowth in place of a lateral branch or leaf.

tomentose Surface covered with matted woolly hairs.

tomentulose Surface covered with matted woolly hairs, more finely than tomentose.

torulose Twisted or knobby, cylindrical, irregularly swollen at close intervals.

trifoliate Leaf with 3 leaflets (e.g., *Rhus lancea*).

trifurcate Having the apex split into 3 lobes.

truncate Ending abruptly, as if cut off at the end.

turbinate Top or turban-shaped.

twig Young woody stem; shoot of a woody plant representing growth of the current season.

two-ranked In 2 side-by-side rows.

umbel Simple inflorescence with pedicels all arising from the same point.

undulate Having a wavy surface or margin in a vertical plane, not inward and outward as in sinuate.

velutinous Surface with visible velvety hairs.

vernal Appearing in spring.

verticillate Arranged in whorls, or seemingly so.

vestiture Covering or surface, referring to condition of hairiness, scaliness, or glandularity.

villous Surface with moderately dense, long, soft, often curly hairs, erect, but not necessarily straight.

viscid Covered with a sticky secretion.

whorled Three or more leaves at a node.

wing Membranous or thin, dry appendage of a seed.

woolly Covered with long matted or tangled hairs.

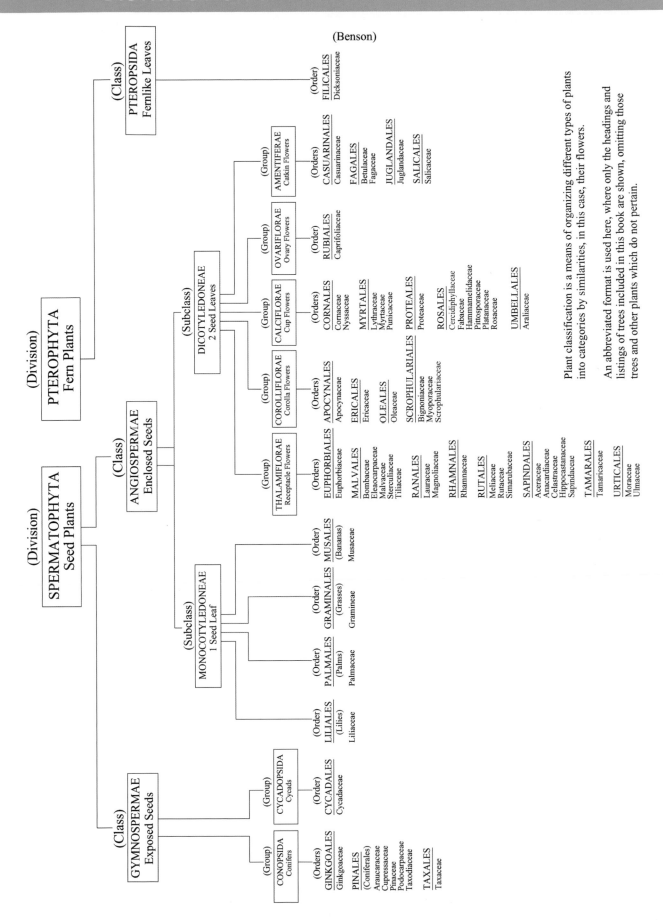

(Benson)

Plant classification is a means of organizing different types of plants into categories by similarities, in this case, their flowers.

An abbreviated format is used here, where only the headings and listings of trees included in this book are shown, omitting those trees and other plants which do not pertain.

Maple Family
1. ACERACEAE
   Acer

Agave Family
2. AGAVACEAE
   Cordyline

Cashew Family
3. ANACARDIACEAE
   Pistacia
   Rhus
   Schinus

Dogbane Family
4. APOCYNACEAE
   Nerium

Aralia Family
5. ARALIACEAE
   Tetrapanax

Araucaria Family
6. ARAUCARIACEAE
   Araucaria
   Agathis

Birch Family
7. BETULACEAE
   Alnus
   Betula
   Carpinus

Bignonia Family
8. BIGNONIACEAE
   Catalpa
   Jacaranda

Bombax Family
9. BOMBACEAE
   Chorisia

Honeysuckle Family
10. CAPRIFOLIACEAE
    Sambucus

Casuarina Family
11. CASUARINACEAE
    Casuarina

Staff Tree Family
12. CELASTRACEAE
    Maytenus

Cercidiphyllum Family
13. CERCIDIPHYLLACEAE
    Cercidiphyllum

Dogwood Family
14. CORNACEAE
    Cornus

Cypress Family
15. CUPRESSACEAE
    Cupressus
    Calocedrus
    Chamaecyparis
    Cupressocyparis
    Juniperus
    Thuja

Cycad Family
16. CYCADACEAE
    Cycas

Tree Fern Family
17. CYCATHACEAE
    Cibotium
    Dicksonia

Eleaocarpus Family
18. ELEAOCARPACEAE
    Crinodendron

Heath Family
19. ERICACEAE
    Arbutus
    Oxydendrum

Spurge Family
20. EUPHORBIACEAE
    Sapium

Pea Family
21. FABACEAE
    Acacia
    Albizia
    Bauhinia
    Ceratonia
    Cercidium
    Cercis
    Cladrastis
    Erythrina
    Gleditsia
    Laburnum
    Prosopis
    Robinia
    Sophora

Beech Family
22. FAGACEAE
    Chrysolepis
    Fagus
    Lithocarpus
    Quercus

Ginkgo Family
23. GINKGOACEAE
    Ginkgo

Grass Family
24. GRAMINEAE
    Phyllostachys

Witch Hazel Family
25. HAMAMELIDACEAE
    Liquidambar

Horsechestnut Family
26. HIPPOCASTANACEAE
    Aesculus

Walnut Family
27. JUGLANDACEAE
    Carya
    Juglans
    Pterocarya

Laurel Family
28. LAURACEAE
    Cinnamomum
    Laurus
    Neolitsea
    Umbellularia

Lily Family
29. LILIACEAE
    Cordyline
    Yucca

Loosestrife Family
30. LYTHRACEAE
    Lagerstroemia

Magnolia Family
31. MAGNOLIACEAE
    Liriodendron
    Magnolia
    Michelia

Mallow Family
32. MALVACEAE
    Lagunaria

Mahogany Family
33. MELIACEAE
    Melia

Mulberry Family
34. MORACEAE
    Broussonetia
    Ficus
    Morus

Banana Family
35. MUSACEAE
    Musa

Myoporum Family
36. MYOPORACEAE
    Myoporum

Myrtle Family
37. MYRTACEAE
    Agonis
    Callistemon
    Eucalyptus
    Eugenia
    Lophostemon
    Melaleuca
    Metrosideros
    Syzygium
    Tristaniopsis

Nyssa Family
38. NYSSACEAE
    Nyssa

Olive Family
39. OLEACEAE
    Fraxinus
    Ligustrum
    Olea

Palm Family
40. PALMACEAE
    Arecastrum
    Butia
    Chamaerops
    Jubaea
    Phoenix
    Trachycarpus
    Washingtonia

Pine Family
41. PINACEAE
    Abies
    Cedrus
    Picea
    Pinus
    Pseudotsuga
    Tsuga

Pittosporum Family
42. PITTOSPORACEAE
    Pittosporum

Plane Family
43. PLATANACEAE
    Platanus

Protea Family
44. PROTEACEAE
    Grevillea
    Stenocarpus

Pomegranate Family
45. PUNICACEAE
    Punica

Buckthorn Family
46. RHAMNACEAE
    Zizyphus

Rose Family
47. ROSACEAE
    Cercocarpus
    Crataegus
    Eriobotrya
    Lyonothamnus
    Malus
    Prunus
    Pyrus
    Quillaja
    Sorbus

Rue Family
48. RUTACEAE
    Geijera

Willow Family
49. SALICACEAE
    Populus
    Salix

Soapberry Family
50. SAPINDACEAE
    Cupaniopsis
    Koelreuteria

Figwort Family
51. SCROPHULARIACEAE
    Paulownia

Quassia Family
52. SIMARUBACEAE
    Ailanthus

Cocoa Family
53. STERCULIACEAE
    Brachychiton
    Firmiana

Tamarix Family
54. TAMARICACEAE
    Tamarix

Yew Family
55. TAXACEAE
    Podocarpus
    Taxus
    Torreya

Redwood Family
56. TAXODIACEAE
    Cryptomeria
    Cunninghamia
    Metasequoia
    Sciadopitys
    Sequoia
    Sequoiadendron
    Taxodium

Linden Family
57. TILIACEAE
    Tilia

Elm Family
58. ULMACEAE
    Celtis
    Ulmus
    Zelkova

1. Reproduction by spores .... **PTERIDOPHYTA**
1. Reproduction by seeds .... **SPERMATOPHYTA**..2
   2. Seeds not enclosed in an ovary, flowers unisexual, leaves fan-shaped, linear, needlelike or scalelike **GYMNOSPERMAE**
   2. Seeds borne in an enclosed ovary which becomes a fruit at maturity .... **ANGIOSPERMAE**..3
3. Linear leaves with parallel veins, flower parts in 3s or 6s, stems without central pith or annual wood layers, embryo with 1 cotyledon .... **MONOCOTYLEDONAE**
3. Leaves generally broad, with netted veins, flower parts usually in 4s or 5s, stems with central pith and annual wood layers, embryo with 2 cotyledons. .... **DICOTYLEDONAE**

PTERIDOPHYTA

1. Leaves fernlike, bearing spores, single basal trunk, or clumped .... *Cyathea* p. 455, *Dicksonia* p.456

SPERMATOPHYTA

GYMNOSPERMAE
Leaves Flattened, Leaflike

1. Leaves flattened, leaflike 2
1. Leaves needlelike or scalelike 6
   2. Leaves simple 3
   2. Leaves pinnately compound, from central trunk, conelike seed fruit .... *Cycas* p.455
3. Deciduous, petioled leaves, twigs with spur shoots, drupelike seed fruit .... *Ginkgo* p.251
3. Leaves persistent, short or no petioles 4
   4. Solitary leaves .... *Podocarpus* spp. p.521
   4. Overlapping leaves 5
5. Leaves rigid and sharply pointed .... *Araucaria araucana, bidwillii* p.148,149,517
5. Leaves flexible and soft to touch .... *Cunninghamia* p.203

Leaves Needlelike

Note: *Refer to Dicotyledoneae, for leaves appearing to be needlelike may in fact be stems or rachii, as in **Casuarina** and **Tamarix**, with tiny vestigial leaves, and **Parkinsonia** with deciduous leaves but rachii may remain, leafless.*

   6. Leaves needlelike 7
   6. Leaves scalelike 19
7. Needles in fascicles of 1-5, enclosed by sheath at base .... *Pinus* spp. p.520
7. Needles single-stalked 8
   8. Needles 3-6" long, slightly thickened, with 2 white stripes on bottom .... *Sciadopitys* p.432
   8. Needles generally 3" or shorter 9
9. Leaves flattish, in flat sprays 10
9. Leaves rounded, in whorled or semi-whorled sprays 15
   10. Seed cones 11
   10. Seed fruit 14
11. Cones with papery scales .... *Tsuga* spp. p.523
11. Cones with hardened projections with peltate flat ends 12
   12. Needles persistent .... *Sequoia* p.433
   12. Needles deciduous, or semi-deciduous 13
13. Leaves and branchlets paired opposite, 2-ranked, long-stemmed cones .... *Metasequoia* p.288
13. Leaves alternate, spirally arranged, appearing 2-ranked, very short-stemmed cones ...... *Taxodium* spp. p.523
   14. Fruit seed partially enclosed in scarlet red fleshy cup, matures 1 year, leaves 1/2-1" .... *Taxus* p.442
   14. Olivelike fruit totally enclosed in purple striped green skin, matures 2 yrs, leaves 1-3" .... *Torreya* spp. p.523
15. Leaf base not persistent, leaving no residual projections 16
15. Leaf base persistent 17
   16. Cones pendulous, scales opening, with trifurcated tail .... *Pseudotsuga* spp. p.521
   16. Cones upright, compressed scales .... *Abies* spp. p.516
17. Terminal and axillary needles evenly spaced, pendulous cones .... *Picea* spp. p.520
17. Terminal growth axillary twigs stout, with tufted needles, upright cones 18
   18. Evergreen, cones to 4" long, scales closed, disintegrate from central spine when ripe .... *Cedrus* spp. p.517
   18. Deciduous, cones to 1-1/2" long, scales open when ripe .... *Larix* p.270

Leaves Scalelike or Awl-like

19. Leaves flattened, scalelike, in flattened sprays 20
19. Leaves rounded, scalelike or awl-like 22
   20. Cones rounded .... *Thujopsis* p.446
   20. Cones trifurcated 21
21. Cones elongated, duckbilled, trifurcated .... *Calocedrus* p.163
21. Cones similar, doubly so .... *Thuja* spp. p.523

ANGIOSPERMAE

MONOCOTYLEDONEAE

Leaves with parallel veins, alternate, simple and persistent

DICOTYLEDONEAE

Leaves with netted veins

Foliage Appearing Needlelike or Featherlike

Opposite-Compound-Persistent

Opposite-Compound-Deciduous

Opposite-Simple-Persistent

18. Dried seed pod fruit 19
18. Soft fleshy fruit 20
19. Elongated linear capsule .... ***Nerium*** p.295
19. Winged achene .... ***Acer paxii*** p.126
20. Composite fleshy fruit .... ***Cornus capitata*** p.193
20. Single fleshy fruit 21
21. Fruit green, turning black when ripe 23
21. Fruit green, color other than black when ripe 22
22. Fruit yellow to orange when ripe .... ***Pittosporum* spp.** p.521
22. Fruit reddish to purplish when ripe .... ***Syzygium*** p.437
23. Leaves ovate, shiny green, drupe seed 1/8-1/4" .... ***Ligustrum*** p.272
23. Leaves elliptical, green with gray underside, olive fruit 1/2-3/4" .... ***Olea*** p.297

Opposite-Simple-Deciduous

24. Leaves linear, elongated seed pods .... ***Chilopsis*** p.188
24. Leaves not linear 25
25. Leaves palmate, paired winged achene seeds .... ***Acer* spp.** p.516
25. Leaves oval, heart-shaped or elliptical 26
26. Leaves heart-shaped 27
26. Leaves oval, lanceolate or elliptical 28
27. Leaves heart-shaped, smooth edges, rough scaly surface, long bean fruit seed pods .... ***Catalpa*** p.169
27. Leaves with coarsely serrated edges, hairy surface, round beaked woody capsule fruit pods .... ***Paulownia*** p.307
28. Long bean fruit seed pods, flowers trumpet-shaped .... ***Chitalpa*** p.189
28. Rounded fruit and seeds 29
29. Flowers with large ovate white or colored bracts, fleshy drupe fruit, turns red .... ***Cornus* spp.** p.517
29. Flowers crinkled, crapelike 30
30. Flowers 1", round woody-shelled 1/4" seed capsules .... ***Lagerstroemia* spp.** p.519
30. Flowers usually larger than 1", fleshy-skinned composite juicy fruit .... ***Punica*** p.372

Alternate-Compound-Persistent

31. Leaves bipinnate .... ***Acacia baileyana, dealbata*** p.516
31. Leaves pinnate, margins entire 32
32. Pinnae lobed .... ***Grevillea*** p.254
32. Pinnae margins not lobed 33
33. Pinnae margins smooth 34
33. Pinnae margins toothed 35
34. Pinnae oblanceolate, gray to brown bark, tiny capsule seed fruit .... ***Cupaniopsis*** p.204
34. Pinnae distinctly oval , brown to black bark, fat legume seed pod fruit .... ***Ceratonia*** p.177
35. Leaves trifoliate .... ***Rhus lancea*** p.412
35. Leaves pinnate (or with few trifoliate) .... ***Schinus* spp.** p.523

Alternate-Compound-Deciduous

36. Leaves bipinnate 37
36. Leaves pinnate 43
37. Usually one or two paired pinnae 38
37. Many paired pinnae 39
38. Pinnae ovate to oval-elliptical .... ***Cercidium*** p.179
38. Pinnae linear .... ***Prosopis* spp.** p.521
39. Pinnae linear 40
39. Pinnae ovate to oval-elliptical 41
40. Flat woody pods; lavender-blue flowers .... ***Jacaranda*** p.256
40. Elongated papery legume pods; pink or yellow flowers .... ***Albizia* spp.** p.516
41. Pinnae oval-elliptical, rounded at ends; leathery legume pods; inconspicuous flowers .... ***Gleditsia* spp.** p.519
41. Pinnae ovate, pointed at ends 42
42. Noticeably toothed margins; lavender-blue flowers; round, fleshy yellow fruit .... ***Melia*** p.287
42. Slightly toothed margins; yellow flowers; bladder pod fruit .... ***Koelreuteria bipinnata*** p.264, 519
43. Pinnae pinnate 45
43. Pinnae trifoliate 44
44. Petiole extended to terminal pinna; red or orange flowers; leathery legume seed pods .... ***Erythrina* spp.** p.518
44. Petiole not extended to terminal pinna; yellow flowers; papery legume seed pods .... ***Laburnum*** p.266
45. Pinnae edges smooth 46
45. Pinnae edges toothed or serrated 53
46. Winged rachis between pinnae, pendulous tassel-like flowers; winged seeds .... ***Pterocarya*** p.371
46. No winged rachis between pinnae 47
47. Pinnae ends rounded 48
47. Pinnae ends pointed 50
48. Fleshy legume pod, yellowish turning brown or black; creamy yellow flowers .... ***Sophora*** p.435
48. Thin, tan, papery seed pods, either legume or key 49
49. Legume seed pods; leaf tips blunt: pendulous racemes of white to pinkish or purple flowers .... ***Robinia* spp.** p.522
49. Keylike seed pods; leaf tips indented; sparse axillary clusters of orange-yellow flowers .... ***Tipuana*** p.450

84. Small yellow flower clusters; pea-sized black fleshy fruit .... *Prunus caroliniana* p.361
84. Small waxy flowers, small fleshy cucumberlike fruit .... *Michelia* p.294

Alternate-Simple-Deciduous

85. Leaf edges entire 86
85. Leaf edges toothed or lobed 92
    86. Leaves narrowly elliptical or linear 87
    86. Leaves ovate, lanceolate or elliptical 88
87. Leaf buds swollen, enlarged; catkin fruit .... *Salix* spp. p.523
87. Leaf buds present, but not swollen, acorn fruit .... *Quercus phellos* p.399
    88. Fruit enlarged, many compartments, cucumberlike appearance when green .... *Magnolia* spp. p.519
    88. Fruit other than above 89
89. Leaves heart-shaped or deltoid 90
89. Leaves elliptical 91
    90. Leaves not sharply pointed, broadly rounded or oval; legume seed pods .... *Cercis* spp. p.517
    90. Leaves with long, pointed ends; small, rounded pealike fruit .... *Sapium* p.428
91. Fruit with long petiole stem, purplish to black when ripe .... *Nyssa* p.296
91. Fruit without petiole, red when ripe .... *Cornus* spp. p.517
    92. Leaf edges toothed, or finely so, or wavy 93
    92. Leaf edges lobed 111
93. Leaves lanceolate to linear .... *Salix* spp. p.523
93. Leaves other than lanceolate to linear 94
    94. Leaf surface not smooth, either textured or noticeably fuzzy 95
    94. Leaf surface smooth 96
95. Leaf surface textured, appearing crinkled .... *Cercidiphyllum* p.178
95. Leaf surface fuzzy, covered with short hairs .... *Broussonetia* p.160
    96. Winged or cottony seeds 97
    96. Fleshy fruit, nuts or berries 101
97. Cottony seeds .... *Populus* spp. p.521
97. Winged seeds 98
    98. Pendulous flowers; round pea-sized single elongated winged fruit .... *Tilia* spp. p.523
    98. Pendulous catkinlike flowers; very tiny seeds 99
99. Seeds with elongated wing .... *Carpinus* spp. p.517
99. Seeds encircled by round papery wing 100
    100. Leaves elliptical .... *Ulmus* spp. p.523
    100. Leaves ovate to deltoid .... *Betula* spp. p.517
101. Fleshy fruit or berries 105
101. Fruit a nut or dry seed fruit 102
    102. Seeds released from woody, husked capsule 103
    102. Seeds rounded 104
103. Seeds flat-sided, released from pendulous husked capsule, with short spurs .... *Fagus* spp. p.518
103. Seeds released from spiny woody-sheathed covering .... *Castanea* p.167
    104. Seeds tiny, smaller than 1/16" .... *Oxydendrum* p.298
    104. Seeds pealike, 1/8" or larger .... *Celtis* spp. p.517
105. Composite berrylike fruit .... *Morus* spp. p.519
105. Single fleshy fruit 106
    106. Single seed drupe fruit 107
    106. Multiple seed fruit 109
107. Showy white or pink flowers .... *Prunus* spp. p.521
107. Insignificant yellowish green flowers 108
    108. Fruit 3/16", egg-shaped, slightly fleshy .... *Zelkova* p.467
    108. Fruit 3/4" or larger, very fleshy .... *Zizyphus* p.468
109. Red pyracantha-like berries .... *Crataegus* spp. p.517
109. Miniature apple or pearlike fruit 110
    110. Fruit smooth, shiny, yellow, orange or red .... *Malus* p.283
    110. Fruit with rough skin, dull brown .... *Pyrus* spp. p.521
111. Palmately lobed 112
111. Pinnately lobed, acorn fruit .... *Quercus* spp. p.521
    112. Leaf apex indented or flat, not pointed 113
    112. Leaf apex tapered, pointed 114
113. Leaf apex emarginate, or indented; flattened legume seed pod fruit .... *Bauhinia* p.152
113. Leaf apex truncated or flat; tapered woody fruit with overlapping flaps .... *Liriodendron* & vars. p.275, 276
    114. Leaf surface fuzzy or tomentose 115
    114. Leaf surface smooth 116
115. Leaves woolly tomentose .... *Platanus* spp. p.521
115. Leaf surface usually with short fuzzy hairs, not woolly .... *Broussonetia* p.160
    116. Leaf surface shiny; edges serrated 117
    116. Leaf surface dull; margins smooth .... *Firmiana* p.241
117. Lobes rounded; fleshy berrylike fruit .... *Morus* spp. p.519
117. Lobes pointed; round woody burrlike fruit .... *Liquidambar* spp. p.519

Abies – *Leaves short, needlelike, flattened or 4-sided, white striped underside; upright cones, scales fall off*
    1. Needles with whitish bands on underside 2
    1. Needles with whitish bands on both sides 8
        2. Needles acute, sharp or semi-sharp 3
        2. Needles blunt or round ended, only slightly sharp 6
    3. Needles very sharp; hairy twigs; non-resinous buds; 2-4" cones, with scale bracts .... *A. bracteata* p.96
    3. Needles prickly-ended 4
        4. Needles stiff, whorled perpendicular to and surrounding twigs; 4-5" cones .... *A. pinsapo* p.104
        4. Needles flexible 5
    5. Needles nearly 2-ranked, flat, underturned at ends; 2-4" cones .... *A. grandis* p.98
    5. Needles outward to each side, v-shaped void down center; 1-2" cones .... *A. nordmanniana* p.103
        6. Needles only slightly flattened along twig .... *A. alba* p.94
        6. Needles lying flat, or nearly so, on lower branches, becoming upright higher up 7
    7. Needles stubby-ended, uneven lengths; thin, fissured bark; 3-6" cones, no bracts .... *A. amabilis* p.95
    7. Needles round-ended; bark thick and fissured; hairless twigs; 3-5" cones .... *A. concolor* p.97
        8. Needles flat in cross-section; 2-4" cones, bracts hidden .... *A. lasiocarpa* p.99
        8. Needles fattened, 3 or 4 sided in cross-section (twirl between fingers) 9
    9. Needles 4 sided; 6-8" cones .... *A. magnifica* & vars. p.100-102
    9. Needles 3 or 4 sided; 4-7" cones, folded scale bracts .... *A. procera* p.105

Acacia – *Leaves simple or pinnately compound; yellow round puffy flowers; flat seed pods*
    1. Leaves simple, margins entire, parallel veined 2
    1. Leaves compound, pinnate 4
        2. Leaves 4-6" long, narrow, linear, with occasional broad-lanceolate leaves .... *A. retinoides* p.110
        2. Leaves 2-4" long, oblong-lanceolate 3
    3. Leaves 2-4" long, 1/4-1/2" wide .... *A. melanoxylon* p.109
    3. Leaves 3-6" long, 1/2" wide .... *A. longifolia* p.108
        4. Leaves 1-1/2-3" long, with 4-6 paired pinnae .... *A. baileyana* p.106
        4. Leaves 4-6" long, with 10-15 paired pinnae .... *A. dealbata* p.107

Acer – *Leaves palmate, simple or compound; pairs of winged seeds*
    1. Leaves compound, 3-5 leaflets; hairy leaf surface 2
    1. Leaves simple 3
        2. Leaf color green to yellow green or variegated; smooth-textured bark .... *A. negundo* & var. p.119, 120
        2. Leaf color dark green to bronzy colored; papery reddish brown peeling bark .... *A. griseum* p.116
    3. Leaves smaller than 3" 4
    3. Leaves 3" or larger 10
        4. Leaves trifurcate, or 3 lobed to ovate 5
        4. Leaves palmate 7
    5. Lobed near base, serrated edges, or doubly so 6
    5. Lobed near apex, finely serrated or mostly smooth margins .... *A. buergerianum* p.111
        6. Flowers in umbels; greenish to grayish brown bark .... *A. tataricum* & ssp. p.132, 133
        6. Flowers on stalks, smooth young bark, green, with distinct white stripes .... *A. capillipes* p.113
    7. Margins wavy to entire, 3-5 lobes 8
    7. Margins serrate 9
        8. Lobes pointed; winged seeds 1/2" long, few or none present .... *A. truncatum* & var. p.133, 134
        8. Lobes rounded; winged seeds 1" long, in abundance .... *A. campestre* p.112
    9. Leaves deeply palmately lobed, green or red to purple, finely serrated .... *A. palmatum* & vars. p.121-123
    9. Leaves lobed in spherical outline, serrated margins 10
        10. Dark green leaves with 5-11 fat pointed lobes .... *A. japonicum* & var. p.117
        10. Light or yellowish green leaves, with 8-11 narrow pointed lobes .... *A. circinatum* p.114
    11. Lobes deeper than halfway to midrib 12
    11. Lobes cut less than halfway to midrib 14
        12. Lobes serrated to double serrated, pointed ends, gray cast underside .... *A. saccharinum* p.130
        12. Lobes and ends smooth, rounded or blunt, light green underside 13
    13. Leaves 2-5" in diameter; seeds with 1/2 to 3/4" long wings .... *A. saccharum* ssp. *grandidentatum* p.115
    13. Leaves 6-12" in diameter; seeds with 2" long wings .... *A. macrophyllum* p.118
        14. Lobes wavy margined, acute to acuminate end; milky juice from cut petiole 15
        14. Lobes serrate to double serrate, acute ends; petiole juice clear 16
    15. Broad oval form; yellow or bronzy-orange fall color .... *A. platanoides* & var. p.124-126
    15. Tall oval crown; glowing yellow-orange fall color .... *A. saccharum* p. 131
        16. Mature height 10-20'; usually multi-trunked; shrublike .... *A. glabrum* p. 115
        16. Mature height to 40' or more, usually single trunked; treelike .... *A. rubrum* & var. p.128, 129

Aesculus – *Leaves palmately compound; spiny skinned chestnut fruit*
    1. Leaflets rather smooth, short petioles 2
    1. Leaflets crinkly, nearly sessile, or with very short petiole 3
        2. White flowers, in drooping 1' long panicles; deciduous in midsummer dryness .... *A. californica* p.135
        2. Red flowers, in short upright panicles .... *A. pavia* p.137

    3. Leaflets 5-7; flowers white .... *A. hippocastanum* p.138
    3. Leaflets usually 5; flowers red or dark pink .... **A. x carnea** p.137
Albizia – *Leaves pinnately compound; red or yellow flowers; flat seed pods.*
    1. Flowers greenish yellow, in fluffy cylindrical spikes; leaves evergreen, 40-60 leaflets .... *A. distachya* p.142
    1. Flowers pink, in puffball clusters; deciduous leaves, 30-40 leaflets .... *A. julibrissin* p.143
Alnus – *Leaves simple; miniature brown conelike fruit, in clusters* ....
    1. Leaves oval, with 5-6 pairs of veins, very finely serrated margins; fat oval cones .... *A. cordata* p.144
    1. Leaves ovate, 10-15 pairs of veins, with roughly serrated edges; elongated cones 2
        2. Leaves wavy-lobed; double-serrate edges, turn under slightly .... *A. oregona* p.145
        2. Leaves serrate to double-serrate .... *A. rhombifolia* p.146
Araucaria – *Leaves short overlapping at base, unappressed; large heavy conelike fruit*
    1. Leaves broadly flat, sharp pointed 2
    1. Leaves round, needlelike or hooklike, whorled and overlapping .... *A. heterophylla* p.149
        2. Leaves whorled and overlapping .... *A. araucana* p.147
        2. Leaves in 2 rows, or semi-whorled .... *A. bidwillii* p.148
Arbutus – *Leaves simple; white or pink bell-shaped flowers; rough, red, spongy, fleshy or pealike fruit*
    1. Leaves 3-6 " long, entire on adult leaves; 1/2" round, orange-red fruit; smooth bark .... *A. menziesii* p.150
    1. Leaves 2-3" long, deeply serrate; 3/4" round juicy red rough fruit; rough dark bark .... *A. unedo* p.151
Betula – *Leaves simple; pendulous; soft, flaky, catkinlike cone fruit*
    1. Bark white, with black or gray markings 2
    1. Bark tan or gray 3
        2. Pronounced weeping habit .... **B. pendula** & vars. p.156, 157
        2. Semi-weeping habit .... **B. papyrifera** p.155
    3. Tan bark, with whitish cast, shedding in thin curling strips; leaves ovate .... **B. nigra** p.153
    3. Gray bark, smooth or cracked, white markings; leaves broadly ovate to oval .... **B. occidentalis** p.154
Brachychiton – *Leaves simple, glossy green; heavy, woody follicle fruit*
    1. Leaves 6-10" long, simple, or often 3-5 lobed; loose racemes of reddish flowers .... **B. acerifolius** p.158
    1. Leaves 1-1/2 to 2-1/2" long, simple, elongated tip; inconspicuous yellowish cup flowers .... **B. populneus** p.159
Callistemon – *Leaves simple; red or yellow fluffy cylindrical flowers with elongated stamens*
    1. Leaves 1/4-1/2" wide, 3" long, densely placed on stiff, erect branches .... *C. citrinus* p.161
    1. Leaves 1/4" wide, 2-4" long, with pliable weeping branches .... *C. viminalis* p.162
Carpinus – *Leaves simple, double-serrate edge; tassel-like flowers; catkin fruit*
    1. Compact, pyramidal or broad conical form .... *C. betulus* 'Compacta' p. 164
    1. Upright, columnar form .... *C. betulus* 'Fastigiata' p.165
Cedrus – *Leaves needlelike, gray or green; fat, vertical "candle" cones of compressed scales*
    1. Branchlets stiff, needles less than 1" long; cones less than 3" long .... *C. atlantica* 'Glauca' p.170
    1. Branchlets drooping, needles 1" or longer; cones 3" or larger .... *C. deodara* & var. p.171, 172
Celtis – *Leaves alternate, simple, deciduous, with tri-veined leaf base; long-stemmed single drupe fruit*
    1. Leaves usually noticeably dark green 2
    1. Leaves usually noticeably light green 3
        2. Leaves doubly serrate, with elongated acuminate tip, dull dark green .... *C. occidentalis* p.175
        2. Leaves with finely crenate margins, acute tailed tip, glossy deep green .... *C. sinensis* p.176
    3. Leaves auriculate at base, very sandpapery texture .... *C. reticulata* p.174
    3. Leaves with finely crenate margins, acute tailed tip, dull green .... *C. australis* p.174
Cercis – *Leaves alternate, simple, round, or distinctly heart-shaped, deciduous; legume seed pods*
    1. Leaves glossy green, reniform or rounded .... *C. reniformis* p.183
    1. Leaves dull green or bluish green, cordate or heart-shaped 2
        2. Mature bark tan and scaly; leaves 3-5", acute apex; height 15' or more .... *C. canadensis* & vars. p.180, 181
        2. Mature bark dark and smooth; leaves 2-3"; obtuse apex, height 15' or less .... *C. occidentalis* p.182
Chamaecyparis – *Leaves flattened, scalelike, green to blue-green; small brown scaly cones*
    1. Foliage gland dotted on back, flat horizontal sprays; 6-10 cone scales .... *C. lawsoniana* & var. p.185, 186
    1. Foliage without glands, sprays in several planes; 4-6 cone scales .... *C. nootkatensis* p.187
Cornus – *Leaves opposite, simple, deciduous; red or orange drupe fruit*
    1. Evergreen, summer-flowering; leaves 2" or less; composite fleshy fruit .... *C. capitata* p.193
    1. Deciduous, spring-flowering; leaves 2" or longer; single separate fruit 2
        2. Height 15' or more; leaves 2" or longer; flowers with large bracts 3
        2. Height 15' or less; leaves 2" or less; yellow stamenlike flowers .... *C. mas* p.196
    3. Flower bracts not notched at ends, white or slight pinkish tinge 4
    3. Flower bracts notched at ends, white, pink or red; bark fissured, squared plates .... *C. florida* & vars. p.194, 195
        4. Flower bracts usually 6, elliptical, with round-pointed ends .... *C. nuttallii* p. 197
        4. Flower bracts usually 4, broadly ovate, tapering to long-pointed ends .... *C. kousa* p. 195
Crataegus – *Leaves alternate, simple, toothed or 3-5 lobed; red pyracantha-like berries*
    1. Leaves simple, toothed 2
    1. Leaves simple, lobed 3
        2. Leaves 1-3" long .... *C. douglasii* p. 198
        2. Leaves 2-4" long .... *C. x lavallei* p. 198
    3. Leaves ovate, 3-5 sharply lobed; white flowers .... *C. phaenopyrum* p. 198
    3. Leaves ovate, 3-5 round-lobed, mitten-like, or palmate; white, pink or red flowers .... *C. laevigata* p. 199
Cryptomeria – *Leaves needlelike, whorled, in tufted sprays*
    1. Needles 1/8" or shorter, dark green; 1/2" round scaled cones .... *C. japonica* p.201

    1. Needles 1/4" or longer, bluish green or bronzy in winter; cones rarely present .... *C. japonica* 'Elegans' p.202

Cupressus – *Leaves scalelike; small round brown scaly cones*
    1. Foliage green to yellow, in flattened, weeping sprays .... *C. funebris* p.207
    1. Foliage greenish to grayish, in rigid sprays, upright or horizontal, not drooping 2
        2. Narrowly vertical form, with vertical branching .... *C. sempervirens* p.212
        2. Oval to conical form, with horizontal branching 3
    3. Bark shedding in thin curling strips, exposing rather shiny reddish underbark 4
    3. Bark dull brown or gray, fibrous, furrowed, cracked or scaly 5
        4. Foliage bluish gray... *C. guadalupensis* p.208
        4. Foliage greenish gray .... *C. arizonica* p.206
    5. Foliage pitted, with 2 shallow grooves on back; height to 60' .... *C. macrocarpa* p.210
    5. Foliage pitted, not grooved; height to 30-40' 6
        6. Green foliage, cones with 6-8 peltate scales, with protruding hook .... *C. macnabiana* p.209
        6. Green foliage, cones with 6-8 peltate scales, with round, pointed protrusion .... *C. sargentii* p.211

Eriobotrya – *Leaves alternate, simple, persistent, leathery, elliptical shape*
    1. Leaves shiny deep green; with reddish cast on new leaves; pink flowers .... *E. deflexa* p.215
    1. Leaves noticeably fuzzy; creamy yellow flowers; yellowish orange 1/2" pome fruit .... *E. japonica* p.216

Erythrina – *Leaves alternate, trifoliate; red legumelike flowers*
    1. Leaflets ovate, with pointed tips; bark smooth, gray; branch ends not twiggy .... *E. caffra* p.217
    1. Leaflets oval-oblong, with pointed ends; bark fissured, brown; twiggy branch ends .... *E. crista-galli* p.218

Eucalyptus – *Leaves alternate, simple, persistent, leathery, with eucalyptol odor*
    1. Juvenile leaves less than 1" long, connate-perfoliate, adult leaves lanceolate, 2-3" long .... *E. gunnii* p.228
    1. Leaves longer than 1" 2
        2. Leaves ovate or spherical, only 3
        2. Leaves lanceolate to elliptical, ovate or spherical juvenile growth may be present 4
    3. Leaves with 1/2" petiole, ovate, bluish gray green, semiglaucous .... *E. baueriana* p.219
    3. Leaves connate-perfoliate, overlapping at stem, glaucous gray-green .... *E. cinerea* p.221
        4. Bark peeling or flaking, to reveal whitish underbark 5
        4. Bark peeling or flaking, with no whitish underbark 8
    5. Gray bark peels in round or irregular plates 6
    5. Bark peels irregularly, in strips or plates 7
        6. Variable ovate to lanceolate gray-green leaves .... *E. polyanthemos* p.231
        6. Lanceolate, gray-green leaves .... *E. leucoxylon* p.229
    7. Foliage tufts sparse .... *E. citriodora* p.222
    7. Foliage fairly dense .... *E. viminalis* p.234
        8. Flowers white or creamy yellow 9
        8. Flowers reddish, pinkish, orange or yellow 14
    9. Leaves elliptical; persistent fused seed capsules with elongated projections .... *E. conferruminata* p.224
    9. Leaves lanceolate; individual clustered seed capsules 10
        10. Lanceolate leaves, 6" long or more, with ovate to connate-perfoliate juvenile growth 11
        10. Lanceolate leaves, 6" long or less, only 12
    11. Mature leaves green with slight gray to whitish pubescence, 3/4" seed capsules .... *E. globulus* p.226
    11. All leaves distinctly gray green, seed capsules 1/2" or smaller .... *E. pulverulenta* p.232
        12. Leaves gray-green, nearly linear; yellow flowers 1/8-1/4" .... *E. nicholii* p.230
        12. Leaves green to gray-green, lanceolate; creamy white flowers 1/4" or larger 13
    13. Long slender lanceolate leaves, medium green; bark mottled tan .... *E. camaldulensis* p.220
    13. Leaves oval-lanceolate, dull gray-green pink/yellow twigs, bark dull brown/white .... *E. cladocalyx* p.223
        14. Leaves grayish cast, lanceolate; red-based flowers 1/2" or smaller .... *E. sideroxylon* 'Rosea' p.233
        14. Leaves thick dark green, broad lanceolate; yellow-based flowers 1/2" or larger .... *E. ficifolia* p.225

Fagus – *Leaves alternate, simple, deciduous, glossy; small woody husklike fruit*
    1. Leaves green to dark bronzy green .... *F. sylvatica* p.235
    1. Leaves purplish, coppery or multi-colored 2
        2. Leaves green, with white variegation, edges with pink .... *F. sylvatica* 'Tricolor' p.237
        2. Leaves reddish to purple 3
    3. Leaves reddish to bronzy purple .... *F. sylvatica* 'Purpurea' p.237
    3. Leaves glossy, deep purple; trees have noticeably upright branching .... *F. sylvatica* 'Riversii' p.236

Ficus – *Leaves alternate, simple, persistent; white milky sap; greenish brown drupe fruit*
    1. Leaves 4" long or less .... *F. microcarpa* 'Nitida' p.239
    1. Leaves 6" long or more 2
        2. Leaves medium green, shiny upperside, lighter underside .... *F. macrophylla* p.238
        2. Leaves dark green, shiny upperside, brown fuzzy underside .... *F. rubiginosa* p.240

Fraxinus – *Leaves opposite, pinnately compound, deciduous; drooping winged seeds*
    1. Leaflets sessile 2
    1. Leaflets stalked 4
        2. 5-7 leaflets, ovate to broad lanceolate .... *F. latifolia* p.245
        2. 7-9 leaflets 3
    3. Leaflets broad-lanceolate, medium to light green...... .... *F. holotricha* 'Moraine' p.244
    3. Leaflets narrow-lanceolate, dark green to purplish cast...... .... *F. angustifolia* 'Raywood' p.242
        4. Height 20' or less, 3-9 elliptical to ovate leaflets, with rounded tips .... *F. dipetala* p.243
        4. Height to 20' or more, leaflets pointed 5

Picea – *Leaves needlelike; narrow elongated cones, usually pendent*

Pinus – *Leaves needlelike, sheathed at base; cones bearing nutlike winged seeds*

27. Needles bluish green, in 2s, somewhat sparse .... *P. brutia* var. *eldarica* p.318
27. Needles shiny dark green, in 2s or 3s, dense 28
    28. Height less than 8' .... *P. mugo* p.333
    28. Height over 8' 29
29. Height to 20' .... *P. contorta* ssp. *contorta* p.321
29. Height to 50' .... *P. contorta* ssp. *murrayana* p.322

<u>**Pittosporum**</u> – *Leaves opposite, simple, persistent*
1. Leaves rhomboid, serrated edges; yellow flowers; orange fruit .... *P. rhombifolium* p.345
1. Leaves light green, very wavy edges .... *P. undulatum* p.346

<u>**Platanus**</u> – *Leaves alternate, simple, lobed, tomentose underside, deciduous; 1" round dry fruit*
1. Leaves palmate, with toothed pointed lobed margins 2
1. Leaves palmate, with finely serrated or smooth lobed margins .... *P. racemosa* p.349
    2. 3-5 lobes, nearly equally long as wide, flaky blotchy older bark .... *P. x acerifolia* p.347
    2. 3-5 lobes, broader than long, dark scaly older bark .... *P. occidentalis* p.348

<u>**Podocarpus**</u> – *Leaves linear, whorled, persistent*
1. Leaves 1-2" long, 1/8" wide, grayish green...... .... *P. gracilior* p.350
1. Leaves 1-3" long, 1/4" wide, dark green...... .... *P. macrophyllus* p.351

<u>**Populus**</u> – *Leaves alternate, simple, deciduous; catkinlike flowers and cottony seeds*
1. Leaves nearly spherical to ovate, with short acute tip .... *P. tremuloides* p.356
1. Leaves nearly deltoid to cordate or ovate 2
    2. Leaves with white tomentose underside 3
    2. Leaves glaucous green 4
3. Broad canopy with horizontal branching .... *P. alba* p.352
3. Columnar upright growth .... *P. alba* 'Pyramidalis' p.353
    4. Ellipsoidal to ovoid form 5
    4. Narrowly columnar upright growth .... *P. nigra* 'Italica' p.355
5. Leaves with slightly lobed margins .... *P. fremontii* p.354
5. Leaves with finely toothed margins .... *P. trichocarpa* p.357

<u>**Prosopis**</u> – *Leaves alternate, bipinnnately compound, deciduous, small spines at leaf base; seed pod fruit*
1. Leaves with 2 paired pinnae, leaflets 1/4 to 1-1/4" long .... *P. glandulosa* p.358
1. Leaves with 1 or 2 pinnae, leaflets 1/4 to 3/8" long .... *P. velutina* p.359

<u>**Prunus**</u> – *Leaves simple, entire, deciduous or persistent; fleshy fruit*
1. Leaves persistent, glossy green; growth often shrublike 2
1. Leaves deciduous; usually tree form 4
    2. Leaves elliptical, pliable .... *P. caroliniana* p.361
    2. Leaves lanceolate to lanceolate-ovate, somewhat stiff 3
3. Leaf usually ovate, margin generally entire, flat .... *P. lyonii* p. 366
3. Leaf widely ovate to round, margin spiny-serrate, somewhat wavy .... *P. ilicifolia* p.365
    4. Leaves 2" or longer, green 5
    4. Leaves 2" or smaller, green or purplish to bronze, slightly crinkled, or textured 8
5. Weeping form .... *P. pendula* p.368
5. Upright form 6
    6. Stiffly upright symmetrical obconical form; double white or pink flowers .... *P.* 'Kwanzan' p.368
    6. Upright, with outward arching branches; semi-double white to light pink flowers 7
7. Regular even branching .... *P. serrulata* p.367
7. Branching angled outward .... *P. x yedoensis* p.368
    8. Leaves green or slightly bronze colored .... *P. cerasifera* p.362
    8. Leaves purplish to bronze colored 9
9. Flowers double, pink, reddish to bronzy cast greenish foliage; trunk often with knots .... *P. x blirieana* p.360
9. Flowers single, white or pink; relatively smooth or fissured bark 10
    10. Flowers white; sets plum fruit; reddish purple foliage .... *P. cerasifera* 'Atropurpurea' p.363
    10. Flowers pink; rarely sets fruit; purple foliage .... *P. cerasifera* 'Krauter Vesuvius' p.364

<u>**Pseudotsuga**</u> – *Needlelike foliage; pendulous cones with scales remaining attached*
1. Lush dark green foliage, appearing somewhat flattened; cone scales semi open .... *P. menziesii* p.370
1. Grayish to bluish green foliage, surrounding branchlets; cone scales open fully .... *P. macrocarpa* p.369

<u>**Pyrus**</u> – *Leaves alternate, simple, deciduous; white flower clusters; small pomelike fruit*
1. Irregular broadly ovate form to semi weeping; semi-deciduous .... *P. taiwanensis* p.375
1. Upright, ovate form symmetrical 2
    2. Long upward-curving branches, bowing outward from stout trunk .... *P. calleryana* 'Bradford' p.374
    2. Upright, vertical branches, with shorter horizontal twiggy ends .... *P. calleryana* 'Aristocrat' p.373

<u>**Quercus**</u> – *Leaves alternate, simple, lobed or entire, deciduous or persistent; acorn nut fruit*
1. Leaves persistent 2
1. Leaves deciduous or semi-deciduous 14
    2. Leaves with entire margins 3
    2. Leaves with serrated , toothed or lobed margins 5
3. Short-stalked acorns; dark green pointed leaves, wavy-edged, whitish woolly underside .... *Q. ilex* p.388
3. Multiple short-stalked acorns from long common stem 4
    4. Shiny green blunt ended leaves, obovate, recurving .... *Q. virginiana* p.408
    4. Dull green pointed leaves, elliptical, wavy edged .... *Q. oblongifolia* p.396
5. Leaves ovate, or shortly elliptical, less than 3 times as long as wide 6

  5. Leaves elongated, elliptical, more than four times as long as wide 10
    6. Dull green to gray-green, elongated spiny toothed edges .... ***Q. turbinella*** p.407
    6. Shiny green leaves, short fine spiny toothed edges 7
  7. Pointed ends, variable entire to sawtoothed edges, finely woolly underside .... ***Q. chrysolepis*** p.380
  7. Rounded ends or only slightly pointed 8
    8. Leaves 1" long or less, trees mostly less than 25' tall, forming thickets, small trunks .... ***Q. dumosa*** p.383
    8. Leaves 1-1/4" long or more, trees usually taller than 25', develop large trunks 9
  9. Densely to sparsely prickly toothed, recurving slightly along margins .... ***Q. agrifolia*** p.377
  9. Variable on same tree, spiny, sparsely toothed, or entire, margins not recurving .... ***Q. wislizeni*** p.409
    10. Leaves 3/4" wide or narrower 11
    10. Leaves 1" wide or wider 12
  11. Light green leaves, pale green underside, finely serrated edges .... ***Q. myrsinifolia*** p.395
  11. Dark green leaves, whitish underside, wavy edges, entire or few sparse serrations .... ***Q. suber*** p.405
    12. Shiny dark green leaves, elliptical, evenly toothed 13
    12. Dull green leaves, obovate, unevenly sparsely toothed wavy edges .... ***Q. rugosa*** p.404
  13. Leaf surface raised between evenly spaced veins, edges roll under evenly .... ***Q. tomentella*** p.406
  13. Leaf surface rather flat, edges roll under only slightly, rather wavy .... ***Q. emoryi*** p.384
    14. Leaf lobes acutely pointed 15
    14. Leaf lobes rounded, not pointed 20
  15. Short-toothed, less than 1/8", narrowly elliptical .... ***Q. acutissima*** p.376
  15. Pointed teeth, 1/8" or longer, ovate 16
    16. Leaves regularly lobed 1/2 or less to midrib, dentate ends 17
    16. Leaves variably lobed 1/2 or more to midrib 18
  17. Leaves 5-8" long, pliable, pointed ends not sharp to touch .... ***Q. rubra*** p.403
  17. Leaves 2-4" long, stiff, pointed ends prickly to touch .... ***Q. morehus*** p.393
    18. Leaves persist dried into winter; deeply lobed leaves, broad rounded sinuses .... ***Q. palustris*** p.398
    18. Leaves normally deciduous 19
  19. Lobes as wide as sinuses, long pointed ends .... ***Q. coccinea*** p.381
  19. Lobes narrower than sinuses, rounded ends with short pointed tips .... ***Q. kelloggii*** p.389
    20. Shallow irregular lobing, 1/4" or less; slightly wavy edged, or nearly entire 21
    20. Deep lobing, with sinuses more than halfway to midrib 27
  21. Leaves linear-elliptical, to pointed tip 22
  21. Leaves ovate, to broad, rounded tip 23
    22. Leaves somewhat sickle-shaped, lobing symmetrical each side of petiole .... ***Q. phellos*** p.399
    22. Oblanceolate, entire or 3-lobed, short-toothed near apex; semi-deciduous .... ***Q. laurifolia*** p.390
  23. Leaves 2" or longer 24
  23. Leaves 2" or shorter 25
    24. Sinuses deeper than wide between lobes .... ***Q. muehlenbergii*** p.394
    24. Sinuses shallower than wide between wide-rounded lobes .... ***Q. robur*** p.401
  25. Shallow lobing, with distinct v-shaped sinuses, flat-angled to petiole .... ***Q. douglasii*** p.382
  25. Very shallow lobing, tapered to petiole 26
    26. Deep bluish green, with distinctly rounded ends .... ***Q. engelmannii*** p.385
    26. Dark green to bluish green, only slightly wider than 26 above .... ***Q. arizonica*** p.379
  27. Leaves up to 6-10" long, lobes constricted at base; wide fat lobes, spatulate end .... ***Q. macrocarpa*** p.392
  27. Leaves 2-6" long 28
    28. Leaves 3-4" long, lobes often nearly to midrib, often club-shaped at ends .... ***Q. lobata*** p.391
    28. Leaves 3-6" long, lobes from halfway, nearly to midrib, widened at ends 29
  29. Height range to 50' or more 30
  29. Height range to 25-35'; leaves 7-9 lobed, acorns 2/3 to 3/4" long .... ***Q. gambelii*** p.386
    30. Height range to 80-100'; leaves 7-9 lobed, acorns 3/4" long or less .... ***Q. alba*** p.378
    30. Height range to 50-70' tall; leaves 5-9 lobed, acorns 1 to 1-3/4" long .... ***Q. garryana*** p.387

<u>Rhus</u> – *Leaves alternate, compound, deciduous or persistent; small panicles of 1/8" round fruit*
  1. Leaves persistent .... ***R. lancea*** p.412
  1. Leaves deciduous 2
    2. Twigs smooth .... ***R. glabra*** p.411
    2. Twigs densely hairy .... ***R. typhina*** p.413

<u>Robinia</u> – *Leaves alternate, pinnately compound, deciduous; legume flowers and pods*
  1. Upright straight branching 2
  1. Upright trunk with contorted twisted branching; white flowers .... ***R. pseudoacacia* 'Tortuosa'** p.415
    2. Green foliage 3
    2. Yellow/green foliage .... ***R. pseudoacacia* 'Frisia'** p.415
  3. Flowers white .... ***R. pseudoacacia*** p.414
  3. Flowers pink or purple 4
    4. Pink flowers .... ***R.* 'Idahoensis'** p.416
    4. Purple flowers .... ***R.* 'Purple Robe'** p.417

<u>Salix</u> – *Leaves opposite, simple, deciduous; catkinlike flowers; cottony seeds*
  1. Leaves blunt-tipped, oblanceolate to obovate 2
  1. Leaves pointed-tipped, elliptical to lanceolate 3
    2. Leaves with smooth edges, turn under slightly .... ***S. lasiolepis*** p.424
    2. Leaves with mostly smooth edges, or barely toothed, with slightly wavy edges .... ***S. scouleriana*** p.425

3. Weeping form 4
3. Irregular, upright, rounded crown or large mounding form 5
    4. Straight branches and leaves, drooping .... ***S. babylonica*** p.418
    4. Twisted corkscrewlike growth and foliage, also drooping .... ***S. matsudana*** 'Tortuosa' p.426
5. Leaves grayish green, with whitish feltlike hairs .... ***S. hindsiana*** p.420
5. Leaves green, without noticeable hairiness 6
    6. Leaves with noticeable stipule at leaf base 7
    6. Leaves leathery, without noticeable stipules .... ***S. hookeriana*** p.421
7. Young twigs reddish .... ***S. laevigata*** p.422
7. Young twigs yellowish 8
    8. Leaves lanceolate; broad mounding shrublike form .... ***S. lasiandra*** p.423
    8. Leaves narrowly lanceolate; rather upright treelike form .... ***S. gooddingii*** p.419

Schinus – *Leaves alternate, pinnately compound; persistent, panicles of tiny round capsules*
1. Leaves pinnate, with 12 or more leaflets; weeping form .... ***S. molle*** p.429
1. Leaves pinnate with 10 or less leaflets; rounded canopy 2
    2. Leaves with 6 or more leaflets, dark glossy green .... ***S. terebinthifolius*** p.431
    2. Leaves with 6 or less leaflets, dull medium green .... ***S. peruviana*** p.430

Tamarix – *Foliage appearing needlelike or featherlike; elongated fluffy pink flower spikes*
1. Evergreen; flowers with 5 petals, in panicled spikes .... ***T. aphylla*** p.438
1. Deciduous; flowers with 3 petals, in elongated spikes .... ***T. parviflora*** p.439

Taxodium – *Leaves needlelike, soft, light green, 2-ranked; globose cones with thick disintegrating scales*
1. Foliage opposite from stems, deciduous; branches spreading, drooping only at ends .... ***T. distichum*** p.440
1. Foliage alternately from stems, evergreen to semi-deciduous; branches drooping .... ***T. mucronatum*** p.441

Thuja – *Leaves needlelike or scalelike*
1. Flattened glandular-pitted leaves; 4-scaled cones 2
1. Flattened grooved leaves; 6-scaled cones .... ***T. occidentalis*** p.443
    2. Tall tree, to 100' or more .... ***T. plicata*** p.444
    2. Small tree or shrub 3
3. Vertical, upright columnar growth habit .... ***T. plicata*** 'Emerald Cone' p.445
3. Rounded, oval growth habit .... ***T. plicata*** 'Gracilis' p.445

Tilia – *Leaves alternate, simple, deciduous*
1. Leaves 3" or larger; fruit hard-shelled .... ***T. tomentosa*** p.449
1. Leaves smaller than 3" 2
    2. Symmetrical upright branching; fruit thin-shelled .... ***T. cordata*** p.447
    2. Irregular branching with twiggy ends; fruit thin-shelled .... ***T. europaea*** p.448

Torreya – *Leaves needlelike, bright shiny green, appearing nearly fir-like*
1. Leaves 1-2" long, straight, pointed tip; cracked furrowed bark somewhat scaly .... ***T. californica*** p.451
1. Leaves 1/2 to 1" long, curved under, dull tip; "smooth" bark with raised "braided" ridges .... ***T. nucifera*** p.452

Tsuga – *Leaves needlelike, 2-ranked, persistent; 1-2" soft papery sessile cones, (eastern species stalked)*
1. Flat needles, groove on back, flatly 2-ranked; 1" light brown cones; north coast .... ***T. heterophylla*** p.459
1. Rounded needles, ridged back, all directions; 1-2" purplish cone; high Sierra .... ***T. mertensiana*** p.460

Ulmus – *Leaves alternate, simple deciduous; seeds 1/8" papery, winged*
1. Leaves 3" or larger, noticeably unequal at base 2
1. Leaves usually 1" or smaller, leather; zigzag twigs hairy; tiny brown buds .... ***U. parvifolia*** p.462
    2. Leaves doubly serrate, hairy, somewhat crinkled texture .... ***U. americana*** p.461
    2. Leaves serrate, shiny and appearing smooth, but sandpapery rough to touch .... ***U. rubra*** p.463

Yucca – *Leaves swordlike, in tufts at ends of thick ascending branches; large, corky trunk; white bell-shaped fls.*
1. Leaves 6-10" long, 1/4-1/2" wide; Mojave Desert .... ***Y. brevifolia*** p.465
1. Leaves 1-2' long, 1-2" wide; southeastern species, common ornamental use .... ***Y. gloriosa*** p.466

INDEX TO TREES

Page	Botanical Name	Common Name(s)	Family	Other Latin Name(s)	Avail.
127	*Acer pseudoplatanus* 'Atropurpureum'	wineleaf maple, pinkleaf maple	Aceraceae		2
128	*Acer rubrum*	red maple, scarlet maple Swamp Maple	Aceraceae		1
129	*Acer rubrum* 'Columnare'	columnar red maple	Aceraceae	*A. r.* 'Pyramidale'	2
130	*Acer saccharinum*	silver maple, white maple	Aceraceae	*A. dasycarpum, A. eriocarpum*	1
131	*Acer saccharum*	sugar maple, rock maple	Aceraceae		2
115	*Acer saccharum* ssp. *grandidentatum*	bigtooth maple, Wasatch maple	Aceraceae		4
133	*Acer tataricum*	Tatarian maple	Aceraceae		4
132	*Acer tataricum* ssp. *ginnala*	amur maple	Aceraceae		4
134	*Acer truncatum*	purpleblow maple	Aceraceae		2
133	*Acer truncatum* 'Pacific Sunset'	Pacific Sunset maple, Shantung maple	Aceraceae		2
135	**Aesculus californica**	California buckeye	Hippocastanaceae		2
137	*Aesculus* x *carnea*	red horsechestnut	Hippocastanaceae		1
136	*Aesculus* x *carnea* 'Briotii'	Briotii red horsechestnut	Hippocastanaceae		1
138	*Aesculus hippocastanum*	horsechestnut	Hippocastanaceae		2
137	*Aesculus pavia*	red buckeye	Hippocastanaceae	*A. discolor, A. austrina, A. rubra*	3
139	*Agathis robusta*	dammar pine, kauri pine, Queensland kauri	Araucariaceae		5
140	*Agonis flexuosa*	peppermint tree Australian willow myrtle	Myrtaceae		1
141	*Ailanthus altissima***	tree-of-heaven	Simarubaceae	*A. glandulosa*	1
142	*Albizia distachya***	plume albizia	Fabaceae	*A. lophantha*	2
143	*Albizia julibrissin*	silk tree	Fabaceae	*Acacia julibrissin, Mimosa julibrissin*	1
144	*Alnus cordata*	Italian alder	Betulaceae	*A. cordifolia*	1
145	**Alnus oregona**	red alder, Oregon alder	Betulaceae	*A. rubra*	1
146	**Alnus rhombifolia**	white alder	Betulaceae		1
147	*Araucaria araucana*	monkey puzzle tree Chilean pine	Pinaceae	*A. imbricata*	4
148	*Araucaria bidwillii*	bunya-bunya	Pinaceae		2
149	*Araucaria heterophylla*	Norfolk Island pine	Pinaceae	*A. excelsa*	2
150	**Arbutus menziesii**	madrone, madrono, Oregon laurel	Ericaceae		4
151	*Arbutus unedo*	strawberry tree	Ericaceae		1
152	*Bauhinia variegata*	orchid tree	Fabaceae	*B. purpurea*	3
157	*Betula jacquemontii*	Jacquemont birch	Betulaceae	*B. utilis* 'Jacquemontii'	1
153	*Betula nigra*	river birch, red birch	Betulaceae	*B. rubra*	1
154	**Betula occidentalis**	western birch, water birch	Betulaceae		5
155	*Betula papyrifera*	paper birch, canoe birch	Betulaceae		1
156	*Betula pendula*	European white birch weeping birch	Betulaceae		1
157	*Betula pendula* 'Crispa'	cutleaf weeping birch	Betulaceae	*B. p.* 'Dalecarlica'	1
157	*Betula pendula* 'Purpurea'	purpleleaf birch	Betulaceae		3
158	*Brachychiton acerifolius*	flame tree	Sterculiaceae		2
159	*Brachychiton populneus*	bottle tree, kurrajong	Sterculiaceae	*Sterculia diversifolia*	4
160	*Broussonetia papyrifera**	paper mulberry	Moraceae	*Morus papyrifera*	4
161	*Callistemon citrinus*	lemon bottlebrush	Myrtaceae	*C. lanceolatus*	1
162	*Callistemon viminalis*	weeping bottlebrush	Myrtaceae		1
163	**Calocedrus decurrens**	incense cedar, western cedar	Pinaceae	*Libocedrus decurrens*	2
164	*Carpinus betulus* 'Compacta'	compact hornbeam	Betulaceae		2
165	*Carpinus betulus* 'Fastigiata'	columnar hornbeam	Betulaceae	*C. b.* 'Pyramidalis', *C. b.* "Erecta'	2
166	*Carya illinoiensis*	pecan, hickory	Juglandaceae	*C. pecan*	2
167	*Castanea sativa*	Spanish chestnut	Juglandaceae		2
168	*Casuarina equisetifolia*	horsetail tree	Casuarinaceae		1

Page	Botanical Name	Common Name(s)	Family	Other Latin Name(s)	Avail.
169	*Catalpa speciosa*	western catalpa	Bignoniaceae		1
170	*Cedrus atlantica* 'Glauca'	blue Atlas cedar	Pinaceae		1
171	*Cedrus deodara*	deodar cedar	Pinaceae		1
172	*Cedrus deodara* 'Descanso Dwarf'	compact deodar cedar	Pinceae	*C. deodara* 'Compacta'	2
173	*Ceiba speciosa*	floss silk tree	Bombaceae	*Chorisia speciosa*	2
174	*Celtis australis*	European hackberry	Ulmaceae		1
175	*Celtis occidentalis*	common hackberry	Ulmaceae		1
174	**Celtis reticulata**	netleaf hackberry, western hackberry	Ulmaceae	*C. douglasii*	5
176	*Celtis sinensis*	Chinese hackberry, Yunnan hackberry	Ulmaceae		1
177	*Ceratonia siliqua*	carob tree, St. John's bread	Fagaceae		2
178	*Cercidiphyllum japonicum*	katsura tree	Cercidiphyllaceae		3
179	**Cercidium floridum**	palo verde	Fabaceae	*C. torreyanum*	4
180	*Cercis canadensis*	eastern redbud	Fabaceae		1
181	*Cercis c.* 'Forest Pansy'	purple-leaf eastern redbud	Fabaceae		2
181	*Cercis c.* 'Silver Cloud'	variegated eastern redbud	Fabaceae		2
182	**Cercis occidentalis**	western redbud	Fabaceae		1
183	*Cercis reniformis*	Texas redbud	Fabaceae	*C. canadensis* var. *texensis*	2
184	**Cercocarpus betuloides**	mountain ironwood hardtack, sweet brush	Rosaceae		4
185	**Chamaecyparis lawsoniana**	Port Orford-cedar, Lawson cypress, Oregon cedar, white cedar, ginger pine	Cupressaceae	*Cupressus lawsoniana*	4
186	*Chamaecyparis l.* 'Stewartii'	Stewart's golden cedar	Cupressaceae		2
187	**Chamaecyparis nootkatensis**	Nootka cypress, Alaska cedar	Cupressaceae	*Thujopsis borealis*, *Cupressus nootkatensis*	4
188	**Chilopsis linearis**	desert-willow	Bignoniaceae		2
189	x *Chitalpa tashkentensis*	chitalpa	Bignoniaceae		2
190	**Chrysolepis chrysophylla**	golden chinquapin	Fagaceae	*Castanopsis chrysophylla*, *Castanea chrysophylla*	4
191	*Cinnamomum camphora*	camphor tree	Lauraceae	*Laurus camphora*, *Camphor camphora*	1
192	*Cladrastis kentukea*	yellowwood, yellow ash, yellow locust	Fabaceae	*C. lutea, C. tinctoria*	2
193	*Cornus capitata*	evergreen dogwood	Cornaceae	*Benthamidia capitata*, *Dendrobenthamia capitata*	3
194	*Cornus florida*	eastern dogwood, flowering dogwood	Cornaceae	*Benthamidia florida*, *Cynoxylon floridum*	1
195	*Cornus f.* 'Rainbow'	variegated dogwood	Cornaceae		3
195	*Cornus f.* 'Cherokee Sunset'	pink variegated dogwood	Cornaceae		3
195	*Cornus kousa*	kousa dogwood	Cornaceae		2
196	*Cornus mas*	cornelian cherry	Cornaceae		2
197	**Cornus nuttallii**	western dogwood, Pacific dogwood	Cornaceae		4
198	**Crataegus douglasii**	black hawthorn, western black hawthorn, black thornberry	Rosaceae	*C. brevispina*, *C. rivularis, C. suksdorfii*	4
199	*Crataegus laevigata*	English hawthorn	Rosaceae	*C. oxyacantha*	1
198	*Crataegus* x *lavallei*	Carriere hawthorn	Rosaceae	*C. carriere*	2
198	*Crataegus phaenopyrum*	Washington thorn	Rosaceae		1
200	*Crinodendron patagua*	lily-of-the-valley tree, lantern tree	Eleaocarpaceae	*C. dependens*, *Tricuspidaria lanceolata*	4
201	*Cryptomeria japonica*	Japanese cryptomeria	Pinaceae		1
202	*Cryptomeria j.* 'Elegans'	plume cedar plume cryptomeria	Pinaceae	*C. gracilis*	2
203	*Cunninghamia lanceolata*	China fir	Pinaceae	*C. sinensis*, *Belis lanceolata*	2
204	*Cupaniopsis anacardioides*	carrot wood, tuckeroo	Sapindaceae	*Cupania anacardioides*	2
205	x *Cupressocyparis leylandii*	Leyland cypress	Cupressaceae		1
206	**Cupressus arizonica**	Arizona cypress	Cupressaceae	*C. arizonica* var. *glabra*	1
207	*Cupressus funebris*	mourning cypress	Cupressaceae		4

Page	Botanical Name	Common Name(s)	Family	Other Latin Name(s)	Avail.
208	*Cupressus guadalupensis*	Guadalupe Island cypress	Cupressaceae		4
209	**Cupressus macnabiana**	McNab cypress	Cupressaceae		4
210	**Cupressus macrocarpa**	Monterey cypress	Cupressaceae		2
214	*Cupressus macrocarpa* 'Lutea'	golden Monterey cypress	Cupressaceae		4
211	**Cupressus sargentii**	Sargent's cypress	Cupressaceae		4
212	*Cupressus sempervirens*	Italian cypress	Cupressaceae		1
213	*Cupressus s.* 'Stricta'	columnar Italian cypress	Cupressaceae		1
214	*Cupressus s.* 'Swane's Golden'	Swane's golden cypress	Cupressaceae		2
215	*Eriobotrya deflexa*	bronze loquat	Rosaceae	*Photinia deflexa*	1
216	*Eriobotrya japonica*	loquat	Rosaceae	*Photinia japonica,* *Mespilus japonica*	1
217	*Erythrina caffra*	kaffirboom coral tree	Fabaceae		1
218	*Erythrina crista-galli*	cockspur coral tree	Fabaceae		1
219	*Eucalyptus baueriana*	blue box	Myrtaceae		1
220	*Eucalyptus camaldulensis*	red gum, river red gum	Myrtaceae	*E. rostrata*	1
221	*Eucalyptus cinerea*	silver dollar tree, ash-colored gum	Myrtaceae		1
222	*Eucalyptus citriodora*	lemon-scented gum	Myrtaceae		1
223	*Eucalyptus cladocalyx*	sugar gum	Myrtaceae	*E. corynocalyx*	1
224	*Eucalyptus conferruminata*	bushy yate	Myrtaceae	*E. lehmannii*	2
225	*Eucalyptus ficifolia*	red-flowering gum, scarlet gum	Myrtaceae		2
226	*Eucalyptus globulus*	blue gum	Myrtaceae		1
227	*Eucalyptus globulus* 'Compacta'	dwarf blue gum	Myrtaceae		1
228	*Eucalyptus gunnii*	cider gum	Myrtaceae	*E. whittinghamensis*	2
229	*Eucalyptus leucoxylon*	white ironbark	Myrtaceae		1
230	*Eucalyptus nicholii*	Nichol's willow-leafed peppermint	Myrtaceae		1
231	*Eucalyptus polyanthemos*	silver dollar gum	Myrtaceae		1
232	*Eucalyptus pulverulenta*	silver mountain gum	Myrtaceae		1
233	*Eucalyptus sideroxylon* 'Rosea'	red ironbark, pink ironbark	Myrtaceae		1
234	*Eucalyptus viminalis*	manna gum	Myrtaceae		1
235	*Fagus sylvatica*	European beech	Fagaceae		2
237	*Fagus sylvatica* 'Purpurea'	purple beech, copper beech	Fagaceae	*F. s.* 'Atropurpurea',	2
236	*Fagus sylvatica* 'Riversii'	Rivers' purple beech	Fagaceae	*F. s.* 'Atropunicea'	2
237	*Fagus sylvatica* 'Tricolor'	tricolor beech	Fagaceae	*F. s.* 'Roseomarginata', *F. s.* 'Purpurea Tricolor'	3
238	*Ficus macrophylla*	Moreton Bay fig	Moraceae		2
239	*Ficus microcarpa* 'Nitida'	little-leaf fig	Moraceae		2
240	*Ficus rubiginosa*	rustyleaf fig	Moraceae		2
241	*Firmiana simplex*	Chinese parasol tree	Sterculiaceae	*F. platanifolia,* *Sterculia platanifolia*	3
242	*Fraxinus angustifolia* 'Raywood'	Raywood ash, claret ash	Oleaceae	*F. oxycarpa* 'Raywood'	1
243	**Fraxinus dipetala**	foothill ash	Oleaceae		4
244	*Fraxinus holotricha* 'Moraine'	Moraine ash.	Oleaceae	*F. angustifolia* 'Moraine'	2
243	*Fraxinus ornus*	flowering ash, manna ash	Oleaeceae	*F. europaea*	3
245	**Fraxinus latifolia**	Oregon ash	Oleaceae	*F. oregona*	2
246	*Fraxinus pennsylvanica*	green ash, red ash	Oleaceae	*F. pubescens*	1
247	*Fraxinus uhdei*	evergreen ash, shamel ash	Oleaceae		1
248	**Fraxinus velutina**	Arizona ash, velvet ash	Oleaceae		1
249	*Fraxinus velutina* 'Modesto'	Modesto ash	Oleaceae		1
250	*Geijera parviflora*	Australian willow, wilga	Rutaceae		2
251	*Ginkgo biloba*	maidenhair tree	Ginkgoaceae		1
252	*Gleditsia triacanthos* 'Inermis'	thornless honey locust	Fabaceae		1
253	*Gleditsia triacanthos* 'Sunburst'	golden honey locust	Fabaceae		1
254	*Grevillea robusta*	silk oak	Proteaceae		1
255	*Hymenosporum flavum*	sweetshade	Pittosporaceae		2
256	*Jacaranda mimosifolia*	jacaranda	Bignoniaceae	*J. acutifolia*	1
257	**Juglans hindsii**	northern California black walnut	Juglandaceae		4
258	**Juglans californica**	southern California black walnut	Juglandaceae		4
258	*Juglans microcarpa*	little walnut, Texas walnut	Juglandaceae		
259	*Juglans regia*	English walnut	Juglandaceae	*J. sinensis*	2
260	**Juniperus californica**	California juniper	Cupressaceae		4

Page	Botanical Name	Common Name(s)	Family	Other Latin Name(s)	Avail.
261	*Juniperus deppeana*	alligator juniper	Cupressaceae		4
262	**Juniperus occidentalis**	western juniper	Cupressaceae	*J. utahensis*	2
263	*Juniperus scopulorum* 'Tolleson's Blue Weeping'	Tolleson's Blue Weeping juniper	Cupressaceae		3
264	*Koelreuteria bipinnata*	Chinese flame tree	Sapindaceae	*K. integrifolia*	2
265	*Koelreuteria paniculata*	goldenrain tree	Sapindaceae	*K. japonica*	1
266	*Laburnum* x *watereri*	goldenchain tree	Fagaceae		2
267	*Lagerstroemia fauriei* hybrids	Japanese crape myrtle	Lythraceae		1
268	*Lagerstroemia indica*	crape myrtle	Lythraceae		1
269	*Lagunaria patersonii*	cow itch tree, primrose tree	Malvaceae		4
270	*Larix kaempferi*	Japanese larch	Pinaceae		2
271	*Laurus nobilis*	Grecian laurel, sweet bay	Lauraceae		1
272	*Ligustrum lucidum*	glossy privet	Oleaceae	*L. macrophyllum*	1
273	*Liquidambar formosana*	Chinese sweet gum, Formosan sweet gum	Hamamelidaceae	*L. acerifolia*	4
274	*Liquidambar styraciflua*	American sweet gum	Hamamelidaceae		2
275	*Liquidambar s.* 'Rotundiloba'	roundleaf sweet gum	Hamamelidaceae		3
276	*Liriodendron tulipifera*	tulip tree, tulip poplar	Magnoliaceae		1
275	*Liriodendron t.* 'Tortuosum'	corkscrew tulip tree	Magnoliaceae		3
275	*Liriodendron t.* 'Aureo-marginatum'	variegated tulip tree	Magnoliaceae	*L. t.* 'Variegata'	3
277	**Lithocarpus densiflorus**	tanbark oak	Fagaceae	*Quercus densiflora*	4
278	*Lophostemon confertus*	Brisbane box	Myrtaceae	*Tristania conferta*	2
279	**Lyonothamnus floribundus ssp. aspleniifolius**	fern-leaved Catalina ironwood	Rosaceae		3
280	*Magnolia grandiflora*	southern magnolia, bull bay	Magnoliacea	*M. foetida*	1
281	*Magnolia* x *soulangeana*	saucer magnolia	Magnoliaceae		1
282	*Magnolia stellata*	star magnolia	Magnoliaceae		1
283	*Malus* x *floribunda*	Japanese flowering crabapple	Rosaceae		1
284	*Maytenus boaria*	mayten	Celastraceae		1
285	*Melaleuca linariifolia*	flaxleaf paperbark	Myrtaceae		2
286	*Melaleuca quinquenervia*	cajeput tree	Myrtaceae		2
287	*Melia azedarach*	chinaberry, bead tree	Meliaceae	*M. australis, M. japonica, M. sempervirens*	1
288	*Metasequoia glyptostroboides*	dawn redwood	Pinaceae		1
289	*Metrosideros excelsus*	New Zealand Christmas tree	Myrtaceae		2
294	*Michelia figo*	banana shrub	Magnoliaceae		3
290	*Morus alba**	white mulberry	Moraceae		1
291	*Morus alba* 'Fruitless'	fruitless white mulberry	Moraceae		1
290	*Morus microphylla***	Texas mulberry	Moraceae		5
292	*Morus nigra**	black mulberry, Persian mulberry	Moraceae	*M. persica*	4
293	*Myoporum laetum*	myoporum	Myoporaceae		1
294	*Neolitsea sericea*	Japanese silver tree	Lauraceae	*N. glauca, Laurus glauca N. latifolia, Litsea sericea*	3
295	*Nerium oleander*	oleander	Apocynaceae		1
296	*Nyssa sylvatica*	tupelo, sour gum, pepperidge gum, black gum	Nyssaceae	*N. biflora, N. villosa, N. multiflora*	1
297	*Olea europaea*	olive	Oleaceae		1
298	*Oxydendrum arboreum*	sorrel tree, sourwood	Ericaceae	*Andromeda arborea*	3
299	PALMS				
	Archontophoenix cunninghamiana	king palm, piccabeen palm, bungalow palm	Palmaceae	*Seaforthia elegans*	2
	Brahea edulis	Guadalupe fan palm	Palmaceae		3
300	PALMS				
	Butia capitata	pindo palm	Palmaceae		3
	Caryota urens	fishtail palm, wine palm, toddy palm	Palmaceae		2
301	PALMS				
	Chamaerops humilis	Mediterranean fan palm	Palmaceae		2

Page	Botanical Name	Common Name(s)	Family	Other Latin Name(s)	Avail.
	Jubaea chilensis	Chilean wine palm	Palmaceae		3
302	PALMS				
	Phoenix canariensis	Canary Island date palm	Palmaceae		1
	Phoenix dactylifera	date palm	Palmaceae		2
303	PALMS				
	Phoenix reclinata	Senegal date palm	Palmaceae		2
	Phoenix roebelenii	pygmy date palm	Palmaceae		3
304	PALMS				
	Syagrus romanzoffianum	queen palm	Palmaceae	*Arecastrum romanzoffianum*	1
	Trachycarpus fortunei	windmill palm	Palmaceae		1
305	PALMS				
	Washingtonia filifera	California fan palm	Palmaceae		1
	Washingtonia robusta	Mexican fan palm	Palmaceae		1
306	*Parkinsonia aculeata*	Mexican palo verde	Fabaceae		4
307	*Paulownia tomentosa**	empress tree	Scrophulariacea	*P. imperialis*	3
308	*Picea abies*	Norway spruce	Pinaceae	*P. excelsa*	2
309	**Picea breweriana**	weeping spruce	Pinaceae		5
310	**Picea engelmannii**	Engelmann spruce	Pinaceae		2
311	*Picea glauca* 'Conica'	dwarf Alberta spruce	Pinaceae		3
311	*Picea glauca* 'Pendula'	weeping white spruce	Pinaceae		3
312	*Picea jezoensis*	yeddo spruce	Pinaceae		5
313	*Picea pungens* 'Glauca'	Colorado blue spruce	Pinaceae		1
314	**Picea sitchensis**	Sitka spruce, tideland spruce	Pinaceae		4
315	**Pinus albicaulis**	whitebark pine	Pinaceae		4
316	**Pinus attenuata**	knobcone pine	Pinaceae		4
317	*Pinus brutia*	Calabrian pine	Pinaceae	*P. halepensis brutia*	4
318	*Pinus brutia* var. *eldarica*	Mondell pine	Pinaceae		1
319	*Pinus bungeana*	lacebark pine	Pinaceae		2
320	*Pinus canariensis*	Canary Island pine	Pinaceae		1
321	**Pinus contorta ssp. contorta**	shore pine, beach pine	Pinaceae		4
322	**Pinus contorta ssp. murrayana**	lodgepole pine	Pinaceae	*P. contorta* ssp. *latifolia*	2
323	**Pinus coulteri**	Coulter pine	Pinaceae		2
324	*Pinus densiflora*	Japanese red pine	Pinaceae		2
325	*Pinus d.* 'Umbraculifera'	tanyosho pine	Pinaceae		2
326	**Pinus edulis**	Colorado pinyon	Pinaceae		4
327	**Pinus flexilis**	limber pine	Pinaceae		2
328	*Pinus halepensis*	Aleppo pine	Pinaceae		1
329	**Pinus jeffreyi**	Jeffrey pine	Pinaceae		2
330	**Pinus lambertiana**	sugar pine	Pinaceae		4
331	**Pinus monophylla**	singleleaf pinyon	Pinaceae		4
332	**Pinus monticola**	western white pine	Pinaceae		4
333	*Pinus mugo*	mugho pine, Swiss mountain pine	Pinaceae		1
334	**Pinus muricata**	Bishop pine	Pinaceae	*P. remorata*	2
335	*Pinus patula*	Jelecote pine	Pinaceae		1
336	*Pinus pinea*	Italian stone pine	Pinaceae		1
337	**Pinus ponderosa**	ponderosa pine, yellow pine	Pinaceae		2
331	**Pinus quadrifolia**	Parry pinyon pine	Pinaceae	*Pinus cembroides* var. *parryana*	4
338	**Pinus radiata**	Monterey pine	Pinaceae		1
339	**Pinus sabiniana**	gray pine, digger pine	Pinaceae		2
340	*Pinus strobus*	eastern white pine	Pinaceae		3
341	*Pinus sylvestris*	Scotch pine	Pinaceae		1
342	*Pinus thunbergii*	Japanese black pine	Pinaceae	*P. thunbergiana*	1
343	**Pinus torreyana**	Torrey pine	Pinaceae		5
344	*Pistacia chinensis*	Chinese pistache	Anacardiaceae		1
345	*Pittosporum rhombifolium*	Queensland pittosporum	Pittosporaceae		1
346	*Pittosporum undulatum*	Victorian box	Pittosporaceae		1
347	*Platanus* x *acerifolia*	London plane	Platanaceae		1
348	*Platanus occidentalis*	American sycamore	Platanaceae		1

Page	Botanical Name	Common Name(s)	Family	Other Latin Name(s)	Avail.
349	*Platanus racemosa*	western sycamore	Platanaceae		1
350	*Podocarpus gracilior*	fern pine	Taxaceae	*P. elongatus*	1
351	*Podocarpus macrophyllus*	yew pine	Taxaceae		1
352	*Populus alba***	white poplar	Salicaceae		
353	*Populus alba* 'Pyramidalis'	Bolleana poplar	Salicaceae	*P. bolleana*	2
354	*Populus fremontii*	Fremont cottonwood, western cottonwood	Salicacae		2
355	*Populus nigra* 'Italica'*	Lombardy poplar	Salicaceae		1
356	*Populus tremuloides*	quaking aspen	Salicaceae		2
357	*Populus trichocarpa*	black cottonwood	Salicaceae	*P. balsamifera* ssp. *trichocarpa*	2
358	*Prosopis glandulosa*	honey mesquite	Fagaceae	*P. chilensis*, *P. g.* var. *torreyana*	4
359	*Prosopis velutina*	Arizona mesquite, velvet mesquite	Fagaceae	*P. glandulosa velutina*	4
360	*Prunus* x *blirieana*	pink flowering plum	Rosaceae		1
361	*Prunus caroliniana**	Carolina cherry laurel	Rosaceae		1
362	*Prunus cerasifera*	cherry plum	Rosaceae		1
363	*Prunus c.* 'Atropurpurea'	purple-leaf plum	Rosaceae		1
364	*Prunus c.* 'Krauter Vesuvius'	pink-flowering purple-leaf plum	Rosaceae		1
365	*Prunus ilicifolia*	hollyleaf cherry, islay	Rosaceae	*P. integrifolia*, *P. ilicifolia* ssp. *integrifolia*, *P. ilicifolia* ssp. *ilicifolia*	1
366	*Prunus lyonii*	Catalina cherry	Rosaceae	*P. ilicifolia* ssp. *lyonii*	1
367	*Prunus serrulata*	Japanese flowering cherry	Rosaceae		1
368	*Prunus* 'Kwanzan'	Kwanzan flowering cherry	Rosaceae	*P.* 'Kanzan', *P.* 'Sekiyama'	2
368	*Prunus pendula*	weeping cherry	Rosaceae	*P.* x *subhirtella* 'Pendula'	1
368	*Prunus* x *yedoensis*	Yoshino flowering cherry	Rosaceae		1
369	*Pseudotsuga macrocarpa*	bigcone Douglas-fir	Pinaceae		4
370	*Pseudotsuga menziesii*	Douglas-fir	Pinaceae	*P. taxifolia*, *P. mucronata*, *P. douglasii*	2
371	*Pterocarya stenoptera*	Chinese wingnut	Juglandaceae	*P. sinensis*, *P. japonica*	2
372	*Punica granatum*	pomegranate	Punicaceae		1
373	*Pyrus calleryana* 'Aristocrat'	Aristocrat pear	Rosaceae		1
374	*Pyrus calleryana* 'Bradford'	Bradford pear	Rosaceae		1
375	*Pyrus taiwanensis*	evergreen pear	Rosaceae	*P. kawakamii*	1
376	*Quercus acutissima*	Chinese oak	Fagaceae	*Q. serrata*	4
377	*Quercus agrifolia*	coast live oak	Fagaceae		1
378	*Quercus alba*	white oak	Fagaceae		2
379	*Quercus arizonica*	Arizona white oak	Fagaceae		4
380	*Quercus chrysolepis*	canyon live oak, maul oak	Fagaceae		4
381	*Quercus coccinea*	scarlet oak	Fagaceae		1
382	*Quercus douglasii*	blue oak	Fagaceae		2
383	*Quercus dumosa*	Nuttall's scrub oak	Fagaceae		4
384	*Quercus emoryi*	Emory oak	Fagaceae		4
385	*Quercus engelmannii*	Engelmann oak	Fagaceae		2
386	*Quercus gambelii*	Rocky Mountain white oak	Fagaceae	*Q. utahensis*	4
387	*Quercus garryana*	Oregon oak, Oregon white oak	Fagaceae		4
388	*Quercus ilex*	holly oak, holm oak	Fagaceae		1
389	*Quercus kelloggii*	California black oak	Fagaceae	*Q. californica*	2
390	*Quercus laurifolia*	laurel oak	Fagaceae	*Q. obtusa*, *Q. rhombica*	2
391	*Quercus lobata*	valley oak	Fagaceae		2
392	*Quercus macrocarpa*	mossycup oak	Fagaceae		3
393	*Quercus morehus*	oracle oak	Fagaceae	*Q. kelloggii* x *Q. wislizeni*	4
394	*Quercus muehlenbergii*	chinquapin oak, yellow oak	Fagaceae	*Q. acuminata*, *Q. castanea*, *Q. prinoides* var. *acuminata*	3
395	*Quercus myrsinifolia*	Japanese live oak	Fagaceae	*Q. myrsinaefolia*, *Q. bambusaefolia*, *Cyclobalanopsis myrsinifolia*	4

Page	Botanical Name	Common Name(s)	Family	Other Latin Name(s)	Avail.
396	*Quercus oblongifolia*	Mexican blue oak	Fagaceae		4
397	**Quercus palmeri**	Palmer's oak, Dunn oak	Fagaceae	*Q. dunnii*	4
398	*Quercus palustris*	pin oak	Fagaceae		1
399	*Quercus phellos*	willow oak	Fagaceae		2
400	*Quercus phillyreoides*	ubame oak	Fagaceae		4
401	*Quercus robur*	English oak	Fagaceae	*Q. pedunculata, Q. femina*	2
402	*Quercus robur* 'Fastigiata'	upright English oak	Fagaceae		2
403	*Quercus rubra*	red oak, Spanish red oak	Fagaceae	*Q. r. maxima, Q. borealis*	1
404	*Quercus rugosa*	netleaf oak	Fagaceae		4
405	*Quercus suber*	cork oak	Fagaceae		1
406	**Quercus tomentella**	island oak	Fagaceae		5
407	**Quercus turbinella**	shrub live oak	Fagaceae		5
408	*Quercus virginiana*	southern live oak	Fagaceae	*Q. virens*	1
409	**Quercus wislizeni**	interior live oak	Fagaceae	*Q. w.* var. *wislizeni*	1
410	*Quillaja saponaria*	soapbark tree	Rosaceae		4
411	*Rhus glabra*	smooth sumac	Anacardiaceae		2
412	*Rhus lancea*	African sumac	Anacardiaceae		1
413	*Rhus typhina*	staghorn sumac	Anacardiaceae	*R. hirta*	2
414	*Robinia pseudoacacia*	black locust	Fabaceae		2
415	*Robinia pseudoacacia* 'Frisia'	golden locust	Fabaceae		3
415	*Robinia p.* 'Tortuosa'	twisted locust	Fabaceae		3
416	*Robinia x ambigua* 'Idahoensis'	Idaho pink locust	Fabaceae		1
417	*Robinia x a.* 'Purple Robe'	purple flowering locust	Fabaceae		1
418	*Salix babylonica*	weeping willow	Salicaceae	*S. napoleonis*	1
419	**Salix gooddingii**	Goodding's black willow	Salicaceae		4
420	**Salix hindsiana**	sandbar willow	Salicaceae	*S. exigua* var. *hindsiana*	4
421	**Salix hookeriana**	Hooker willow, coastal willow, beach willow	Salicaceae	*S. ampliflora*	4
422	**Salix laevigata**	red willow	Salicaceae		4
423	**Salix lasiandra**	yellow willow	Salicaceae	*S. lucida* ssp. *lasiandra*	4
424	**Salix lasiolepis**	arroyo willow	Salicaceae		4
425	**Salix scouleriana**	Scouler's willow western pussy willow	Salicaceae	*S. flavescens, S. brachystachys, S. nuttallii*	4
426	*Salix matsudana* 'Tortuosa'	corkscrew willow, twisted Hankow willow	Salicaceae		2
427	**Sambucus mexicana**	blue elderberry	Caprifoliaceae	*S. glauca, S. caerulea*	2
428	*Sapium sebiferum***	Chinese tallow tree	Euphorbiaceae		1
429	*Schinus molle*	pepper tree	Anacardiaceae		1
430	*Schinus peruviana*	Peruvian pepper	Anacardiaceae		3
431	*Schinus terebinthifolius*	Brazilian pepper	Anacardiaceae		2
432	*Sciadopitys verticillata*	Japanese umbrella pine	Pinaceae		3
433	**Sequoia sempervirens**	coast redwood	Taxodiaceae		1
434	**Sequoiadendron giganteum**	giant sequoia, big tree	Taxodiaceae	*Sequoia gigantea*	1
435	*Sophora japonica*	Japanese pagoda tree, Chinese scholar tree	Fagaceae		1
436	*Sorbus aucuparia*	European mountain ash, rowan tree	Rosaceae	*Styphnolobium japonicum*	2
437	*Syzygium paniculatum*	brush cherry	Myrtaceae	*Eugenia myrtifolia*	1
438	*Tamarix aphylla**	athel tree	Tamarixaceae		4
439	*Tamarix parviflora***	tamarisk	Tamarixaceae	*T. tetrandra*	4
440	*Taxodium distichum*	bald cypress	Taxodiaceae		2
441	*Taxodium mucronatum*	Montezuma cypress	Taxodiaceae		3
442	**Taxus brevifolia**	western yew, Pacific yew	Taxaceae		4
443	*Thuja occidentalis*	American arborvitae	Pinaceae		2
444	**Thuja plicata**	western red cedar, giant arborvitae	Pinaceae	*T. menziesii, T. gigantea*	2
445	*Thuja plicata* 'Emerald Cone'	columnar red cedar	Pinaceae		3
445	*Thuja plicata* 'Gracilis'	giant arborvitae	Pinaceae		5
446	*Thujopsis dolobrata*	deerhorn cedar	Pinaceae		5

Page	Botanical Name	Common Name(s)	Family	Other Latin Name(s)	Avail.
447	*Tilia cordata*	little-leaf linden	Tiliaceae	*T. parvifolia*	1
448	*Tilia europaea*	European linden	Tiliaceae		1
449	*Tilia tomentosa*	silver linden	Tiliaceae	*T. argentea*	1
450	*Tipuana tipu*	tipu tree	Fagaceae		4
451	**Torreya californica**	California nutmeg	Taxaceae	*Tumion californicum*	4
452	*Torreya nucifera*	Japanese nutmeg	Taxaceae		4
453	*Tristaniopsis laurina*	laurel leaf box, water gum	Myrtaceae	*Tristania laurina*	2
454	TROPICALS				
	Cordyline australis*	green dracaena dracaena palm, grass palm	Liliaceae	*Dracaena australis, Dracaena indivisa*	3
455	TROPICALS				
	Cyathea cooperi	Australian tree fern	Dicksoniaceae		1
	Cycas revoluta	sago palm	Cycadaceae		2
456	TROPICALS				
	Dicksonia antarctica	Tasmanian tree fern	Dicksoniaceae		2
	Phyllostachys bambusoides	timber bamboo	Gramineae		3
457	TROPICALS				
	Musa x *paradisiaca*	banana	Musaceae		3
458	TROPICALS				
	Stenocarpus sinuatus	firewheel tree	Proteaceae		3
	Tetrapanax papyriferus	rice paper plant	Araliaceae	*Aralia papyrifera*	3
459	**Tsuga heterophylla**	western hemlock	Pinaceae		4
460	**Tsuga mertensiana**	mountain hemlock	Pinaceae		4
461	*Ulmus americana*	American elm	Ulmaceae	*U. alba*	4
462	*Ulmus parvifolia*	Chinese elm, evergreen elm	Ulmaceae	*U. chinensis*	1
463	*Ulmus rubra*	slippery elm	Ulmaceae	*U. fulva*	4
464	**Umbellularia californica**	California bay	Lauraceae	*Laurus regia*	2
465	**Yucca brevifolia**	Joshua tree	Liliaceae		4
466	*Yucca gloriosa*	Spanish dagger, soft-tip yucca	Liliaceae		2
467	*Zelkova serrata*	sawtooth zelkova	Ulmaceae	*Z. acuminata*	1
468	*Zizyphus jujuba*	Chinese jujube	Rhamnaceae	*Z. zizyphus, Z. vulgaris, Rhamnus zizyphus*	2

<u>Key to Availability Codes</u>

(Generally the higher the code, the less readily available)

1   Readily available, commonly used.
2   Limited availablity, or must be shipped from out of state, or natives available in small sizes, or as seedling stock.
3   Specialty nursery item.
4   Seldom grown as a nursery item.
5   Rarely available.

<u>Notes:</u>

California native species are in bold type.
* Indicates invasive tendencies; use is cautioned.
** Indicates species that may become very invasive.

Common Name	Botanical Name	Page
cork oak	Quercus suber	405
corkscrew willow	Salix matsudana 'Tortuosa'	426
cornelian cherry	Cornus mas	196
cottonwood	Populus fremontii	354
Coulter pine	Pinus coulteri	323
cow itch tree	Lagunaria patersonii	269
crape myrtle	Lagerstroemia indica	268
cutleaf weeping birch	Betula pendula 'Crispa'	157
cutleaf sumac	Rhus typhina 'Laciniata'	413
cycad	Cycas revoluta	455
dammar pine	Agathis robusta	139
date palm	Phoenix dactylifera	302
dawn redwood	Metasequoia glyptostroboides	288
deodar cedar	Cedrus deodara	171
desert-willow	Chilopsis linearis	188
digger pine	Pinus sabiniana	339
dogwood	Cornus florida	194
Douglas-fir	Pseudotsuga menziesii	370
dracaena	Cordyline australis	454
dunn oak	Quercus palmeri	397
dwarf blue gum	Eucalyptus globulus 'Compacta'	227
dwarf maple	Acer glabrum	115
eastern redbud	Cercis canadensis	180
elderberry	Sambucus mexicana	427
Emory oak	Quercus emoryi	384
empress tree	Paulownia tomentosa	307
Engelmann spruce	Picea engelmannii	310
English hawthorn	Crataegus laevigata	199
English oak	Quercus robur 'Fastigiata'	402
English walnut	Juglans regia	259
eugenia	Syzygium paniculatum	437
European hackberry	Celtis australis	174
European linden	Tilia europaea	448
European mountain ash	Sorbus aucuparia	436
European silver fir	Abies alba	94
evergreen ash	Fraxinus uhdei	247
evergreen elm	Ulmus parvifolia	462
evergreen pear	Pyrus taiwanensis	375
fern pine	Podocarpus gracilior	350
fern-leaved Catalina ironwood	Lyonothamnus floribundus ssp. aspleniifolius	279
firewheel tree	Stenocarpus sinuatus	458
flame tree	Koelreuteria bipinnata	264
flaxleaf paperbark	Melaleuca linariifolia	285
floss silk tree	Ceiba speciosa	173
flowering cherry	Prunus serrulata	367
flowering crabapple	Malus x floribunda	283
flowering locust	Robinia pseudoacacia	414
flowering plum	Prunus x blirieana	360
Formosan sweet gum	Liquidambar formosana	273
Fremont cottonwood	Populus fremontii	354
fishtail palm	Caryota urens	300
fruitless mulberry	Morus alba 'Fruitless'	291
fullmoon maple	Acer japonicum	117
giant arborvitae	Thuja plicata	444
giant bamboo	Phyllostachys bambusoides	456
giant sequoia	Sequoiadendron giganteum	434
ginger pine	Chamaecyparis lawsoniana	185
glossy privet	Ligustrum lucidum	272
golden chinquapin	Chrysolepis chrysophylla	190
golden honey locust	Gleditsia triacanthos 'Sunburst'	253
golden Monterey cypress	Cupressus macrocarpa 'Lutea'	214
goldenchain tree	Laburnum x watereri	266
goldenrain tree	Koelreuteria paniculata	265
grand fir	Abies grandis	98
grass palm	Cordyline australis	454
gray pine	Pinus sabiniana	339
Grecian laurel	Laurus nobilis	271
green acacia	Acacia dealbata	107
green dracaena	Cordyline australis	454
Guadalupe fan palm	Brahea edulis	299
Guadalupe Island cypress	Cupressus guadalupensis	208
hackberry	Celtis occidentalis	175
hardtack	Cercocarpus betuloides	184
hawthorn	Crataegus laevigata	199
hedge maple	Acer campestre	112
hemlock	Tsuga heterophylla	459
hickory	Carya illinoiensis	166
holly oak	Quercus ilex	388
Holm oak	Quercus ilex	388
honey locust	Gleditsia triacanthos 'Inermis'	252
honey mesquite	Prosopis glandulosa	358
horsechestnut	Aesculus hippocastanum	138
horsetail tree	Casuarina equisetifolia	168
Idaho pink locust	Robinia x ambigua 'Idahoensis'	416
incense cedar	Calocedrus decurrens	163
Indian laurel fig	Ficus microcarpa	239
interior live oak	Quercus wislizeni	409
ironwood	Cercocarpus betuloides	184
ironwood	Lyonothamnus floribundus	279
islay	Prunus ilicifolia	365
Italian alder	Alnus cordata	144
Italian cypress	Cupressus sempervirens	212
Italian stone pine	Pinus pinea	336
Japanese black pine	Pinus thunbergii	342
Japanese crape myrtle	Lagerstroemeria fauriei hybrids	267
Japanese cryptomeria	Cryptomeria japonica	201
Japanese flowering cherry	Prunus serrulata	367
Japanese flowering crabapple	Malus x floribunda	283
Japanese larch	Larix kaempferi	270
Japanese live oak	Quercus myrsinifolia	395
Japanese maple	Acer palmatum	121
Japanese pagoda tree	Sophora japonica	435
Japanese red pine	Pinus densiflora	324
Japanese silver tree	Neolitsea sericea	294
Japanese umbrella pine	Sciadopitys verticillata	432
Jeffrey pine	Pinus jeffreyi	329
Jelecote pine	Pinus patula	335
Joshua tree	Yucca brevifolia	465
jujube	Zizyphus jujuba	468
kaffirboom coral tree	Erythrina caffra	217
katsura tree	Cercidiphyllum japonicum	178
kauri pine	Agathis robusta	139
king palm	Archontophoenix cunninghamiana	299
knobcone pine	Pinus attenuata	316
kurrajong	Brachychiton populneus	159
Kwanzan flowering cherry	Prunus 'Kwanzan'	368
lacebark pine	Pinus bungeana	319
laceleaf Japanese maple	Acer palmatum 'Dissectum'	123
lantern tree	Crinodendron patagua	200
laurel leaf box	Tristaniopsis laurina	453

Common Name	Botanical Name	Page
laurel oak	*Quercus laurifolia*	390
Lawson cypress	*Chamaecyparis lawsoniana*	185
lemon bottlebrush	*Callistemon citrinus*	161
lemon-scented gum	*Eucalyptus citriodora*	222
Leyland cypress	x *Cupressocyparis leylandii*	205
libocedrus	*Calocedrus decurrens*	163
lily-of-the-valley tree	*Crinodendron patagua*	200
little-leaf linden	*Tilia cordata*	447
little leaf fig	*Ficus microcarpa* 'Nitida'	239
live oak	*Quercus agrifolia*	377
live oak	*Quercus chrysolepis*	380
live oak	*Quercus myrsinifolia*	395
live oak	*Quercus virginiana*	408
live oak	*Quercus wislizeni*	409
locust	*Gleditsia triacanthos* 'Inermis'	252
locust	*Robinia pseudoacacia*	414
lodgepole pine	*Pinus contorta* ssp. *murrayana*	322
Lombardy poplar	*Populus nigra* 'Italica'	355
London plane tree	*Platanus* x *acerifolia*	347
loquat	*Eriobotrya japonica*	216
lovely fir	*Abies amabilis*	95
madrone	*Arbutus menziesii*	150
maidenhair tree	*Ginkgo biloba*	251
manna gum	*Eucalyptus viminalis*	234
maul oak	*Quercus chrysolepis*	380
mayten	*Maytenus boaria*	284
McNab cypress	*Cupressus macnabiana*	209
Mediterranean fan palm	*Chamaerops humilis*	301
Mexican blue oak	*Quercus oblongifolia*	396
Mexican fan palm	*Washingtonia robusta*	305
Mexican palo verde	*Parkinsonia aculeata*	306
Modesto ash	*Fraxinus velutina* 'Modesto'	249
monkey puzzle tree	*Araucaria araucana*	147
Monterey cypress	*Cupressus macrocarpa*	210
Monterey pine	*Pinus radiata*	338
Montezuma cypress	*Taxodium mucronatum*	441
Moraine ash	*Fraxinus holotricha* 'Moraine'	244
mossycup oak	*Quercus macrocarpa*	392
mountain ash	*Sorbus aucuparia*	436
mountain hemlock	*Tsuga mertensiana*	460
mountain ironwood	*Cercocarpus betuloides*	184
mountain maple	*Acer glabrum*	115
mourning cypress	*Cupressus funebris*	207
mugho pine	*Pinus mugo*	333
mulberry	*Morus alba*	290
myrobalan	*Prunus cerasifera*	362
netleaf oak	*Quercus rugosa*	404
New Zealand Christmas tree	*Metrosideros excelsus*	289
Nootka cypress	*Chamaecyparis nootkatensis*	187
Nordmann fir	*Abies nordmanniana*	103
Norfolk Island pine	*Araucaria heterophylla*	149
Norway maple	*Acer platanoides*	124
Nuttall's scrub oak	*Quercus dumosa*	383
oleander	*Nerium oleander*	295
oracle oak	*Quercus morehus*	393
orchid tree	*Bauhinia variegata*	152
Oregon ash	*Fraxinus latifolia*	245
Oregon laurel	*Arbutus menziesii*	150
Oregon white oak	*Quercus garryana*	387
Pacific dogwood	*Cornus nuttallii*	197
Pacific silver fir	*Abies amabilis*	95

Common Name	Botanical Name	Page
pagoda tree	*Sophora japonica*	435
Palmer oak	*Quercus palmeri*	397
palo verde	*Cercidium floridum*	179
paper birch	*Betula papyrifera*	155
parasol tree	*Firmiana simplex*	241
pear	*Pyrus taiwanenesis*	375
pecan	*Carya illinoiensis*	166
pepper tree	*Schinus molle*	429
pepperidge	*Nyssa sylvatica*	296
peppermint tree	*Agonis flexuosa*	140
Persian mulberry	*Morus nigra*	292
Peruvian pepper	*Schinus peruviana*	430
piccabeen palm	*Archontophoenix cunninghamiana*	299
pin oak	*Quercus palustris*	398
pindo palm	*Butia capitata*	300
pink ironbark	*Eucalyptus sideroxylon* 'Rosea'	233
pink locust	*Robinia* x *ambigua* 'Idahoensis'	416
pistache	*Pistacia chinensis*	344
plane tree	*Platanus* x *acerifolia*	347
plume albizia	*Albizia distachya*	142
plume cedar	*Cryptomeria japonica*	201
pomegranate	*Punica granatum*	372
ponderosa pine	*Pinus ponderosa*	337
poplar	*Populus nigra* 'Italica'	355
Port Orford-cedar	*Chamaecyparis lawsoniana*	185
primrose tree	*Lagunaria patersonii*	269
privet	*Ligustrum lucidum*	272
purple beech	*Fagus sylvatica* 'Purpurea'	237
purpleblow maple	*Acer truncatum*	134
purple flowering locust	*Robinia* 'Purple Robe'	417
purple-leaf plum	*Prunus cerasifera* 'Atropurpurea'	363
pygmy date palm	*Phoenix roebelenii*	303
quaking aspen	*Populus tremuloides*	356
queen palm	*Syagrus romanzoffianum*	304
Queensland kauri	*Agathis robusta*	139
Queensland pittosporum	*Pittosporum rhombifolium*	345
Raywood ash	*Fraxinus angustifolia* 'Raywood'	242
redbud	*Cercis canadensis*	180
red birch	*Betula nigra*	153
red fir	*Abies magnifica*	100
red-flowering gum	*Eucalyptus ficifolia*	225
red gum	*Eucalyptus camaldulensis*	220
red ironbark	*Eucalyptus sideroxylon* 'Rosea'	233
red Japanese maple	*Acer palmatum* 'Atropurpureum'	122
red maple	*Acer rubrum*	128
red oak	*Quercus rubra*	403
redwood	*Sequoia sempervirens*	433
rice paper plant	*Tetrapanax papyriferus*	458
river birch	*Betula nigra*	153
river red gum	*Eucalyptus camaldulensis*	220
Rivers' purple beech	*Fagus sylvatica* 'Riversii'	236
rock maple	*Acer saccharum*	131
Rocky Mountain white oak	*Quercus gambelii*	386
roundleaf sweet gum	*Liquidambar styraciflua* 'Rotundiloba'	275
rowan tree	*Sorbus aucuparia*	436
rustyleaf fig	*Ficus rubiginosa*	240
St. John's bread	*Ceratonia siliqua*	177
sandbar willow	*Salix hindsiana*	420

Common Name	Botanical Name	Page
Santa Lucia fir	*Abies bracteata*	96
Sargent's cypress	*Cupressus sargentii*	211
saucer magnolia	*Magnolia x soulangeana*	281
sawtooth zelkova	*Zelkova serrata*	467
scarlet maple	*Acer rubrum*	128
scarlet oak	*Quercus coccinea*	381
scholar tree	*Sophora japonica*	435
Schwedler maple	*Acer platanoides* 'Schwedleri'	125
Scotch pine	*Pinus sylvestris*	341
shamel ash	*Fraxinus uhdei*	247
shore pine	*Pinus contorta* ssp. *contorta*	321
shrub live oak	*Quercus turbinella*	407
Sidney golden wattle	*Acacia longifolia*	108
silk oak	*Grevillea robusta*	254
silver dollar gum	*Eucalyptus polyanthemos*	231
silver linden	*Tilia tomentosa*	449
silver maple	*Acer saccharinun*	130
silvertip	*Abies magnifica*	100
Sitka spruce	*Picea sitchensis*	314
slippery elm	*Ulmus rubra*	463
smooth sumac	*Rhus glabra*	411
snakebark maple	*Acer capillipes*	113
soft-tip yucca	*Yucca gloriosa*	466
sour gum	*Nyssa sylvatica*	296
sourwood	*Oxydendrum arboreum*	298
southern live oak	*Quercus virginiana*	408
southern magnolia	*Magnolia grandiflora*	280
Spanish chestnut	*Castanea sativa*	167
Spanish dagger	*Yucca gloriosa*	466
Spanish fir	*Abies pinsapo*	104
Spanish red oak	*Quercus rubra*	403
spruce	*Picea abies*	308
staghorn sumac	*Rhus typhina*	413
star magnolia	*Magnolia stellata*	282
sterculia	*Brachychiton populneus*	159
stinking cedar	*Torreya californica*	451
stone pine	*Pinus pinea*	336
strawberry tree	*Arbutus unedo*	151
sugar gum	*Eucalyptus cladocalyx*	223
sugar maple	*Acer saccharum*	131
sugar pine	*Pinus lambertiana*	330
sumac	*Rhus glabra*	411
swamp maple	*Acer rubrum*	128
sweet bay	*Laurus nobilis*	271
sweet brush	*Cercocarpus betuloides*	184
sweet gum	*Liquidambar styraciflua*	274
sweetshade	*Hymenosporum flavum*	255
sycamore	*Platanus racemosa*	349
tallow tree	*Sapium sebiferum*	428
tamarisk	*Tamarix parviflora*	439
tanbark oak	*Lithocarpus densiflorus*	277
Tanyosho pine	*Pinus densiflora* 'Umbraculifera'	325
Tasmanian tree fern	*Dicksonia antarctica*	456
Texas redbud	*Cercis reniformis*	183
Texas walnut	*Juglans microcarpa*	258
tideland spruce	*Picea sitchensis*	314
timber bamboo	*Phyllostachys bambusoides*	456
tipu tree	*Tipuana tipu*	450
toddy palm	*Caryota urens*	300
Torrey pine	*Pinus torreyana*	343
tree fern	*Cyathea cooperi*	455
tree-of-heaven	*Ailanthus altissima*	141

Common Name	Botanical Name	Page
tree willow	*Salix lasiandra*	423
trident maple	*Acer buergerianum*	111
tuckeroo	*Cupaniopsis anacardioides*	204
tulip poplar	*Liriodendron tulipifera*	276
tulip tree	*Liriodendron tulipifera*	276
tupelo	*Nyssa sylvatica*	296
twisted Hankow willow	*Salix matsudana* 'Tortuosa'	426
twisted locust	*Robinia pseudoacacia* 'Tortuosa'	415
ubame oak	*Quercus phillyreoides*	400
umbrella pine	*Sciadopitys verticillata*	432
upright European hornbeam	*Carpinus betulus* 'Fastigiata'	165
upright English oak	*Quercus robur* 'Fastigiata'	402
valley oak	*Quercus lobata*	391
variegated box elder	*Acer negundo* 'Variegatum'	120
variegated tulip tree	*Liriodendron tulipifera* 'Aureo-marginatum'	275
velvet mesquite	*Prosopis velutina*	359
Victorian box	*Pittosporum undulatum*	346
vine maple	*Acer circinatum*	114
walnut	*Juglans regia*	259
Wasatch maple	*Acer saccharum* ssp. *grandidentatum*	115
water birch	*Betula occidentalis*	154
weeping birch	*Betula pendula* 'Crispa'	157
weeping bottlebrush	*Callistemon viminalis*	162
weeping cherry	*Prunus pendula*	368
weeping spruce	*Picea breweriana*	309
weeping white spruce	*Picea glauca* 'Pendula'	311
weeping willow	*Salix babylonica*	418
western catalpa	*Catalpa speciosa*	169
western cedar	*Calocedrus decurrens*	163
western cottonwood	*Populus fremontii*	354
western dogwood	*Cornus nuttallii*	197
western hackberry	*Celtis reticulata*	174
western hemlock	*Tsuga heterophylla*	459
western pussy willow	*Salix scouleriana*	425
western red cedar	*Thuja plicata*	444
western redbud	*Cercis occidentalis*	182
western yew	*Taxus brevifolia*	442
white alder	*Alnus rhombifolia*	146
white fir	*Abies concolor*	97
white ironbark	*Eucalyptus leucoxylon*	229
white maple	*Acer saccharinum*	130
white mulberry	*Morus alba*	290
white poplar	*Populus alba*	352
whitebark pine	*Pinus albicaulis*	315
wilga	*Geijera parviflora*	250
willow-leafed peppermint	*Eucalyptus nicholii*	230
willow oak	*Quercus phellos*	399
willow myrtle	*Agonis flexuosa*	140
windmill palm	*Trachycarpus fortunei*	304
wine palm	*Caryota urens*	300
wingnut	*Pterocarya stenoptera*	371
yeddo spruce	*Picea jezoensis*	312
yellow oak	*Quercus muehlenbergii*	394
yellow pine	*Pinus ponderosa*	337
yellow tree willow	*Salix lasiandra*	423
yellowwood	*Cladrastis kentukea*	192
yew pine	*Podocarpus macrophyllus*	351
Yoshino flowering cherry	*Prunus x yedoensis*	368
Yunnan hackberry	*Celtis sinensis*	176

An arboretum is defined as a collection of trees and other woody plants used for scientific or other educational purposes, a type of botanical garden. Usually plants are arranged in some sort of order, such as by family, climatic zone, or horticultural use. Some parks can serve this purpose, as do many riparian areas and greenways, as examples of restoring vegetation in native habitats. Arboretums are good places to view many different trees in close proximity grown in ways that display their natural growth habits and cultural requirements and facilitate comparison with other species and varieties.

NORTHERN CALIFORNIA AND OREGON

Arcata Marsh and Wildlife Sanctuary
   736 F St, Arcata
Bidwell Park
   E. 8th St, Chico
California State University Arboretum
   6000 J St, Sacramento
Capitol Park
   State Capitol grounds, Sacramento
East Bay Regional Parks Botanic Garden, Tilden Park
   Grizzly Peak Blvd, Berkeley
Hoyt Arboretum
   4000 SW Fairview Blvd, Portland
Institute of Forest Genetics
   2480 Carson Rd, Placerville
Mt. Tamalpais State Park
   801 Panoramic Hwy, Mill Valley
San Francisco Botanical Garden at Strybing Arboretum
   9th Avenue at Lincoln Way, San Francisco
Stanford University Arboretum
   Stanford University campus, Palo Alto
University of California, Berkeley, Botanical Garden
   200 Centennial Dr, Berkeley
University of California, Davis, Arboretum
   La Rue Rd, Davis
University of California, Santa Cruz, Arboretum
   Western Dr and High St, Santa Cruz

SOUTHERN CALIFORNIA

Balboa Park
   Park Blvd, San Diego
Descanso Gardens
   1418 Descanso Dr, La Cañada-Flintridge
Fullerton Arboretum
   California State University, Fullerton
Huntington Library, Art Collection, and Botanical Gardens
   1151 Oxford Rd, San Marino
Los Angeles County Arboretum
   301 N. Baldwin Ave, Arcadia
Quail Botanical Gardens
   230 Quail Gardens Dr, Encinitas
Rancho Santa Ana Botanic Garden
   1500 N. College Ave, Claremont
Santa Barbara Botanic Garden
   1212 Mission Canyon Road, Santa Barbara
South Coast Botanical Garden
   26300 Crenshaw Blvd, Palos Verdes Peninsula
University of California, Irvine, Arboretum
   Jamboree Rd and Campus Dr, Irvine
University of California, Riverside, Arboretum
   Campus Dr, Riverside

Ruth Risdon Storer Garden, UC Davis Arboretum.

California State University Arboretum, Sacramento.

BIBLIOGRAPHY

Barbour, Michael G. and Jack Major (eds.)
1977, *Terrestrial vegetation of California*, Wiley-Interscience, reprinted by California Native Plant Society (1988), Sacramento CA

Barbour, Michael G., Todd Keeler-Wolf, and Allan A. Schoenherr (eds.)
2007, *Terrestrial vegetation of California* (3rd ed.), University of California Press, Berkeley CA

Bass, Ronald E., Albert I. Herson, and Kenneth M. Bogdan
1999, *CEQA Deskbook* (2nd ed.), Solano Press Books, Point Arena CA

Benson, Lyman
1979, *Plant classification* (2nd ed.), D.C. Heath, Lexington MA

Boltz, Howard O.
1963, *The landscape use of shrubs and vines*, Educational Publishers, St. Louis MI.

Bradley, Gordon A.
1995, *Urban forest landscapes*, University of Washington Press, Seattle WA

Brooker, M.I.H. and D.A. Kleinig
1983, *Field guide to eucalypts, south eastern*, Inkata Press, Melbourne and Sydney, Australia

Cheatham N.H. and J.R. Haller
1975, *An annotated list of California habitat types*, University of California Natural Land and Water Reserves System, Berkeley CA

CNDDB/Holland
1986, revision of Cheatham and Haller by the California Natural Diversity Database

Costello, L.R. and K.S. Jones
1994, *WUCOLS: water use classification of landscape species: a guide to the water needs of landscape plants*, University of California Cooperative Extension, San Francisco and San Mateo County Office, Half Moon Bay CA

Cowardin, Lewis M. *et al.*
1979, *Classification of wetlands and deepwater habitats of the United States*, US Fish and Wildlife Service, FWS/OBS 79/31, Washington DC

Critchfield, W.B.
1971, *Profiles of California vegetation*, USDA Forest Service Research Paper PSW-76, Pacific Southwest Forest and Range Experiment Station, Berkeley CA

Dirr, Michael A.
1998, *Manual of woody landscape plants* (5th ed.), Stipes Publishing, Champaign IL

Druse, Ken with Margaret Roach
1994, *The natural habitat garden*, Clarkson Potter, New York NY

East Bay Municipal Utility District
1990, *Water-conserving plants and landscapes for the Bay Area*, EBMUD, Oakland CA

East Bay Municipal Utility District
2004, *Plants and landscapes for summer-dry climates*, EBMUD, Oakland CA

Eckbo, Garrett
1969, *The landscape we see*, McGraw-Hill, New York NY

Elias, Thomas S.
1980 (2000), *The complete trees of North America*, Van Nostrand Reinhold, New York NY

Everett, Thomas H.
1960, *New illustrated encyclopedia of gardening*, Greystone Press, New York NY

Everett, Thomas H.
1986, *Living trees of the world*, Doubleday, New York NY

Eyre, F.H. (ed.)
1980, *Forest cover types of the United States and Canada*, Society of American Foresters, Washington DC

Faber, Phyllis M. and Robert F. Holland
1988, *Common riparian plants of California*, Pickleweed Press, Mill Valley CA

Francis, Mark and Andreas Reimann
1999, *The California landscape garden: ecology, culture and design*, University of California Press, Berkeley CA

French, Jere S.
1993, *The California garden and the landscape architects who shaped it*, American Society of Landscape Architects, Washington, DC

Gilman, Edward F.
1997, *Trees for urban and suburban landscapes*, Delmar Publishers, Albany NY

Griffin, James R. and William B. Critchfield
1972 *The distribution of forest trees in California*, USDA Forest Service, Research Paper PSW-82 (rev. with suppl. 1976), Pacific Southwest Forest and Range Experiment Station, Berkeley, CA.

Grillo, Paul J.
1975, *Form, function, and design*, Dover Publications, Mineola NY

Heritage Oaks Committee
1988, *Native oaks: our valley heritage*, Heritage Oaks Committee, Sacramento CA

Hickman, James C. (ed.)
1993, *The Jepson manual: higher plants of California*, University of California Press, Berkeley CA

Hogan, Sean, chief consultant
2003, *Flora: a gardener's encyclopedia*, Timber Press, Portland OR

Holland, Robert F.
1986, *Preliminary descriptions of the terrestrial natural communities of California*, California Resources Agency, Department of Fish and Game, Sacramento CA

Holland, V.L. *et al.*
1986, *Major plant communities of California*, El Corral Bookstore, California Polytechnic State University, San Luis Obispo CA

Holliday, Ivan and Ron Hill
1984, *A field guide to Australian trees*, Rigby, Adelaide, Australia

Hora, Baynard (ed.)
1981, *The Oxford encyclopedia of trees of the world*, Oxford University Press, New York NY

Hoyt, Roland Stewart
1978, *Ornamental plants for subtropical regions*, Livingston Press, Anaheim CA

Hunter, Serena C. and Timothy E. Payson
1986, *Vegetation classification system for California: a user's guide*, USDA. Forest Service Research Paper PSW-82, Pacific Southwest Forest and Range Experiment Station, Berkeley CA

Jacobsen, Arthur L.
1996, *North American landscape trees*, Ten Speed Press, Berkeley CA

Jellicoe, Geoffrey and Susan Jellicoe
1999, *The landscape of man* (3rd ed.), Thames and Hudson, New York NY

Jensen, Herbert A.
1947, "A system for classifying vegetation in California," *California Fish and Game*, 33(4):199-266.

Jepson, Willis L.
1923, *Trees of California*, University of California, Berkeley CA

Krussmann, Gerd
1985, *Manual of cultivated conifers*, Timber Press, Portland OR

Kuchler, A.W.
1964, *Potential natural vegetation of the coterminous United States*, American Geographical Society, Special Publication No. 36, New York NY

Kuchler, A.W.
1977, *Map of natural vegetation of California*, in: Barbour and Major, *op. cit.*, pp. 909-938

Lanner, Ronald M.
1999, *Conifers of California*, Cachuma Press, Los Olivos CA

Latting, June (ed.)
1976, *Plant communities of Southern California*, California Native Plant Society Special Publication No. 2, Sacramento CA

Little, Elbert L.
1980, *National Audubon Society field guide to trees, western region*, Alfred A. Knopf, New York NY

Lynch, Kevin and Gary Hack
1984, *Site planning*, MIT Press, Cambridge MA

Lyle, John T.
1996, *Regenerative design for sustainable development*, John Wiley, New York NY

Maino, Evelyn and Frances Howard
1955 (1962), *Ornamental trees: an illustrated guide to their selection and care*, University of California Press, Berkeley CA

Mathias, Mildred E. and Elizabeth McClintock
1963, *A checklist of woody ornamental plants of California*, University of California College of Agriculture, Agricultural Experiment Station Extension Service, Manual 32, Berkeley, CA

Mayer, Kenneth E. and William F. Laudenslayer, Jr.
1988, *A guide to wildlife habitats of California: California Wildlife Habitat Relationships (WHR) System*, California Resources Agency, Department of Forestry and Fire Protection, Sacramento, CA

McClintock, Elizabeth
2001, *Trees of Golden Gate Park and San Francisco*, Heyday Books, Berkeley CA

McHarg, Ian L.
1971, *Design with nature*, Doubleday, New York NY

McMinn, Howard E. and Evelyn Maino
1963, *Pacific coast trees* (2nd ed.), University of California Press, Berkeley CA

Menninger, Edwin A.
1962, *Flowering trees of the world for tropics and warm climates*, Hearthside Press, New York NY

Metcalf, Woodbridge
1968, *Introduced trees of central California*, University of California Press, Berkeley CA

Mohlenbrock, Robert H. and John W. Thieret
1987, *Trees: a quick reference guide to trees of North America*, McMillan, New York NY

Mueller-Dombois, Dieter and Heinz Ellenberg
1974, *Aims and methods of vegetation ecology*, John Wiley, New York NY

Munz, Philip A. and David D. Keck
1973, *A California flora with supplement*, University of California Press, Berkeley CA

Ornduff, Robert, Phyllis M. Faber, and Todd Keeler-Wolf
2003, *Introduction to California plant life*, University of California Press, Berkeley CA

Parker, I., and W. J. Matyas.
1981, *CALVEG: a classification of Californian vegetation*, USDA Forest Service, Regional Ecology Group, San Francisco CA

Pavlik, Bruce M. *et al.*
1991, *Oaks of California*, Cachuma Press and California Oak Foundation, Los Olivos CA.

Paysen, Timothy E. and Jeanine A. Derby
1982, *A vegetation classification system for use in California: its conceptual basis*, USDA. Forest Service Research Paper PSW-63, Pacific Southwest Forest and Range Experiment Station, Berkeley CA

Paysen *et al.*
1980, *A vegetation system applied to Southern California*, USDA Forest Service General Technical Report PSW-45, Pacific Southwest Forest and Range Experiment Station, Berkeley CA

Perry, Bob
1981, *Trees and shrubs for dry California landscapes*, Land Design Publishing, San Dimas CA

Perry, Bob
1992, *Landscape plants for western regions*, Land Design Publishing, Claremont CA

Phillips, Roger
1978, *Trees of North America and Europe*, Random House, New York NY

Preston, Richard J., Jr. and Richard R. Braham
1948 (1989), *North American trees*, Iowa State University Press, Ames IA

Proctor, C.M. *et al.*
1980, *An ecological characterization of the Pacific Northwest Coastal Region*, US Department of the Interior, Fish and Wildlife Service, FWS/OBS 79/11-79/15, Portland OR

Rehder, Alfred
1927 (1977), *Manual of cultivated trees and shrubs*, MacMillan, New York NY

Sargent, Charles S.
1905 (1965), *Manual of the trees of North America*, Dover Publications, New York NY

Sawyer, John O. and Todd Keeler-Wolf
1995, *A manual of California vegetation*, California Native Plant Society, Sacramento CA

Schuster, Max J.
1982, *Capitol Park trees: a users guide*, Peterson Publishing, Sacramento CA

Seymour, E.L.D.
1970, *The wise garden encyclopedia*, Grosset & Dunlap, New York NY

Simonds, John O.
1961 (1997), *Landscape architecture: a manual of site planning and design,* McGraw-Hill, New York NY

Street Tree Seminar, Inc.
1994, *Street trees recommended for Southern California*, Street Tree Seminar Inc., Anaheim, CA

Stuart, John D. and John O. Sawyer
2001, *Trees and shrubs of California*, University of California Press, Berkeley CA

Sudworth, George B.
1967, *Forest trees of the Pacific slope*, Dover Publications, New York NY

Sunset Books
2007, *Sunset western garden book* (8th ed.), Sunset Books, Menlo Park CA

Thorne, Robert F.
1976, "The vascular plant communities of California," in: Latting, *op. cit.*, pp. 1-31

University of California, Berkeley, Department of Environmental Science, Policy, and Management
2005, *Wieslander Vegetation Type Mapping*, http://vtm.berkeley.edu/

University of California Cooperative Extension and California Department of Water Resources
2000, *A guide to estimating irrigation water needs of landscape plantings in California: the landscape coefficient method and WUCOLS III* (1999 ed.), California Department of Water Resources, Sacramento CA

Walker, Peter and Leah Levy
1997, *Minimalist gardens*, Spacemaker Press, Washington DC

Wieslander, A.E.
1935, "A vegetation type map of California," *Madroño*, 3(3): 140-44

Wieslander, A. E. and Herbert A. Jensen
1946, *Forest areas, timber volumes and vegetation types in California*, California Forest and Range Experiment Station, USDA Forest Survey Release No. 4., Berkeley CA

Wyman, Donald
1951 (1972), *Trees for American gardens*, McMillan, New York NY

Zion, Robert L.
1968, *Trees for architecture and landscape*, Van Nostrand Reinhold, New York NY